Photometric and Spectroscopic Binary Systems

NATO ADVANCED STUDY INSTITUTES SERIES

Proceedings of the Advanced Study Institute Programme, which aims at the dissemination of advanced knowledge and the formation of contacts among scientists from different countries

The series is published by an international board of publishers in conjunction with NATO Scientific Affairs Division

A	Life Sciences	Plenum Publishing Corporation
B	Physics	London and New York
C	Mathematical and Physical Sciences	D. Reidel Publishing Company Dordrecht, Boston and London
D	Behavioural and Social Sciences	Sijthoff & Noordhoff International Publishers
E	Applied Sciences	Alphen aan den Rijn and Germantown U.S.A.

Series C – Mathematical and Physical Sciences

Volume 69 – *Photometric and Spectroscopic Binary Systems*

Photometric and Spectroscopic Binary Systems

*Proceedings of the NATO Advanced Study Institute
held at Maratea, Italy, June 1-14, 1980*

edited by

ELLEN B. CARLING

and

ZDENĚK KOPAL

Department of Astronomy, The University of Manchester, England

D. Reidel Publishing Company

Dordrecht : Holland / Boston : U.S.A. / London : England

Published in cooperation with NATO Scientific Affairs Division

Library of Congress Cataloging in Publication Data
NATO advanced study institute (1980: Maratea, Italy)

    Photometric and spectroscopic binary systems.
    (NATO advanced study institutes series. Series C: Mathematical and
physical sciences; v. 69).
    'Published in cooperation with NATO scientific affairs division'
    Bibliography: p.
    Includes index.
    1. Stars, double—Congresses. I. Carling, Ellen B., 1933–
II. Kopal, Zdeněk, 1914–   . III. North Atlantic Treaty Organization.
Division of scientific affairs. IV. Title. V. Series.
QB821.N37   1980   523.8'41   81–4588
ISBN 90-277-1281-6   AACR2

Published by D. Reidel Publishing Company
P.O. Box 17, 3300 AA Dordrecht, Holland

Sold and distributed in the U.S.A. and Canada
by Kluwer Boston Inc.,
190 Old Derby Street, Hingham, MA 02043, U.S.A.

In all other countries, sold and distributed
by Kluwer Academic Publishers Group,
P.O. Box 322, 3300 AH Dordrecht, Holland

D. Reidel Publishing Company is a member of the Kluwer Group

All Rights Reserved
Copyright © 1981 by D. Reidel Publishing Company, Dordrecht, Holland
No part of the material protected by this copyright notice may be reproduced or utilized
in any form or by any means, electronic or mechanical, including photocopying,
recording or by any informational storage and retrieval system,
without written permission from the copyright owner

Printed in The Netherlands

TABLE OF CONTENTS

INTRODUCTION   vii

    PART I: LIGHT CHANGES OF ECLIPSING VARIABLES AND
            THEIR ANALYSIS

Z. KOPAL / The Royal Road of Eclipses   1

I. JURKEVICH, A. F. PETTY and I. R. GOODMAN / Error Analysis
    of the Elements of an Eclipsing Binary System
    in Frequency Domain   17

P. B. ETZEL / A Simple Synthesis Method for Solving the
    Elements of Well-Detached Eclipsing Systems   111

B. NELSON / The Uniqueness Criterion: The Analysis of
    Eclipsing Binary Light Curves   121

O. DEMIRCAN / Light Curve Analysis for the Basic Spherical-
    Model of Eclipsing Variables in the Frequency
    Domain   125

A. GIMÉNEZ and J. M. GARCÍA-PELAYO / A Computer Program
    for the Frequency-Domain Analysis of the Light
    Curves of Eclipsing Variables   155

F. ZAFIROPOULOS / A Study of Special Functions in the Theory
    of Eclipsing Binary Systems   161

H. ROVITHIS-LIVANIOU / Photometric Perturbations of Close
    Binaries in the Frequency-Domain   181

P. G. NIARCHOS / An Analysis of the Light Changes of
    W UMa-type Systems in the Frequency Domain   199

    PART II: ECLIPSING VARIABLES: PHOTOMETRIC OBSERVATIONS

G. SEDMAK / Data Acquisition in Astronomical Photoelectric
    Photometry   217

G. LONGO and M. RIGUTTI / Detection of Close Binary Systems
    by Means of Lunar Occultations   253

M. RODONÒ / Progress and Problems in RS CVn Star Research 285

P. VIVEKANANDA RAO and M. B. K. SARMA / Observations
of RS CVn-type Binary Stars 305

S. MANCUSO, L. MILANO, A. VITTORE, E. BUDDING and
D. M. Z. JASSUR / The System RT And and the
Transit/Occultation Question 313

L. MILANO / The Short-Period Group of Eclipsing Binaries
with Properties Similar to the Classical
RS CVn Group 331

P. VIVEKANANDA RAO and M. B. K. SARMA / Photoelectric
Study of the RS CVn-type Binary TY Pyxidis 361

M. KURUTAÇ and C. IBANOĞLU / BV Photometry of Selected
RS CVn-type Eclipsing Binaries 389

E. BUDDING / Observations and Analysis of UU Sge 405

O. DEMIRCAN and N. GÜDÜR / The Light and Period Variations
of OO Aquilae 413

P. G. NIARCHOS / Photoelectric Observations of Eclipsing
Binary Systems at Kryonerion Observatory, Greece 441

P. ROVITHIS / Stepanian's Star 447

K. OTMIANOWSKA / Note on the Eclipsing Binary
AW Ursae Maioris 451

PART III: SPECTROSCOPIC STUDIES OF CLOSE BINARY SYSTEMS

M. HACK / Ultraviolet Spectra of Interacting Binaries 453

A. H. BATTEN / U Cephei Viewed from Maratea 465

E. BUDDING / Certain Aspects of the Problem of Mass Transfer
in Semi-Detached Binaries 473

A. GIMÉNEZ / The Study of Apsidal Motion in Eclipsing
Binaries 511

T. IIJIMA / Temperature Determination of Exciting Stars
in Highly Excited Planetary Nebulae and
Symbiotic Stars 517

Z. KOPAL / Gas Streams in Close Binary Systems 535

NAME INDEX 559

INDEX OF INDIVIDUAL ECLIPSING VARIABLES 571

INTRODUCTION

Our conference - opening today - has two aims in view: first, to commemorate some milestones in the development of the studies of close binary systems whose anniversaries fall in these years, as well as to take stock of our present knowledge accumulated throughout preceding decades, in order to consider where do we go from here.

This summer, 310 years will have elapsed since the first eclipsing binary - Algol - was discovered in Bologna by Geminiano Montanari (1633-1687) to be a variable star; and 198 years have gone by since John Goodricke of York (1764-1786) established the fact that Algol's light changes were periodic. Moreover, it is almost exactly (to a month) now 100 years since Edward Charles Pickering (1846-1919) of Harvard Observatory in the United States took the first steps towards the development of systematic methods of analysis of the light changes of Algol and related systems - a topic which will constitute the major part of the programme of our present conference. The three dates recalled above illustrate that the discoverers of such celestial objects and observers of their light changes have been systematically ahead of the theoreticians endeavouring to understand the significance of the observed data by decades and centuries in the past - a fact which, incidentally, continues to hold good (albeit with a diminishing lead-time) up to the present.

Pickering (1880) based his pioneer analysis of the light changes of Algol on the following four simplifying assumptions:

1) both components constituting Algol are spherical;

2) the star undergoing eclipse is uniformly bright;

3) the eclipsing component is completely dark; and

4) the eclipses in the Algol system are grazing.

Under these conditions Pickering deduced (from Schönfeld's light curve based on visual estimates made by Argelander's method) a set of the elements including the fractional radii r_1 = 0.171,

$r_2 = 0.223$; the orbital inclination $i = 87°$, and the fractional luminosities of the two components $L_1 = 1$ and $L_2 = 0$ (assumed) - values which do not differ too much from the most up-to-date set of the elements ($r_1 = 0.227$, $r_2 = 0.239$; $i = 82°0$, $L_1 = 0.966$ and $L_2 = 0.034$) based on modern photoelectric observations. Moreover, soon after Pickering, J. Harting (1889) freed the reduction procedures from Pickering's assumption 4 (grazing eclipses) as unnecessarily restrictive; and E. Hartwig (1900) admitted (for Z Herculis) that $L_2 > 0$, and thus freed the procedure from Pickering's assumption 3. Pickering's assumption 2 (of uniformly bright discs) was removed independently and simultaneously by the (then) young astronomers A. Pannekoek (1902) and C. Rödiger (1902), who were the first to introduce in our work the "limb-darkening" in their Ph.D. theses - a problem to which Antonie Pannekoek (1873-1960) devoted also one of the last papers of his lifetime (cf. Pannekoek and van Dien, 1937).

The abandonment of the spherical shape of the components (Pickering's "assumption 1") was proposed first by Johannes Plassman in 1888, and replaced by distorted (ellipsoidal) shape to account for the observed light changes of the eclipsing system of β Lyrae (discovered also by Goodricke in 1785) - a proposal which was elaborated mathematically by G. W. Myers in his doctoral dissertation (1896); and subsequently in one of the earliest volumes of the then newly founded Astrophysical Journal (1898). The second and less conspicuous source of the light changes of close binary systems - the "reflection effect" - was detected somewhat later by R. S. Dugan (1908) in RT Persei, and J. Stebbins (1910) for Algol. Another landmark in subsequent development of our subject - namely, a realization that in Algol and other binaries whose components are unequal in mass or size - the shape of the constituent stars may be very different - goes back to K. Walter (1931), the senior participant of our conference, whom we have the pleasure to welcome among us today.

So much for a brief outline of past milestones in the evolution of our subject. An Advanced Study Institute, such as ours, cannot of course limit itself to a survey of what was accomplished by our predecessors. Its main aim should be to take stock of our present knowledge in its field, and to chart our course for the future. Since the work of the "founding fathers" of our subject, briefly alluded to before, its further outgrowth has proceeded on a very massive scale, and its results would have exceeded the fondest hopes of our ancestors.

This was due very largely to two reasons: first, to a greatly increased precision of underlying observations (which followed in the wake of the replacement of earlier visual or photographic photometry by photoelectric work); and, secondly, by an outgrowth of our knowledge of the physical structure of the stars, as well as

INTRODUCTION

our ability to base our analysis of the observations on physically sounder (though more complicated) models of close binary systems with the aid of automatic computers, which emerged on the scene in the latter 1940's and whose impact on the study of our subject cannot be overestimated. This impact has provided, in fact, the motivation for a large part of the programme of the first part (and to only somewhat lesser extent) also for the second part of our Conference.

* * * *

As is evident from the Table of Contents of the present volume, the subject matter for discussion in the days ahead splits up naturally in two parts. In the first part, we shall be concerned mainly with the present state, and immediate prospects, of the analysis of the light changes of eclipsing binary systems; with particular attention to the Fourier techniques developed in recent years. The observed light changes constitute, however, only one part of the observational evidence bearing on the astronomy and physics of close binary systems. Another is traditionally based on spectroscopic observations of the radial velocities of the components of such binaries, or of the gaseous medium surrounding the entire system; and a discussion of the phenomena arising in this connection will constitute the second part of this volume.

A word of explanation may be in order concerning the editorial structure of this book. According to the practice followed by the previous volumes of the ASI series, the contributors were expected to provide the text of their lectures for the press in camera-ready form. Although not all contributors failed to do so, the fact that several of them did provide their text in this (though not necessarily the expected) form prevented us from attaining such uniformity in style, notations and references as could be expected in a letter-press edition of the entire material. To re-type all papers in a uniform style would have delayed their publication unacceptably, and any thought of it had to be given up at the outset. We can, therefore, only hope that the surviving stylistic differences will in no way impair the legibility and general usefulness of this volume, not only to the participants of the conference, but also for a wider circle of interested readers.

* * * * *

In conclusion, it is a pleasure to acknowledge the encouragement and support received from several sources to render our conference a success. First of all, our sincere thanks are due to the Scientific Affairs Division of the North Atlantic Treaty Organization in Brussels, Belgium, which sponsored the entire conference and provided us with the principal means for organizational purposes. Additional support (in the form of travel funds) was received from the National Science Foundation of the United States of America and other branches of the U.S. Government; and also from the Goto Telescope Manufacturing Company of Tokyo, Japan. Our Italian colleagues and hosts at Maratea - in particular, Dr. A. Guzzardi, manager of the Hotel Villa del Mare where our conference was held, and to his able secretary Signora Susy Bisogno for all the care they took of the conferees, and all the help which they willingly extended to us in the course of the Conference. And last, but not least, our sincere thanks are also due to the D. Reidel Publishing Company for their cooperation and care with which they produced the present book.

Ellen B. Carling
Zdeněk Kopal

REFERENCES

Dugan, R. S.: 1908, Publ. Amer. Astr. Soc., 1, p.311.
Harting, J.: 1889, "Untersuchung über den Lichtwechsel des Sternes β Persei", Diss. München.
Hartwig, E.: 1900, Astr. Nachr., 152, p.309.
Myers, G. W.: 1896, "Untersuchung über den Lichtwechsel des Sternes β Lyrae", Diss. München.
Myers, G. W.: 1898, Astrophys. J., 7, p.1.
Pannekoek, A.: 1902, "Untersuchungen über den Lichtwechsel Algols", Diss. Leiden.
Pannekoek, A. and van Dien, E.: 1937, Bull. Astr. Inst. Netherlands, 8, p.141.
Pickering, E. C.: 1880, Proc. Amer. Acad. Sci., 16, p.1.
Plassmann, J.: 1888, *Die Veränderlichen Sterne*, Köln.
Rödiger, C.: 1902, "Untersuchungen über das Doppelsternsystem Algol", Diss. Königsberg.
Stebbins, J.: 1910, Astrophys. J., 32, p.213.
Walter, K.: 1931, Königsberg Veröff., No.2.

<u>Sitting in front</u> (from left to right): Güdür, Demircan, Lorenzi, Mammano, Mrs. Zafiropoulos, Niarchos, Milano, Mantegazza, Awadalla, Hamzaoglu, Iijima.

<u>Standing</u>: Moutsoulas, Kizilirmak, Jurkevich, Mrs. Petty, Giménez, Mrs. Jurkevich, Petty, Rao, Etzel, Rovithis, Mrs. Carling, Robertson, Mrs. Rovithis-Livaniou, Mrs. Budding, Tsouroplis, Mrs. Hack, Longo, Walter, Kopal, Bonifazi, Pastori, Scaltriti, Sedmak, Burchi, Nelson, Miss Cerrutti-Sola, Ibanoğlu, Miss Romeo, Mahdy, Kurutaç, Miss Otmianowska, Miss Balucinska, Budding.

THE ROYAL ROAD OF ECLIPSES

Zdeněk Kopal

Department of Astronomy,
University of Manchester, England.

More than a third of a century has elapsed since 28 December 1946, when the late Professor Henry Norris Russell (1877-1957) delivered an address, under the above title, before the American Astronomical Society in Cambridge, Massachusetts - as the first of a series of annual Russell Lectures sponsored by the Society to commemorate the name of that distinguished scholar of the past generation (for its published text see the volume of the *Centennial Symposia*, Harvard Observatory Monograph No. 7, pp.181-209; 1948). I am the only one in this room who was present at that occasion, and who can recall Russell's enthusiasm and inimitable style exhibited not only in the delivery of his lecture, but also in the course of discussion which started in the large auditorium of the Radcliffe College, and continued to reverberate through the snow-clad streets of a crisp winter evening as we accompanied the aged master to his local Cambridge home - through weather so different from the greenery and mellow air which our hosts have prepared for us at Maratea today.

If I have chosen for my introductory talk to this Conference the same title which Russell used in 1946, it is to underline the continuity of purpose with our scientific predecessors one or two generations ago; though the advances made in the past 30 years in our understanding of a wide range of phenomena exhibited by close binary systems, as well as new methods of their analysis, introduced in practical use since that time, would have astonished Russell and his contemporaries if they could be with us today. Indeed, the importance of our subject has greatly increased since that time; and progressed from an ad-hoc treatment of isolated curious phenomena to one whose results became one of the cornerstones of modern stellar astronomy, and which provided almost the entire evidence we possess today empirically to document the story of stellar evolution.

Whenever we study the properties of any celestial body - be it a planet or a star - all information we wish to gain can reach us through two different channels: their gravitational attraction, and their light. Gravitational interaction between our Earth and its celestial neighbours is, however, measurable only at distances of the order of the dimensions of our solar system; and the only means of communication with the realm of the stars are their nimble-footed photons reaching us - with appropriate time-lag - across the intervening gaps of space.

As long as a star is single and emits constant light, it does not constitute a very revealing source of information. Spectrometry of its light can disclose, to be sure, the temperature (colour, or ionization) of the star's semi-transparent outer layers, their chemical composition, and prevalent pressure (through Stark effect) or magnetic field (Zeeman effect), it can disclose even some information about its absolute luminosity or rate of spin. It cannot, however, tell us anything about what we should like to know most - namely, the mass or size (i.e., density) of the respective configuration; its absolute dimensions, or its internal structure.

In order to disclose its mass, the star must be made to "step on the scales" by entering into gravitational partnership with another star to form a "binary system". A certain amount of information on masses of the stars can be obtained from observations of nearby "visual" binaries which are within measurable distances, and close enough to exhibit absolute motions about their common centre of gravity within 1-2 centuries of their observation. The number of such systems is, however, limited by their required proximity; and their available supply is not copious. Binary systems situated at greater distance can be discovered spectroscopically (from periodic variations of Doppler shifts of spectral lines of their components) the more easily, the closer they happen to be; and although their spectroscopic observations can furnish absolute values for the lower bounds of their masses, they can say nothing about (absolute or relative) dimensions of their components.

If, however, the system happens to be sufficiently close for the dimensions of its components to represent not too small a fraction of their separation, and if their orbit is not inclined too much to the line of sight, each component may eclipse - partly or wholly - its made in the course of each orbit; and the system becomes thus an *eclipsing variable* - recognizable as such not only by fluctuating Doppler shifts in the spectra of each component, but also by the *variation of light* of the system within eclipses. The characteristic variations of light due to this cause represent an even more eloquent (and more easily observable) tell-tale feature of such systems; and their interpretation will constitute the principal objective of this conference.

A study of the phenomena exhibited by eclipsing binary systems occupies a very important position in contemporary stellar astronomy for several reasons. First, because of a truly prodigous abundance of the objects of its study. Surveys of stars in our neighbourhood disclose that at least 0.1 per cent (probably more) of them form systems which happen to eclipse; and if stars with masses greater than that of the Sun were considered alone, their percentage would be much higher. If, moreover, the foregoing conservative estimate of their percentage were to apply to our whole galactic system, the total number of eclipsing variables within it should be of the order of 10^9. Only a minute fraction (a few thousands) of these have been identified so far, and their periods determined; but their total number in our Galaxy is beyond the hope of individual discovery. Eclipsing variables are, therefore, manifestly no exceptional or uncommon phenomena!

The significance of eclipsing variables is further emphasized by the fact that they constitute the only class of double stars that can be discovered in more distant parts of the Universe. In the neighbourhood of the Sun - up to distances of the order of 100 parsecs - at least wide binaries can be recognized by their orbital motion (or, for very wide pairs, by common proper motions of their components). Spectroscopic binaries can be discovered as such up to distances of the order of one or two thousand parsecs (depending on their absolute brightness) with the aid of modern powerful reflectors capable of decomposing their light into spectra of sufficient dispersion. Beyond that limit close binaries can, however, be detected if, and only if, they happen to be eclipsing variables. Their characteristic light variations can be measured photometrically almost down to the limit attainable by our telescopes - not only in our Galaxy, but in any system resolvable by them into stars. At present dozens of them are known in external galaxies, down to distances exceeding one million parsecs.

But the significance of eclipsing variables in astronomy is not based only on their ubiquitous presence and enormous numbers; it rests also as much (or more) on the unique nature of the information which they - and they alone - can provide. We mentioned above that spectrographic observations alone can furnish only the minimum values of the masses of the components of close binary systems or of dimensions of their orbits. The missing clue - necessary to convert their lower bounds into actual values - is represented by the inclination of the orbital plane to the line of sight; and it can be obtained from an analysis of the light changes if the respective binary happens to be an eclipsing variable. As will be shown later in this book, an analysis of their observed light curves can specify not only the orbital inclination, but also fractional dimensions of the components - which, on combination with spectroscopic data, can furnish the masses, densities and absolute dimensions of the constituent stars.

Astrophysical data which can be deduced from the observed light changes of eclipsing variables transcend, moreover, information on the absolute dimensions of their components or the characteristics of their orbits; for even their internal constitution may not (under certain conditions) remain concealed from us. Even though the interiors of the stars are concealed from view by enormous opacity of the overlying material, a gravitational field emanates from them which the overlying layers - opaque as they may be - cannot appreciably modify; and the distribution of brightness over the exposed surfaces of the components is governed by the energy flux originating in the deep interior.

These and other possibilities opening up by the studies of eclipsing variables have long attracted due attention on the part of the observers. Largely because of the recurrent nature of the phenomena which they exhibit, eclipsing variables have always been favourites for pioneers of accurate photometry of any kind - visual, photographic or photoelectric - and the total number of observations made in this field runs into several millions. Observations alone are, however, insufficient to disclose to inspection a wealth of information which they contain. To develop this information calls for introduction of systematic methods rooted in physically sound models of the phenomena we observe to decipher what observations have to say. This sets the tasks which we are going to discuss at our meetings.

The problem at issue is indeed one of astronomical cryptography: the messages these stars send out on waves of light are encoded by processes responsible for them; while the task of the analyst is to decode the photometric evidence to yield the information which it contains. To do so requires some knowledge of the code; and to provide it is the task of the theoretician. The decoding process constitutes an essentially mathematical problem; but the identification of the code is primarily one for the astrophysicist; and the extent to which this code can be set up determines the gain to be expected from its use. Eclipsing variables do not represent by any means the only sources of encoded information reaching us from the stars: indeed, virtually all stars to the right or left of the Main Sequence in the Hertzsprung-Russell diagram are known to be variable (though not necessarily periodic). The reason, however, why their light variations tell us (at best) only a part of their story is the fact that - because of physical complexities - we lack as yet a code to decipher some (or most) part of it. In a physical sense, the quest for a code to unlock the information contained in the light changes of eclipsing variables will not confront us with unsurmountable problems. However, their gravity should also not be underestimated; for unless the theoretical basis of an analysis of their light changes is well understood - and due care exercised in their interpretation - not only do we fail to do justice to the skill and perseverance of the observers, but, worse still, we run the risk that a considerable part of stellar astronomy may rest on inaccurate empirical foundations.

It is, incidentally, astonishing to contemplate the range of inspiration which astronomy (and physical science in general) received from the phenomena of eclipses. Not later than in the 4th century B.C., the observed features of the terrestrial shadow cast on the Moon by the Earth enabled Aristotle (384-322 B.C.) to formulate the first scientific proof that the Earth was a sphere; and only somewhat later, the eclipses of the Sun provided Aristarchos (in the early part of the 3rd century B.C.) and Hipparchos (2nd half of the 3rd century) with the geometric means to gauge the distance which separates us from the Sun. In the 17th century A.D., the eclipses of the satellites of Jupiter furnished Roemer (in 1676) with the first experimental proof of the fact that the velocity of light was finite. Total eclipses of the Sun have, in particular, proved a veritable godsend to students of many branches of science as well as humanities - from the historians of ancient times (whom they enabled to straighten out many an obscure date of early chronology of their subject), to the geophysicists (who used the old eclipse records to detect secular irregularities in the length of the day), or to the chemists who learned (in 1868) of the existence of at least one new element (helium) from the solar flash spectra before it was identified also in the laboratory. Students of the motion of the Moon in the sky found the records of ancient eclipses invaluable for identification of its "secular acceleration"; and a disappearance ("occultation") of the stars of known positions behind the Moon's limb valuable for accurate location of the instantaneous positions of our satellite; while high-speed photometry of such occultations led in several cases to a more precise determination of the angular diameters of such stars than would (at present) be possible by any other alternative optical method. In recent decades astrophysicists have also taken advantage of the fleeting minutes of total eclipses to measure the extent of light deflection in the gravitational field of the Sun, and thus to study the metric properties of space in the neighbourhood of large masses. In the face of these facts, the choice of the subject of this conference needs scarcely any further explanation or apology!

As we mentioned already in the introduction, the history of eclipsing binary systems goes back to Montanari in 1670, and the first attempt to develop the actual information contained in their light changes to Pickering in 1880. For another 30 years, work along the line of approach initiated by Pickering made only slow progress - limited largely to applications to individual specific systems. The first general method, applicable to an analysis of the light changes due to eclipses of any type was not launched till 1912 by H. N. Russell (1912), who in collaboration with H. Shapley (cf. Russell and Shapley, 1912) developed numerical procedures designed to cope with observations of limited accuracy available at that time - methods which were re-stated later by Russell and Merrill (1952) without, however, injection of any essentially new ideas into the problem.

The principal aspects in which the methods proposed in 1912 by Russell and Shapley (together with their subsequent modifications) were subsequently found to be wanting go back to their use of the observational data; or, more specifically, to the replacement of actual observations - consisting of discrete measurements of the star's brightness at different times - by continuous light curves drawn by free hand to follow the course of individual observations. A graphical interpolation of this nature may have been justified, on grounds of expediency, at a time when the accuracy of individual photometric observations amounted to 1-10% of the measured light flux. With the gradual increase in precision of photometric observations throughout the first half of this century the time was bound to come when an empirical interpolation or smoothing of the observed data (as represented by a drawing of free-hand curves) was bound to turn into a liability; for a substitution of such curves for actual observations amounts to an unwarranted and risky interference with the data provided by the observer, which may vitiate all the results based upon them in a systematic manner.

For once we replace the actual observations by graphical inferences in the form of free-hand curves, and stake our subsequent analysis solely upon them, we not only give up any possibility to specify the *uncertainty* within which the results are defined by available data, but - worse still - we have no means to ascertain the *uniqueness*; or to investigate the extent to which any elements are defined at all by the given photometric data.

This last point is one of cardinal importance; though it was not considered - let alone solved - by the founding fathers of our subject in the first half a century or more after 1880. Instead, they burdened our souls with a "hereditary sin" of a mistaken notion that the sufficient aim of the analyst should be to represent the observed light changes by *some* set of elements of the system within the limits of observational errors - without stopping to inquire whether the set which can do so is unique! A satisfactory agreement of the theoretical light curve (i.e., one computed from the elements resulting from an analysis) with the observations is indeed a *necessary, but not sufficient* condition for the correctness of the elements in question; for the set (or sets) which may satisfy this requirement need not be the only one which can do so. We now know that photometric observations of certain types of eclipsing systems (for example, those exhibiting shallow partial eclipses;

or those consisting of two components of comparable luminosity) may not define *any* set of elements uniquely - no matter how precise the available observations may be - or can do so only with wide limits. And should the solution border on such an indeterminacy, nothing is easier than to represent the respective light changes by a "right combination of wrong elements" - as only too many investigators found in the past the hard way to their sorrow!

These facts came only gradually to light in the 1940's, following the introduction of the "iterative" methods by Kopal (1941, 1946, 1948) and Piotrowski (1947, 1948); and caused more than one unpleasant surprise to the users of earlier "direct" methods which always led to some solution by a restriction of the inherent degrees of freedom (for instance, by the adoption of certain "fixed points" on the light curves, admitting no subsequent improvement). The iterative process - in which the intrinsic indeterminacy of a case manifests itself by a failure of successive iterations to converge to a definite answer - turned out also to be more laborious of application; and for this reason it was greeted by some investigators (e.g., Russell, 1942) with somewhat less than enthusiasm. Although it was impossible at that time to foresee the full extent to which the actual burden of numerical work could eventually be relegated to automatic computers working at electronic speed, the success with which the iterative methods have since been automated (cf.,e.g.,Huffer and Collins,1962. Jurkevich,1970; Linnell and Proctor, 1970, 1971; Söderhjelm, 1974; etc.) should not, however, make us lose sight of the *limitations* inherent in *all* methods mentioned so far - direct as well as iterative - which are inherent in their roots and are not affected by a different strategy of approach to a solution of the respective equations of the problem. All methods proposed for their solution between 1880-1960 have been based on the use of the actual observations of the star's brightness made at different phases of the eclipse - direct use in iterative methods, or indirect (through free-hand light curves drawn to follow the course of individual observations) in Russell's method. Each approach aims basically at the same end: namely, *to interpret the observed light changes in the time-domain.* Such an interpretation does not, however, represent by any means the only avenue of approach to the solution of the underlying problem. The alternative approach - equally valid in principle, is to interpret the observed light changes in the *frequency domain* - i.e., to subject to an appropriate analysis, not the light curve of eclipsing variables as a plot of brightness versus time, but its *Fourier transform.* And it is to the discussion of this new approach that a major part of this conference will be devoted.

"Mere opinions end, and real knowledge begins in astronomy only in such subjects which can be treated mathematically. This is the case with size and shape of celestial bodies, their distances, relative positions and their changes - or, in effect, their motions

... " wrote the great Carl Friedrich Gauss in a preamble to physical astronomy almost 200 years ago. Since that time, we could adduce other branches of our science which have matured far enough to merit Gauss's approval - such as the theory of radiative transfer in stellar atmospheres, of internal structure of the stars; or - I claim with some pride on behalf of us all - of the theory of light changes of close binary systems. And which more powerful technique is known to mathematical science to this end than the Fourier analysis?

Earlier in this chapter we outlined already the range of information concealed in the light changes of eclipsing binary systems. This information is, of course, not sent out "en clair", but encoded in a language to which we must obtain first a clue - or "code" - before we can reap the full benefits of its contents. This code is represented by a physically reasonable model of the system, consistent with our knowledge of contemporary astrophysics; it is furnished to us partly by a theory of radiative transfer in stellar atmospheres, and partly by the equilibrium theory of tides. To set up such a code is primarily a task for the astrophysicist; while decoding of the message constitutes a challenge to the mathematician. The problem at issue is indeed another example of "celestial cryptography", similar to those encountered in other branches of science or engineering - regardless of whether the actual carriers of information are sound waves, optical, or electrical signals. Indeed, any such information can be regarded as the "language" of the source; and its decoding depends on the internal logic of this language rather than the technical aspects of its external transmission.

Now the logic of the language by which the eclipsing binary systems have been communicating to us the principal features of their structure is such as to lead the celestial cryptographer into the frequency domain, offering the most equitable avenue of approach to their secrets. This was first noted by the present speaker - also in Italy (at the congress to commemorate the 50th anniversary of the death of G. V. Schiaparelli) - almost 20 years ago (cf. Kopal, 1960); although fuller details of the underlying procedures were not worked out more fully by the Manchester school of double-star astronomy until since 1974.

The results of this work obtained through 1978 have been summarized in my recent book on *The Language of the Stars* which appeared last year (Kopal, 1979). As copies of this book are available in this room for your inspection (and since some of you have already seen it), its contents need not be repeated at this time; and only a few remarks of more general nature may be in order - as a preface to more specific individual contributions by several speakers who will follow me on the roster in the first part of our conference.

As regards the more formal aspects of the analysis of the light changes of eclipsing variables in the frequency-domain, the legitimacy of the underlying change of mathematical scenario needs no elaboration. The Fourier (or, indeed, any other integral) transform of the light changes of eclipsing variables contains exactly the same amount of information as their original light curves in the time-domain; and the latter can be synthesized from the former in every respect - indeed, a "spectral analysis" and "synthesis" of any function constitute complementary operations covered by the Fourier theorem. In which respect can an approach to the solution of our problem via the frequency-domain offer advantage over that through the time-domain? A more detailed answer to this question has been given in Chapters V and VI of my *Language of the Stars*; here we wish only to summarize the salient features.

As we already mentioned, the relation between the elements of the eclipses and their observed characteristics are *transcendental*, and not capable of any type of analytic inversion. In the frequency domain, however, these relations become *algebraic* (albeit nonlinear); and can be inverted to yield the desired results. This difference entails, in turn, a different extent of insight into the nature of our problem, and permits us to obtain a deeper understanding of the relations between the observational data and the unknowns of the problem.

A much more fundamental difference concerns, however, the behaviour of the photometric *proximity effects* in the time- and frequency-domains. In the time-domain, the light changes of close eclipsing systems (and, in view of the probability of their discovery, a large majority of known eclipsing variables are bound to constitute close systems) consist of a superposition of photometric proximity effects (due to mutual distortion as well as irradiation of both components), which are *continuous* in time, upon *discontinuous* changes arising from the eclipses.

In the time-domain - if we plot the observed brightness of the system against the time (or the phase angle), the two phenomena are difficult to separate by inspection, and may merge indistinguishably with each other for really close systems. In the frequency-domain, on the other hand, *the two causes of light variation are clearly distinguished by their frequency-spectra* - the proximity effects being characterized by *discrete* spectra; while those arising from the eclipses, by *continuous* spectra of the fundamental frequency range.

In the time-domain, the customary procedure in the past had been to "rectify" the light curve - i.e., attempt to remove the photometric proximity effect algebraically from the light curve *before* solving the latter for the elements of the eclipse; and to do so with the aid of the "rectification constants", deduced from the

light changes between eclipses. Apart from practical difficulties
inherent in such a scheme (which become the greater, the closer
the system; for the range of the light curve which is to furnish
these "rectification constants" shrinks as the proximity effects
grow in magnitude), the procedure itself can be theoretically justified only under severely simplified assumptions which are unlikely
to be met in reality. In order to fulfil these assumptions, the
constituent components must be similar ellipsoids, seen in projection as stars with isophotes constantly symmetrical around the
centre of the apparent discs. The latter part can be fulfilled only
if the discs in question were either uniformly bright; or for combinations of limb- and gravity-darkening so unlikely on physical
grounds as to be of no more than geometrical significance. And
if any one of these conditions is not fulfilled, the "rectified" light
curve within minima does *not* become equivalent to one arising
from eclipses of circular discs (to which standard methods of solution could then be applied), but becomes an artifact whose interpretation would be even more complicated than that of the light
changes directly observed. A general treatment of these phenomena by Kopal (1945; cf. also 1959, Section V.1) showed that as
soon as spherical-harmonic distortion of order higher than the
second is taken into account, a symmetry of the isophotes required
for rectification is irretrievably lost for *any* distribution of brightness over the apparent discs of such stars; and no amount of limb-
or gravity-darkening can help to restore it.

In the face of this situation - now known for more than thirty
years - the reaction of the investigators was two-fold. An increasing fraction of more critical students of close eclipsing systems preferred to abstain from any solution which would have to be based
on manifestly shaky premises, and limited themselves to publishing
their observations as they stood - leaving the task of their analysis to the future. Others - less patient to wait - persisted in
the use of discredited techniques because nothing better could be
put in its place; and in the hope that errors committed by the use
of "rectified" light curves as a basis for their solution for the elements of the eclipses (as if the latter were caused by mutual eclipses of spherical stars) may not have too serious an effect on the
outcome. This obviously unsatisfactory situation continued to obtain as long as our operations remained restricted to the time-domain. In the frequency-domain, however, no "rectification" of old
style is needed for decoding of the light curve at all; for - as was
shown in Chapter VI of Kopal (1979) - the photometric proximity
effects (due to ellipticity or reflection) can be removed from the
observed light changes by a suitable "modulation" of the light curves.

In order to illustrate this, let it be stressed that the situation
facing the investigator of the phenomena exhibited by eclipsing
variables can be compared with the function of a television camera

which optically scans a three-dimensional image and transforms what it sees sequentially, at high speed, into a series of linear elements which are reassembled by the receiver into a two-dimensional picture. An eclipsing variable represents, in principle, an analogous source of information; for as one component proceeds to eclipse its mate, a scanning takes place (by the eclipsing limb) giving rise to characteristic changes of light that can be observed at a distance. Our eclipsing variable can, accordingly, be regarded as an elementary one-channel television system - transmitting continuous light signals which our detector (i.e., photometer) records sequentially; and the aim of the receiver must be to reconstruct, from time-variations of the intensity of one picture-point, a time-independent two-dimensional picture of the system. Needless to say, the simplicity of the receiver is bound to add to the complexity of the interpretation of its messages: the task is more complicated, but it can be done.

Or - to use another parallel - in the preceding part of this introduction we stressed that, in close binary systems, the photometric proximity effects between eclipses are caused by a superposition of partial tides ("diurnal", "semidiurnal", etc.) which sweep around in front of our view with discrete characteristic frequencies. A spectroscopist would say that the variations of light arising from this cause are "multiplexed" by Nature, and reach us along the same (optical) channel - i.e., our photometer - with different frequencies. The task of the analyst is to separate ("unscramble") these frequencies by a suitable (mathematical) procedure, in order to reconstruct their message.

The technique of "modulation" of the light changes for the removal of photometric proximity effects - so familiar in the field of communication engineering - was applied to our problem already several years ago (cf. Kopal, 1976). Let us add to what was said there a comment that the *observational errors* - which are responsible for the dispersion of individual observations around the mean light curve in the time-domain - will manifest themselves in the frequency-domain by affecting the high-frequency tail of their (empirical) Fourier transforms in a way which a communication engineer would describe as "noise". And while, in the time-domain, we minimize the cumulative effects of observational errors by least-squares solutions of the respective equations of condition, in the frequency-domain this noise will be filtered out by a disregard of the high-frequency end of the respective transform (to which a Fourier series terminating after a certain number of terms will provide a suitable approximation).

A sequential receipt and registration of all information contained in periodically varying light signals from the stars is forced upon us by a very low frequency at which these signals are emitted. Since the orbital periods of close binary systems are, on the average, of the order of 10^5 seconds, periodic information from them on the waves of light will reach us at frequencies generally of the order of 10^{-5} Hertz. The messages which these systems keep transmitting with such insistence are, therefore, played in a very low key - in fact, some 23-24 octaves below that of audible sound. Moreover, the photometric eclipse phenomena translated into the frequency domain produce a dissonant cacophony (characterized as they are by continuous frequency spectra); while - in contrast - the "signature tune" of the proximity effects consist of harmonic chords of individual tones whose wavelengths bear integral ratios to each other. It is the combined frequencies of all these tones that add up to (*sit venia verbo*) the "music of the spheres" with which we shall be concerned at this conference.

In all these instances an analogy between astronomical and terrestrial phenomena encountered in other branches of science or engineering is manifest; and so should be among their treatment. As it happened so often in the history of science in the past, a dissolution of interdisciplinary barriers - and a resort to methods developed in one branch to assist another - is invariably found to be beneficial to both.

The extent to which this may apply to the subject of our present conference will be for you to judge; and for the readers of this book. Before stepping down from the roster and making room for speakers who follow I should, however, like to make one final observation: and this concerns the construction of "synthetic" light curves as a tool for light-change analysis. What can such techniques as were developed in recent years under the code names of WINK (Wood), EBOP (Etzel) and others - contribute to the ends which we have in mind?

In order to place the significance of these methods in proper perspective, let us recall the well-known fact that the complete Fourier theorem can be split up in two consecutive stages: namely, an "analysis" of the input data (in our case, of photometric observations of an eclipsing variable) by a decomposition in their natural "frequency-spectrum" (whose role is played, in our case, by the elements characteristic of the respective eclipsing system); and a subsequent "synthesis" of this spectrum which should give us back the original data. The Fourier analysis of the light changes of eclipsing variables, as developed by the present speaker and his school from 1974 on, is rooted in the first part of the Fourier theorem; while the synthetic methods operate (in effect) under the protection of the second.

In carrying out the Fourier "analysis", we depart directly from the photometric data provided by the observer, and by a series of purely algebraic operations are led to a specification of the most probable values of the elements of the system - a process which can yield the uncertainty within which these elements are specified by the available data as a logical by-product. It may, perhaps, be superfluous to add that this - like any other - process of reduction of the observations can only *lose* information encoded in the underlying data (i.e., increase its "entropy"). The most important consideration for the analyst should be to minimize this increase of entropy (loss of information), e.g., by least-squares techniques in the time-domain, suppression ("filtering out") of high-frequency noise in the frequency-domain, etc.

A "synthesis" of the light curves as practised so far is, on the other hand, nothing else but a numerical construction of theoretical light curves based on assumed (trial) values of the elements - and, as such, it corresponds (theoretically) to the second stage of the Fourier theorem. This stage does not, however, depart from the observations themselves, but rather from an anticipation of what we think the system should look like. And the need to depart from *assumed* values of its elements is bound to introduce into synthetic work a *bias*, the consequences of which can be removed only by a *very large* number of trials - an effort that may entail numerical work 100 times as onerous as that required for light curve analysis.

Let me explain what I mean in more specific terms. First, the number of the elements characterizing an eclipsing system which we seem to extract from its observed light changes is generally large: four (i.e., r_1, r_2, i and L_1) in the simplest possible cases; five (or six) if the coefficients of (linear) limb-darkening are to be included among the unknowns of the problem; and, of course, many more for close eclipsing systems (including coefficients of gravity-darkening, mass-ratio, etc.). To compute a "synthetic" light curve for any one set of assumed elements cannot be done in practice without the aid of an automatic computer - an exercise belonging to that branch of modern astronomy now generally referred to as "computer game" - a game which can be practised only by those having access to such computers, and able to pay for the requisite computer time.

As most astronomers are an impecunious lot (unable to pay for computer time from their own pockets even if one exists in their neighbourhood), the costs of the computations have, in general, to be paid from the public purse; and those who disburse funds for this purpose are likely to insist on economies. Which way to lessen the costs more effectively exists other than a temptation to minimize the number of necessary combinations of desired elements from prior knowledge? Which investigator, caught in this predicament, would not be tempted to depart in his trial-and-error synthesis process from the results already obtained by his predecessors, and to regard his own task only as an improvememnt of a previously accepted picture of the system?

Such a strategy can indeed reduce numerical work, but at the same time opens also the door to a perpetuation of errors which may have been committed by previous investigators. In the full sense it continues to labour under the spell of the "hereditary sin", implying that the duty of the analyst is satisfactorily discharged if he can reproduce the observed light changes of an eclipsing system by a certain set of the elements. But are these necessarily the elements of this system? For no *finite* amount of numerical "modelling" of the light curve can guarantee the *uniqueness* of the representation of the observations by *any* set of assumed elements. Such work can furnish *particular* sets of the elements capable of matching the observed light changes, but will not readily disclose the range of other combinations of the elements by which these light changes can be similarly represented. In contemplating the outcome of such an effort, we find ourselves face-to-face with a situation characterized succinctly by the great Italian poet Dante Alighieri (1265-1321) by the words "Se non è vero, è molto bene trovato"* - cf. Canto 35 of his *Divina Comedia* (Part I: "Inferno", where many astronomers still suffer for their sins - both personal and hereditary).

In point of fact, the smaller the intrinsic determinacy of the case, the easier it is to reproduce the observed evidence by, not one, but a whole range of combinations of the elements (cf., in particular, a high degree of correlation between the degree of limb-darkening of the component undergoing eclipse, and its fractional radius r_1!). Therefore, to find *some* set of the elements which can do so should be the easier, the more indeterminate the case: in such cases (encountered only too often) the first trial set may very often do - but to establish it may constitute only a hollow success, engendering spurious confidence in the outcome. In cases bordering on indeterminacy, merely to establish that a certain set of the elements can reproduce the observed light curve within the limits of observational errors contributes very little to our actual knowledge of the respective eclipsing system - in fact, if can lull us into a false belief that we know something about the system, while, in reality, this is not the case.

As a corollary, we should also add that the computation of a limited number of "synthetic" light curves cannot teach us anything about the limits of *uncertainty* within which these are defined by the available photometric data. To connect the errors of individual observations with the uncertainry of the elements determined from these observations by any kind of analytical process, we must de-

* Dr. L. Milano of the Naples Observatory kindly pointed out to me that not all medievalists are in agreement about the exact reading of the above phrase. Some believe that the word "trovato" really means "inventato" - which, if true would have anticipated even more closely the situation we have in mind.

part from the observed data rather than from any postulated model – the importance of this message cannot be overestimated!

A notion is, to be sure, sometimes voiced that – no matter how uncertain a set of preliminary elements (obtained by whichever method) may be – it should always be possible to arrive at the correct answer by resorting (repeatedly if need be) to a least-squares fit involving differential corrections; with an implication that the uncertainty of so corrected elements is identical with those of the respective differential corrections. Such methods have indeed been worked out in some detail by a number of investigators in the past (Wyse, 1939; Kopal, 1943; Irwin, 1947; Otrebski, 1948); but practical results of their application proved to be rather disappointing: namely, in almost every case to which these methods have been applied, the magnitude of the differential corrections resulting from the least-squares fit to the light curves turned out to be very small; and the sum of the squares of the residuals was reduced to a barely significant extent. An optimist may be inclined to interpret so (apparently) an auspicious outcome as an indication that his preliminary elements (adopted to serve as a basis for a least-squares fit) were very near the truth; but a more thoughtful investigator may ponder on chances that this should happen time and again with every system on hand.

After a lot of beating about the bush it was eventually discovered that the real cause of deceptive success goes back to the nonlinearity of the problem. The coefficients of differential corrections in the variational equations are, not constant, but functions of the elements which our procedure is trying to improve; and some of them depend so sensitively on the outcome that the convergence of successive approximations becomes very slow – if convergent at all. In other words, a multi-dimensional surface whose coordinates are all elements of an eclipsing system is of so complex a structure that *a* minimum of the residuals can always be located in the proximity of any point – without, however, any assurance that it represents *the* minimum.

Accordingly, the method of differential corrections can be relied upon to result in a significant improvement (and establishment of a genuine uncertainty) of preliminary elements only if the latter are very near their actual values; but it cannot automatically iron out blunders (or even indicate that we may be belabouring an entirely wrong solution). It can be invoked with profit to verify the final results already arrived at; it can demonstrate the quality of the fit they afford and distribution of the residuals; as well as provide quantitative expressions for the uncertainty within which the final elements are defined by available observational data in the least-squares sense. But it does not provide sufficient safeguards against major blunders committed at previous stsges of analysis; and may, in fact, turn an impassively blind eye on them.

The moral to be drawn from all this is simple: avoid blunders to begin with, and do not entertain vain hopes that differential corrections will automatically take care of them. Instead, embark from the outset on the path of righteousness avoiding risky short-cuts taken for the same of mere convenience, and do not stray from a course which you can justify to yourself as well as to anyone else. The principal aim of our meetings should indeed be to provide a compass which should enable you to chart just such a course.

REFERENCES.

Huffer, C. M. and Collins, G. W.: 1962, Astrophys. J. Suppl., 7, p.351.
Irwin, J. B.: 1947, Astrophys. J., 106, p.380.
Jurkevich, I.: 1970, in *Vistas in Astronomy* (ed. A. Beer, Pergamon Press, London, vol. 12, pp.63ff.
Kopal, Z.: 1941, Astrophys. J., 94, p.145.
Kopal, Z.: 1943, Proc. Amer. Phil. Soc., 86, p.342.
Kopal, Z.: 1945, Proc. Amer. Phil. Soc., 89, p.517.
Kopal, Z.: 1946, *An Introduction to the Study of Eclipsing Variables* (Harvard Obs. Mono. No. 6), Harvard Univ. Press.
Kopal, Z.: 1948, Astrophys. J., 108, p.46.
Kopal, Z.: 1959, *Close Binary Systems*, Chapman-Hall and John Wiley, London and New York.
Kopal, Z.: 1960, in "Atti del Convegno per le Celebrazioni di G. V. Schiaparelli" (ed. F. Zagar), Milano, pp.156-161.
Kopal, Z.: 1976, Astrophys. Space Sci., 45, p.269.
Kopal, Z.: 1979, *Language of the Stars* (Astrophys. Space Sci. Library, vol. 77), D. Reidel Publ. Co., Dordrecht and Boston.
Linnell, A. P. and Proctor, D. D.: 1970, Astrophys. J., 161, p. 1045; 162, p.683.
Linnell, A. P. and Proctor, D. D.: 1971, Astrophys. J., 164, p.131
Otrebski, A.: 1948, Acta Astron. (a)4, p.139.
Piotrowski, S. L.: 1947, Astrophys. J., 106, p.472.
Piotrowski, S. L.: 1948, Astrophys. J., 108, p.510.
Russell, H. N.: 1912, Astrophys. J., 35, p.315; 36, p.54.
Russell, H. N.: 1942, Astrophys. J., 95, p.345.
Russell, H. N. and Shapley, H.: 1912, Astrophys. J., 36, p.239, p.285.
Russell, H. N. and Merrill, J. E.: 1952, Princeton Obs. Contr., No. 26.
Söderhjelm, S.: 1974, Astron. Astrophys., 34, p.59.
Wyse, A. B.: 1939, Lick Obs. Bull., No.494.

ERROR ANALYSIS OF THE ELEMENTS OF AN ECLIPSING BINARY SYSTEM
IN FREQUENCY DOMAIN

I. Jurkevich, A. F. Petty and I. R. Goodman*

Naval Research Laboratory
Washington, D.C., U.S.A.

1. Introduction

A. GENERAL REMARKS

The fundamental ideas involved in the frequency domain approach to the analysis of light curves were described by Kopal (1975) in the first of a sequence of papers devoted to the subject. Details of the procedure have been elaborated upon ever since and the current state of its development is summarized in Kopal's monograph (1979a) "Language of the Stars."

For reasons fully discussed in the above references, one of the most analytically tractable methods of transforming the problem from the time domain into the frequency domain is to consider the loss of light to be a function of even powers of the sine of the phase angle and evaluate the quantities

$$A_{2m} = \int_0^{\theta'} (1 - \ell) \, d(\sin^{2m}\theta) \quad ,$$

in which θ' is the phase angle at external contact. These quantities, subsequently called moments, can be related to the system elements by means of expressions whose complexity depends on numerous factors such as, the type of eclipse, presence of proximity effects, etc. System elements are determined by solving these relations. In this lecture we shall ignore all details of how to accomplish this objective and refer the reader to Kopal (1979a) and other lectures of this summer school. Let it be assumed that the elements of an eclipsing binary system have been evaluated. Having done

*Contributed Appendix B

that, it remains to establish their reliability or the uncertainty with which they can be computed from the observations.

By way of a historical digression let us note that the need for doing this was, with some exceptions, consciously ignored from the first attempts to study the variation of light produced by mutual eclipses of stars until the decade of the 1970's. Early investigators justified this attitude by noting that visual and then photographic observations simply did not warrant the horrendous amount of hand labor involved in estimating the errors in the elements. This became an unstated article of faith which carried into the age of photoelectric photometry and automatic computation. Although a few investigators e.g. Wyse (1939), Kron (1942a, b), Tsessevich (1940, 1947), Kopal (a series of papers starting in 1941 and culminating in the monograph of 1959), recognized the need and developed the necessary procedures, the practitioners of eclipsing binary work paid no heed. For instance, Tsessevich (1947) stated that he could not understand why an observer would readily invest many months of his time to obtain the best observations of a light curve he possibly could, but would not spend an additional fraction of this time to establish whether the resulting elements could be trusted. The last decade saw a change in this attitude, yet the current literature still contains many papers on eclipsing binaries in which no mention of the reliability of the derived elements is made.

The most up-to-date summary of these questions can be found in Kopal's "Language of the Stars" and we shall consider that discussion more than sufficient justification for our subsequent work. We shall also limit the discussion to error analysis in the frequency domain.

B. RELATION OF OBSERVATIONAL ERROR TO ERROR IN THE ELEMENTS

It is clear from equation (1) that errors in observations of light intensity ℓ translate into errors in the moments A_μ ($\mu = 2m$) which in turn, via a long indirect chain of intermediate quantities, produce uncertainties in the system elements.

In order to use the frequency domain method it is, first of all, necessary to evaluate the moments A_μ from the observations. This can be accomplished by either graphical or analytical quadrature methods. In the graphical approach the observed intensities describing the light curve are plotted as functions of $\sin^2 \Theta$, $\sin^4 \Theta$, $\sin^6 \Theta$, etc., smooth curves are drawn through the points, and the areas under the respective curves are obtained by planimetry. Although the use of a free hand drawn curve is hardly the best procedure, it should not be too objectionable because the moments are derived from the entire curve, and integration is by its very nature a smoothing procedure which reduces the effects of random errors in the data points. Consequently, the contribution of any inaccuracy in the way the curve is drawn to the total error in the computed elements should be relatively small.

One quickly concludes that if planimetry is to be used, a separate plot of ℓ versus $\sin^\mu \Theta$ must be prepared for each m since there is a strong 'compresssive' effect on the abscissa with an increasing value of m. Thus, for instance, a plot of $\ell(\sin^6 \Theta)$ on the abscissa scale of $\ell(\sin^2 \Theta)$ becomes a "spike" and such a plot is no longer suitable for graphical integration. In fact, unless the external tangency occurs at a phase angle of at least 20 to 25 degrees, it is impractical to include even two successive plots on the same abscissa scale. Typical plots are shown in Figures 1 to 4.

The basic problem with this approach is that while it permits the determination of relatively good numerical values of the first four even moments, it has little to offer in the way of their uncertainty.

Integrals in Equation (1) can be evaluated numerically by any of the standard numerical procedures. Most of these methods require data at equally spaced intervals of the abscissa, a condition which normally does not occur in the observational data.

Even if one had intensities at equally spaced values of time or phase, the transformation of these variables to even powers of $\sin^2 \Theta$ makes the spacing unequal and consequently resampling would be necessary.

For the higher values of m, only the wings of the transformed curve remain discernible and, in fact, dominate the determination of the respective moments. Clearly then, if the shoulder of a minimum is sparsely observed and, perhaps, less carefully than other parts of the eclipse, the problem of both drawing a free-hand curve and selecting appropriate ordinates from it for numerical integration becomes difficult, prone to errors, and eventually leads to computational difficulties. Some impartial procedure is needed therefore to resolve this difficulty. Such a procedure must account not only for the observational errors, but if possible provide a measure of its performance in interpolating and smoothing the data.

One elementary approach to this problem is to form normal points. The use of this common procedure to reduce the number of data points in a light curve is of questionable validity since it does violence not only to both variables involved, but to their statistical descriptors as well. The problem is that in practice the number of observations going into the making of a normal point is never large enough to rescue the statistics. Furthermore, the problem of resampling for each moment would still be with us.

Another technique to obtain A_μ's which is potentially very promising is based on concepts embodied in the so-called Kalman filter. The attractiveness of this procedure is rooted in such powerful concepts as a recursive estimation in the minimum variance sense, optimal filtering, smoothing, etc.

Complete mathematical developments of Kalman's concepts have been discussed in numerous papers and monographs. For example, those of Bryson and Ho (1969) and Bucy and Joseph (1969) are particularly relevant to the present discussion. A variant of this procedure was used by Jurkevich et al. (1976) to evaluate the A_μ's.

Numerical values for the moments produced by the graphical approach and the Kalman filter are in excellent agreement and we shall use Kalman values as our standard of comparison later in the discussion. Although it is theoretically possible with the Kalman filter formalism to compute the variance of the errors in the estimates of the moments, A_μ, numerical experience has shown that the variances obtained in this way are much too small. The reason for this is not too clear at the present time. A guess is that the stochastic process chosen by us to represent the fluctuations of the light curve, while a reasonable approximation for the purpose of estimating moments is unrealistic for the purpose of determining their probable errors. Only further work will show whether there is a solution to this problem. In the meantime we appear to have reached an impasse. We have a number of ways of obtaining the moments, but not a reasonably impartial estimate of their errors.

This vexing problem has been under continuous discussion between Kopal, Jurkevich and Petty since Dr. Kopal's first stay at the Naval Research Laboratory in Washington. In due course it became apparent that the moments of a light curve defined by equation (1) are related to the local values of the Fourier transform of the light changes and that these, in turn, are expressible as functions of the amplitudes of the Fourier expansion of the corresponding section of the light curve. Theoretical background, in complete detail, is given by Kopal (Paper XXV, 1979b; or 1979a).

As far as the error analysis is concerned the merit of this development resides in the fact that the amplitudes of the Fourier expansion are readily computable by any of the conventional estimation techniques such as, for instance, the method of least squares. These methods also yield estimates of uncertainties in the estimated quantities. Thus, aside from the numerical detail of the procedure, we have a way of connecting the errors in the observations to uncertainties in the A_μ's.

In the subsequent discussion we shall explore this approach in considerable detail. The objectives will be to

- Determine the Fourier coefficients of the light changes. Considerations will be given to the statistical and computational aspects of the problem.

- Develop insight into practical bounds of the mathematical formulations.

- Compile expressions for estimating errors.

- Provide examples illustrating procedures used.

- Provide algorithms used in the computations.

For numerical examples we shall use Ebbinghausen's (1966) observations of EK Cephei. There is no partiuclar reason for this except that the observations are good and the authors have used them in some of their previous work.

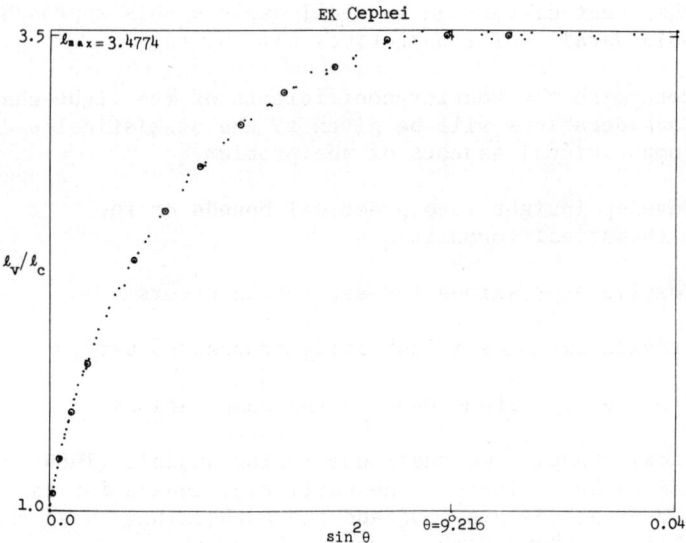

Fig. 1

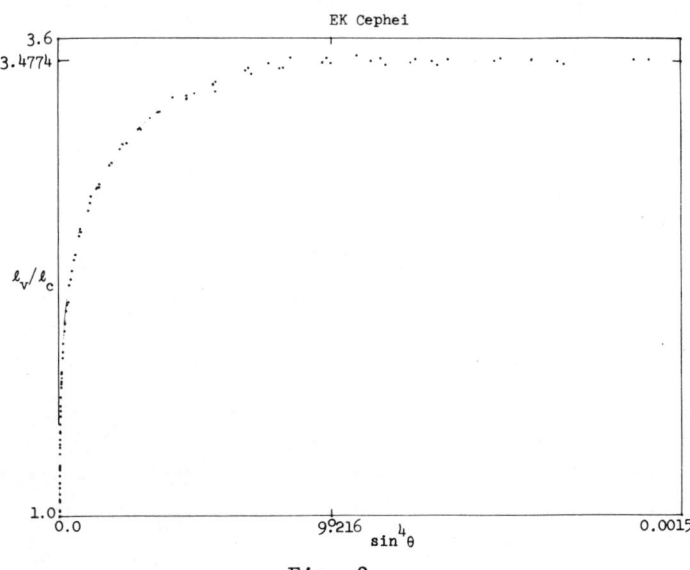

Fig. 2

Fig. 1 and 2. Plots of ℓ_v/ℓ_c for EK Cep (B) as functions of $\sin^2\theta$ and $\sin^4\theta$. Circled points in Fig. 1 represent the intensities filtered and smoothed by the Kalman filter.

ELEMENTS OF AN ECLIPSING BINARY SYSTEM IN FREQUENCY DOMAIN

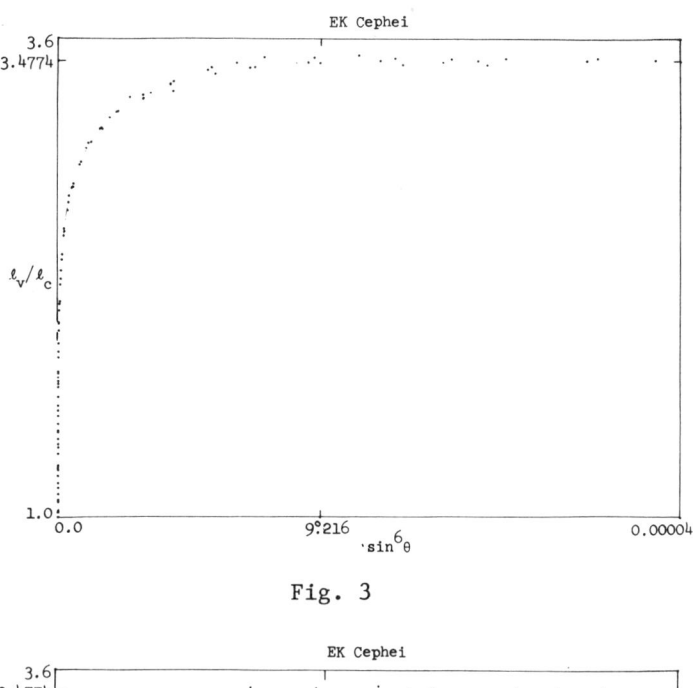

Fig. 3

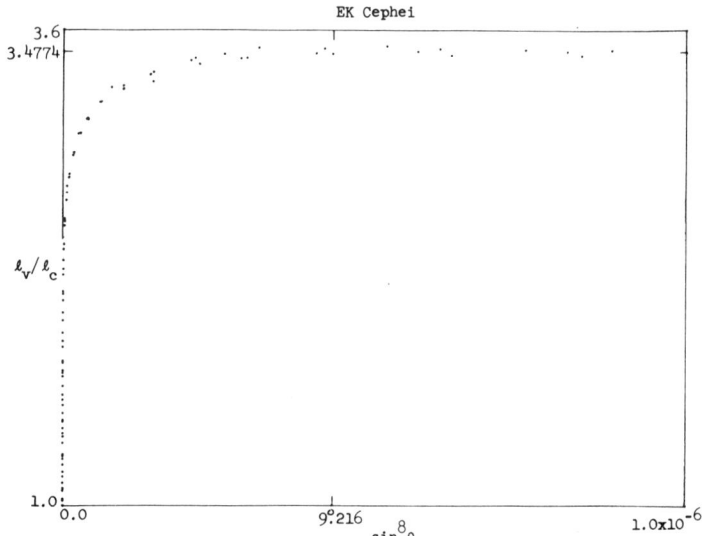

Fig. 4

Fig. 3 and 4. Plots of ℓ_v/ℓ_c for EK Cep (B) as functions of $\sin^6\theta$ and $\sin^8\theta$.

2. Determination of Fourier Coefficients

A. EQUATIONS OF CONDITION AND NORMAL EQUATIONS

Let the variation of light during eclipse be described as shown in Fig. 5.

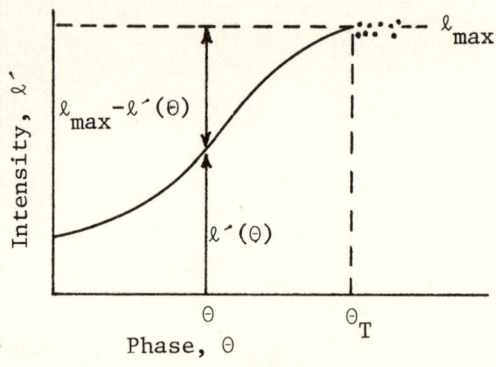

Figure 5

The loss of light at any phase, Θ, $\ell_{max} - \ell´(\Theta)$, if expressed as a Fourier series, can be written as

$$\ell_{max} - \ell´(\Theta) = \tfrac{1}{2}a´_o + \sum_{i=1}^{\infty} a´_n \cos i\, \pi(\Theta/\Theta_T) \qquad (2)$$

At this point assume that the light changes are symmetric with respect to zero phase and that no complications are present. The adopted value of ℓ_{max} is the mean of the intensities observed outside the eclipse. If, as is customary, e.g. Kopal (1959, 1979b) Piotrowski (1948), it is assumed that a systematic error is present in ℓ_{max}, the latter is written as $\ell_{maxTR} + \Delta\ell_{max}$ where ℓ_{maxTR} is the true value of ℓ_{max}. If all quantities are now normalized with respect to ℓ_{maxTR}, equation (2) becomes

$$1 + \Delta u - \ell(\Theta) = \tfrac{1}{2}a_o + \sum_{i=1}^{\infty} a_n \cos i\, \pi(\Theta/\Theta_T) \qquad (3)$$

Equation (3) will serve as the equation of condition for the determination of coefficients a_o, a_n. Although this equation contains Δu, it is not possible to determine the latter, because a_o and Δu will not generate independent normal equations. Consequently, equation (3) as it stands can yield an estimate of only a combination of a_o and Δu. To decouple these quantities we can adjoin to equation (3) an equation of the form

$$\sqrt{M}\,\Delta u = 0,$$

where $\sqrt{M} = \varepsilon/\sigma$, M is the number of observations defining ℓ_{max}, ε is the mean error of a single observation of unit weight, and σ is the standard deviation of observations from their mean. The rationale behind this relation can be found in Kopal (1959, 1979b) or Piotrowski (1948). Note, however, that even with this equation in place, no numerical value for Δu can be determined since, by definition, it will always be zero. What this procedure will do is give some idea about how uncertain that zero is and, furthermore, it will couple this uncertainty into the coefficients a_n.

Since the treatment of the above estimation problem is of fundamental importance to the subsequent analysis we give its detailed analysis in Appendix B.

At this point rewrite equation (3) as

$$\tfrac{1}{2}a_o + \sum_{i=1}^{\infty} a_n \cos i\,\pi(\Theta_k/\Theta_T) - \Delta u - (1-\ell(\Theta_k)) = 0. \qquad (4)$$

If, furthermore, we define

$$x_j = a_{j-1} \qquad j = 1, 2, 3 \ldots n-1$$

$$x_n = \Delta u$$

$$f_{1,k} = \tfrac{1}{2}$$

$$f_{i,k} = \cos i\,\pi(\Theta_k/\Theta_T) \qquad n = 1, 2, 3, \ldots n-1$$

$$\begin{aligned} f_{n,k} &= -1 & k &\leq m \\ &= \sqrt{M} & k &= m+1 \end{aligned}$$

$$\begin{aligned} g_{n+1,k} &= -\{1-\ell(\Theta_k)\} & k &= 1, 2, 3 \ldots, m \\ &= 0 & k &= m+1 \end{aligned}$$

in which k denotes the k'th observation, x_j denotes the quantities to be estimated and the f's and g's the regression functions and the free terms, the complete set of equations of condition, becomes, in matrix form,

$$\underline{F}\underline{X} - \underline{G} = 0. \qquad (5)$$

In the simplest variant of the least squares problem the normal equations are formed by left multiplying equation (5) by the transpose of F; namely,

$$\underline{F}^T \underline{FX} = \underline{F}^T \underline{G} \quad . \tag{6}$$

The solution of (6) is

$$\underline{X} = \frac{\underline{F}^T \underline{G}}{\underline{F}^T \underline{F}} \stackrel{d}{=} \underline{R}^{-1} \underline{G} \quad . \tag{7}$$

Let us designate the elements of the array $\underline{F}^T \underline{F}$ by R_{ij}. Then equation (6) can be written as,

$$\begin{pmatrix} R_{11} & R_{12} & R_{13} & \cdots & R_{1n} \\ R_{21} & R_{22} & R_{23} & \cdots & R_{2n} \\ \cdot & \cdot & \cdot & \cdots & \cdot \\ \cdot & \cdot & \cdot & \cdots & \cdot \\ R_{n1} & R_{n2} & R_{n3} & \cdots & R_{nn} \end{pmatrix} \begin{pmatrix} x_1 \\ x_2 \\ \cdot \\ \cdot \\ x_n \end{pmatrix} + \begin{pmatrix} R_{1,n+1} \\ R_{2,n+1} \\ \cdot \\ \cdot \\ R_{n,n+1} \end{pmatrix} = 0 \tag{8}$$

The array of R_{ij}'s is a symmetric one, that is, $R_{ij} = R_{ji}$. Written explicitly, the elements become

$$R_{11} = \frac{m}{4} ,$$

$$R_{i1} = \sum_{k=1}^{m} \tfrac{1}{2} \cos(i-1)\pi (\Theta_k/\Theta_T) ,$$

$$R_{ij} = \sum_{k=1}^{m} \cos(i-1) \pi (\Theta_k/\Theta_T) \cos(i-1) (\Theta_k/\Theta_T) ,$$

for (1 < i < n and 1 < j < n).

Coefficients of the Δu terms become

$$R_{in} = \sum_{k=1}^{m} (-\tfrac{1}{2}) = -\frac{m}{2} \qquad \text{if } i = 1 ,$$

$$= \sum_{k=1}^{m} (-1)\cos(i-1)\pi(\Theta_k/\Theta_T) \qquad \text{if } n > i > 1 ,$$

$$R_{nn} = m+M \qquad \text{if } i = n ;$$

and

ELEMENTS OF AN ECLIPSING BINARY SYSTEM IN FREQUENCY DOMAIN

Absolute or free terms represented by

$$R_{n+1,1} = -\tfrac{1}{2} \sum_{k=1}^{m} (1-\ell_i),$$

$$R_{n+1,i} = - \sum_{k=1}^{m} (1-\ell_i) \cos (j-1)\pi(\Theta_k/\Theta_T) \text{ for } n > j > 1,$$

$$R_{n+1,n} = \sum_{k=1}^{m} (1-\ell_i) \qquad \text{for } n = j.$$

For instance, if we limit ourselves to a second harmonic approximation the equations of condition become

$$\tfrac{1}{2}a_o + a_1 \cos\pi (\Theta_i/\Theta_T) + a_2 \cos 2\pi(\Theta_i/\Theta_T) - \Delta u(1-\ell_i) = 0,$$

$$\sqrt{M}\, \Delta u = 0\ ;$$

and the normal equations would be given by

$$a_o \tfrac{1}{2}m + a_i \tfrac{1}{2}\sum \cos\pi(\Theta_i/\Theta_T) + a_2 \tfrac{1}{2}\sum \cos 2\pi (\Theta_i/\Theta_T) - m\Delta u - \tfrac{1}{2}\sum(1-\ell_i) = 0,$$

$$\tfrac{1}{2}a_o \sum \cos\pi(\Theta_i/\Theta_T) + a_1 \sum \cos^2 \pi (\Theta_i/\Theta_T) + a_2 \sum \cos \pi(\Theta_i/\Theta_T)\cos 2\pi(\Theta_i/\Theta_T)$$
$$- \Delta u \sum \cos\pi(\Theta_i/\Theta_T) - \sum(1-\ell_i)\cos(\Theta_i/\Theta_T) = 0,$$

$$\tfrac{1}{2}a_o \sum \cos 2\pi(\Theta_i/\Theta_T) + a_1 \sum \cos\pi(\Theta_i/\Theta_T)\cos 2\pi(\Theta_i/\Theta_T) + a_2 \sum \cos^2 2\pi(\Theta_i/\Theta_T)$$
$$- \Delta u \sum \cos 2\pi(\Theta_i/\Theta_T) - \sum(1-\ell_i)\cos 2\pi(\Theta_i/\Theta_T) = 0,$$

$$-\tfrac{1}{2}ma_o - a_1 \sum \cos\pi(\Theta_i/\Theta_T) - a_2 \sum \cos 2\pi(\Theta_i/\Theta_T) + \Delta u(M+m) + \sum(1-\ell_i) = 0.$$

It should be mentioned at this point that Θ_T is assumed known. In reality the point of external tangency is only an estimate and the ultimate effect of its uncertainty on system elements must be ascertained. The discussion of this question is given by Dr. Giménez in another chapter appearing in the proceedings. Another open question is the number of harmonics to use in the regression relation. This point will be discussed later.

B. SOLUTION OF THE NORMAL EQUATIONS

In principle the solution of the normal equations is given by expression (7). There are numerous standard procedures for solving n equations in n unknowns and a specific choice is governed by considerations such as, minimization of arithmetic operations, preservation of accuracy, ease with which the solution yields estimates of errors, and, to some extent the labor involved.

If we carry out the necessary computations with the matrix formalism and if we designate elements of the inverse matrix R^{-1} by R^{ij} it can be shown that the standard errors and covariances of estimated quantities are related to R^{ij} by

$$SE(x_i) = \sigma\sqrt{R^{ij}}, \qquad (9)$$

$$COV(x_i x_j) = \sigma^2 R^{ij}.$$

The true value of the error variance, σ^2, of the observed quantities is not known a priori. It can be estimated, however, from the residual fluctuations of the observed light changes about the values predicted by the regression relation. In our previous notation the residual sum of the squares, v^2, will be defined as

$$v^2 = \sum_k (g_{(n+1,k)obs} - g_{(n+1,k)com})^2 = \sum_k g^2_{(n+1,k)obs}$$
$$- x_1 \sum_k f_{1,k} \, g_{(n+1,k)obs} - x_2 \sum_k f_{2,k} \, g_{(n+1,k)obs} - \cdots$$
$$- x_n \sum_k f_{2,k} \, g_{(n+1,k)obs},$$

in which $g_{(n+1,k)obs} = 1 - \ell(\Theta_k)$ as actually observed and

$g_{(n+1,k)com}$ = the value predicted by equation (4).

The quantity $(x_1 \sum_k f_{1,k} g_{(n+1,k)obs} + \cdots x_n \sum_k f_{n,k} g_{(n+1)obs})$ is referred to as the sum of squares due to regression. In terms of the above quantities the unbiased estimate, s^2, of the error variance, σ^2, is given by

$$s^2 = \frac{v^2}{m-t+1},$$

in which m continues to be the number of independent observations and t is the number of parameters estimated.

Apart from the numerical problems which can arise in inverting a specific matrix we should point out that if the near dependencies between estimated quantities are of interest, then the above approach permits easy detection of correlations only between pairs of variables. Correlations between more than two variables are practically impossible to detect by inspection of the variance-covariance matrix (9).

The general problem of fitting observational data to an appropriate theory and the associated problem of error analysis has given rise to a vast amount of literature. A brief historical review can be found in Kopal's 1979a monograph in which a case is made for the use of a method proposed by Banachiewicz (1938, 1942). The valuable feature of this approach is that as a by-product of solving for the unknowns in the normal equations, the method yields all elements of the conventional inverse matrix which are required to compute weights of any linear combination of the estimated parameters.

Historically, the method has its origins in the work of the French geodesist A. L. Cholesky (1916, described in 1924 by Benoit). Banachiewicz, in the course of developing his "Cracovian" calculus proposed a scheme for solving normal equations by means of a "Cracovian Sqare Root." The Cholesky and Banachiewicz algorithms are in effect identical. However, Banachiewicz generalized the algorithm to solve normal equations by the method of undetermined multipliers. In the literature this approach appears under various names, e.g. the method of Cholesky, method of square roots, Banachiewicz algorithm, and the "Cracovian" method.

We propose to use this method in our subsequent discussion for the following reasons:

(a) It is relatively easy to mechanize and it is modest in its requirements for machine memory.

(b) Compared to elimination schemes (e.g. Gauss-Jordan), elements of the "square root matrices" are generally numerically larger than those obtained for elimination equations. Therefore, the relative effect of round-off errors is smaller and, for a fixed precision, the unknowns are determined with higher precision.

(c) The effect of correlations is easier to account for.

In what follows we give a brief account of this method.

C. ALGORITHM OF BANACHIEWICZ

A symmetric matrix, such as that comprising the coefficients of the normal equations, can be decomposed into a product of two mutually transposed triangular matrices, namely

$$\underline{R} = \underline{r}^T \underline{r}, \qquad (10)$$

where

$$\underline{r} = \begin{pmatrix} r_{11} & r_{12} & r_{13} & \cdots & r_{1n} \\ 0 & r_{22} & r_{23} & \cdots & r_{2n} \\ \cdot & \cdot & \cdot & & \cdot \\ 0 & 0 & 0 & \cdots & r_{nn} \end{pmatrix}.$$

The elements of $\underline{R}$ are related to elements of $\underline{r}$ by

$$R_{ij} = r_{1i} r_{1j} + r_{2i} r_{2j} + \cdots r_{ii} r_{ij} \quad i < j,$$

$$R_{ii} = r_{1i}^2 + r_{2i}^2 + \cdots r_{ii}^2 \quad i = j.$$

Consequently

$$r_{11} = \sqrt{R_{11}} \quad \text{or} \quad r_{11} = R_{11}/r_{11},$$

$$r_{1j} = R_{1j}/r_{11},$$

$$r_{ii} = (R_{ii} - \sum_{k=1}^{i-1} r_{ki}^2)^{\frac{1}{2}} \quad i > 1,$$

$$r_{ij} = (R_{ij} - \sum_{k=1}^{i-1} r_{ki} r_{kj})/r_{ii} \quad j > i,$$

$$r_{ij} = 0 \quad i > j.$$

The solution of the original system

$$\underline{R}\underline{X} = \underline{G}$$

is reduced now to the solution of two triangular systems

ELEMENTS OF AN ECLIPSING BINARY SYSTEM IN FREQUENCY DOMAIN

$$\underline{r}^T \underline{K} = \underline{G} \quad \text{and} \quad \underline{r}\underline{X} = \underline{K} \ .$$

The first of these relations yields elements of the vector $\underline{K}$:

$$k_1 = \frac{g_1}{r_{11}} \quad \text{and} \quad k_i = (g_i - \sum_{j=1}^{i-1} r_{ji} k_j)/r_{ii}, \quad i > 1 \ . \tag{11}$$

The second produces the element of $\underline{X}$:

$$x_n = \frac{k_n}{r_{nn}}, \quad x_i = (k_i - \sum_{j=i+1}^{n} r_{ij} x_j)/r_{ii} \quad i < n \ . \tag{12}$$

To apply this to the specific problem formulated in (8), interchange column and row indices (in keeping with the "Cracovian" formalism), include the absolute terms in the $\underline{R}$ matrix and operate on them in exactly the same way as the rest of the elements. This will, in effect, generate the elements of the $\underline{K}$ vector directly. Next compute the inverse of the $\underline{r}$ matrix explicitly.

Let the matrix of normal equations be given by

$$\begin{pmatrix} R_{11} & R_{21} & R_{31} & \cdots & R_{n1} & \vdots & R_{n+1,1} \\ & R_{22} & R_{32} & \cdots & R_{n2} & \vdots & R_{n+1,2} \\ & & \cdots & & \cdot & & \cdots \\ & & & & R_{nn} & \vdots & R_{n+1,n} \end{pmatrix}$$

and $R_{ij} = R_{ji}$.

The decomposition of R follows the discussion just outlined. The elements of the upper triangular matrix $\underline{r}$ will be designated by

$$\begin{pmatrix} r_{11} & r_{21} & r_{31} & \cdots & r_{n+1,1} \\ 0 & r_{22} & r_{32} & \cdots & r_{n+1,2} \\ 0 & 0 & r_{33} & \cdots & r_{n+1,3} \\ \cdot & \cdot & \cdot & \cdots & \cdot \\ 0 & 0 & 0 & \cdots & r_{n+1,n} \\ 0 & \cdot & \cdot & \cdots & 1 \end{pmatrix} \tag{13}$$

$r_{n+1,n+1} = 1$ if the absolute terms are transposed to the left side of the normal equations.

An algorithm for the inversion of a triangular matrix is derived from the following considerations. Suppose we have a matrix relation

$$\underline{u} = \underline{r}\underline{p} \; ;$$

that is,

$$u_1 = r_{11}p_1 \, ,$$
$$u_2 = r_{21}p_1 + r_{22}p_2 \, ,$$
$$\cdot$$
$$\cdot \quad \cdot \cdot \cdot$$
$$\cdot$$
$$u_n = r_{n1}p_1 + r_{n2}p_2 \cdots + r_{nn}p_n \, .$$

Solving this system by successively eliminating each additional unknown in terms of the quantities already computed we find the elements of vector $\underline{p}$ in terms of the elements of vector $\underline{u}$; namely,

$$p_1 = q_{11}u_1 \, ,$$
$$p_2 = q_{21}u_1 + q_{22}u_2 \, ,$$
$$\cdots$$
$$p_n = q_{n1}u_1 + q_{n2}u_2 + \cdots q_{n1n}u_n \, .$$

It is easily shown that the diagonal elements q_{ii} are given by

$$q_{ii} = 1/r_{ii} \, .$$

The remaining elements are obtained from

$$q_{i,i+k} = (-\sum_{j=0}^{k-1} q_{i,i+j} r_{i+k,i+j})/ r_{i+k,i+k} .$$

The product of $\underline{r}^T\underline{q}$, taken column by column, yields a unitary matrix I. This can serve as a control on the numerical calculations. In this case, since the matrix $\underline{R}$ includes the absolute terms, the last row of the q matrix represents the desired unknowns. Their uncertainties are specified by the sum of the squares of the remaining elements of the respective column. Thus,

$$x_i = q_{i,n+1} \pm \varepsilon \left(\sum_{j=1}^{n} q_{ij}^2 \right)^{1/2}, \quad i = 1, 2, \ldots, n.$$

The estimate of ε, the mean error of a single equation of condition, is given by S^2, defined earlier in this discussion. In our current notation the sum of the squares due to regression is given by

$$\sum_{j=1}^{n} r_{n+1,j}^2 ;$$

and, therefore, the residual sum of the squares becomes

$$v^2 = \sum_{k=1}^{m} g_{(n+1,k)obs}^2 - \sum_{j=1}^{n} r_{n+1,j}^2$$

$$= \sum_{k=1}^{m} (1 - \ell(\Theta_k))^2 - \sum_{j=1}^{n} r_{n+1,j}^2 .$$

This completes the discussion on computing the a_i's. Keep in mind, however, that we have assumed a value of Θ_T and that we terminated our approximation at the n'th harmonic. We have sidestepped the question of what to do about Θ_T. The question of selecting the highest harmonic to use will, for the time being, be handled as follows. Start computations with an arbitrary harmonic, say, the first or second, carry out the computations; increment the highest harmonic, repeat computations, and so on. At each stage monitor the value of S^2 and terminate the computation either when it reaches a sufficiently small value or when it ceases to change significantly. To do this properly requires setting up a statistical significance test for the effect of the added terms. For now, ignore this requirement, although Appendix B contains a discussion of the necessary procedure.

D. MECHANIZATION OF THE COMPUTATIONS

Certain sections of the discussion will be accompanied by actual computer codes. These codes are reasonably well annotated, but a brief discussion is attached to each code. The codes are written in FORTRAN IV tailored for Data General Corp. (DGC) machines.

Except for the input/output operations and a few minor deviations this version is basically standard ANSI FORTRAN IV. The codes as given, are working codes, checked on a NOVA 800 and on an ECLIPSE S/200 computers. These are exploratory codes and so no attempt was made to optimize their performance. As far as the machines are concerned, they represent an older and a newer version of typical minicomputers of modest size (memory of 32 thousand 16 bit words). At our installation a set of magnetic tapes and disk drives is attached to them. The operating system is a fairly sophisticated disk oriented system. It is useful to point out that programs discussed here could be run easily on desk-type computers such as a Hewlett-Packard System 45. Finally the codes, as given, are interactive in nature.

The enclosed codes contain certain common features which will be discussed in this section

(a) Input-Output (I/O)

Input/output activity is always specific to a given machine and its operating system. Normally there would be no reason to describe these for our machine. However, we hope to give you enough understanding of our specific code, so that you can adapt it to your own installation.

Since our operating system is disk oriented, we find it convenient to keep programs, data, etc. on the disk in units called "files." Each file is suitably organized and has a user assigned name. Access to a file is made via its name. For instance, in the attached programs observations of EK Ceph are stored in a file named "DATA." Communication with this file is established by a call to a system subroutine OPEN. Call parameters signify that the communication will be via channel $\emptyset$, with the file "DATA," and that the file is open for both reading and writing. Should an error occur during this procedure, its code will be contained in IER$\emptyset$. The file "DATA" has been filled with observations of phase T, and intensity EL, by a separate program. It was done in such a manner that a full set of T's was stored in 4 disk blocks (256-16 bit words/block) followed by the corresponding set of EL's also stored in 4 blocks.

To call the observations into memory, the use is made of a system routine which performs direct block I/O. To read the disk, the name is RDBLK, to write to the disk the name is WRBLK. Unfortunately, in our version of the FORTRAN IV libraries these names are interchanged and, therefore, in the programs you see WRBLK even though we read the disk. The call parameters signify that communication takes place on channel $\emptyset$. starting block is $\emptyset$ or 4, the arrays in memory which accept the data are IT and IBR, 4 blocks at a time are transferred into memory and that, should errors occur, the error codes will be contained in IER1 and IER2. Two calls are made due to the nature of the data file "DATA."

A problem peculiar to DGC FORTRAN should be mentioned. Direct block I/O requires integer arrays, yet our observations are real numbers. To enable the direct block I/O of real quantities we equivalence arrays T and IT and EL and IBR. Also, since a real number is stored in two words whereas an integer requires one word the real arrays must have a dimension twice as larger as the corresponding integer arrays. This is not a very efficient use of memory, but that is how life is if one wants direct block I/O. In this system one can equivalence only those arrays whose elements are in COMMON area. Once the data have been read into the memory the communication channel, $\emptyset$, is broken by a call to a system subroutine CLOSE.

For a different computer arrays IT and IBR, COMMON and EQUIVALENCE statements, and statements from CALL OPEN through CALL CLOSE will have to be altered or replaced altogether.

The user interaction with the program is via a system terminal. In response to an ACCEPT statement the user keys in the appropriate parameters in free format. TYPE statements produce output on the terminal. Most of the output is to logical unit 12 which is the system line printer.

(b) Precision and the Run-time Stack

Precision of the floating point quantities in a 16 bit computer is about 6 significant places. Since the inversion of matrices is notorious for a loss of precision we have decided to carry out all calculations in double precision (64 bits). Command COMPILER DOUBLE PRECISION accomplishes this objective.

In our computer, quantities not in COMMON storage are put on the so-called run-time stack, a dynamically allocated section of memory. It is a great memory saver provided you do not re-enter the program. If you do you will not find certain quantities that you loaded, like input parameters and data. The stack is a dynamically changing entity and the required quantities have been overwritten by something else. Since this is an interative computation this problem is present on each return to the beginning. The simplest way to avoid this is to allocate permanent storage to all quantities by the use of a command COMPILER NOSTACK.

Finally certain lines in the attached programs are labeled with an X in the first column. The DGC FORTRAN compiler allows us to either include or exclude such lines from the program at the compilation time. This feature is valuable in the course of program development when a look at intermediate results is desirable. For instance, BINMOMENTS permits us to examine the variability of $\varphi_i(\mu)$. Once established, however, there is no need to print them on each iteration. Therefore, the corresponding lines of the code are labeled with an X.

(c) Computation of the Fourier Amplitudes, a_i.

The program which implements the computation of the Fourier coefficients, a_i, is listed in Appendix A under the name BINMOMENTS.

The program is annotated in reasonable detail. The input quantities are

    NMAX = the highest harmonic used
      MI = number of observations in the minimum
    NMAX = number of observations outside the minimum
    ISWITCH1 is initially set to Ø to read data from the disk.
        On subsequent iterations it is set to 1
    IEND is initially set to Ø. To terminate the session, set to 1 on return for the next iteration.

Typical output for NMAX = 2 and 3 is attached in the appendix. It consists of

    RU   Elements of matrix $\underline{R}$
    RL   Elements of matrix $\underline{r}$
    Q   Elements of matrix $\underline{q}$

each accompanied by their column and row indices j,i. The quantities labeled A(I) and AER(I) are

    a_o error in a_o
    a_1 error in a_1
    ...   ...
    s^2 estimate of error variance

Subroutines used: CRACOVIAN
Call Parameters: RU = $\underline{R}$ matrix
                    NP1 = number of columns in $\underline{R}$ matrix
                    RL = $\underline{r}$ matrix, returned
                     Q = $\underline{q}$ matrix, returned.

We have found that with the Ebbinghausen's data for EK Cephei one can go up to the 8'th harmonic before a decision should be made to terminate the computation.

3. Computation of the A_μ's from the a_i's

Equation (3.17) of Kopal's paper XXV (1979b) or equation (3.37), p. 241 of Kopal(1979a) show that A_μ is a linear combination of a_i's e.g.

$$A_\mu = \sum_{i=0}^{N} \varphi_i(\mu) \, a_i, \qquad (14)$$

in which

$$\varphi_i(\mu) = \frac{\Gamma(\mu+1)}{2^\mu} \sum_{j=0}^{\infty} \frac{(-1)^{i+j}[(2j+1)\Theta_T]^2}{\Gamma(\frac{\mu+3}{2}+j)\Gamma(\frac{\mu+1}{2}-j)\{[(2j+1)\Theta_T]^2 - (i\pi)^2\}} \sin(2j+1)\Theta_T$$

$$= \frac{\Gamma(\mu+1)}{2^\mu} \sum_{j=0}^{\infty} \varphi_{i(\mu)}^{(j)}, \qquad (15)$$

It is claimed that this expression is valid for $\mu \geq 0$, integral even or odd, or fractional. The heart of the above expression is the Fourier expansion of $\sin^\mu(\Theta)$ in terms of its odd harmonics as given by Oberhettinger (1973),

$$\sin^\mu \Theta = \frac{\Gamma(\mu+1)}{2^{\mu-1}} \sum_{j=0}^{\infty} \frac{(-1)^j}{\Gamma(\frac{\mu+3}{2}+j)\Gamma(\frac{\mu+1}{2}-j)} \sin(2j+1)\Theta$$

If we limit ourselves to integral moments, we note that for odd moments the series in $\varphi_i(\mu)$ terminates in a finite number of terms since $\Gamma(\frac{\mu+1}{2}-j) \to \infty$ for $\frac{\mu+1}{2} = j$. For even moments one must decide on the desired precision in evaluating φ_i, that is, on the number of terms in the series. It is reasonable to assume that φ_i should be at least as accurate as the precision with which the corresponding a_i's have been computed. For our numerical example it ought to be good to about 1 part in 10^3. At this point we encounter the following difficulty. Figures 6 and 7 show the behavior of the dominant term of $\varphi_i(0)$ for $i = 0, 1, 2, 3$ and the increasing values of j. With the exception of $\varphi_o^{(0)}(0)$,

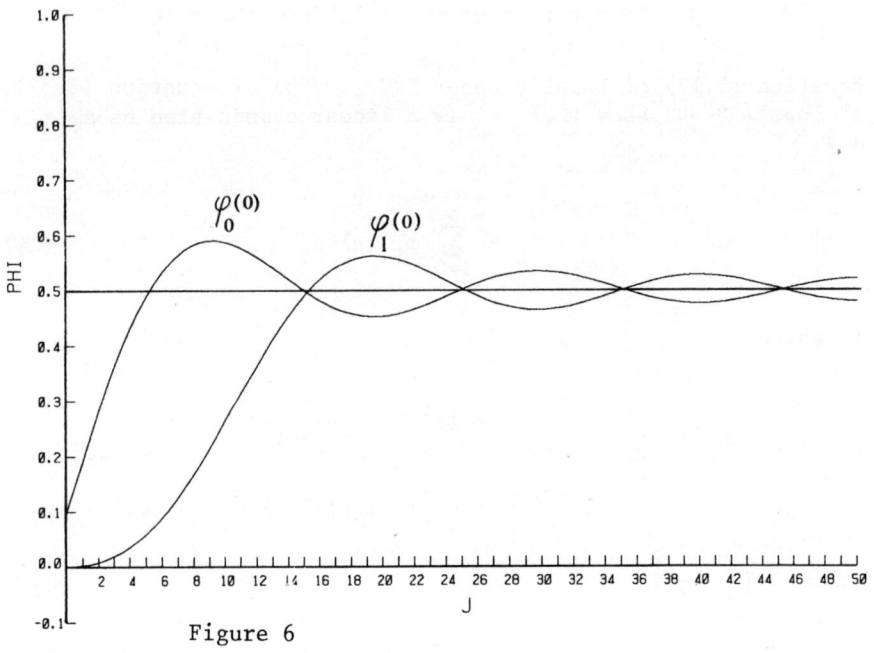

Figure 6

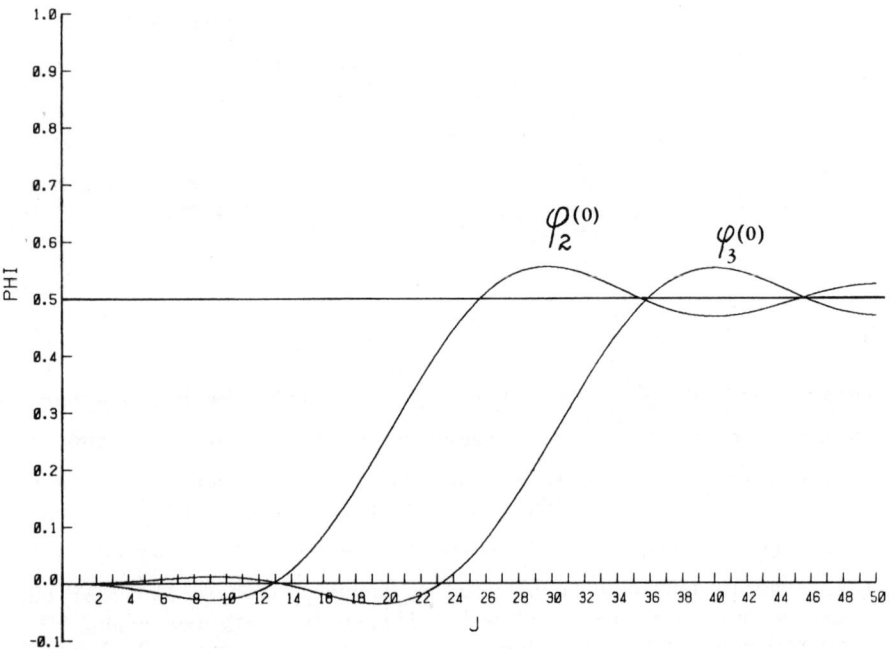

Figure 7

the first term of each is a small positive or negative value, oscillates for a while around zero, then builds up in an oscillatory fashion to a value of 0.5. It is easy enough to show that in the limit of large j, $\varphi_o(0) = 0.5$. What is surprising is that 0.5 seems to be the limit regardless of the value of i and Θ_T. We have not been able to prove that this is so, and for the time being we shall consider this to be a conjecture. There are two other numerical difficulties. The first of these is an excessively slow convergence of $\varphi_i(0)$. To achieve the required precision the value of j had to exceed 50. Unfortunately with our 16 bit machine $\Gamma(x)$ functions exceed at this point the largest floating point number allowed. A solution to this difficulty will be mentioned shortly. The second difficulty is associated with the factor $[(2j+1)^2 \Theta_T^2 - (i\pi)^2]^{-1}$ of equation (15). It is conceivable that for a specific eclipsing system this quantity could become nearly singular. By pure bad luck, that is precisely what happens for EK Cephei for which $\Theta_T = 8°.78$ (.1533 rad). For $i = 2$ and $j = 20$, $(2j+1)\Theta_T = 6.285 \approx 2\pi$ to within 3 places in the third decimal. Consequently $\varphi_2^{(20)}(\mu)$ is always of questionable value. Let us return for a moment to the problem of computing $\varphi_i^{(j)}(0)$ for $j > 50$. If an asymptotic approximation to $\Gamma(x)$ is used then

$$[\Gamma(\tfrac{\mu+3}{2}+j)\Gamma(\tfrac{\mu+1}{2}-j)]^{-1} \sim \frac{(j-\tfrac{\mu+1}{2j})^{j-\mu/2}}{(j-\tfrac{\mu-1}{2j})^{j+1+\mu/2}} \frac{e^{(1+\mu)}}{\pi} \sin(j-\tfrac{\mu-1}{2})\pi \ . \tag{16}$$

This form permits our machine to carry computations for j equal to several tens of thousands without causing floating number overflow. What we have done then to study the behavior of $\varphi_i(0)$ is to compute $\varphi_i^{(j)}(0)$ from equation (15) for $j \leq 50$, and for $j > 50$ we use equation (16). With machines having a larger word size the onset of such problems will occur for much larger values of j than discussed here and, consequently, may not require special handling.

Moreover, for even values of $\mu \geq 2$ the problem disappears entirely. For instance, for $\mu = 2$, $\varphi_o(2)$ is computable to one part in 10^5 with about 20 terms. As μ and i increase the number of terms required to reach a given precision rapidly decreases. The problem then is the computation of $\varphi_i(0)$, that is, of A_o.

Yet even A_o is apparently computable if we take j large enough. It is curious, however, that for $\mu = 0$ all φ_i appear to converge to 0.5, so that

$$A_o = 0.5 \sum a_i . \qquad (17)$$

Now A_o is supposed to represent the ordinate of the light curve at zero phase. For EK Cep the observed ordinate and that given by (17) are different by 26%. The discrepancy is not due to an inadequate approximation of the light curve by a_i's, since with 6 to 8 harmonics the approximation is good to several parts in 10^3 at any observation. The problem is burried in $\varphi_i(\mu)$. As it turns out the difficulty extends over all values of μ, as shown in Table 1. It can be seen that even in the case of odd moments, when the series in (15) terminates, equation (14) does not yield the correct values of A_μ.

Table 1

Comparison of A_μ Computed by Kalman Filter and Fourier Expansion (NMAX = 6)

Moment	Kalman Filter	$\Sigma \varphi_i a_i$
A_1	5.203×10^{-2}	5.198×10^{-2}
A_2	4.831×10^{-3}	6.401×10^{-3}
A_3	5.104×10^{-4}	8.612×10^{-4}
A_4	5.855×10^{-5}	
A_5	7.070×10^{-6}	17.676×10^{-6}
A_6	8.916×10^{-7}	
A_7	1.149×10^{-7}	
A_8	1.529×10^{-8}	

Originally when we undertook the computation of the even moments, the above difficulties were attributed to convergence problems, particulary in view of our experience with A_o. It was then thought reasonable to compute odd moments for which the convergence question does not arise, and then use the recursion relations suggested by Kopal (1979a, L. of S.) to arrive at the even moments. The coefficients of these relations were generated by the code AMOM. At the outset it became clear that this approach cannot generate A_o. But it also failed for all values of μ. When the Kalman filter values were inserted into the recursion relations based on at least the 5th order Legendre polynomial, the correct Kalman even moments were generated. But if the odd moments obtained from the φ_i functions were used, incorrect even moments were generated. This clearly indicates that there is a fatal flaw of some sort in φ_i.

The code with which the behavior of $\varphi_i(\mu)$ was studied is shown in the appendix under the names PHICALC and PHINEW. A subroutine DGAMMA to compute the required values of $\Gamma(x)$ is included.

The code is sufficiently annotated and, hopefully, requires no further discussion. A typical output is also attached.

Essentially, expression (15) as it stands does not yield a reasonable value of A_μ for any value of μ. Assuming that there is no flaw in the theoretical development, it is reasonable to question the various approximations used in arriving at equation (15).

Recall that in the defining expression for A_μ, $\sin^\mu\theta$ in the differential element was replaced by its Fourier Series expansion in terms of the odd sine harmonics. How good is this approximation, and, what is more important, does its derivative represent the actual derivative $d(\sin^\mu\theta)$? At the very outset we ought to point out a discrepancy between Oberhettinger's and Kopal's statement about the validity of the Fourier expansion of $\sin^\mu\theta$ at $\theta = 0$. Oberhettinger seems to imply that it is not valid, Kopal says it is. We shall side with the latter author. It is not too difficult to convince ourselves of this. For instance, plotting $\sin^2\theta$ and its counterpart from the Fourier Series note that the discrepancy between the two is barely discernible even for 5 terms, practically indistinguishable for 10 terms, and for our purposes the difference is practically zero for 40 terms. The agreement for $\sin^4\theta$ is even better. It should be pointed out that whatever discrepancies exist they are most noticeable for small values of θ, say $\theta \leq 10^0$. Carrying out a similar analysis for the derivatives discloses a much poorer agreement. Thus, for $d(\sin^2\theta)$ the discrepancy does not entirely disappear even after

40 terms. For $d(\sin^4 \theta)$ the agreement is dramatically better for as few as 5 terms. Again the region of greatest error occurs for $\theta < 10^0$. Therefore the first few moments, say A_0, A_1, A_2, and A_3, will be affected the most by our approximations. Also, the error will be greater the smaller the moment of external contact and since the behavior of our approximations near $\theta = 0$ is questionable in any event, the Fourier expansion of $\sin^\mu \theta$ may not be valid for $\mu = 0$ and, consequently, A_0 may not be computable at all. Yet if a sufficiently large number of terms is taken, all errors tend to diminish. This indicates that for the series in question all conditions for differentiability are satisfied. What is then the problem? Smoothing the approximation by a sinx/x filter, that is, multiplying each term in the series for φ_i by sinx/x in which $x = (2j+1)\pi/(2N+1)$ and N is the number of terms in the sum will make things even worse. The reason being that for any φ_i, the first term $\varphi_i^{(0)}$ is considerably larger than the expected value of the corresponding A_μ, and the second and succeeding terms are too small to decrease it to any great extent. The filter in question will not affect the first term since $\lim_{x \to 0}(\sin x/x) = 1$, and it will only further diminish the contribution of the other terms.

At this point the only remaining possibility is that equation (15) is in error.

    a. Re-examination of Equation (15).

A re-examination of the derivation of equation (15) indicates that the problem lies in the relation between the local value of the Fourier transforms, F_1, of the light curve and the Fourier amplitudes in the corresponding series expansion of the light changes.

The definition of F_1 (e.g. Kopal L. of S., p. 238) is

$$F_1(\nu) = 2 \int_0^{\theta_T} (1-\ell) \cos h\theta \, d\theta .$$

Consider now Oberhettinger's expansion for $\cos \mu\theta$ in terms of cosine harmonics (Oberhettinger, 1.60, p. 13). If in this expression we identify

ELEMENTS OF AN ECLIPSING BINARY SYSTEM IN FREQUENCY DOMAIN

$$\mu = \frac{\pi b}{a \ell}, \quad \ell = \Theta_T,$$

we can write

$$\cos h\Theta = \frac{h\Theta_T}{\pi^2} \sin\mu\Theta_T \sum_{n=0}^{\infty} \frac{(-1)^n \varepsilon_n \cos(n\pi\Theta/\Theta_T)}{(h\Theta_T/\pi)^2 - n^2}.$$

Following the same interpolation procedure used by Kopal to arrive at the local value of the Fourier transform we obtain

$$F_1\left(\frac{2j+1}{2\pi}\right) = \frac{\sin(2j+1)\Theta_T}{(2j+1)} \sum_{n=0}^{\infty} \frac{(-1)^{n+1} \varepsilon_n a_n}{\left[\frac{n\pi}{(2j+1)\Theta_T}\right]^2 - 1}.$$

This is the same expression as Kopal's eq. 3.32, p. 240 (L. of S.) except for the Neumann's number ε_n. This factor must also appear in equations 3.36 and 3.37, p. 241, of L. of S. The implication then is that our equation (14) ought to read

$$A_\mu = \sum_{i=0}^{\infty} \varphi_i(\mu) \, \varepsilon_i a_i \tag{18}$$

in which $\varepsilon_o = 1$ and $\varepsilon_i = 2$, $i = 1, 2, \ldots$.

If we use equation (18) to evaluate the A_μ's we get

$$A_0 = 0.7077,$$
$$A_1 = 5.204 \times 10^{-2},$$
$$A_2 = 4.839 \times 10^{-3},$$
$$A_3 = 5.156 \times 10^{-4},$$
$$A_4 = 5.864 \times 10^{-5},$$
$$A_5 = 7.138 \times 10^{-6}.$$

The above values are based on a regression with six harmonics, but the actual sums contain only 3 to 4 terms. The displayed quantities compare very well indeed with those obtained by the Kalman filter. In fact, filtering the series with a sin x/x filter is useful only to decrease the number of terms in the series, but otherwise there is little need for it since it contributes little to the precision of the sum. Having overcome the above hurdle, consider now the errors of the moments A_μ.

4. Errors in Moments A_μ

It is useful, at this point, to recall the form of the inverse matrix, q, of the normal equations, namely,

$$\underline{q} = \begin{pmatrix} q_{11} & 0 & 0 & \cdots & 0 & 0 & | & 0 \\ q_{12} & q_{22} & 0 & \cdots & 0 & 0 & | & 0 \\ q_{13} & q_{23} & q_{33} & \cdots & 0 & 0 & | & 0 \\ \cdot & \cdot & \cdot & \cdots & \cdot & \cdot & | & \cdot \\ \cdot & \cdot & \cdot & \cdots & \cdot & \cdot & | & \cdot \\ q_{1,n} & q_{2,n} & q_{3,n} & \cdots & q_{n-1,n} & q_{n,n} & | & 0 \\ \hline q_{1,n+1} & q_{2,n+1} & q_{3,n+1} & \cdots & q_{n-1,n+1} & q_{n,n+1} & | & 1 \end{pmatrix},$$

Recall that $q_{i,n+1}$, the elements of the last row are the unknowns (a_i's) of the problem and that the moments A_μ are a linear combination of a_i's. If we multiply the above matrix by a column matrix (vector) $(f_1\ f_2\ f_3\ \ldots f_n\ 0)$ we obtain the following column matrix

$$\begin{pmatrix} q_{11}f_1 \\ q_{12}f_1 + q_{22}f_2 \\ q_{13}f_1 + q_{23}f_2 + q_{33}f_3 \\ \cdots \\ q_{1,n}f_1 + q_{2,n}f_2 + q_{3,n}f_3 \cdots q_{n,n}f_n \\ q_{1,n+1}f_1 + q_{2,n+1}f_2 + q_{3,n+1}f_3 \cdots q_{n,n+1}f_n\quad 0 \end{pmatrix}.$$

The last element of which is a linear combination of $q_{i,n+1}$'s and the error of this combination is given by the sum of the squares of the remaining elements. For instance, identifying f_i with $\varphi_i(\mu)$ and $q_{i,n+1}$ with a_i the error in

ELEMENTS OF AN ECLIPSING BINARY SYSTEM IN FREQUENCY DOMAIN

$$A_\mu = \sum_{i=0}^{n} \varphi_i(\mu) \, a_i \varepsilon_i$$

is given by

$$\varepsilon_{A_\mu}/\varepsilon^2 = (\varphi_0 q_{11})^2 +$$
$$+ (\varphi_0 q_{12} + \varphi_1 q_{22})^2 +$$
$$+ (\varphi_0 q_{13} + \varphi_1 q_{23} + \varphi_2 q_{33})^2 + \ldots$$
$$\ldots$$
$$+ (\varphi_0 q_{1n} + \varphi_1 q_{2n} + \varphi_2 q_{3n} + \ldots \varphi_{n-2} q_{n-1,n})^2$$
$$= (\underline{q} \cdot \underline{\varphi}(\mu))^T (\underline{q} \cdot \underline{\varphi}(\mu))$$

in which the last row of $\underline{q}$ is omitted.

Let us now summarize the results obtained for EK Cephei (observations given in the appendix) for the case of NMAX = 8. These are shown in TABLE 2.

The zero'th moment A_o, as estimated from the regression expression is equal to .70437 and is within .2% of the value shown in the table. For the above to work properly an additional word of warning is in order with regard to the function φ_i. Estimates of A_μ, mentioned previously, were obtained with φ_i computed to a tolerance of 10^{-5}. An attempt to extend these estimates to A_6, A_7, and A_8 produced significant deviations from the corresponding Kalman filter values. The reason for this is as follows. Recall that $\varphi_i^{(j)}(\mu)$ varies with j in an oscillatory fashion. For higher values of μ the very first term $\varphi_i^{(0)}(\mu)$ is quite small and could be accepted as an adequate representation of the function. Yet the function may not stabilize to its true value for quite a few terms. For instance, for EK Ceph $\varphi_8^{(0)}(8)$ is of the order 10^{-5} and the function does not show any sign of stability until approximately 15 terms are summed. To force this stability with our program we need a tolerance of at least 10^{-10}. The discrepancy between the stable value and the first few terms can be 3 to 4 orders of magnitude. Clearly, if we use the "unstable" values to estimate A_μ, we run into disastrous consequences with the higher order moments. As far as their errors are concerned, all ε_A would be affected.

It is therefore suggested that prior to the estimation of the moments, the behavior of $\varphi_i(\mu)$ be studied to establish their stable values.

TABLE 2

MOMENT	KALMAN FILTER (A_k)	$\Sigma\varphi_i a_i$	$A_\mu = \Sigma\varphi_i \epsilon_i a_i$	ϵ_μ	$\left|\frac{A_k - A_\mu}{A_k}\right| \times 10^2$
A_0			.70628	1.113×10^{-3}	
A_1	5.203×10^{-2}	5.198×10^{-2}	5.203×10^{-2}	7.972×10^{-5}	.01
A_2	4.831×10^{-3}	6.401×10^{-3}	4.831×10^{-3}	1.293×10^{-5}	.01
A_3	5.104×10^{-4}	8.612×10^{-4}	5.112×10^{-4}	2.068×10^{-6}	.16
A_4	5.855×10^{-5}		5.864×10^{-5}	3.277×10^{-7}	.15
A_5	7.070×10^{-6}	17.676×10^{-6}	7.105×10^{-6}	5.186×10^{-8}	.50
A_6	8.916×10^{-7}		8.951×10^{-7}	8.185×10^{-9}	.40
A_7	1.140×10^{-7}		$1.161 \cdot 10^{-7}$	1.289×10^{-9}	1.04
A_8	1.529×10^{-8}		1.542×10^{-8}	2.026×10^{-10}	.85

The code which implements the computation of $\varphi_i(\mu)$, A_μ, and errors in A_μ is included in BINMOMENTS and starts at Statement 102.

A complete run with a considerable amount of intermediate output is given in Appendix A for NMAX = 8.

5. Transition From The Errors In The A_μ's To The Errors In The Elements

In the previous section we have obtained the values of A_μ and their errors.

In the Fourier domain formulation the A_μ's are related to the conventional elements via two auxiliary quantitites, a and c_o, defined by

$$a = r_1/(r_1 + r_2) \qquad c_o = \cos i/(r_1 + r_2)$$

in which r_1 and r_2 are the radii of the component stars and i is the inclination of the relative orbit.

The first task will be to compute these quantities in terms of the moments as well as to derive their errors. The second task is to compute the elements and their errors.

A. COMPUTATION OF THE AUXILIARY CONSTANTS a AND c_o

Recall (e.g. Kopal, L.o.S., equation 2.8, Chapter V) that the relation between A_μ, a, and c_o is given by

$$A_{2m} = r(m+1)L_1 \sin^{2m}(\theta_T) \sum_{\ell=0}^{\infty} \frac{C^{(\ell)} (1-c_o^2)^{\nu+1}}{\nu \, \Gamma(\nu+1)\Gamma(\nu+1+m)} \sum_{n=0}^{\infty} \frac{\nu+2n+2}{n+1}(n+1+\nu) \times$$

$$\times \left\{ \frac{\Gamma(n+1+\nu)}{\Gamma(n+1)} G_{n+1}(\nu,\nu+1,a) \right\}^2 G_n(\nu+2, \nu+m+2, 1-c_o^2)$$

$$\overset{d}{\equiv} L_1 \sin^{2m}(\theta_T) f_{2m}(a, c_o) \, . \tag{19}$$

In the above, quantities $C^{(\ell)}$ are functions of the limb-darkening coefficients, and G_n represent Jacobi polynomials.

The unknowns are L_1, θ_T, a, and c_o. The phase of the first contact, θ_T, continues to exhibit a "split personality" in the sense that we assume it known, although the degree of certainty with which we know it may be weak.

If we restrict ourselves to even moments then, a linear ratio of two A_μ's will eliminate L_1,

$$A_{2m}/A_{2\mu} = \sin^{2(m-\mu)} \theta_T \; (f_{2m}/f_{2\mu}) \; .$$

Quadratic ratios,

$$(A_{2m}/A_{2\mu}) \cdot (A_{2m'}/A_{2\mu'}) = \sin^{2(m+m'-\mu-\mu')} \theta_T \; (f_{2m}f_{2m'}/f_{2\mu}f_{2\mu'})$$

permit elimination of θ_T by selecting

$$m+m' \stackrel{d}{\equiv} \mu+\mu' \; .$$

For instance, setting $m=1$, $\mu=0$, and $m'=1$, $\mu'=2$, we obtain

$$\frac{A_2^2}{A_0 A_4} = \frac{f_2^2}{f_0 f_4} \stackrel{d}{\equiv} g_2(a,c_0) \; . \tag{20}$$

Similarly, setting $m=2$, $\mu=1$, and $m'=2$, $\mu'=3$, we obtain

$$\frac{A_4^2}{A_2 A_6} = \frac{f_4^2}{f_2 f_6} \stackrel{d}{\equiv} g_4(a,c_0) \; . \tag{21}$$

Equations (20) and (21) are two equations in the unknown quantities, a and c_0. Unfortunately these relations are highly transcendental, complicating not only their solution but the establishment of differential relations between A_μ's, a and c_0.

To obtain a and c_0 it is necessary to explore the (a,c_0) plane. Procedures to do this were studied by Demircan in his doctoral dissertation, and will be discussed by him later during this lecture series.

B. COMPUTATION OF ERRORS IN a AND c_0

To proceed with computation of the error, we define the logarithmic relations

$$G_{2,4} \stackrel{d}{\equiv} \log g_{2,4} = 2 \log f_{2,4} - \log f_{0,4} - \log f_{4,6} \; . \tag{22}$$

Straightforward evaluation of the total differential change yields

$$\begin{pmatrix} dG_2 \\ dG_4 \end{pmatrix} = \begin{pmatrix} \dfrac{\partial G_2}{\partial a} & \dfrac{\partial G_2}{\partial c_o} \\ \dfrac{\partial G_4}{\partial a} & \dfrac{\partial G_4}{\partial c_o} \end{pmatrix} \begin{pmatrix} da \\ dc_o \end{pmatrix} = \begin{pmatrix} \dfrac{1}{g_2}\dfrac{\partial g_2}{\partial a} & \dfrac{1}{g_2}\dfrac{\partial g_2}{\partial c_o} \\ \dfrac{1}{g_4}\dfrac{\partial g_4}{\partial a} & \dfrac{1}{g_4}\dfrac{\partial g_4}{\partial c_o} \end{pmatrix} \begin{pmatrix} da \\ dc_o \end{pmatrix} =$$

$$= \begin{pmatrix} 2\dfrac{dA_2}{A_2} - \dfrac{dA_o}{A_o} - \dfrac{dA_4}{A_4} \\ 2\dfrac{dA_4}{A_4} - \dfrac{dA_2}{A_2} - \dfrac{dA_6}{A_6} \end{pmatrix},$$

$$\begin{pmatrix} da \\ dc_o \end{pmatrix} = \dfrac{1}{\left(\dfrac{\partial G_2}{\partial a}\dfrac{\partial G_4}{\partial c_o} - \dfrac{\partial G_2}{\partial c_o}\dfrac{\partial G_4}{\partial a}\right)} \begin{pmatrix} \dfrac{\partial G_4}{\partial c_o} & \dfrac{\partial G_2}{\partial c_o} \\ \dfrac{\partial G_4}{\partial a} & \dfrac{\partial G_2}{\partial a} \end{pmatrix} \begin{pmatrix} dG_2 \\ dc_o \end{pmatrix}. \qquad (23)$$

The denominator of this expression is the Jacobian determinant, J. If it is not equal to zero, the inverse transformation exists, that is,

$$a = a(G_2, G_4),$$
$$c_o = c_o(G_2, G_4),$$

and

$$\begin{pmatrix} da \\ dc_o \end{pmatrix} = \begin{pmatrix} \dfrac{\partial a}{\partial G_2} & \dfrac{\partial a}{\partial G_4} \\ \dfrac{\partial c_o}{\partial G_2} & \dfrac{\partial c_o}{\partial G_4} \end{pmatrix} \begin{pmatrix} dG_2 \\ dG_4 \end{pmatrix}. \qquad (24)$$

Since (23) and (24) represent identities, we identify

$$\dfrac{\partial a}{\partial G_2} = \dfrac{1}{J}\dfrac{\partial G_4}{\partial c_o}, \quad \dfrac{\partial a}{\partial G_4} = -\dfrac{1}{J}\dfrac{\partial G_2}{\partial c_o},$$

$$\dfrac{\partial c_o}{\partial G_2} = -\dfrac{1}{J}\dfrac{\partial G_4}{\partial a}, \quad \dfrac{\partial c_o}{\partial G_4} = \dfrac{1}{J}\dfrac{\partial G_2}{\partial a}.$$

ELEMENTS OF AN ECLIPSING BINARY SYSTEM IN FREQUENCY DOMAIN

In terms of g_{2m} or f_{2m} the Jacobian determinant can be written as

$$J = \frac{1}{g_2 g_4} \frac{\partial(g_2, g_4)}{\partial(a, c_o)} = \frac{1}{f_o f_2} \frac{\partial(f_o, f_2)}{\partial(a, c_o)} - \frac{2}{f_o f_4} \frac{\partial(f_o, f_4)}{\partial(a, c_o)} - \frac{2}{f_2 f_6} \frac{\partial(f_2, f_6)}{\partial(a, c_o)} + \frac{3}{f_2 f_4} \frac{\partial(f_2, f_4)}{\partial(a, c_o)} + \frac{1}{f_o f_6} \frac{\partial(f_o, f_6)}{\partial(a, c_o)} + \frac{1}{f_4 f_6} \frac{\partial(f_4, f_6)}{\partial(a, c_o)}.$$

Let us replace in (24) the vector $(dG_2 \, dG_4)^T$ by its equivalent in terms of the A_μ's and then express da and dc_o as linear combinations of dA_μ/A_μ; namely,

$$da = \sum_{k=0}^{3} \alpha_{2k} \frac{dA_{2k}}{A_{2k}},$$

$$dc_o = \sum_{k=0}^{3} \gamma_{2k} \frac{dA_{2k}}{A_{2k}},$$

(25)

in which

$$\alpha_o = -\frac{\partial a}{\partial G_2}, \qquad \gamma_o = -\frac{\partial c_o}{\partial G_2},$$

$$\alpha_2 = 2\frac{\partial a}{\partial G_2} - \frac{\partial a}{\partial G_4}, \qquad \gamma_2 = 2\frac{\partial c_o}{\partial G_2} - \frac{\partial c_o}{\partial G_4},$$

$$\alpha_4 = 2\frac{\partial a}{\partial G_4} - \frac{\partial a}{\partial G_2}, \qquad \gamma_4 = 2\frac{\partial c_o}{\partial G_4} - \frac{\partial c_o}{\partial G_2},$$

$$\alpha_6 = -\frac{\partial a}{\partial G_4}, \qquad \gamma_6 = -\frac{\partial c_o}{\partial G_4}.$$

At this point, let us make a reasonable assumption that the errors ε_{dA_μ}, of dA_μ are identifiable with the errors

ε_{A_μ}, of A_μ. Under these conditions

$$dA_\mu = \sum_{i=1}^{N} \varphi_i(\mu) \varepsilon_{a_i} , \qquad (26)$$

in which ε_{a_i} are the errors of the Fourier coefficients, a_i. Combining (25) and (26), we obtain

$$da = \sum_{i=1}^{N} \sum_{k=0}^{3} \varphi_i(2k) \alpha_{2k} \varepsilon_{a_i} \stackrel{d}{\equiv} \sum_{i=1}^{N} B_i \varepsilon_{a_i} ,$$

$$dc_o = \sum_{i=1}^{N} \sum_{k=0}^{3} \varphi_i(2k) \gamma_{2k} \varepsilon_{a_i} \stackrel{d}{\equiv} \sum_{i=1}^{N} D_i \varepsilon_{a_i} , \qquad (27)$$

in which μ is set equal to $2k$. Combining matrix $\underline{q}$, which accounts for the errors in a_i, with (27) we have

$$(\varepsilon_a/\varepsilon)^2 = (\underline{q} \cdot B)^T \cdot (\underline{q} \cdot B) \quad \text{and}$$

$$(\varepsilon_{c_o}/\varepsilon)^2 = (\underline{q} \cdot D)^T \cdot (\underline{q} \cdot D) \qquad (28)$$

in which superscript T denotes the transpose of the respective matrix.

In the expanded form the first of (28) will appear as follows:

$$\begin{aligned}(\varepsilon_a/\varepsilon)^2 = & (B_1 q_{11})^2 \\ & + (B_1 q_{12} + B_2 q_{22})^2 \\ & + (B_1 q_{13} + B_2 q_{23} + B_3 q_{33})^2 \\ & + \cdots \\ & + (B_1 q_{1N} + B_2 q_{2N} + B_3 q_{3N} + \cdots B_{N-1} q_{N-1,N})^2 ;\end{aligned}$$

and a similar expression obtains for $(\varepsilon_{c_o}/\varepsilon)^2$

C. NUMERICAL ASPECTS OF EVALUATING a, da, c_o, AND dc_o

At this point it is of some importance to examine what is involved in carrying out the above calculations

(i) DIFFERENTIABILITY OF g_{2m}.

The functions g_{2m} involve three different values of the functions f_{2m} which, in turn, are enormously complicated functions of a and c_o. The expressions for f_{2m} are themselves Fourier expansions obtained from $A_{2\mu}$.

In principle, f_{2m} defined by equation (19) can be differentiated with respect to a and c_o, but we have no evidence that the resulting series represents the true value of the derivative. It is a problem similar to that raised earlier in evaluating functions $\varphi_i(\mu)$. A cursory look at $\partial f_{2m}/\partial a$ and $\partial f_{2m}/\partial c_o$ indicates that in this instance the problem is a serious one and cannot be ignored.

There are two practical ways of obtaining the required derivatives.

One way is to differentiate the series, but filter it by $\sin x/x$, a procedure mentioned earlier in this discussion.

The alternate way is to obtain estimates of the required derivatives by finite differences; namely,

$$\frac{\partial f_{2m}}{\partial a} = \frac{f_{2m}(a+\Delta a, c_o) - f_{2m}(a-\Delta a, c_o)}{2\Delta a} \; ;$$

and similarly for $\partial f_{2m}/\partial c_o$

The smoothing effect of this approach depends on a suitable choice of the increments Δa and Δc_o. It must be large enough to attenuate oscillations in the Fourier series, but small enough to represent the derivative with sufficient accuracy. As of the moment we have not examined the performance of either approach.

(ii) DETERMINANCY OF da AND dc_o.

The transformation from dG_2 and dG_4 to da and dc_o depends on the nature of the Jacobian determinant, J. If it is equal to zero the inverse transformation (24) does not exist. If the J is small, da and dc_o are relatively large and subsequent analysis may not be of much value. The unanswered question at the moment is what is the magnitude of the J required to produce an adequate solution? We have little feel for what is meant by "small" or "large enough". Additional comments on this problem will be made by Dr. Demircan.

D. UNCERTAINTIES IN THE GEOMETRIC ELEMENTS

(i) UNCERTAINTY IN r_1

The point of departure for evaluation of uncertainties in the geometric elements will be equation (3.14), Chapter V of L.o.S.

$$Xr_1^2 = a^2 A_2 f_o(a,c_o) , \qquad (29)$$

in which

$$X = (1-c_o^2) A_o f_2 + c_o^2 A_2 f_o . \qquad (30)$$

The logarithmic derivatives of both sides of equation (29) yield

$$\frac{dr_1^2}{r_1^2} = 2\frac{da}{a} + \frac{dA_2}{A_2} + \frac{df_o}{f_o} - \frac{dX}{X} ,$$

in which dX/X is obtained by differentiation of (30),

$$dX = (1-c_o^2)f_2 dA_o + c_o^2 f_o dA_2 + (1-c_o^2)A_o df_2 + c_o^2 A_2 df_o + (A_2 f_o - A_o f_2)dc_o^2 ,$$

and

$$\frac{dX}{X} = \frac{r_1^2(1-c_o^2)}{a}\frac{f_2}{f_o}\frac{dA_o}{A_2} + \frac{r_1^2 c_o^2}{a}\frac{dA_2}{A_2} + \frac{r_1^2(1-c_o^2)}{a}\frac{A_o}{A_2}\frac{df_2}{f_o}$$

$$+ \frac{r_1^2 c_o^2}{a}\frac{df_o}{f_o} - \frac{r_1^2(A_2 f_o - A_o f_2)}{a^2 f_o A_2}dc_o . \qquad (31)$$

We note that, by definition of a and c_o, we have

$$\frac{r_1^2 c_o^2}{a} = \cos^2 i \qquad \frac{r_1^2(1-c_o^2)}{a^2} = \sin^2 i \sin^2 \theta_T ;$$

and that θ_T can be defined by one of the following relations :

ELEMENTS OF AN ECLIPSING BINARY SYSTEM IN FREQUENCY DOMAIN

$$\sin^2\theta_T \sin^2 i + \cos^2 i = (r_1+r_2)^2,$$

$$\sin^2\theta_T = \frac{f_o A_2}{f_2 A_o}.$$

Substitution of these relations into equation (31) results in the following expression for dX/X,

$$\frac{dX}{X} = \frac{dA_o}{A_o} + \frac{df_2}{f_2} \sin^2 i + \frac{dA_2}{A_2} + \frac{df_o}{f_o} \cos^2 i - \frac{dc_o^2}{c_o^2} \cos^2 i \cot^2\theta_T.$$

(32)

For eclipsing systems the values of i and θ_T are typically

$$i > 80° \quad \text{and} \quad \theta_T < 20°,$$

so that $\cos^2 i$ and $\cos^2 i \cot^2 \theta_T$ are numerically on the order of a few hundredths and a few tenths, respectively. If the terms associated with them are neglected, equation (32) becomes

$$\frac{dX}{X} = \frac{dA_o}{A_o} + \frac{df_2}{f_2}.$$

Insertion of the above approximations into the expression for dr_1^2/r_1^2 yields

$$\frac{dr^2}{r_1^2} = \frac{da^2}{a^2} - \frac{dA_o}{A_o} + \frac{dA_2}{A_2} + \frac{df_o}{f_o} - \frac{df_2}{f_2}.$$

Furthermore, since

$$df_{2m} = \frac{\partial f_{2m}}{\partial a^2} da^2 + \frac{\partial f_{2m}}{\partial c_o^2} dc_o^2,$$

the final expression for dr_1^2/r_1^2 can be written as

$$\frac{dr_1^2}{r_1^2} = \left\{1 + \frac{a^2}{f_o}\frac{\partial f_o}{\partial a^2} - \frac{a^2}{f_2}\frac{\partial f_2}{\partial a^2}\right\}\frac{da^2}{a^2} + \left\{\frac{c_o^2}{f_o}\frac{\partial f_o}{\partial c_o^2} - \frac{c_o^2}{f_2}\frac{\partial f_2}{\partial c_o^2}\right\}\frac{dc_o^2}{c_o^2}$$

$$+ \frac{dA_2}{A_2} - \frac{dA_o}{A_o}.$$

(ii) UNCERTAINTY IN r_2

The fractional radius r_2 of the eclipsing component is defined by

$$r_2 \stackrel{d}{=} \frac{1-a}{a} r_1 \,.$$

Differentiating logarithmically we obtain

$$\frac{dr_2^2}{r_2^2} = \frac{dr_1^2}{r_1^2} - \frac{1}{1-a} \frac{da^2}{a^2} \,;$$

and, furthermore, since the first term on the right hand side is already known, the above can be written as

$$\frac{dr_2^2}{r_2^2} = \frac{da^2}{a(1-a)} - \frac{dA_o}{A_o} + \frac{dA_2}{A_2} + \frac{df_o}{f_o} - \frac{df_2}{f_2}$$

$$= \left\{ \frac{a}{1-a} + \frac{a^2}{f_o} \frac{\partial f_o}{\partial a^2} - \frac{a^2}{f_2} \frac{\partial f_2}{\partial a^2} \right\} \frac{da^2}{a^2} + \left\{ \frac{c_o^2}{f_o} \frac{\partial f_o}{\partial c_o^2} \right. \qquad (34)$$

$$\left. - \frac{c_o^2}{f_2} \frac{\partial f_2}{\partial c_o^2} \right\} \frac{dc_o^2}{c_o^2} + \frac{dA_2}{A_2} - \frac{dA_o}{A_o} \,.$$

(iii) UNCERTAINTY IN i

Following the same procedure as above, we obtain an expression for $d(\cos^2 i)/\cos^2 i$

$$\frac{d\cos^2(i)}{\cos^2(i)} = \frac{dc_o^2}{c_o^2} + \frac{dr_1^2}{r_1^2} - \frac{da^2}{a^2}$$

$$= \left\{ \frac{a^2}{f_o} \frac{\partial f_o}{\partial a^2} - \frac{a^2}{f_2} \frac{\partial f_2}{\partial a^2} \right\} \frac{da^2}{a^2} \qquad (35)$$

$$+ \left\{ 1 + \frac{c_o^2}{f_o} \frac{\partial f_o}{\partial c_o^2} - \frac{c_o^2}{f_2} \frac{\partial f_2}{\partial c_o^2} \right\} \frac{dc_o^2}{c_o^2} + \frac{dA_2}{A_2} - \frac{dA_o}{A_o} ,$$

(iv) UNCERTAINTY IN THE FRACTIONAL LUMINOSITY L_1.

The error in the fractional luminosity L_1, of the star being eclipsed is arrived at by a logarithmic differentiation of the following relation

$$L_1 \equiv \frac{dA_o}{f_o} ,$$

leading to

$$\frac{dL_1}{L_1} = \frac{dA_o}{A_o} - \frac{df_o}{f_o} = \frac{dA_o}{A_o} - \frac{a^2}{f_o} \frac{\partial f_o}{\partial a^2} \frac{da^2}{a^2} - \frac{c_o^2}{f_o} \frac{\partial f_o}{\partial c_o^2} \frac{dc_o^2}{c_o^2} . \quad (36)$$

Note that for total eclipses $f_o = 1$ and, therefore, the error in L_1 is identical to the error in A_o.

(v) GENERAL REMARKS

Equations (33), (34), (35), and (36) are linear functions of dA_o, dA_2, dA_3, dA_6, da^2, and dc_o^2.

The errors in dA_μ have already been evaluated in equation (26) as functions of the errors in the Fourier coefficients a_i. The corresponding errors in da and dc_o are given by (27) or (28).

If the errors in the differential changes dr_1, dr_2, $d(\cos i)$ and dL_1 are identified as with the errors of the respective elements and equations (33) - (36) are rewritten in terms of errors in a_i, we obtain an expression of the form

$$(\varepsilon_{e\ell}/\varepsilon)^2 = (\underline{q} \cdot \underline{\beta})^T \cdot (\underline{q} \cdot \underline{\beta}) ,$$

in which $\underline{q}$ continues to be the inverse triangular matrix of the normal equations, defining a_i's. Quantities $\underline{\beta}$ represent a column vector defining the weight to be attached to the elements of $\underline{q}$.

References

Banachiewicz, Th.: 1938, Bull. Acad. Polonaise (Kraków), A. pp. 134 - 135 and 393-404.

Banachiewicz, Th.: 1942, Astron. J. $\underline{50}$, pp. 38-41.

Benoit, Commandant: 1924, Bulletin Géodésique (Toulouse) No. 2, pp. 67-77.

Bryson, A.R. Jr. and Ho, Y.C.: 1969, Applied Optimal Control, Blaisdell Publishing Co., Waltham, Mass.

Bucy, R.S. and Joseph, P.D.: 1968, Filtering for Stochastic Processes with Applications in Guidance, Interscience Publishers, New York.

Deutsch, R.: 1965, Estimation Theory, Prentice-Hall, Inc.

Ebbinghausen, E.G.: 1966, Astron. J. $\underline{71}$, pp. 642-646.

Jenkins, G. and Watts, D., 1968: Spectral Analysis and Its Applications, John Wiley and Sons, Inc.

Jurkevich, I., Willman, W.W., and Petty, A.F.: 1976, Astrophys and Space Sci., $\underline{44}$, pp. 63-83.

Kopal, Z.: 1959, "Close Binary Systems, Chapman-Hall and John Wiley, London and New York.

Kopal, Z.: 1975, Astrophys. and Space Sci. pp. $\underline{34}$, 431-457.

Kopal, Z.: 1979a, "Language of the Stars," D. Reidel Publishing Co., Dordrecht:Holland, Boston: USA, London: England.

Kopal, Z.: 1979b, Astrophys. and Space Sci. $\underline{66}$, pp. 91-101.

Kron, G.E.: 1942a, Astrophys. J. $\underline{96}$, p. 173.

Kron, G.E.: 1942b, Lick Observ. Contributions, Series II, No. 5, p. 87.

Lehmann, E.: 1959, Testing Statistical Hypotheses, John Wiley and Sons, Inc.

Oberhettinger, F.: 1973, Fourier Expansions, Academic Press, New York.

Piotrowski, S.L.: 1948, Astrophys. J. 108, No. 3, pp. 510-518.

Plackett, R.: 1968, Principles of Regression Analysis, Oxford Press.

Rao, C.R.: 1973, Linear Statistical Inference and Its Applications, John Wiley and Sons, Inc.: 2nd Edition.

Scheffé, M.: 1964, The Analysis of Variance, John Wiley and Sons, Inc.

Tsessevich, V.P.: 1940, Bull. Astr. Inst. USSR, Acad. of Sci., No. 50.

Tsessevich, V.P.: 1947, Metody Izuchenia Peremennykh Zviozd, Vol. 3, Gosudarstrennoe Izdatel'stvo Tekhniko-Teoreticheskoi Literatury (OGIZ), Moscow-Leningrad.

Wyse, A.B.: 1939, Lick Observatory Bulletin No. 494.

Appendix A
Computer Code

```
●●●●●●●●●●●●●●●●●●●●●●●●●●●●●●●●●●●●●●●
                        BINMOMENTS
 05/28/80                          10:51:05
 CURRENT DIR.  KOPAL
 NASHUA NO.  -4-  ARDOS 6.3
●●●●●●●●●●●●●●●●●●●●●●●●●●●●●●●●●●●●●●●●

       COMPILER DOUBLE PRECISION
       COMPILER NOSTACK
C**** PROGRAM TO ESTIMATE MOMENTS OF BINARY LIGHT CURVES FROM
C**** FOURIER COSINE SERIES APPROXIMATION OF LIGHT LOSS
C
C
       DIMENSION T(400),EL(400),IT(800),IBR(800)
       DIMENSION RU(13,13),RL(13,13),Q(13,13),AER(13),CS(13)
       DIMENSION SN(13),X(13),APHI(9,9),AMOM(13),AMOMERR(13),EPS(13)
       COMMON IT,IBR
       EQUIVALENCE (T,IT),(EL,IBR)
C**** DIMENSION OF 13 ON RU AND RL IMPLIES 13 TERMS IN THE REGRESSION
C**** EQUATION:10 HARMONICS,A0,(-DELU),AND THE ABSOLUTE TERM
C
     4 ACCEPT "NMAX,MI,MMAX,ISWITCH1,IEND=",NMAX,MI,MMAX,ISWITCH1,IEND
       ACCEPT "I2,I3=",I2,I3
       ACCEPT "THETAT=",THETAT
C**** THETA IS THE POINT OF EXTERNAL CONTACT IN RADIANS
C**** IF I2=1 DO NO COMPUTE INDIVIDUAL RESIDUALS FIRST TIME AROUND
C**** IF I3=1 (AUTOMATICALLY SETS I1=0) COMPUTE INDIVIDUAL RESIDUALS
C**** NMAX=HIGHEST HARMONIC USED, MI=NUMBER OF OBSERVATIONS WITHIN
C**** ECLIPSE, MMAX=NUMBER OF OBSERVATIONS OUTSIDE ECLIPSE
C
       IF(IEND.NE.0) GO TO 102
C**** IF IEND NOT EQUAL TO    TERMINATE THE COMPUTATION
       WRITE(12,800)
   800 FORMAT(1H1)
       WRITE(12,120)NMAX,MI,MMAX,ISWITCH1,IEND
   120 FORMAT(1X,5HNMAX=I2,2X,3HMI=I3,2X,5HMMAX=I3,2X,9HISWITCH1=I1,
      12X,5HIEND=I1)
       WRITE(12,121)I2,I3
   121 FORMAT(1X,5H  I2=I2,2X,3HI3=I2)
       WRITE(12,122)THETAT
   122 FORMAT(1X,7HTHETAT=F8.6//)
       IF(ISWITCH1.EQ.1)GO TO 3
C**** IF ISWITCH1=0 READ A NEW SET OF OBSERVATIONS, IF "1" REPEAT
C**** WITH THE SAME SET
       CALL OPEN(0,"DATA",2,IER0)
       CALL WRBLK(0,0,IT,4,IER1)
C**** ACTUALLY READS BLOCKS, SEE COMMENT BELOW
       IF(IER1.EQ.0)GO TO 100
       IF(IER1.GE.3)GO TO 100
       CALL WRBLK(0,4,IBR,4,IER2)
       IF(IER2.EQ.0)GO TO 101
       IF(IER2.GE.3)GO TO 101
C
C**** NOTE:FORT 5.2 LIBRARY HAS A BUG IN WHICH NAMES WRBLK AND RDBLK
C**** ARE REVERSED
C
C**** DEFINE CONSTANTS
       PI=3.1415926535897930D0
C
     3 IMAX=NMAX+2
       JMAX=NMAX+3
       REFANG=PI/THETAT
C
C**** COMPUTE LOSS OF LIGHT
       DO 53 I=1,MI
       EL(I)=1.-EL(I)
    53 CONTINUE
C
C**** COMPUTE SUM OF SQUARES OF LIGHT LOSS
```

ELEMENTS OF AN ECLIPSING BINARY SYSTEM IN FREQUENCY DOMAIN

```
         DO 52 I=1,MI
         TRY1=TRY1+EL(I)*EL(I)
      52 CONTINUE
C
C**** ZERO OUT RU MATRIX
         DO 1 J=1,IMAX
         DO 2 I=1,J
       2 RU(J,I)=0.
         RU(JMAX,J)=0.
       1 CONTINUE
C
C**** COMPUTATION OF COSINES FOR K'TH OBSERVATION
      11 DO 10 M=1,MI
         ANG=REFANG*T(M)
         SSN=SIN(ANG)
         CSN=COS(ANG)
         SN(1)=SSN
         CS(1)=CSN
         DO 200 J=2,NMAX
         SN(J)=SN(J-1)*CSN+CS(J-1)*SSN
     200 CS(J)=CS(J-1)*CSN-SN(J-1)*SSN
C
C**** COMPUTATION OF INDIVIDUAL RESIDUALS IF NEEDED
         IF(I2-1)206,207,102
     206 Y=0.
         DO 501 I=2,NMAX
     501 Y=Y+Q(I,JMAX)*CS(I-1)
         Y=Y+0.5*Q(1,JMAX)-Q(IMAX,JMAX)
         S1=EL(M)-Y
         S3=S3+S1*S1
         WRITE(12,128)T(M),EL(M),Y,S1
     128 FORMAT(10X,D23.16,3(3XD23.16))
         GO TO 10
C
C**** COMPUTATION OF COEFFICIENTS OF NORMAL EQUATIONS
     207 X(1)=0.5
         NU=NMAX+1
         DO 201 J=2,NU
         X(J)=CS(J-1)
     201 CONTINUE
         X(JMAX)=-EL(M)
         DO 203 J=1,NU
         DO 202 I=1,J
     202 RU(J,I)=RU(J,I)+X(I)*X(J)
         RU(JMAX,J)=RU(JMAX,J)+X(J)*X(JMAX)
     203 CONTINUE
      10 CONTINUE
         IF(I2.EQ.1)GO TO 41
         WRITE(12,127)S3
     127 FORMAT(7X,3HS3=D23.16)
         GO TO 4
      41 DO 500 J=1,IMAX
         RU(IMAX,J)=-2.*RU(J,1)
     500 CONTINUE
         RU(IMAX,IMAX)=RU(IMAX,IMAX)+MMAX
         RU(JMAX,IMAX)=-2.*RU(JMAX,1)
C
C**** SOLUTION OF NORMAL EQUATIONS
         NP1=JMAX
         CALL CRACOVIAN(RU,NP1,RL,Q)
         WRITE(12,124)
     124 FORMAT(//)
```

```
      125 FORMAT(2X,1HJ,3X,1HI,12X,2HRU,24X,2HRL,18X,1HI,3X,1HJ,12X,1HQ/)
          DO 505 I=1,JMAX
          DO 504 J=I,JMAX
          WRITE(12,123)J,I,RU(J,I),RL(J,I),I,J,Q(I,J)
      123 FORMAT(1X,I2,2X,I2,3X,D23.16,3X,D23.16,5X,I2,2X,I2,3X,D23.16)
      504 CONTINUE
          WRITE(12,506)
      506 FORMAT(/)
      505 CONTINUE
C**** COMPUTE THE SECOND TERM OF EQUATION (15)
          TRY2=0.
          DO 36 I=1,IMAX
          TRY2=TRY2+RL(JMAX,I)*RL(JMAX,I)
       36 CONTINUE
          DEN=MI+1-IMAX
          S2=(TRY1-TRY2)/DEN
          ERRMEAN=SQRT(S2)
C
C**** COMPUTE ERRORS OF UNKNOWNS
          DO 33 I=1,IMAX
          AER(I)=0.
       33 CONTINUE
          DO 35 I=1,IMAX
          DO 32 J=I,IMAX
       32 AER(I)=AER(I)+Q(I,J)*Q(I,J)
          AER(I)=ERRMEAN*SQRT(AER(I))
       35 CONTINUE
C
C**** PRINT UNKNOWNS AND THEIR ERRORS
          WRITE(12,124)
          WRITE(12,38)
       38 FORMAT(6X,1HI,12X,4HA(I),22X,6HAER(I)/)
          DO 37 I=1,IMAX
          I1=I-1
          WRITE(12,39)I1,Q(I,JMAX),AER(I)
       39 FORMAT(5X,I2,2(3X,D23.16))
       37 CONTINUE
          WRITE(12,126)S2
      126 FORMAT(/7X,3HS2=D23.16//)
          DO 618 L=1,2
          WRITE(12,800)
      618 CONTINUE
C
C**** DECISION WHETHER TO COMPUTE INDIVIDUAL RESIDUALS
          IF(I3-1)4,208,102
      208 I2=0
          S3=0.
          WRITE(12,110)
      110 FORMAT(20X,4HT(M),20X,5HEL(M),22X,1HY,25X,2HS1)
          GO TO 11
C
C**** ERROR RETURNS
      100 TYPE "IER1=",IER1
          GO TO 4
      101 TYPE "IER2=",IER2
          GO TO 4
      102 ACCEPT"COMPUTE MOMENTS AND THEIR ERRORS ?(0=YES,1=NO)",IYN
          IF(IYN.EQ.1) STOP

          ACCEPT "THETA,MUSTART,MUTOP,ISTART,ITOP=",THETA,MUIN,MUFIN,IIN,IFIN
          ACCEPT "EPS(1),KSIGMA,N=",EPS(1),KSIGMA,N
          DO 617 K=2,9
          EPS(K)=EPS(K-1)*0.1D0
```

```
      617 CONTINUE
C
C**** NOTE THAT MU AND I ARE OFFSET. TO RUN MU=0 CASE MUTOP MUST BE1 1.
C**** THETA IS IN RADIANS
C
      DO 12 J=MUIN,MUFIN
      TOL=EPS(J)
      DO 12 I=IIN,IFIN
C**** BELOW J1 IS MU,I1 IS THE RUNNING INDEX IN PHI(I,MU)
      J1=J-1
      I1=I-1
X     WRITE(12,103)J1,I1,EPS(J)
X 103 FORMAT(1X,3HMU=I3,3X,2HI=I3,3X,4HEPS=D23.16/)
      CALL PHI(I1,J1,THETA,TOL,JJ,KSIGMA,N,PHI1)
X     WRITE(12,104)J1,I1,JJ,PHI1
X 104 FORMAT(1X,3HMU=I3,3X,2HI=I3,3X,3HJJ=I6,3X,10HPHI(I,MU)=D23.16//)
      APHI(I,J)=PHI1
   12 CONTINUE
C
C**** PRINT TABLE OF PHI'S
      WRITE(12,800)
      WRITE(12,614)
  614 FORMAT(2X,2HMU,2X,1HI,9X,10HAPHI(I,MU),11X,7HEPS(MU)/)
      DO 616 J=1,MUFIN
      DO 619 I=1,IFIN
      J1=J-1
      I1=I-1
      WRITE(12,615)J1,I1,APHI(I,J),EPS(J)
  615 FORMAT(2X,I2,2X,I1,3X,D23.16,3X,D8.2)
  619 CONTINUE
      WRITE(12,506)
  616 CONTINUE
C**** COMPUTE MOMENTS FROM ASUBMU=SUM(APHI*NEUMANNNO*ASUBI)
      IFIN=IMAX-1
      DO 300 J=1,MUFIN
      I=1
      SUM=APHI(I,J)*Q(I,JMAX)
      DO 400 I=2,IFIN
      SUM=SUM+2.*APHI(I,J)*Q(I,JMAX)
  400 CONTINUE
      AMOM(J)=SUM
  300 CONTINUE
C
C**** COMPUTE ERRORS IN AMOM(I)
      DO 600 J=1,MUFIN
      SUM1=0.
      DO 602 I=1,IFIN
      SUM=0.
      DO 601 K=1,I
  601 SUM=SUM+APHI(K,J)*Q(K,I)
      SUM1=SUM1+SUM*SUM
  602 CONTINUE
      AMOMERR(J)=ERRMEAN*SQRT(SUM1)
  600 CONTINUE
C
C**** PRINT A TABLE OF MOMENTS AND THEIR ERRORS
      WRITE(12,800)
      WRITE(12,610)
  610 FORMAT(10X,2HMU,12X,6HMOMENT,22X,5HERROR/)
      DO 612 J=1,MUFIN
      MU=J-1
      WRITE(12,611)MU,AMOM(J),AMOMERR(J)
  611 FORMAT(10X,I2,4X,D23.16,4X,D23.16)
  612 CONTINUE
      STOP
      END
```

```
************************************
             CRACOVIAN

       05/28/80                10:51:17
       CURRENT DIR.  KOPAL
       NASHUA NO. -4- ARDOS 6.3
************************************
```

```
X      COMPILER DOUBLE PRECISION
       SUBROUTINE CRACOVIAN(RU,NP1,RL,Q)
       DIMENSION RU(13,13),RL(13,13),Q(13,13)
       INTEGER DI
       N=NP1-1
       RL(NP1,NP1)=1.
       RL(1,1)=SQRT(RU(1,1))
       DO 50 K=1,N
       KM1=K-1
       DO 40 I=K,NP1
       SUM=0.
       IF(KM1.NE.0)GO TO 10
       RL(I,K)=RU(I,K)/RL(1,1)
       GO TO 40
10     DO 30 J=1,KM1
       SUM=SUM+RL(I,J)*RL(K,J)
30     CONTINUE
       IF(K.EQ.I)GO TO 60
       RL(I,K)=(RU(I,K)-SUM)/RL(K,K)
       GO TO 70
60     RL(I,K)=SQRT(RU(I,K)-SUM)
70     CONTINUE
40     CONTINUE
50     CONTINUE
       DO 100 I=1,N
       Q(I,I)=1/RL(I,I)
100    CONTINUE
       Q(NP1,NP1)=1.
       DO 130 I=1,N
       K2=NP1-I
       DO 120 K=1,K2
       SUM=0
       DO 110 J=1,K
       DI=J-1
       SUM=SUM+RL(I+K,I+DI)*Q(I,I+DI)
110      CONTINUE
       Q(I,I+K)=-SUM/RL(I+K,I+K)
120    CONTINUE
130    CONTINUE
       RETURN
       END
```

ELEMENTS OF AN ECLIPSING BINARY SYSTEM IN FREQUENCY DOMAIN   65

```
                  *******************************************
                                    AMOM

                        05/26/80                  14:31:55
                        CURRENT DIR.  KOPAL
                        NASHUA NO. -4- ARDOS 6.3
                  *******************************************
      COMPILER DOUBLE PRECISION
C**** PROGRAM TO GENERATE RECURSIVE RELATIONS BETWEEN MOMENTS
C**** N=ORDER OF LEGENDRE POLYNOMIAL, MU=POWER OF THE APPROXIMATED
C**** SINE, THETA1=POINT OF EXTERNAL TANGENCY, THETA=RUNNING PHASE
C**** AM=COMMON FACTOR IN LEGENDRE POLYNOMIALS, NTERMS=NUMBER OF
C**** TERMS IN THE APPROXIMATING POLYNOMIAL, A(I) ARRAY OF COEFFICIENTS
C**** OF LEGENDRE POLYNOMIALS ORTHOGONAL IN /X/=1, C(I) COEFFICIENTS
C**** A(I)/(AJ-AMU) IN EQUATION 3.47,P.243 OF THE MONOGRAPH,
C**** B(I) - COEFFICIENTS OF SIN(THETA)**MU IN THE FINAL APPROXIMATING
C**** POLYNOMIAL
C
C**** P0=1
C**** P1=X
C**** P2=(1/2)*(-1+3X**2)
C**** P3=(1/2)(-3X+5X**3)
C**** P4=(1/8)(3-30X**2+35X**4)
C**** P5=(1/8)(15X-70X**3+63X**5)
C**** P6=(1/16)(-5+105X**2-315X**4+231X**6)
C**** P7=(1/16)(-35X+315X**3-693X**5+429X**7)
C**** P8=(1/128)(35-1260X**2+6930X**4-12012X**6+6435X**8)
C**** P9=(1/128)(315X-4620X**3+18018X**5-25740X**7+12155X**9)
C**** ETC.
      DIMENSION A(15),B(15),C(15)
C**** THIS WILL ALLOW USE OF 15'TH ORDER POLYNOMIAL
      COMMON/AA/A
C**** THIS PROGRAM WILL RUN PROPERLY WITH THE ABOVE STATEMENT DELETED
   10 ACCEPT "MU,THETA=",MU,THETA
C**** THETA IS INPUT IN DEGREES
      ACCEPT "THETA1=",THETA1
C**** THETA1 IS INPUT IN RADIANS
      ACCEPT "N,M=",N,AM
      IN=N/2
      NTERMS=IN+1
      ACCEPT "A(I)=",(A(I),I=1,NTERMS)
C
      DO 1 I=1,N
      A(I)=A(I)/AM
    1 CONTINUE
C
      PI=3.14159265359D0
      CONV=PI/180.
      SNTHETA=SIN(THETA*CONV)
      SNTHETA1=SIN(THETA1)
      X1=1/SNTHETA1
C
C**** DETERMINE WHETHER LEGENDRE POLYNOMIAL IS EVEN OR ODD
      AMU=FLOAT(MU)
      Y=FLOAT(N)
      Y=Y/2.
      AIN=FLOAT(IN)
      TEMP=AIN-Y
      IF(ABS(TEMP).NE.0.) GO TO 2
C**** REORTHOGONALIZE THE RANGE
      K=2
      GO TO 4
```

```
      2 K=1
      4 DO 3 I=1,NTERMS
        J=2*I-K
        A(I)=A(I)*(X1**J)
      3 CONTINUE
C
C**** DETERMINE THE DENOMINATOR OF (3.47)
C
     11 SUM1=0.
        DO 5 I=1,NTERMS
        J=2*I-K
        AJ=FLOAT(J)
        IF(J.NE.MU) GO TO 9
        DO 12 LL=1,NTERMS
        C(LL)=0.
     12 CONTINUE
        C(I)=A(I)
        SUM1=C(I)*(SNTHETA1**MU)
        GO TO 13
      9 C(I)=A(I)/(AJ-AMU)
        SUM1=SUM1+C(I)*(SNTHETA1**J)
      5 CONTINUE
C
C**** EVALUATE COEFFICIENTS OF THE NUMERATOR OF (3.47)
     13 DO 6 I=1,NTERMS
        B(I)=C(I)*(SNTHETA1**MU)/SUM1
      6 CONTINUE
C
C**** PRINT OUT THE COEFFICIENTS OF THE APPROXIMATION TO SIN(THETA)**MU
        WRITE(12,102)THETA,THETA1
    102 FORMAT(//1X,11HTHETA(DEG)=D23.16,3X,12HTHETA1(RAD)=D23.16/)
        WRITE(12,100)N,(B(I),I=1,NTERMS)
    100 FORMAT(1X,2HN=I2/(1X,D23.16))
C
C**** EVALUATIONS OF THE QUALITY OF APPROXIMATION
        SUM=0.
        DO 7 I=1,NTERMS
        J=2*I-K
        SUM=SUM+B(I)*(SNTHETA**J)
      7 CONTINUE
C
C**** ACTUAL VALUE OF (SIN(THETA))**MU
        X=SNTHETA**MU
C
C**** WRITE OUT THE RESULT
        WRITE(12,101) MU,X,SUM
    101 FORMAT(/1X,3HMU=,I2,3X,2HX=D23.16,3X,4HSUM=D23.16)
        ACCEPT"TO ALTER MU OR THETA RESPOND WITH 1,OTHERWISE 0=",L
        IF(L.EQ.1) GO TO 8
        GO TO 10
      8 ACCEPT"NEW MU AND/OR THETA=",MU,THETA
        SNTHETA=SIN(THETA*CONV)
        AMU=FLOAT(MU)
        GO TO 11
        STOP
        END
```

ELEMENTS OF AN ECLIPSING BINARY SYSTEM IN FREQUENCY DOMAIN

```
****************************************
                    PHICALC
            05/27/80             01:45:50
            CURRENT DIR. KOPAL
            NASHUA NO. -4- ARDOS 6.3
****************************************
      COMPILER DOUBLE PRECISION
      ACCEPT"THETA,MUSTART,MUTOP,ISTART,ITOP=",THETA,MUIN,MUFIN,IIN,IFIN
      ACCEPT"EPS=",EPS
      TOL=EPS
C**** NOTE THAT MU AND I ARE OFFSET. TO RUN MU=0 CASE MUTOP MUST BE 1.
C**** THETA IS IN RADIANS
      DO 10 J=MUIN,MUFIN
      DO 10 I=IIN,IFIN
C**** J1 IS MU, I1 IS THE RUNNING INDEX IN PHI(I,MU)
      J1=J-1
      I1=I-1
      WRITE(12,101)J1,I1,EPS
  101 FORMAT(1X,3HMU=I3,3X,2HI=I3,3X,4HEPS=D23.16/)
      CALL PHI(I1,J1,THETA,TOL,JJ,PHI1)
      WRITE(12,100)J1,I1,JJ,PHI1
  100 FORMAT(1X,3HMU=I3,3X,2HI=I3,3X,3HJJ=I6,3X,10HPHI(I,MU)=D23.16//)
   10 CONTINUE
      STOP
      END
```

```
****************************************
                    PHINEW
            05/27/80             01:45:55
            CURRENT DIR. KOPAL
            NASHUA NO. -4- ARDOS 6.3
****************************************
      COMPILER DOUBLE PRECISION
      SUBROUTINE PHI(I,MU,THETA,TOL,JJ,PHI1)
      PI=3.14159265359D0
      E=2.71828182845D0
      ONE=FLOAT(-1)
      AMU=FLOAT(MU)
      FACT=(E**(1+MU))/PI
      CALL GAMMA(1.+AMU,GAM,IER)
      TS1=GAM/(2.D0**MU)
      AI=FLOAT(I)
      AIPI=AI*PI
      TS3=(AMU+3.D0)/2.D0
      TS4=(AMU+1.D0)/2.D0
      SUM=0.D0
      J=-1
C
C**** IS MU ODD OR EVEN ?
      IODD=1
```

```
          Y=AMU
          Y=Y/2.D0
          IY=IFIX(Y)
          DIFF=Y-FLOAT(IY)
          IF(ABS(DIFF).LE.0.1D-10)IODD=0
C
C**** GENERAL FORM OF SERIES  FOLLOWS
C
   20     J=J+1
          ITS=MU+1-2*J
          IF(ITS.EQ.0)GO TO 10
          AJ=FLOAT(J)
          TS2=2.*AJ+1.
          ARG=TS2*THETA
          ARGSQ=ARG*ARG
          TEMPNUM=(ONE**(I+J))*ARGSQ*SIN(ARG)/(ARGSQ-AIPI*AIPI)
          IF((MU.GE.0).AND.(J.GT.50))GO TO 21
          CALL GAMMA(TS3+AJ,GAM1,IER)
          CALL GAMMA(TS4-AJ,GAM2,IER)
          FACT1=1./(GAM1*GAM2)
          GO TO 30
C
C**** ASSYMPTOTIC SECTION FOR J GREATER THAN 50
C
   21     A1=(AMU+1.)/(2.*AJ)
          A2=(AMU-1.)/(2.*AJ)
          ANUM=(1.-A1)**(J-MU/2)
          DENA=(1.-A2)**(J+1+MU/2)
          DENB=AJ**(1+MU)
          TEMP1=ANUM/(DENA*DENB)
          AMPL=FACT*TEMP1
          ANG=(AJ-AMU/2.+0.5D0)*PI
          FACT1=AMPL*SIN(ANG)
C
C**** COMMON SECTION
C
   30     TEMP=TEMPNUM*FACT1
          SUM=SUM+TEMP
X         WRITE(12)J,SUM
          IF(IODD.EQ.1)GO TO 20
          IF(ABS(TEMP).LE.TOL)GO TO 10
          GO TO 20
   10     CONTINUE
          JJ=J
          PHI1=TS1*SUM
X         WRITE(12)J,PHI1
          RETURN
          END
```

```
****************************************
                DGAMMA
          05/27/80              01:49:50
          CURRENT DIR. KOPAL
          NASHUA NO. -4- ARDOS 6.3
****************************************
      COMPILER DOUBLE PRECISION
      SUBROUTINE GAMMA(XX,GX,IER)
C     SUBROUTINE TO COMPUTE VALUES OF GAMMA FUNCTION
C     PARAMETERS:XX - THE ARGUMENT OF THE GAMMA FUNCTION
C                GX - THE RESULTANT GAMMA FUNCTION VALUES
C                IER- TOLERANCE CODE:
C                       IER=0 NO TOLERANCE
C                          =1 XX IS WITHIN .000001 OF
C                             BEING NEGATIVE INTEGER
C                          =2 XX.GT.34.5,GX IS SET TO 1.E38
      IF(XX.LE.34.5D0) GO TO 6
      IER=2
      GX=1.0D+38
    6 X=XX
      TOL=1.0 D-06
      IER=0
      GX=1.D0
      IF(X-2.D0)50,50,15
   10 IF(X.LE.2.D0) GO TO 110
   15 X=X-1.D0
      GX=GX*X
      GO TO 10
   50 IF(X-1.D0)60,120,110
C     SEE IF X IS NEAR NEGATIVE INTEGER OR ZERO
   60 IF(X.GT.TOL)GO TO 80
      K=X
      Y=FLOAT(K)-X
      IF(ABS(Y).LE.TOL) GO TO 130
      IF((1.D0-Y).LE.TOL) GO TO 130
C     X NOT NEAR A NEGATIVE INTEGER OR ZERO
   70 IF(X.GT.1.D0) GO TO 110
   80 GX=GX/X
      X=X+1.D0
      GO TO 70
  110 Y=X-1.D0
      GY=1.D0+Y*(-0.5771017D0+Y*(+0.9858540D0+Y*(-0.8764218D0+
     1Y*(0.8328212D0+Y*(-0.5684729D0+Y*(0.2548205D0+Y*(-0.0514993D0)))))))
      GX=GX*GY
  120 RETURN
  130 IER=1
      RETURN
      END
```

****
                       MOMTESTEVEN

              05/27/80   01:52:06
              CURRENT DIR. KOPAL
              NASHUA NO. -4- ARDOS 6.3
****

```
C     PROGRAM TO ESTIMATE MOMENTS OF BINARIES BY KALMAN FILTER
C     EVEN MOMENTS A(2M)
      COMPILER DOUBLE PRECISION
      DIMENSION F(400),EL(400),IT(1600),IBR(1600)
      COMMON IT,IBR
      EQUIVALENCE (F,IT),(EL,IBR)
      REAL M,M12,M14,M16,M18
    4 ACCEPT"T,X,N,ISWITCH1,IEND=",T,X,N,ISWITCH1,IEND
      IF(IEND.NE.0)GO TO 102
      WRITE(10,201)T,X,N,ISWITCH1
  201 FORMAT(4H  T=F7.1,2X2HX=F5.3,2X2HN=I4,2X9HISWITCH1=I1)
      IF(ISWITCH1.EQ.1) GO TO 3
      CALL OPEN(0,"DATA",2,IER0)
      CALL RDBLK(0,0,IT,4,IER1)
      IF(IER1.EQ.0) GO TO 100
      IF(IER1.GE.3) GO TO 100
      CALL RDBLK(0,4,IBR,4,IER2)
      IF(IER2.EQ.0) GO TO 101
      IF(IER2.GE.3) GO TO 101
      CALL CLOSE(0)
    3 P=1.
      P12=0.
      P14=0.
      P16=0.
      P18=0.
      X2=0.
      X4=0.
      X6=0.
      X8=0.
      B=0.
      DO 2 I=1,N
      Z=1.-EL(I)
      DIFZX=Z-X
      A=B
      B=F(I)
      SA=SIN(A)
      SB=SIN(B)
      C=B-A
      M=P+T*C
      SB2=SB*SB
      SA2=SA*SA
      SB4=SB2*SB2
      SA4=SA2*SA2
      SB6=SB2*SB4
      SB8=SB4*SB4
      DIF2=(SB+SA)*(SB-SA)
      SIG=SB2+SA2
      DIF4=SIG*DIF2
      SBSA=SB*SA
      DIF6=DIF2*(SIG-SBSA)*(SIG+SBSA)
      DIF8=(SB4+SA4)*SIG*DIF2
      M12=P12+P*DIF2+T*(C*SB2+S2(A)-S2(B))
      M14=P14+P*DIF4+T*(C*SB4+S4(A)-S4(B))
      M16=P16+P*DIF6+T*(C*SB6+S6(A)-S6(B))
```

```
      M18=P18+P*DIF8+T*(C*SB8+S8(A)-S8(B))
      DM=1./(1.+M)
      P=M*DM
      P12=M12*DM
      P14=M14*DM
      P16=M16*DM
      P18=M18*DM
      X2=X2+X*DIF2+P12*DIFZX
      X4=X4+X*DIF4+P14*DIFZX
      X6=X6+X*DIF6+P16*DIFZX
      X8=X8+X*DIF8+P18*DIFZX
    2 X=X+P*DIFZX
      WRITE(10,300)X2,X4,X6,X8
  300 FORMAT(1X,3HA2=E11.5,2X3HA4=E11.5,2X3HA6=E11.5,2X3HA8=E11.5//)
      GO TO 4
  100 TYPE"IER1=",IER1
      GO TO 4
  101 TYPE"IER2=",IER2
      GO TO 4
  102 STOP
      END
```

```
******************************************
                    MOMTESTODD

                    05/27/80                01:52:12
                    CURRENT DIR.  KOPAL
                    NASHUA NO.  -4-  ARDOS 6.3
******************************************

C       PROGRAM TO ESTIMATE MOMENTS OF BINARIES BY KALMAN FILTER
C       ODD MOMENTS A(2M+1)
        COMPILER DOUBLE PRECISION
        DIMENSION F(400),EL(400),IT(1600),IBR(1600)
        COMMON IT,IBR
        EQUIVALENCE (F,IT),(EL,IBR)
        REAL M,M11,M13,M15,M17
      4 ACCEPT"T,X,N,ISWITCH1,IEND=",T,X,N,ISWITCH1,IEND
C
C****   T=Q/R INTENSITY PARAMETER OF WHITE NOISE/VARIANCE OF VARIABLE
C****   X= ESTIMATED VALUE OF 1.-EL AT ZERO PHASE
C****   N= NUMBER OF OBSERVATIONS
C****   IEND =0 TO STOP COMPUTATIONS
C****   ISWITCH1 EQ 0 READ OBSERVATIONS FROM FILE "DATA",OTHERWISE
C****   SET EQUAL TO 1 AND REPEAT WITH THE SAME DATA SET
        IF(IEND.NE.0)GO TO 102
        WRITE(10,201)T,X,N,ISWITCH1
    201 FORMAT(4H  T=F7.1,2X2HX=F5.3,2X2HN=I4,2X9HISWITCH1=I1)
        IF(ISWITCH1.EQ.1) GO TO 3
        CALL OPEN(0,"DATA",2,IER0)
        CALL WRBLK(0,0,IT,4,IER1)
        IF(IER1.EQ.0) GO TO 100
        IF(IER1.GE.3) GO TO 100
        CALL WRBLK(0,4,IBR,4,IER2)
        IF(IER2.EQ.0) GO TO 101
        IF(IER2.GE.3) GO TO 101
        CALL CLOSE(0)
      3 P=1.
        P11=0.
        P13=0.
        P15=0.
        P17=0.
        X1=0.
        X3=0.
        X5=0.
        X7=0.
        B=0.
        DO 2 I=1,N
        Z=1.-EL(I)
        DIFZX=Z-X
        A=B
        B=F(I)
        SA=SIN(A)
        SB=SIN(B)
        C=B-A
        M=P+T*C
        SB1=SB
        SA1=SA
        SB2=SB*SB
        SA2=SA*SA
        SB3=SB1*SB2
        SA3=SA1*SA2
```

```
      SB5=SB2*SB3
      SA5=SA2*SA3
      SB7=SB5*SB2
      SA7=SA5*SA2
      DIF1=SB-SA
      DIF3=SB3-SA3
      DIF5=SB5-SA5
      DIF7=SB7-SA7
      M11=P11+P*DIF1+T*(C*SB1+S1(A)-S1(B))
      M13=P13+P*DIF3+T*(C*SB3+S3(A)-S3(B))
      M15=P15+P*DIF5+T*(C*SB5+S5(A)-S5(B))
      M17=P17+P*DIF7+T*(C*SB7+S7(A)-S7(B))
      DM=1./(1.+M)
      P=M*DM
      P11=M11*DM
      P13=M13*DM
      P15=M15*DM
      P17=M17*DM
      X1=X1+X*DIF1+P11*DIFZX
      X3=X3+X*DIF3+P13*DIFZX
      X5=X5+X*DIF5+P15*DIFZX
      X7=X7+X*DIF7+P17*DIFZX
    2 X=X+P*DIFZX
      WRITE(10,300)X1,X3,X5,X7
  300 FORMAT(1X,3HA1=E11.5,2X3HA3=E11.5,2X3HA5=E11.5,2X3HA7=E11.5//)
      GO TO 4
  100 TYPE"IER1=",IER1
      GO TO 4
  101 TYPE"IER2=",IER2
      GO TO 4
  102 STOP
      END
```

FILE "DATA" (EK Ceph, Ebbinghausen)

PHASE	INTENSITY
0.6283187107347654D -3	0.2956278536051284D 0
0.1256637421469727D -2	0.2945407241391754D 0
0.5654868396613769D -2	0.2978141649062478D 0
0.5654868396613777D -2	0.2972660759045793D 0
0.6911505818083530D -2	0.2972660759045793D 0
0.1256637421469709D -1	0.3101311218996468D 0
0.1319469292543207D -1	0.3112757944467506D 0
0.1382301163616699D -1	0.3147352239647666D 0
0.1633628647910645D -1	0.3229574818948305D 0
0.1822124261131108D -1	0.3323114842704690D 0
0.2010619874351560D -1	0.3397389578193154D 0
0.2261947358645508D -1	0.3515162044282315D 0
0.2324779229718994D -1	0.3603672786971346D 0
0.2513274842939439D -1	0.3674052435643858D 0
0.2576106714012940D -1	0.3653804869179679D 0
0.2827434198306885D -1	0.3843663416766671D 0
0.2827434198306889D -1	0.3854298537465664D 0
0.3078761682600830D -1	0.3954989640022581D 0
0.3141593553674316D -1	0.4006317372017321D 0
0.3267257295821289D -1	0.4088324412802680D 0
0.3455752909041748D -1	0.4191267214748171D 0
0.3518584780115242D -1	0.4214493129797871D 0
0.3644248522262217D -1	0.4320612822795520D 0
0.3895576006556153D -1	0.4449849780930806D 0
0.3958407877629639D -1	0.4536755247309336D 0
0.4146903490850096D -1	0.4608348878131417D 0
0.4272567232997071D -1	0.4746192516398110D 0
0.4586726588364502D -1	0.4897171915447026D 0
0.4649558459437988D -1	0.4951597673508964D 0
0.4838054072658447D -1	0.4992813573075913D 0
0.4838054072658449D -1	0.5109111141092815D 0
0.5026549685878901D -1	0.5137423289963912D 0
0.5340709041246338D -1	0.5359760086116637D 0
0.5466372783393311D -1	0.5449358126023893D 0
0.5592036525540278D -1	0.5576289982788432D 0
0.6031859623054687D -1	0.5780233068136232D 0
0.6157523365201660D -1	0.5903986424568555D 0
0.6471682720569083D -1	0.6091795983345321D 0
0.6660178333789550D -1	0.6193631321280773D 0
0.6911505818083497D -1	0.6233692140161312D 0
0.7100001431303960D -1	0.6497485967162184D 0
0.7351328915597900D -1	0.6587875000770038D 0
0.7539824528818359D -1	0.6722725059742015D 0
0.7853983884185789D -1	0.6892001292369697D 0
0.7979647626332763D -1	0.6975018281214187D 0
0.8419470723847168D -1	0.7263490121446943D 0
0.8545134465994142D -1	0.7371321446672510D 0
0.8607966337067627D -1	0.7330698368059814D 0
0.9236285047802490D -1	0.7662056501193123D 0
0.9361948789949462D -1	0.7782969888533047D 0
0.9424780661022947D -1	0.7883977028607242D 0
0.9864603758537353D -1	0.8023158099178310D 0
0.9990267500684325D -1	0.8023158099178310D 0
0.9990267500684326D -1	0.8075052418808986D 0
0.1055575434034570D 0	0.8370383384329113D 0
0.1068141808249268D 0	0.8416767879389165D 0
0.1118407305108057D 0	0.8708539984660319D 0
0.1137256866430101D 0	0.8716564541278106D 0

0.1181239176181543D	0	0.8936045063520407D	0
0.1187522363288892D	0	0.8952521037116878D	0
0.1193805550396239D	0	0.8927818450935628D	0
0.1250354234362378D	0	0.9203337535476270D	0
0.1256637421469726D	0	0.9322778162500355D	0
0.1256637421469727D	0	0.9211818027413622D	0
0.1294336544113818D	0	0.9435074858029534D	0
0.1332035666757909D	0	0.9400378709589688D	0
0.1332035666757910D	0	0.9417710805647293D	0
0.1332035666757910D	0	0.9461180983628737D	0
0.1394867537831397D	0	0.9645957635954727D	0
0.1394867537831397D	0	0.9637077470745611D	0
0.1401150724938745D	0	0.9522376530815871D	0
0.1401150724938745D	0	0.9681560199247218D	0
0.1463982596012232D	0	0.9861553841880718D	0
0.1470265783119580D	0	0.9897952155875916D	0
0.1476548970226928D	0	0.9798178571903070D	0
0.1507964905763672D	0	0.9971152310707185D	0
0.1526814467085718D	0	0.9897952155875916D	0
0.1533097654193066D	0	0.9907072706233942D	0
0.1545664028407763D	0	0.1005414976714505D	1
0.1595929525266553D	0	0.9980340311956851D	0
0.1602212712373901D	0	0.1005414976714505D	1
0.1608495899481250D	0	0.9971152310707185D	0
0.1646195022125341D	0	0.1009125893939589D	1
0.1665044583447388D	0	0.1000795514507977D	1
0.1677610957662085D	0	0.1003564638631950D	1
0.1683894144769434D	0	0.9934484813533680D	0
0.1721593267413526D	0	0.1003564638631950D	1
0.1740442828735571D	0	0.1000795514507977D	1
0.1746726015842920D	0	0.9934484813533680D	0
0.1759292390057617D	0	0.1002640746974386D	1
0.1809557886916406D	0	0.1000795514507977D	1
0.1815841074023755D	0	0.1004489381617169D	1
0.1847257009560497D	0	0.1000795514507977D	1
0.1847257009560498D	0	0.1001717705860734D	1
0.1872389757989893D	0	0.9989536779566709D	0
0.1878672945097241D	0	0.9952801676118729D	0
0.1922655254848680D	0	0.1002640746974386D	1
0.1941504816170728D	0	0.1002640746974386D	1
0.1941504816170728D	0	0.1001717705860734D	1
0.1954071190385425D	0	0.1002640746974386D	1
0.1998053500136865D	0	0.1006341424709143D	1
0.2010619874351563D	0	0.9934484813533680D	0
0.2054602184103003D	0	0.1001717705860734D	1
0.2067168558317700D	0	0.1012850507903801D	1
0.2067168558317700D	0	0.9916201660798969D	0
0.2086018119639745D	0	0.9961972768023490D	0
0.2136283616498535D	0	0.9971152310707185D	0
0.2155133177820581D	0	0.1004489381617169D	1
0.2186549113357324D	0	0.1004489381617169D	1
0.2205398674679370D	0	0.9980340311956851D	0

PHICALC

$\Theta_T = .153310$

MU= 2 I= 0 EPS= 0.10000000000000000D -6
MU= 2 I= 0 JJ= 20   PHI(I,MU)= 0.11708526959062660D -1

MU= 2 I= 1 EPS= 0.10000000000000000D -6
MU= 2 I= 1 JJ= 61   PHI(I,MU)=-0.4739878684789743D -2

MU= 2 I= 2 EPS= 0.10000000000000000D -6
MU= 2 I= 2 JJ= 61   PHI(I,MU)= 0.2045052525764569D -4

MU= 2 I= 3 EPS= 0.10000000000000000D -6
MU= 2 I= 3 JJ= 20   PHI(I,MU)=-0.2087593509625364D -3

MU= 2 I= 4 EPS= 0.10000000000000000D -6
MU= 2 I= 4 JJ= 20   PHI(I,MU)= 0.1229128101232320D -3

MU= 2 I= 5 EPS= 0.10000000000000000D -6
MU= 2 I= 5 JJ= 20   PHI(I,MU)=-0.7988587052978869D -4

MU= 2 I= 6 EPS= 0.10000000000000000D -6
MU= 2 I= 6 JJ= 20   PHI(I,MU)= 0.5588127643744050D -4

MU= 2 I= 7 EPS= 0.10000000000000000D -6
MU= 2 I= 7 JJ= 20   PHI(I,MU)=-0.4122275844833574D -4

MU= 2 I= 8 EPS= 0.10000000000000000D -6
MU= 2 I= 8 JJ= 20   PHI(I,MU)= 0.3164069890166272D -4

BINMOMENTS

```
NMAX= 2   MI= 77   MMAX= 33   ISWITCH1=0   IEND=0
  I2= 1   I2= 0
THETAT=0.153310
```

J	I	RU		RL		I	J	Q	
1	1	0.1925000000000000D	2	0.4387482193696061D	1	1	1	0.2279211529192759D	0
2	1	0.2982478695479562D	1	0.6797699828308797D	0	1	2	-0.2430609190411476D	-1
3	1	0.2593597106395147D	1	0.5911356426976798D	0	1	3	-0.2234346799710381D	-1
4	1	-0.3850000000000000D	2	-0.8774964387392122D	1	1	4	0.3481553119113957D	0
5	1	-0.1407195063561071D	2	-0.3207295212691531D	1	1	5	0.6773185275430524D	0
2	2	0.4109359710639516D	2	0.6374285048288730D	1	2	2	0.1568803391163790D	0
3	2	0.3475259601256188D	0	-0.8520219154034286D	-2	2	3	0.2213252986481471D	-3
4	2	-0.5964957390959124D	1	-0.3483443291960130D	-16	2	4	0.9513061281779564D	-18
5	2	-0.1549618070575176D	2	-0.2089011958707704D	1	2	5	0.3277537909766670D	0
3	3	0.3682293098084355D	2	0.6039322564546583D	1	3	3	0.1655814852265765D	0
4	3	-0.5187194212790294D	1	0.0000000000000000D	0	3	4	0.0000000000000000D	0
5	3	-0.2666375457674629D	1	-0.1305159265924712D	0	3	5	0.2161102097090421D	-1
4	4	0.1100000000000000D	3	0.5744562646538029D	1	4	4	0.1740776559556978D	0
5	4	0.2814390127122143D	2	0.0000000000000000D	0	4	5	0.0000000000000000D	0
5	5	0.0000000000000000D	0	0.1000000000000000D	1	5	5	0.1000000000000000D	1

```
       0.6773185275430524D  0     0.5377874109324596D -2
       0.3277537909766670D  0     0.2021123972687732D -2
       0.2161102097090421D -1     0.2133220632155132D -2
       0.0000000000000000D  0     0.2242678562604745D -2
  S2=  0.1659770354605073D -3
```

```
NMAX= 3   MI= 77   MMAX= 33   ISWITCH1=0   IEND=0
  I2= 1   I2= 0
THETAT=0.153310
```

J	I	RU		RL		I	J	Q	
1	1	0.1925000000000000D	2	0.4387482193696061D	1	1	1	0.2279211529192759D	0
2	1	0.2982478695479562D	1	0.6797699828308797D	0	1	2	-0.2430609190411476D	-1
3	1	0.2593597106395147D	1	0.5911356426976798D	0	1	3	-0.2234346799710381D	-1
4	1	-0.2634952735353944D	1	-0.6005614653296705D	0	1	4	0.2410604360115167D	-1
5	1	-0.3850000000000000D	2	-0.8774964387392122D	1	1	5	0.3481553119113957D	0
6	1	-0.1407195063561071D	2	-0.3207295212691531D	1	1	6	0.6797638292074426D	0
2	2	0.4109359710639516D	2	0.6374285048288730D	1	2	2	0.1568803391163790D	0
3	2	0.3475259601256188D	0	-0.8520219154034286D	-2	2	3	0.2213252986481471D	-3
4	2	0.9165280872386859D	0	0.2078306404842027D	0	2	4	-0.5369668870933032D	-2
5	2	-0.5964957390959124D	1	-0.3483443291960130D	-16	2	5	0.9854401616128597D	-18
6	2	-0.1549618070575176D	2	-0.2089011958707704D	1	2	6	0.3272090951964483D	0
3	3	0.3682293098084355D	2	0.6039322564546583D	1	3	3	0.1655814852265765D	0
4	3	0.8989752032630489D	0	0.2079304823070486D	0	3	4	-0.5622205766954680D	-2
5	3	-0.5187194212790294D	1	0.0000000000000000D	0	3	5	0.3599373278433412D	-19
6	3	-0.2666375457674629D	1	-0.1305159265924712D	0	3	6	0.2103664893252974D	-1
4	4	0.3742023991261920D	2	0.6080554019033444D	1	4	4	0.1644586984787545D	0
5	4	0.5269905470707887D	1	0.3651716673019989D	-16	4	5	-0.1045434801933897D	-17
6	4	0.8480714642150588D	0	-0.1014393612178362D	0	4	6	0.1668258532040159D	-1
5	5	0.1100000000000000D	3	0.5744562646538029D	1	5	5	0.1740776559556978D	0
6	5	0.2814390127122143D	2	0.0000000000000000D	0	5	6	0.0000000000000000D	0
6	6	0.0000000000000000D	0	0.1000000000000000D	1	6	6	0.1000000000000000D	1

```
       0.6797638292074426D  0     0.2184397601352468D -2
       0.3272090951964483D  0     0.8200593939550927D -3
       0.2103664893252974D -1     0.8655410267119118D -3
       0.1668258532040159D -1     0.8591697097269593D -3
       0.0000000000000000D  0     0.9094213350881376D -3
  S2=  0.2729255643554520D -4
```

NMAX= 8 MI= 77 MMAX= 33 ISWITCH1=0 IEND=0
I2= 1 I3= 0
THETAT=0.153310

J	I	RU		RL		I	J	Q	
1	1	0.1925000000000000D	2	0.4387482193696061D	1	1	1	0.2279211529192759D	0
2	1	0.2982488653756032D	1	0.6797722525327337D	0	1	2	-0.2430617641135752D	-1
3	1	0.2593588988748557D	1	0.5911337925143100D	0	1	3	-0.2234338712867665D	-1
4	1	-0.2634393718242982D	1	-0.6005584984547325D	0	1	4	0.2410593498900548D	-1
5	1	-0.1677081995992137D	1	-0.3822424620666882D	0	1	5	0.9924196896244063D	-2
6	1	-0.2083490585841638D	1	-0.4748715764214837D	0	1	6	0.1270288397091137D	-1
7	1	-0.1079776568998019D	1	-0.2461039205012486D	0	1	7	0.7214548337851271D	-2
8	1	0.1121759575412805D	1	0.2556727357263240D	0	1	8	-0.1223416115100045D	-1
9	1	0.2169130045626902D	0	0.4943906208311249D	-1	1	9	0.1429519564228440D	-2
10	1	-0.3850000000000000D	2	-0.8774964387392122D	1	1	10	0.3481553119113957D	0
11	1	-0.1407195063561071D	2	-0.3207295212691531D	1	1	11	0.6799388053771923D	0
2	2	0.4109358898874855D	2	0.6374284169491906D	1	2	2	0.1568803607448380D	0
3	2	0.3475489355130484D	0	-0.8516629119305332D	-2	2	3	0.2212321052212887D	-3
4	2	0.9165069927564212D	0	0.2078272572737006D	0	2	4	-0.5369580185268942D	-2
5	2	-0.4718430304084622D	1	-0.6994655974032744D	0	2	5	0.1846152990710171D	-1
6	2	-0.2756858564990155D	1	-0.3818552764672823D	0	2	6	0.8510366328587110D	-2
7	2	-0.9617310104288328D	0	-0.1246314680846566D	0	2	7	0.6740022767084610D	-3
8	2	-0.8628635644353289D	0	-0.1626320339930043D	0	2	8	0.4494712361136679D	-2
9	2	0.1972422651512595D	1	0.3041620513557674D	0	2	9	-0.8263293987484083D	-2
10	2	-0.5964977307512063D	1	-0.3483443772208394D	-16	2	10	0.8736721030434347D	-18
11	2	-0.1549618604724529D	2	-0.2089011942659897D	1	2	11	0.3275438295252922D	0
3	3	0.3682291800400788D	2	0.6039321676346109D	1	3	3	0.1655815095785752D	0
4	3	0.8989980679143925D	0	0.2079336961503742D	0	3	4	-0.5662309910467870D	-2
5	3	0.1513812419750539D	1	0.2870871696945634D	0	3	5	-0.7153267301093646D	-2
6	3	-0.1513180142830178D	1	-0.2046123211998619D	0	3	6	0.6455865755972492D	-2
7	3	-0.1460168991429447D	1	-0.2178638857232103D	0	3	7	0.5753844065167910D	-2
8	3	-0.1232875097418480D	1	-0.2293882416946778D	0	3	8	0.4247728853366059D	-2
9	3	-0.2597236194844968D	0	-0.4741563700606273D	-1	3	9	0.4835522653019427D	-3
10	3	-0.5187177977497114D	1	-0.3676648087726502D	-16	3	10	0.1140670334948201D	-17
11	3	-0.2666378892297098D	1	-0.1305162552500580D	0	3	11	0.2108849115718726D	-1
4	4	0.3742022343100199D	2	0.6080552962525293D	1	4	4	0.1644587270537799D	0
5	4	0.4104248229168836D	1	0.6513161245146616D	0	4	5	-0.1747429704907870D	-1
6	4	0.2810501993311247D	1	0.4353583596328059D	0	4	6	-0.1009380670044405D	-1
7	4	-0.1784276642143193D	1	-0.3060368679252923D	0	4	7	0.1025303093650914D	-1
8	4	-0.8570290464786147D	0	-0.1022909761093011D	0	4	8	0.1379888753287127D	-2
9	4	-0.2350915574603272D	1	-0.3905201379858496D	0	4	9	0.9589483686442622D	-2
10	4	0.5269879436485964D	1	0.3651717307513012D	-16	4	10	-0.9389949671360016D	-18
11	4	0.8480690784454751D	0	-0.1014392889222656D	0	4	11	0.1621785034496449D	-1
5	5	0.3871691300456270D	2	0.6129838484474868D	1	5	5	0.1631364354106091D	0
6	5	0.3833151729855821D	1	0.5154665548205738D	0	5	6	-0.1356107411664779D	-1
7	5	0.3413641938262079D	1	0.5700423962952634D	0	5	7	-0.1405234680045615D	-1
8	5	-0.2902364707004615D	1	-0.4544839260065123D	0	5	8	0.1540932924339488D	-1
9	5	-0.3953461394093705D	0	0.1700973760920484D	-1	5	9	-0.1059037586591898D	-2
10	5	0.3354163991984273D	1	0.3622356534962658D	-16	5	10	-0.1028277710476846D	-17
11	5	0.2557238122658745D	1	-0.4303489510191008D	-2	5	11	0.6274847274349225D	-3
6	6	0.3932005294951351D	2	0.6200937640359818D	1	6	6	0.1612659339599444D	0
7	6	0.2715063664994399D	1	0.3782371552209684D	0	6	7	-0.9755238239384037D	-2
8	6	0.3875324845331324D	1	0.6719150768192158D	0	6	8	-0.1675200987201918D	-1
9	6	-0.1125530047406167D	1	-0.1345539293906965D	0	6	9	0.5158528539280764D	-2
10	6	0.4166981171683276D	1	0.0000000000000000D	0	6	10	-0.2016029381148286D	-20
11	6	0.2175528920078114D	1	-0.2024679821826035D	-1	6	11	0.3261437929070908D	-2
7	7	0.3978173585658277D	2	0.6252719472170818D	1	7	7	0.1599304117913386D	0
8	7	0.4491898324582844D	1	0.7130022342142720D	0	7	8	-0.1883941944625080D	-1
9	7	0.2054844849766146D	1	0.3224634461569316D	0	7	9	-0.7061729717829006D	-2
10	7	0.2159553137996038D	1	0.0000000000000000D	0	7	10	0.2759828599272660D	-20
11	7	0.1161540884812416D	1	0.9993836414739304D	-2	7	11	-0.1595177798015684D	-2
8	8	0.3796125586101759D	2	0.6053094913370117D	1	8	8	0.1652047447316898D	0
9	8	0.2690534610840211D	1	0.4204064123827731D	0	8	9	-0.1123568602321137D	-1
10	8	-0.2243519150825610D	1	0.0000000000000000D	0	8	10	0.4391072564136571D	-20
11	8	-0.4641002775517399D	0	-0.3240520798610869D	-2	8	11	0.6374750226185437D	-3
9	9	0.3875948845373374D	2	0.6181476938549279D	1	9	9	0.1617736359677641D	0
10	9	-0.4338260091253803D	0	0.2245060192859251D	-17	9	10	-0.6322353375051255D	-19
11	9	-0.6874639340073391D	0	0.9089397040405564D	-2	9	11	-0.1470424807981042D	-2
10	10	0.1100000000000000D	3	0.5744562646538029D	1	10	10	0.1740776559556978D	0
11	10	0.2814390127122143D	2	0.0000000000000000D	0	10	11	0.0000000000000000D	0
11	11	0.0000000000000000D	0	0.1000000000000000D	1	11	11	0.1000000000000000D	1

I	A(I)		AER(I)	
0	0.6799388053771923D	0	0.1879896080936504D	-2
1	0.3275438295252922D	0	0.7119567520582355D	-3
2	0.2108849115718726D	-1	0.7458522743771322D	-3
3	0.1621785034496449D	-1	0.7466541841554968D	-3
4	0.6274847274349225D	-3	0.7409696475849779D	-3
5	0.3261437929070908D	-2	0.7296692525562373D	-3
6	-0.1595177798015684D	-2	0.7237516771491161D	-3
7	0.6374750226185437D	-3	0.7434877161087649D	-3
8	-0.1470424807981042D	-2	0.7263683954468329D	-3
9	0.0000000000000000D	0	0.7816138079809374D	-3

S2= 0.2016036477927324D -4

MU	I	APHI(I,MU)	EPS(MU)
0	0	0.50000000000000000D 0	0.10D -5
0	1	0.50000000000000000D 0	0.10D -5
0	2	0.50000000000000000D 0	0.10D -5
0	3	0.50000000000000000D 0	0.10D -5
0	4	0.50000000000000000D 0	0.10D -5
0	5	0.50000000000000000D 0	0.10D -5
0	6	0.50000000000000000D 0	0.10D -5
0	7	0.50000000000000000D 0	0.10D -5
0	8	0.50000000000000000D 0	0.10D -5
1	0	0.7635507006825518D -1	0.10D -6
1	1	0.1822697447505469D -3	0.10D -6
1	2	-0.4548600032167906D -4	0.10D -6
1	3	0.2020931179454311D -4	0.10D -6
1	4	-0.1136642170388743D -4	0.10D -6
1	5	0.7274120066239159D -5	0.10D -6
1	6	-0.5051325227339040D -5	0.10D -6
1	7	0.3711112582379761D -5	0.10D -6
1	8	-0.2841288204601814D -5	0.10D -6
2	0	0.1168607688026026D -1	0.10D -7
2	1	-0.4723414435737678D -2	0.10D -7
2	2	0.2474384302895605D -5	0.10D -7
2	3	-0.5560281612235868D -3	0.10D -7
2	4	0.5524311974601161D -4	0.10D -7
2	5	-0.5066104419487127D -3	0.10D -7
2	6	-0.5043013481466072D -3	0.10D -7
2	7	-0.4122275844833574D -4	0.10D -7
2	8	0.3164069890166272D -4	0.10D -7
3	0	0.1780629775807466D -2	0.10D -8
3	1	-0.1078571220908499D -2	0.10D -8
3	2	0.2647937592427742D -3	0.10D -8
3	3	-0.1172947237491624D -3	0.10D -8
3	4	0.6590154052299572D -4	0.10D -8
3	5	-0.4215428795868279D -4	0.10D -8
3	6	0.2926525519469661D -4	0.10D -8
3	7	-0.2149721519240700D -4	0.10D -8
3	8	0.1645692323448398D -4	0.10D -8
4	0	0.2716541648676528D -3	0.10D -9
4	1	-0.1958376344778814D -3	0.10D -9
4	2	0.8137972849538839D -4	0.10D -9
4	3	-0.3398091725783665D -4	0.10D -9
4	4	0.2009384318851745D -4	0.10D -9
4	5	-0.1289827657701711D -4	0.10D -9
4	6	0.8972036130113968D -5	0.10D -9
4	7	-0.6598403261679510D -5	0.10D -9
4	8	0.5055262366295827D -5	0.10D -9
5	0	0.4152497529839590D -4	0.10D-10
5	1	-0.3292640143492883D -4	0.10D-10
5	2	0.1769473764814489D -4	0.10D-10
5	3	-0.8584644187068336D -5	0.10D-10
5	4	0.4968394023143014D -5	0.10D-10

```
5  5   -0.3220852785178122D -5    0.10D-10
5  6    0.2252153843161562D -5    0.10D-10
5  7   -0.1661476641552383D -5    0.10D-10
5  8    0.1275459959268478D -5    0.10D-10

6  0    0.6345045035629471D -5    0.10D-11
6  1   -0.5333717755180693D -5    0.10D-11
6  2    0.3333250142349797D -5    0.10D-11
6  3   -0.1831473990741430D -5    0.10D-11
6  4    0.1095117738350550D -5    0.10D-11
6  5   -0.7201525364965361D -6    0.10D-11
6  6    0.5073910308725893D -6    0.10D-11
6  7   -0.3760080389534175D -6    0.10D-11
6  8    0.2894878698851033D -6    0.10D-11

7  0    0.9683784899902998D -6    0.10D-12
7  1   -0.8457239283582058D -6    0.10D-12
7  2    0.5846867761299686D -6    0.10D-12
7  3   -0.3581681842845573D -6    0.10D-12
7  4    0.2230907330750817D -6    0.10D-12
7  5   -0.1494258500002199D -6    0.10D-12
7  6    0.1063196174024407D -6    0.10D-12
7  7   -0.7925363064779722D -7    0.10D-12
7  8    0.6124900998388793D -7    0.10D-12

8  0    0.1477848716237346D -6    0.10D-13
8  1   -0.1325497112524552D -6    0.10D-13
8  2    0.9841201315024618D -7    0.10D-13
8  3   -0.6548797972975600D -7    0.10D-13
8  4    0.4275804906407704D -7    0.10D-13
8  5   -0.2928323331073077D -7    0.10D-13
8  6    0.2108905693769593D -7    0.10D-13
8  7   -0.1583523846065344D -7    0.10D-13
8  8    0.1229571317123244D -7    0.10D-13
```

MU	MOMENT	ERROR
0	0.7062803687891677D 0	0.1112514252430851D -2
1	0.5203497721454133D -1	0.7972066602049386D -4
2	0.4831863911667200D -2	0.1292774010792516D -4
3	0.5111628169564173D -3	0.2067520253713816D -5
4	0.5863672723582121D -4	0.3277128883503605D -6
5	0.7104796100799799D -5	0.5185971505535716D -7
6	0.8950983375753760D -6	0.8185139295189667D -8
7	0.1161426128522142D -6	0.1288655779627542D -8
8	0.1541858081906828D -7	0.2026672208036972D -9

AMOM

THETA(DEG)= 0.00000000000000000D 0   THETA1(RAD)= 0.15331000000000000D 0

N= 5
-0.13911170123495790D -3
 0.83513357887154100D -1
 0.32230178006517760D 1

MU= 4   X= 0.00000000000000000D 0   SUM= 0.00000000000000000D 0

THETA(DEG)= 0.10000000000000000D 1   THETA1(RAD)= 0.15331000000000000D 0

N= 5
-0.13911170123495790D -3
 0.83513357887154100D -1
 0.32230178006517760D 1

MU= 4   X= 0.92772930165972930D -7   SUM=-0.19786779560890970D -5

THETA(DEG)= 0.20000000000000000D 1   THETA1(RAD)= 0.15331000000000000D 0

N= 5
-0.13911170123495790D -3
 0.83513357887154100D -1
 0.32230178006517760D 1

MU= 4   X= 0.14834627841659690D -5   SUM=-0.11381878623127910D -5

THETA(DEG)= 0.30000000000000000D 1   THETA1(RAD)= 0.15331000000000000D 0

N= 5
-0.13911170123495790D -3
 0.83513357887154100D -1
 0.32230178006517760D 1

MU= 4   X= 0.75024075890382530D -5   SUM= 0.59566769570973770D -5

THETA(DEG)= 0.40000000000000000D 1   THETA1(RAD)= 0.15331000000000000D 0

N= 5
-0.13911170123495790D -3
 0.83513357887154100D -1
 0.32230178006517760D 1

MU= 4   X= 0.23677621504706420D -4   SUM= 0.23966568148418340D -4

THETA(DEG)= 0.50000000000000000D 1   THETA1(RAD)= 0.15331000000000000D 0

N= 5
-0.13911170123495790D -3
 0.83513357887154100D -1
 0.32230178006517760D 1

MU= 4   X= 0.57701092134533480D -4   SUM= 0.59373779973947870D -4

THETA(DEG)= 0.60000000000000000D 1   THETA1(RAD)= 0.15331000000000000D 0

N= 5
-0.13911170123495790D -3
 0.83513357887154100D -1
 0.32230178006517760D 1

MU= 4   X= 0.11938183842232430D -3   SUM= 0.12105877312769190D -3

THETA(DEG)= 0.70000000000000000D 1   THETA1(RAD)= 0.15331000000000000D 0

N= 5
-0.13911170123495790D -3
 0.83513357887154100D -1
 0.32230178006517760D 1

MU= 4   X= 0.22058596936768930D -3   SUM= 0.22085072542462650D -3

THETA(DEG)= 0.80000000000000000D 1   THETA1(RAD)= 0.15331000000000000D 0

ELEMENTS OF AN ECLIPSING BINARY SYSTEM IN FREQUENCY DOMAIN

Appendix B

A Regression Problem Arising In Fourier Domain Analysis of Eclipsing Binaries

I. Introduction

In this appendix we consider in detail a problem which arises from discrete measurements of the light intensity changes of a binary system whose components undergo mutual eclipses. For the purposes of the following discussion we shall assume the following model of the light variation:

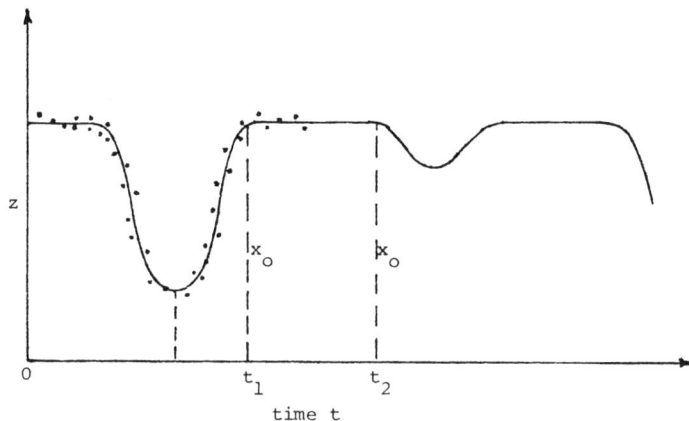

Figure B-1

in which

- z represents theoretical light intensity-solid line
- • indicates a measurement of intensity
- x_o is the unknown maximal constant intensity level over known interval $[t_1, t_2]$.

The basic problem is to determine as accurately as possible from measurements of light intensity the unknown curve $z = z(t)$ vs. t. Often the available measurements will not be in the initial form as depicted in Fig. B-1, but rather will be in normalized form, being divided by a 'best' estimate of x_0. This must also be taken into account in addressing the problem.

II. ANALYSIS

Assume in Fig. B-1, that the nonconstant part of the curve $z = z(t)$ (over the interval $[0, t_1]$) is exactly represented by a sufficiently large Fourier expansion (of size 2M+1)

$$z = z(t) \equiv \sum_{j=0}^{2M} a_j \cdot \cos(j\omega t), \tag{1}$$

for $0 \le t \le t_1$.

$\omega > 0$ is known, integer M is known (pre-chosen), while the a_j's are unknown and to be estimated from the measurements.

The constant part of the curve $z = z(t)$ in Figure B-1 (over the interval $[t_1, t_2]$ is given by the relation

$$z = z(t) \equiv x_0 \tag{2}$$

for $t_1 \le t \le t_2$. x_0 is also unknown.

For convenience choose

$$\omega = \pi/2t_1 . \tag{3}$$

In addition, the boundary condition for continuity of $z(t)$ at $t=0$ and $t=t_1$ requires

$$z(0) = z(t_1) = x_0 \qquad (4)$$

That is, the unknown a_j's and x_0 are subject to the two constraints:

$$\sum_{j=0}^{2M} a_j - x_0 = 0$$

$$\sum_{j=0}^{M} (-1)^j a_{2j} - x_0 = 0 . \qquad (5)$$

Initial measurements $y(t)$, of $z(t)$ are made at:

$$0 \leq t_{11} < t_{12} < \cdots < t_{1m} \leq t_1 < t_{21} < t_{22} < \cdots < t_{2n} \leq t_2 \qquad (6)$$

These measurements relate to z by

$$y(t) = z(t) + \varepsilon(t) , \qquad (7)$$

for all t as in (6), where unobserved measurement error $\varepsilon(t)$ is Gaussian distributed $n(0, \sigma^2(t))$, with $\sigma^2(t) > 0$ known, and where, for any $t' \neq t''$, $\varepsilon(t')$ is statistically independent of $\varepsilon(t'')$.

In compact matrix form, eqs. (5) – (7) reduce to:

$$Y = B \cdot X + \varepsilon, \qquad (8)$$

with

$$C \cdot X = D, \qquad (9)$$

where,

$$D \equiv 0_2 \quad \text{and}$$

$$Y = \begin{pmatrix} Y_1 \\ --- \\ Y_2 \end{pmatrix}, \quad \varepsilon = \begin{pmatrix} \varepsilon_1 \\ -- \\ \varepsilon_2 \end{pmatrix}, \quad (10)$$

(m+n by 1)   (m+n by 1)

$$Y_1 = \begin{pmatrix} y(t_{11}) \\ \vdots \\ y(t_{1m}) \end{pmatrix}, \quad Y_2 = \begin{pmatrix} y(t_{21}) \\ \vdots \\ y(t_{2n}) \end{pmatrix}, \quad (11)$$

(m by 1)   (n by 1)

$$\varepsilon_1 = \begin{pmatrix} \varepsilon(t_{11}) \\ \vdots \\ \varepsilon(t_{1m}) \end{pmatrix}, \quad \varepsilon_2 = \begin{pmatrix} \varepsilon(t_{21}) \\ \vdots \\ \varepsilon(t_{2n}) \end{pmatrix}, \quad (12)$$

(m by 1)   (n by 1)

$$B = \begin{pmatrix} B_1 & | & 0_m \\ ---- & + & --- \\ 0_{m,n} & | & \underset{\sim}{1}_n \end{pmatrix} \quad \text{(m+n by 2M+2)}, \quad (13)$$

$$B_1 = \begin{pmatrix} 1 & \cos(\omega t_{11}) & \cos(2\omega t_{11}) & \cdots & \cos(2M\omega t_{11}) \\ 1 & \cos(\omega t_{12}) & \cos(2\omega t_{12}) & \cdots & \cos(2M\omega t_{12}) \\ \vdots & \vdots & \vdots & \cdots & \vdots \\ 1 & \cos(\omega t_{1m}) & \cos(2\omega t_{1m}) & \cdots & \cos(2M\omega t_{1m}) \end{pmatrix}, \quad (14)$$

(m by 2M+1 of rank 2M+1)

0_m is an m by 1 column vector of zeroes, $0_{m,n}$ is an m by n matrix of zeroes $\underset{\sim}{1}_n$ is an n by 1 vector of ones.

$$X = \begin{pmatrix} X_1 \\ --- \\ x_o \end{pmatrix} \quad, \quad X_1 = \begin{pmatrix} a_o \\ a_1 \\ \vdots \\ a_{2M} \end{pmatrix} \quad \text{(2M+1 by 1)}. \tag{15}$$

x_o is a (1 by 1) scalar.

X is unknown.

$$C = \begin{pmatrix} \mathbf{1}_{2M+1}^T & \vdots & -1 \\ ----- & + & ---- \\ \mathbf{J}_{2M+1}^T & \vdots & -1 \end{pmatrix}, \tag{16'}$$

(2 by 2M+2 of rank 2)

$$\mathbf{J}_{2M+1}^T = (1, 0, -1, 0, 1, 0, -1, \ldots, 0, \pm 1), \tag{16''}$$

the last entry being +1, if M is even, and -1, if M is odd.

Now it easily follows that if it is assumed that

$$m+n \geq 2M+2 \tag{17}$$

then the weighted least squares - and also minimal expected normed squared absolutely unbiased-estimator of X subject to the linear constraint (9) is

$$\hat{\hat{X}} = \hat{X} + K \cdot (D - C \cdot \hat{X}), \tag{18}$$

$$K = \Lambda^{-1} C^T (C \Lambda^{-1} C^T)^{-1}, \tag{19}$$

$$\Lambda = B^T \Sigma^{-1} B, \tag{20}$$

$\hat{X}$ = unconstrained weighted least squares estimator of X

 = minimal expected normed squared absolutely unbiased estimator of X

$$= \Lambda^{-1} \cdot B^T \Sigma^{-1} Y, \tag{21}$$

$$\Sigma = \text{Cov}(\varepsilon) = \left(\begin{array}{c|c} \Sigma_1 & O_{m,n} \\ \hline O_{n,m} & \Sigma_2 \end{array}\right) \quad ,$$

(m+n by m+n positive definite diagonal) (22)

$$\Sigma_1 = \text{Cov}(\varepsilon_1) = \begin{pmatrix} \sigma^2(t_{11}) & & 0 \\ & \ddots & \\ 0 & & \sigma^2(t_{1m}) \end{pmatrix} \quad , \quad (23)$$

$$\Sigma_2 = \text{Cov}(\varepsilon_2) = \begin{pmatrix} \sigma^2(t_{21}) & & 0 \\ & \ddots & \\ 0 & & \sigma^2(t_{2n}) \end{pmatrix} \quad , \quad (24)$$

(For related proofs, basic properties, and general background, see, e.g. Rao, Deutsch, Plackett, or Sheffé. See also Appendix B-1 for a direct proof by matrix orderings with respect to positive semidefiniteness.)

Note that as a check, $C \cdot \hat{\hat{X}} = D$ (=0 here).

Now, by matrix partitioning considerations,

$$\hat{X} = \begin{pmatrix} \hat{X}_1 \\ \hat{X}_0 \end{pmatrix} \quad , \quad (25)$$

$$\hat{X}_1 = \Lambda_1^{-1} \cdot B_1^T \cdot \Sigma_1^{-1} Y_1 \quad , \quad (26)$$

$$\Lambda_1 = B_1^T \Sigma_1^{-1} B_1 \quad (27)$$

$$= (\lambda_{ij}) \quad 0 \leq i,j \leq 2M \quad ,$$

$$\Lambda_1^{-1} \equiv (\lambda^{(ij)}) \quad 0 \leq i,j < 2M \quad ,$$

$$\lambda_{ij} = \sum_{k=1}^{m} (1/\sigma^2(t_{1k})) \cos(i\omega t_{1k}) \cos(j\omega t_{1k})$$

$$= \lambda_{ji} \quad (28)$$

ELEMENTS OF AN ECLIPSING BINARY SYSTEM IN FREQUENCY DOMAIN

for all $0 \leq i, j \leq 2M$

$$B_1^T \Sigma_1^{-1} Y_1 = \begin{pmatrix} \mu_0 \\ \mu_1 \\ \vdots \\ \mu_{2M} \end{pmatrix} , \quad (29)$$

where

$$\mu_j = \sum_{k=1}^{m} (1/\sigma^2(t_{1k})) \cdot y_{1k} \cdot \cos(j\omega t_{1k}) , \quad (30)$$

for $0 \leq j \leq 2M$

Combining (26) - (30) yields

$$\hat{x}_1 = \begin{pmatrix} \hat{a}_0 \\ \hat{a}_1 \\ \vdots \\ \hat{a}_{2M} \end{pmatrix} , \quad (31)$$

where

$$\hat{a}_i = \sum_{j=0}^{2M} \lambda^{(ij)} \cdot \mu_j , \quad (32)$$

with the $\lambda^{(ij)}$'s obtained from the positive definite $2M+1$ by $2M+1$ matrix Λ_1 by usual inversion procedures applied to $B_1^T \Sigma^{-1} B_1$. (See e.g., Rao, pp. 74, 75.)

Next, consider $\hat{x}_0$. Again by matrix partitionings,

$$\hat{x}_0 = \bar{x}_0 = \Lambda_2^{-1} \cdot \underset{\sim}{1}_n^T \cdot \Sigma_2^{-1} \cdot Y_2 , \quad (33)$$

$$\Lambda_2 = \underset{\sim}{1}_n^T \cdot \Sigma_2^{-1} \cdot \underset{\sim}{1}_n$$

$$= \sum_{i=1}^{n} 1/\sigma^2(t_{2i}) , \quad (34)$$

$$\lambda_n^T \cdot \Sigma_2^{-1} \cdot y_2 = \sum_{i=1}^{n} (1/\sigma^2(t_{2i})) \cdot y_{2i} \tag{35}$$

yielding an immediate evaluation of the scalar quantity $\hat{x}_o$.

Eq. (20) reduces to

$$\Lambda = \begin{pmatrix} \Lambda_1 & 0_{2M+1} \\ 0_{2M+1}^T & \Lambda_2 \end{pmatrix} \tag{36}$$

Hence, eq. (19) yields by matrix partitioning multiplication, etc.

$$C\Lambda^{-1}C^T = \begin{pmatrix} g_{11} & g_{12} \\ g_{12} & g_{22} \end{pmatrix} \tag{37}$$

(2 by 2 positive definite matrix).

$$g_{11} = \underline{1}_{2M+1}^T \cdot \Lambda_1^{-1} \cdot \underline{1}_{2M+1} + \Lambda_2^{-1}$$

$$= \sum_{0 \le i,j \le 2M} \lambda^{(ij)} + \Lambda_2^{-1} \quad , \tag{38}$$

$$g_{12} = \underline{1}_{2M+1}^T \cdot \Lambda_1^{-1} \cdot \underline{J}_{2M+1} + \Lambda_2^{-1}$$

$$= \sum_{\substack{0 \le i,j \le 2M, \\ \text{such that} \\ i \text{ is even \&} \\ \text{divisible} \\ \text{by 4}}} \sum_{j=0}^{2M} \lambda^{(i,j)} - \sum_{\substack{0 \le i \le 2M \\ \text{such that} \\ i \text{ is even} \\ \text{\& not divisible by 4}}} \sum_{j=0}^{2M} \lambda^{(i,j)} \quad , \tag{39}$$

$$g_{22} = \Bigg(\sum_{\substack{0 \le i,j \le 2M \\ \text{such that} \\ i\&j \text{ are even} \\ \text{and } i+j \text{ is divisible by 4}}} \lambda^{(i,j)} - \sum_{\substack{0 \le i,j \le 2M \\ \text{such that} \\ i\&j \text{ are even} \\ \text{and } i+j \text{ is not} \\ \text{divisible by 4}}} \lambda^{(i,j)} \Bigg) + \Lambda_2^{-1} \tag{40}$$

Hence,

$$K = \begin{pmatrix} K_{11} & K_{12} \\ \hline K_{21} & K_{22} \end{pmatrix} \quad (2M+2 \text{ by } 2) \quad , \tag{41}$$

$$K_{11} = \Lambda_1^{-1} \cdot \underline{1}_{2M+1} \cdot g^{(11)} + \Lambda_1^{-1} \cdot \underline{J}_{2M+1} \cdot g^{(12)}$$

$$= \begin{pmatrix} k_{11,0} \\ \vdots \\ k_{11,2M} \end{pmatrix} \quad , \tag{42}$$

$$k_{11,i} = g^{(11)} \cdot \sum_{j=0}^{2M} \lambda^{(i,j)} + g^{(12)} \cdot \left(\sum_{\substack{0 \le j \le 2M, \\ j \text{ div. by } 4}} \lambda^{(i,j)} - \sum_{\substack{0 \le j \le 2M \\ j \text{ not div.} \\ \text{by } 4}} \lambda^{(i,j)} \right) ,$$

for $0 \le i \le 2M$; \hfill (43)

$$K_{12} = \Lambda_1^{-1} \cdot \underline{1}_{2M+1} \cdot g^{(12)} + \Lambda_1^{-1} \cdot \underline{J}_{2M+1} \cdot g^{(22)}$$

$$= \begin{pmatrix} k_{12,0} \\ \vdots \\ k_{12,2M} \end{pmatrix} \quad ; \tag{44}$$

$$k_{12,i} = g^{(12)} \cdot \sum_{j=0}^{2M} \lambda^{(i,j)} + g^{(22)} \cdot \left(\sum_{\substack{0 \le j \le 2M, \\ j \text{ div.} \\ \text{by } 4}} \lambda^{(i,j)} - \sum_{\substack{0 \le j \le 2M, \\ j \text{ not div.} \\ \text{by } 4}} \lambda^{(i,j)} \right)$$

for $0 \le i \le 2M$. \hfill (45)

$$k_{21} = -\Lambda_2^{-1} (g^{(11)} + g^{(12)}) \tag{46}$$

$$k_{22} = -\Lambda_2^{-1} (g^{(12)} + g^{(22)}) \quad , \tag{47}$$

where

$$(C\Lambda^{-1}C^T)^{-1} = \begin{pmatrix} g^{(11)} & g^{(12)} \\ \hline g^{(12)} & g^{(22)} \end{pmatrix} , \tag{48}$$

$$g^{(11)} = (1/g_0) \cdot g_{22} \tag{49}$$

$$g^{(22)} = (1/g_o) \cdot g_{11} \quad , \tag{50}$$

$$g^{(12)} = (1/g_o) \cdot (-g_{12}) \quad . \tag{51}$$

Thus, from the above equations, (18) yields,

$$\hat{\hat{x}} = \begin{pmatrix} \hat{\hat{x}}_1 \\ \hat{\hat{x}}_o \end{pmatrix} \quad , \tag{52}$$

$$\hat{\hat{x}}_1 = \hat{x}_1 - \{K_{11}(\underset{\sim}{1}^T_{2M+1} \cdot \hat{x}_1 - \overline{x}_o)$$

$$+ K_{12} \cdot (\underset{\sim}{J}^T_{2M+1} \cdot \hat{x}_1 - \overline{x}_o)\}$$

$$= \begin{pmatrix} \hat{\hat{a}}_o \\ \hat{\hat{a}}_1 \\ \vdots \\ \hat{\hat{a}}_{2M} \end{pmatrix} \quad , \tag{53}$$

where

$$\hat{\hat{a}}_i = \hat{a}_i - \{k_{11,i} \cdot (\sum_{k=0}^{2M} \hat{a}_k - \overline{x}_o)$$

$$+ k_{12,i} \cdot (\underset{\substack{0 \leq k \leq 2M, \\ k \text{ even} \\ \& \text{ div.} \\ \text{by 4}}}{\sum} \hat{a}_k - \underset{\substack{0 \leq k \leq 2M, \\ k \text{ even} \\ \& \text{ not} \\ \text{div. by 4}}}{\sum} \hat{a}_k - \overline{x}_o)\} \quad , \tag{54}$$

for $0 \leq i \leq 2M$.

$$\hat{\hat{x}}_o = \overline{x}_o - \{K_{21} \cdot (\underset{\sim}{1}^T_{2M+1} \cdot \hat{x}_1 - \overline{x}_o)$$

$$+ K_{22} \cdot (\underset{\sim}{J}^T_{2M+1} \cdot \hat{x}_1 - \overline{x}_o)\}$$

ELEMENTS OF AN ECLIPSING BINARY SYSTEM IN FREQUENCY DOMAIN 93

$$= \bar{x}_o - \left\{ K_{21} \cdot \left(\sum_{k=0}^{2M} \hat{a}_k - \bar{x}_o \right) \right.$$
$$\left. + K_{22} \cdot \left(\sum_{\substack{0 \le k \le 2M \\ k \text{ even} \\ \& \text{ div.} \\ \text{by } 4}} \hat{a}_k - \sum_{\substack{0 \le k \le 2M, \\ k \text{ even} \\ \& \text{ not div.} \\ \text{by } 4}} \hat{a}_k - \bar{x}_o \right) \right\}. \tag{55}$$

The optimal estimator of curve $z = z(t)$ in eq. (1) is given by $\hat{\hat{z}}$, where

$$\hat{\hat{z}}(t) = \begin{cases} \sum_{j=0}^{2M} \hat{\hat{a}}_j \cdot \cos(j\omega t); & 0 \le t \le t_1 \\ \hat{\hat{x}}_o & ; t_1 \le t \le t_2 \end{cases} \tag{56}$$

Often, data appears in a normalized form where Y_1 is replaced by

$$Y_1' = (1/\bar{x}_o) \cdot Y_1 \equiv \begin{pmatrix} y_{11}' \\ \vdots \\ y_{1m}' \end{pmatrix}, \tag{57}$$

Consequently, in terms of Y_1', for all calculations involving y_{ik} — as in eqs. (30) and (35) — replace y_{ik} by $\bar{x}_o \cdot y_{ik}'$, etc.

$(\hat{\hat{x}}|X)$, $(\hat{\hat{x}}_1|X)$ $(\hat{\hat{x}}_o|X)$, $(\hat{x}|X)$, $(\hat{x}_o|X)$, as well as all errors such as $(\hat{\hat{x}}-X|X)$, are all Gausssian distributed with computable moments. It readily follows that the basic error covariance matrix, e.g., is

$$\text{Cov}(\hat{\hat{x}}-X|X) = \Lambda^{-1} - \Lambda^{-1} C^T (C \Lambda^{-1} C^T)^{-1} C \Lambda^{-1}$$
$$= (I - K \cdot C) \cdot \Lambda^{-1}, \tag{58}$$

with the absolute unbiasedness of $\hat{X}$ yielding zero expected error

$$E(\hat{X}-X|X) = 0_{2M+2} \,, \tag{59}$$

for all X satisfying eq. (9).

Similarly, for <u>all</u> X ,

$$\text{Cov}(\hat{X}-X|X) = \Lambda^{-1} \,, \tag{60}$$

$$E(\hat{X}-X|X) = 0_{2M+2} \,, \tag{61}$$

$$\text{Cov}(\hat{X}_1-X_1|X_1) = \Lambda_1^{-1} \,, \tag{62}$$

$$E(\hat{X}_1-X_1|X_1) = 0_{2M+1} \,, \tag{63}$$

$$\text{var}(\hat{x}_0-x_0|x_0) = \Lambda_2^{-1} \,, \tag{64}$$

$$E(\hat{x}_0-x_0|x_0) = 0 \,, \tag{65}$$

$$(\hat{x}_0 \equiv \bar{x}_0)$$

$$\text{Cov}(\hat{X}_1-X_1|X_1)$$

$$= \Lambda^{-1} - \left(K_{11} \cdot \underset{\sim}{1}_{2M+1}^T + K_{12} \cdot \underset{\sim}{J}_{2M+1}^T\right) \cdot \Lambda_1^{-1}$$

$$= (\gamma_{ij})_{0 \leq i,j \leq 2M} \,, \tag{66}$$

for X_1 satisfying eq. (9), where

$$\gamma_{ij} = \lambda^{(i,j)} - \left(k_{11,i} \cdot \sum_{\ell=0}^{2M} \lambda^{(\ell,j)} \right.$$
$$\left. + k_{12,i} \cdot \left(\sum_{\substack{0 < \ell < 2M, \\ \ell \text{ even} \\ \text{\& div.} \\ \text{by } 4}} \lambda^{(\ell,j)} - \sum_{\substack{0 < \ell < 2M, \\ \ell \text{ even \&} \\ \text{not div by } 4}} \lambda^{(\ell,j)} \right) \right) \,, \tag{67}$$

for $0 \leq i,j \leq 2M$,

ELEMENTS OF AN ECLIPSING BINARY SYSTEM IN FREQUENCY DOMAIN

$$E(\hat{x}_1 - X_1 | X_1) = 0_{2M+1} \quad , \tag{68}$$

$$\text{var}(\hat{x}_0 - x_0 | x_0) = (1 + K_{21} + K_{22})\Lambda_2^{-1} \quad , \tag{69}$$

and again, by absolute unbiasedness,

$$E(\hat{x}_0 - x_0 | x_0) = 0 \quad , \tag{70}$$

for x_0 (and X_1) satisfying eq. (9).

Note that the <u>goodness of fit</u> in estimating $y(t)$ by $\hat{z}(t)$ in (56) is, also Gaussian distributed, where for t at the sampling times $t = t_{11}, t_{12}, \ldots, t_{1m}, t_{21}, \ldots, t_{2n}$,

$$\left(\hat{z}(t) - y(t)\right)_{t = t_{11}, \ldots, t_{2n}}$$

$$= B \cdot \hat{X} - Y$$

$$= B \cdot (\hat{X} - X) + \varepsilon \quad , \tag{71}$$

with expected mean (for X satisfying (9))

$$E(B \cdot \hat{X} - Y | X) \equiv 0 \tag{72}$$

(absolutely unbiased) and with covariance matrix of error

$$\text{Cov}(B \cdot \hat{X} - Y | X)$$

$$= B \cdot \text{Cov}(\hat{X} - X | X) B^T + \Sigma \quad , \tag{73}$$

and expected normed square error

$$E(\|B\hat{X} - Y\|^2 | X)$$

$$= \text{tr } \text{Cov}(B\hat{X} - Y | X)$$

$$= \text{tr}(B_1 \cdot \text{Cov}(\hat{x}_1 - X_1 | X_1) B_1^T) + \text{tr } \Sigma_1$$
$$+ \text{var}(\hat{x}_0 - x_0 | x_0) \cdot n + \text{tr } \Sigma_2 \quad . \tag{74}$$

Additionally, note that the mean squared error, for all X satisfying eq. (9)

$$E(\|\hat{\hat{x}}-x\|^2 | X)$$
$$= \text{tr Cov}(\hat{\hat{x}}-x | X)$$
$$= \text{tr}(\Lambda^{-1} - \Lambda^{-1} C^T (C\Lambda^{-1} C^T)^{-1} C\Lambda^{-1}), \qquad (75)$$

etc., from eq. (58),

$$E(\|\hat{x}-x\|^2 | X)$$
$$= \text{tr Cov}(\hat{x}-x | X)$$
$$= \text{tr}(\Lambda^{-1}), \qquad (76)$$

etc., from eq. (60).

The difference between the two basic estimators is, from (18)

$$\hat{\hat{x}}-\hat{x} = K \cdot (D-C\hat{x})$$
$$= KC \cdot (X-\hat{x}). \qquad (77)$$

Hence,

$$\text{Cov}(\hat{\hat{x}}-\hat{x} | X) = KC \cdot \text{Cov}(\hat{x}-X | X) C^T K^T$$
$$= KC\Lambda^{-1} C^T K^T$$
$$= \Lambda^{-1} C^T (C\Lambda^{-1} C^T)^{-1} C\Lambda^{-1}, \qquad (78)$$

and

$$E(\|\hat{\hat{x}}-\hat{x}\|^2 | X) = \text{tr Cov}(\hat{\hat{x}}-\hat{x} | X), \qquad (79)$$

etc.

Thus, eqs. (58), (76), (78), (79) imply

$$E(\|\hat{\hat{x}}-x\|^2 | X) = E(\|\hat{x}-x\|^2 | X)$$
$$- E(\|\hat{\hat{x}}-\hat{x}\|^2 | X), \qquad (80)$$

showing the explicit improvement in accuracy of $\hat{\hat{x}}$ over $\hat{x}$.

The <u>quadratic</u> <u>form</u> <u>weighted</u> <u>goodness</u> <u>of</u> <u>fit</u> of $\hat{\hat{z}}$ to y is

ELEMENTS OF AN ECLIPSING BINARY SYSTEM IN FREQUENCY DOMAIN

$$(B\hat{\hat{X}}-Y)^T \cdot \Sigma^{-1} \cdot (B\hat{\hat{X}}-Y)$$

$$= (B\hat{X}-Y)^T \, \Sigma^{-1} (B\hat{X}-Y)$$

$$+ (D-C\hat{X})^T \cdot (C\Lambda^{-1}C^T)^{-1} \cdot (D-C\hat{X})$$

$$= (B\hat{X}-Y)^T \, \Sigma^{-1} (B\hat{X}-Y)$$

$$+ (\hat{X}-X)^T C^T (C\Lambda^{-1}C^T)^{-1} C(\hat{X}-X) \quad , \qquad (81)$$

for all X such that $D = C \cdot X$.

Since, $(B\hat{X}-Y|X)$ and $(\hat{X}-X|X)$ are all normally distributed and $\text{Cov}(B\hat{X}-Y, \hat{X}-X|X) = O_{m+n, \, 2M+2}$, then the two quantities above are statistically independent, yield by considering standardized quadratic forms (see, [1], Chapter 3)

$$((B\hat{X}-Y)^T \cdot \Sigma^{-1} \cdot (B\hat{X}-Y)|X)$$

$$= (Y^T \cdot (\Sigma^{-1} - \Sigma^{-1} \cdot B(B^T \Sigma^{-1} B)^{-1} \cdot B^T \Sigma^{-1}) \cdot Y | X)$$

is distributed as $\chi^2_{m+n-(2M+2)}$;

$$((\hat{X}-X)^T C^T (C\Lambda^{-1}C^T)^{-1} C(\hat{X}-X)|X)$$

is distributed as χ^2_2 ;

$$((B\hat{\hat{X}}-Y)^T \cdot \Sigma^{-1} \cdot (B\hat{\hat{X}}-Y)|X)$$

is distributed as $\chi^2_{m+n-(2M+2)+2} = \chi^2_{m+n-2m}$

Hence, the mean normed differences become

$$E(B\hat{X}-Y)^T \Sigma^{-1} (B\hat{X}-Y)|X) = m+n - (2M+2),$$
$$E(\hat{X}-X)^T C^T (C\Lambda^{-1}C^T)^{-1} C(\hat{X}-X)|X) = 2 \quad ,$$
$$E((B\hat{\hat{X}}-Y)^T \Sigma^{-1} (B\hat{\hat{X}}-Y)|X) = m+n - 2M \qquad (82)$$

Note the special case of equal interval constant error variance sampling over $[0, t_1]$:

In this case, the critically important quantity Λ_1 becomes:

$$\Lambda_1 = (1/\sigma^2_{(1)}) \cdot B_1^T \cdot B_1 , \tag{83}$$

where

$$\sigma^2_{(1)} = \sigma^2(t_{11}) = \cdots = \sigma^2(t_{1m}) , \tag{84a}$$

$$t_{1k} = (k-1) \cdot t_1/m ; \quad k = 1,\ldots,m, \tag{84b}$$

$$B_1^T B_1 = (b_{ij}) , \tag{85}$$
$$0 \leq i, j \leq 2M$$

$$b_{ij} = \sum_{k=1}^{m} \cos(i\omega t_{1k}) \cos(j\omega t_{1k})$$

$$= \sum_{k=0}^{m-1} \cos(i\pi k/2m) \cos(j\pi k/2m)$$

$$= (\tfrac{1}{2}) \cdot \sum_{k=0}^{m-1} \cos((i+j)\pi k/2m)$$

$$+ (\tfrac{1}{2}) \cdot \sum_{k=0}^{m-1} \cos((i-j)\pi k/2m) \tag{86}$$

using the basic identity

$$\cos\alpha \cos\beta = (\tfrac{1}{2}) (\cos(\alpha+\beta) + \cos(\alpha-\beta)) . \tag{87}$$

Now, for any θ,

$$\sum_{k=0}^{m-1} \cos(\theta k) = \sum_{k=0}^{m-1} (\tfrac{1}{2}) (e^{i\theta k} + e^{-i\theta k}) ,$$

where $i = \sqrt{-1}$,

$$= (\tfrac{1}{2}) \left(\frac{1-e^{i\theta m}}{1-e^{i\theta}} + \frac{1-e^{-i\theta m}}{1-e^{-i\theta}} \right)$$

$$= (\tfrac{1}{2}) \cdot (1 + \frac{\cos(\theta \cdot (m-1)) - \cos(\theta m)}{1-\cos\theta})$$

$$= (\tfrac{1}{2}) \cdot (1 - \cos(\theta m) + \frac{\sin(\theta)\sin(\theta m)}{1-\cos\theta}). \tag{88}$$

Combining (86) and (88) yields for $0 \leq i,j \leq 2M$,

$$b_{ij} = \begin{cases} (\tfrac{1}{4}) \cdot (2m + (1-\cos(i\pi))) & ; \quad \text{if } i=j \neq 0 \\ \underline{m} & ; \quad \underline{\text{if } i=j=0} \\ (\tfrac{1}{4}) \cdot \{ \dfrac{\sin((i+j) \cdot \tfrac{\pi}{2}) \cdot \sin((i+j)\tfrac{\pi}{2m})}{1-\cos((i+j)\pi/2m)} \\ \quad + \dfrac{\sin((i-j)\tfrac{\pi}{2}) \cdot \sin((i-j)\tfrac{\pi}{2m})}{1-\cos((i-j)\pi/2m)} \\ \quad + 2(1-\cos(\pi i/2) \cdot \cos(\pi j/2)) \} & ; \quad \text{if } i \neq j \end{cases} \tag{89}$$

The quantities in most of the previous matrix equations can be computed out in scalar forms by straightforward operations.

It should also be noted that constraint (9), for sufficiently large sample sizes m and n, can be ignored, i. e., $\hat{\hat{X}} \approx \hat{X}$ ($\approx X$, as a consistent statistic). This is shown in detail in Appendix B-2.

As a final remark: the quantities M, t_1 and t_2 were assumed known throughout the analysis. In actuality, given M, t_1 and t_2 could be estimated (at least theoretically):

Given t_2, t_1 could be estimated by a least squares procedure, whereby m, the sample size up to t_1 is determined, for m+n, total sample, held fixed (The functional dependency on t through m by the partitioning of the forms into the first and second sampling intervals is shown):

Choose m, $1 \leq m \leq$ m+n, which minimizes

$$(Y(m) - B(m) \cdot \hat{X}(m))^T \cdot (\Sigma(m))^{-1} \cdot (Y(m) - B(m) \cdot \hat{X}(m)) \tag{90}$$

Determination of t_2 is more complicated involving periodogram analysis and sampling beyond the one period considered here. (See e.g., Jenkins and Watts, 1968.)

Choice of M can be determined by a nested chi-square significance test approach in which a value M is tested vs. M+1. Thus, e.g., one can use the test statistic

$$t = \hat{X}_2^T \cdot \text{Cov}^{-1}(\hat{X}_2 - X_2 | X_2) \cdot \hat{X}_2, \tag{91}$$

where

$$X_2 \equiv \begin{pmatrix} a_{2M+1} \\ a_{2M+2} \end{pmatrix}$$

is the proposed additional vector in eq. (8), with X_1 replaced by

$\begin{pmatrix} X_1 \\ X_2 \end{pmatrix}$, B_1 by $(B_1 | B_2)$, where

$$B_2 = \begin{pmatrix} \cos((2M+1)\omega t_{11}) & \cos((2M+2)\omega t_{11}) \\ \vdots & \vdots \\ \cos((2M+1)\omega t_{lm}) & \cos((2M+2)\omega t_{lm}) \end{pmatrix}. \tag{92}$$

$\hat{X}_2$ and $\text{Cov}(\hat{X}_2 - X_2 | X_2)$ are obtained, by partitioning considerations, from the new model resulting from the above substitution, in eq. (8) and all following results.

Under the null hypothesis $H_o: X_2 \equiv 0_2$, it follows that t is distributed as χ_2^2 and the test can be set up, for level α, $0 < \alpha \ll 1$, as:

Accept H_o, i.e., M remains as is, if $t \leq t_\alpha$;

Reject H_o, i.e., M should be replaced by $M+1$ (and X_1 by $\frac{X_1}{X_2}$, etc.), if $t > t_\alpha$, where t_α is determined from

$$1-\alpha = \Pr(\text{Reject } H_o | H_o \text{ true})$$
$$= \Pr(t > t_\alpha | H_o \text{ true})$$
$$= \Pr(\chi_2^2 > t_\alpha).$$

If H_o is rejected, then, $M+1$ can next be tested analogously vs. $M+2$, etc.

The above test can be shown to be the uniformly most powerful α-level test among all tests with power function $\Pr(\text{Reject } H_o | X_2)$ depending on X_2 through $X_2^T \cdot \text{Cov}^{-1}(\hat{X}_2 - X_2 | X_2) \cdot X_2 \equiv \psi^2$. Indeed, the power function here is,

$$\Pr(\chi_{2;\psi^2}^2 > t_\alpha) \tag{94}$$

(See, e.g. Rao, Sheffé, Lehman, for related results on the 'General Linear Hypothesis'.)

III.

Verification that $\hat{\hat{X}}$ is Optimal

Consult Rao, Deutsch, Placket, and Sheffe, for general background. See, e.g., Rao, page 70 for matrix ordering definitions and basic properties. Briefly, in review, if A_1 and A_2 are any two given square matrices of the same size, we write $A_1 \leq A_2$, iff $A_2 - A_1$ is at least positive semi-definite, and $A_1 < A_2$, iff $A_2 - A_1$ is positive definite.

Let $\tilde{X}(Y)$ be <u>any</u> estimator of X relative to the models in eqs. (8) and (9), where Y, B, X, Σ, C, D are general of full rank where required (in the case of B, rank (B) = dim (col. (B))).

Then $C\tilde{X} = D$ from eq. (9), and

$$(Y-B\cdot\tilde{X})^T \cdot \Sigma^{-1} \cdot (Y-B\tilde{X})$$
$$= (Y-B\cdot\hat{\hat{X}} + B\cdot(\hat{\hat{X}}-\tilde{X}))^T \cdot \Sigma^{-1} \cdot$$
$$\cdot (Y-B\hat{\hat{X}} + B(\hat{\hat{X}}-\tilde{X})) \qquad (B-1,1)$$
$$= (Y-B\hat{\hat{X}})^T \cdot \Sigma^{-1} \cdot (Y-B\hat{\hat{X}})$$
$$+ (\hat{\hat{X}}-\tilde{X})^T \cdot B^T \Sigma^{-1} B \cdot (\hat{\hat{X}}-\tilde{X}) + y,$$

where

$$y = 2 \cdot (Y-B\hat{\hat{X}})^T \Sigma^{-1} \cdot B \cdot (\hat{\hat{X}}-\tilde{X})$$
$$= 2 \cdot [(Y-B\hat{X}) - BK(D-C\hat{X})]^T \cdot \Sigma^{-1} \cdot B \cdot$$
$$\cdot [X+K\cdot(D-C\cdot\hat{X}) - \tilde{X}]$$
$$= 2 \cdot [(Y-B\hat{X})^T \cdot \Sigma^{-1}B - (D-C\hat{X})^T K^T \cdot B^T \Sigma^{-1}B] \cdot$$
$$\cdot [\hat{X} + K\cdot(D-C\hat{X}) - \tilde{X}]$$
$$= 2 \cdot [0 - (D-C\hat{X})^T \cdot (C\Lambda^{-1}C^T)^{-1}C] \cdot$$
$$\cdot [\hat{X} + K\cdot(D-C\hat{X}) - \tilde{X}]$$
$$= -2 (D-C\hat{X})^T \cdot [C\Lambda^{-1}C^T]^{-1}C\hat{X} +$$
$$+ (C\Lambda^{-1}C^T)^{-1}C(K(D-C\hat{X}) - \tilde{X})]$$

$$= -2 (D-C\hat{X})^T \cdot [(C\Lambda^{-1}C^T)^{-1}C\hat{X} +$$

$$(C\Lambda^{-1}C^T)^{-1} \cdot (D-C\hat{X}-D)]$$

$$= 0 \qquad (B-1,2)$$

Hence, (A-1) and (A-2) imply

$$\min_{\text{over all}} (Y-B\cdot\tilde{X})^T \cdot \Sigma^{-1} \cdot (Y-B\tilde{X}) \qquad (B-1,3)$$
$$\tilde{X} = \tilde{X}(Y)$$
such that
$$C\cdot\tilde{X} = D$$

occurs for $\tilde{X} = \hat{X}$, with minimal value

$$(Y-B\hat{X})^T \Sigma^{-1} (Y-B\hat{X}) \qquad (B-1,4)$$
$$= (Y-B\cdot[\hat{X}+K\cdot(D-C\hat{X})])^T \cdot \Sigma^{-1} \cdot$$
$$\cdot (Y-B[\hat{X}+K\cdot(D-C\hat{X})])$$
$$= (Y-B\hat{X})^T \cdot \Sigma^{-1}(Y-B\hat{X}) + (D-C\hat{X})^T \cdot (C\Lambda^{-1}C^T)^{-1} (D-C\hat{X}) \quad .$$

Thus $\hat{X}$ enjoys the minimal weighted least squares estimator property of X subject to eq. (9).

Next, consider the following:

This time, let $\tilde{X} = \tilde{X}(Y)$ be any estimator of X satisfying eq. (9) which is absolutely unbiased in X, i.e., $E(\tilde{X}|X) = X$ for all X satisfying eq. (9).

Then:

$$\text{Cov}(\tilde{X}-X|X) \qquad (B-1,5)$$
$$= E(\tilde{X}-\hat{X} + \hat{X}-X)(\tilde{X}-\hat{X} + \hat{X}-X)^T|X)$$
$$= \text{Cov}(\tilde{X}-\hat{X}|X) + \text{Cov}(\hat{X}-X|X) + z+z^T \quad ,$$

where

$$z = \text{Cov}(\tilde{x}-\hat{\hat{x}}, \hat{x}-x|x)$$

$$= \text{Cov}(\tilde{x}-\hat{x} - K(D-C\hat{x}),$$

$$\hat{x}+K(D-C\hat{x}) - x|x)$$

$$= \text{Cov}((KC-I)(\hat{x}-\tilde{x}), KC\cdot(\tilde{x}-\hat{x}) + \hat{x}-x|x)$$

$$= (KC-I)\cdot((-\text{Cov}(\hat{x}-\tilde{x})\cdot c^T K^T + \text{Cov}(\hat{x}-\tilde{x}, \hat{x}|x)) \quad (B-1,6)$$

Now, from the Rao reference Section 5a.2, since $\hat{x}$ is the minimal expected normed squared absolutely unbiased estimator of X, and since from absolute unbiasedness in X for $\hat{x}$ and $\tilde{x}$, $E(\hat{x}-\tilde{x}) = 0$,

$$\text{Cov}(\hat{x}-\tilde{x}, \hat{x}|x) = 0 \quad (B-1,7)$$

Also, since

$$C\cdot(\hat{x}-\tilde{x}) = 0,$$

$$0 = \text{Cov}(C\cdot(\hat{x}-\tilde{x}))$$

$$= C\cdot\text{Cov}(\hat{x}-\tilde{x})C^T,$$

$$= (C\cdot\text{Cov}^{\frac{1}{2}}(\hat{x}-\tilde{x}))(C\cdot\text{Cov}^{\frac{1}{2}}(\hat{x}-\tilde{x}))^T,$$

implying

$$0 = C\cdot\text{Cov}^{\frac{1}{2}}(\hat{x}-\tilde{x}|x), \text{ and by postmultiplying by } \text{Cov}^{\frac{1}{2}}(\hat{x}-\tilde{x}|x),$$

$$0 = C\cdot\text{Cov}(\hat{x}-\tilde{x}|x). \quad (B-1,8)$$

Substituting (B-1,7) and (B-1,8) into (B-1,6) yields $z = 0$.

Thus (B-1,5) implies, by positive semidefinite matrix ordering,

$$\text{Cov}(\tilde{x}-x|x) = \text{Cov}(\tilde{x}-\hat{x}|x) + \text{Cov}(\hat{x}-x|x)$$

$$\geq \text{Cov}(\hat{x}-x|x). \quad (B-1,9)$$

Hence, $\hat{x}$ is also (in a positive semidefinite matrix ordering sense) the minimal expected normed square absolutely unbiased estimator of X for any estimator satisfying eq. (9).

By taking traces, (B-1,9) yields,

$$E(\|\tilde{x}-x\|^2|\hat{x}) = E(\|\tilde{x}-\hat{x}\|^2|x) + E(\|\hat{x}-x\|^2|x)$$

$$\geq E(\|\hat{x}-x\|^2|x). \quad (B-1,10)$$

ELEMENTS OF AN ECLIPSING BINARY SYSTEM IN FREQUENCY DOMAIN 105

IV.

Asymptotic Properties of $\hat{\hat{x}}$ and $\hat{x}$

Equation (70) implies

$$E(\|\hat{\hat{x}}-\hat{x}\|^2 | X)$$
$$= E(\|K \cdot (D-C\hat{x})\|^2 | X)$$
$$= tr(\Lambda^{-1} C^T (C\Lambda^{-1} C^T)^{-1} C\Lambda^{-1})$$
$$= tr(\Lambda^{-\frac{1}{2}} \cdot L \cdot \Lambda^{-\frac{1}{2}}) \quad, \tag{B-2,1}$$

where

$$L = \Lambda^{-\frac{1}{2}} C^T (C\Lambda^{-1} C^T)^{-1} C\Lambda^{-\frac{1}{2}} \quad, \tag{B-2,2}$$

an <u>idempotent</u> matrix ($L^2 = L$) with maxeig (L) = 1. (maxeig ≡ maximum eigenvalue).

Hence, in a matrix ordering sense (by positive semidefiniteness - see Appendix A).

$$0 \preceq L \preceq I \tag{B-2,3}$$

and

$$0 \preceq \Lambda^{-\frac{1}{2}} L \Lambda^{-\frac{1}{2}} \preceq \Lambda^{-\frac{1}{2}} \cdot I \cdot \Lambda^{-\frac{1}{2}}$$
$$= \Lambda^{-1} \quad. \tag{B-2,4}$$

Thus, substituting (B-2,4) into (B-2,1),

$$E(\|\hat{\hat{x}}-\hat{x}\|^2 | X) \leq tr(\Lambda^{-1}) \quad. \tag{B-2,5}$$

Now,

$$tr(\Lambda^{-1}) = tr(\Lambda_1^{-1}) + tr(\Lambda_2^{-1}) \tag{B-2,6}$$

But since

$$0_{m,m} \prec \Sigma_1 \preceq \sigma_1^2 \cdot I_m \quad ;$$

$$0_{n,n} \prec \Sigma_2 \preceq \sigma_2^2 \cdot I_n \quad , \tag{B-2,7}$$

where

$$\sigma_1^2 = \max_{1 \leq k \leq m} (\sigma^2(t_{1,k}))$$
$$\sigma_2^2 = \max_{1 \leq k \leq n} (\sigma^2(t_{2,k})) \quad , \tag{B-2,8}$$

by basic matrix ordering properties,

$$O_{2M+1,2M+1} < \Lambda_1^{-1} < \sigma_1^2 \cdot (B_1^{(m)^T} \cdot B_1^{(m)})^{-1} \quad ,$$
$$O \qquad\qquad < \Lambda_2^{-1} < \sigma_2^2 \cdot (1/n) \quad . \tag{B-2,9}$$

Hence, (B-2,6) - (B-2,9) yields

$$\text{tr}(\Lambda^{-1}) \leq \sigma_1^2 \text{tr}((B_1^{(m)^T} \cdot B_1^{(m)})^{-1}) + \sigma_2^2/n \quad , \tag{B-2,10}$$

indicating the functional dependence of B_1 on m by $B_1^{(m)}$.

Consider now the data sampling over the interval $[0, t_1]$, and assume

$$\lim_{m \to +\infty} \left(\max_{1 \leq k \leq m+1} \left| t_1 - m \cdot (t_{1,k} - t_{1,k-1}) \right| \right) = 0 \quad , \tag{B-2,11}$$

where

$$t_{1,o} \equiv 0 \quad , \quad t_{1,m+1} \equiv t_1 \quad .$$

If equal interval sampling is done, i.e., $t_{1k} - t_{1,k-1} = t_1/m$, then (B-2,11) is trivially satisfied.

Now,

$$B_1^{(m)^T} \cdot B_1^{(m)} = \left(b_{ij}^{(m)} \right)_{0 \leq i,j \leq 2M} \quad , \tag{B-2,12}$$

$$b_{ij}^{(m)} = \sum_{k=1}^{m} \cos(i\omega t_{1k}) \cos j\omega t_{1k} \quad . \tag{B-2,13}$$

Define

$$I_m^{(i,j)} = \sum_{k=1}^{m+1} \cos(i\omega t_{1k}) \cos(j\omega t_{1k}) \cdot (t_{1,k} - t_{1,k-1}) \tag{B-2,14}$$

ELEMENTS OF AN ECLIPSING BINARY SYSTEM IN FREQUENCY DOMAIN

Then

$$|(t_1 \cdot b_{ij}^{(m)}/m) - I_m^{(i,j)}| \leq (m+1)|(t_{1k} - t_{k,k-1})|$$

$$< |1 - m(t_{1,k} - t_{1,k-1})|$$

$$+ |(t_1/m) - (t_{1,k} - t_{1,k-1})| \quad . \qquad (B-2,15)$$

Thus, using (B-2,11) in (B-2,15),

$$\lim_{m \to \infty} |t_i \cdot b_{ij}^{(m)}/m - I_m^{(i,j)}| = 0 \quad . \qquad (B-2,16)$$

But, by definition

$$\lim_{m \to \infty} I_m^{(i,j)} = \int_{x=0}^{t_1} \cos(i\omega x)\cos(j\omega x)dx$$

$$= (2t_1/\pi) \int_{t=0}^{\pi/2} \cos(it)\cos(jt)dt \qquad (B-2,17)$$

using eq. (3).

Then using the simple trigonometric identity from eq. (87), in (B-2,17) yields

$$g_{ij} = \int_{t=0}^{\pi/2} \cos(it)\cos(jt)dt = \begin{cases} (\tfrac{1}{2}) \cdot \left(\dfrac{\sin((i+j)\tfrac{\pi}{2})}{i+j} + \dfrac{\sin((i-j)\tfrac{\pi}{2})}{i-j} \right), \\ \qquad \text{for } i \neq j \quad , \\ \pi/4, \quad \text{for } i = j \neq 0 \\ \pi/2, \quad \text{for } i = j = 0 \quad , \end{cases} \qquad (B-2,18)$$

$0 \leq i, j \leq 2M$.

Thus, combining (B-2,12) and (B-2,16) - (B-2,18),

$$\lim_{m \to +\infty} ((1/m) B_1^{(m)^T} \cdot B_1^{(m)}) = (2/\pi) \cdot G_o \quad , \qquad (B-2,19)$$

where

$$G_o = (g_{ij})_{0 \leq i,j \leq 2M}$$

$$= \int_{t=0}^{\pi/2} f(t) \cdot f^T(t) dt \quad , \qquad (B-2,20)$$

where

$$f(t) \equiv \begin{pmatrix} \cos(0 \cdot t) \\ \cos(1 \cdot t) \\ \vdots \\ \cos(2M \cdot t) \end{pmatrix} \quad (2M+1 \text{ by } 1) \ .$$

(B-2,21)

If V is an arbitrary 2M+1 by 1 fixed vector such that $V^T G_o V = 0$, then

$$0 = V^T G_o V$$

$$= \int_{t=0}^{\pi/2} V^T f(t) f(t)^T V \, dt$$

$$= \int_{t=0}^{\pi/2} (V^T f(t))^2 \, dt \ .$$

Thus,

$V^t \cdot f(t) = 0$, almost every where over $[0,\pi/2]$. But this directly contradicts the linear independence of the elements of $f(t)$ in t, unless $V = 0_{2M+1}$.

Hence, if $V \neq 0_{2M+1}$, $V^T G_o V \neq 0$.

Since G_o from its structure in (B-2,20) is clearly positive semi-definite, it follows that $V^T G_o V > 0$ and hence G_o is positive definite.

Then, combining (B-2,19) and (B-2,10), using the positive definiteness of G_o, and employing the notation $\Lambda(m,n)$ to indicate functional dependence on m and n, etc.,

$$\lim_{m,n \to +\infty} \left(tr(\min(m,n) \cdot \Lambda^{-1}(m,n)) \right)$$

$$\leq \sigma_1^2 \cdot \lim_{m \to +\infty} \left((tr(m \cdot (B_1(m)^T \cdot B_1(m))^{-1})) \right) + \sigma_2^2$$

$$= \sigma_1^2 \cdot (\pi/2) \cdot tr(G_o^{-1}) + \sigma_2^2$$

$$\overset{df}{=\!=} \kappa_o < +\infty \qquad (B\text{-}2,22)$$

Thus (B-2,22) implies,

$$\lim_{n,n \to +\infty} tr(\Lambda^{-1}(m,n)) = 0, \quad \text{with, for } m,n \text{ sufficiently large,}$$

$$tr \Lambda^{-1}(m,n) \approx \kappa_o/\min(m,n) \qquad (B\text{-}2,23)$$

Hence, combining (B-2,5) (B-2,22) it follows that

$$\lim_{m,n \to +\infty} \left(E(\|\hat{\hat{x}} - \hat{x}\|^2 | X) \right) = 0, \qquad (B\text{-}2,24)$$

with, for m, n sufficiently large,

$$E(\|\hat{\hat{x}} - \hat{x}\|^2 | X) \leq \kappa_o/\min(m,n) \qquad (B\text{-}2,25)$$

Note also from eq. (75)

$$E(\|\hat{\hat{x}} - \hat{x}\|^2 | X)$$
$$= tr(\Lambda^{-1} - \Lambda^{-1} C^T (C \Lambda^{-1} C^T)^{-1} C \Lambda^{-1})$$
$$= tr(\Lambda^{-\frac{1}{2}} Q \Lambda^{-\frac{1}{2}}) \qquad (B\text{-}2,26)$$

where

$$Q = I - \Lambda^{-\frac{1}{2}} C^T (C \Lambda^{-1} C^T)^{-1} C \Lambda^{-\frac{1}{2}}$$

is idempotent, and hence, as in eq. (B-2,3),

$$0 \leq Q \leq I, \qquad (B\text{-}2,27)$$

yielding from (B-2,26)

$$E(\|\hat{\hat{x}} - \hat{x}\|^2 | X) \leq tr(\Lambda^{-1}) \qquad (B\text{-}2,28)$$

Hence, summarizing

$$E(\|\hat{\hat{x}} - \hat{x}\|^2 | X), \quad E(\|\hat{x} - x\|^2 | X), \quad E(\|\hat{\hat{x}} - X\|^2 | X)$$

are all

$$\leq \mathrm{tr}\Lambda^{-1}(m,n) \approx k_o/\min(m,n) \qquad (B-2,29)$$

for m,n sufficiently large, yielding

$$\lim_{m,n \to +\infty} \left(E(\|\hat{\hat{x}}-\hat{x}\|^2|x)\right) = 0$$

$$\lim_{m,n \to +\infty} \left(E(\|\hat{x}-x\|^2|x)\right) = 0$$

$$\lim_{m,n \to +\infty} \left(E(\|\hat{\hat{x}}-x\|^2|x)\right) = 0 \qquad (B-2,30)$$

Thus, for sample sizes m and n sufficiently large, $\hat{x}$ can be used in place of $\hat{\hat{x}}$, i.e., the constraint in eq. (9) can be ignored (and K can be set formally equal to zero).

A SIMPLE SYNTHESIS METHOD FOR SOLVING THE ELEMENTS OF WELL-DETACHED ECLIPSING SYSTEMS

Paul B. Etzel

University of California, Los Angeles
Los Angeles, California 90024   USA

ABSTRACT

A simple synthesis method for analyzing the light curves of detached eclipsing binaries exhibiting only small oblateness is outlined. The resulting computer program is flexible to use and faster than D.B. Wood's widely-used program WINK, which was designed for systems with moderately distorted components. The basic indeterminacies of light curve solutions are mentioned.

I. INTRODUCTION

We have heard several papers at this NATO ASI concerned with the development and application of the method of Fourier analysis of light curves in the frequency domain, as recently stated in the monograph by Kopal (1979). The ultimate goal of this method is to find a unique closed-form analytical solution for the photometric orbital elements of a binary system from an integral function of the observations. This function is not directly based upon a physical model, but is equated or transformed to an equivalent spherical limb-darkened model, with adequate perturbations included for distorted stars exhibiting proximity effects.

Another approach, which is popular and easily implemented, is based upon the method of synthesis. A physical model is chosen as a basis to fit the observed data, usually by some optimizing scheme or by differential-correction least squares. Two models currently in wide use are by Wilson and Devinney (1971), based upon Roche equipotentials defining the stellar surfaces, and by Wood (1971, 1972), defined by a triaxial ellipsoid approximation

to the stellar surfaces. Both models describe the stellar surface
brightness distributions due to limb darkening and to the local
temperature as complicated by surface gravity variations and the
reflection effect. The Wilson-Devinney model has been applied to
the entire realm of detached, semi-detached, contact, and overcontact systems, with different degrees of reliability. The Wood
model has been applied mostly to detached systems with varying
amounts of photometric distortion.

The end result of synthesis methods is a model which, it is
hoped, uniquely describes the nature of the system. Regardless of
the method, often one must draw from theory to specify certain
parameters not solvable from the observations. Wilson (1968) has
pointed out that solutions for limb darkening for partial eclipses
are largely indeterminate. Decisions, such as the nature of the
primary eclipse, frequently need the assistance of spectroscopic
information, as has been emphasized by Popper (1976). Computer
programs cannot assure unique solutions in themselves. Even the
Fourier frequency-domain analysis, which was designed to yield a
unique closed-form analytical solution, can fall prey to these
problems when applied to distorted stars, as pointed out by Kopal
(1979, p. 213).

The model to be described here, as formalized in the computer
program, EBOP, has recently been used by Popper and Etzel (1980)
for the analysis of seven well-detached eclipsing binaries. It is
a refinement of the model developed by Nelson and Davis (1972).
The main advantages of this model are that it is fast, inexpensive,
and flexible to use. The disadvantage is that it should only be
used on well-detached eclipsing binaries. Since binaries of this
class, as a whole, are more suitable for the determination of fundamental parameters than semi-detached or contact systems, this
later restriction poses no obstacle to the primary motivation to
our research. The results from this model were compared to results
from similar computer runs with Wood's program WINK (with Status
Reports through number 6 included), using the same data, and they
were found to be in excellent agreement.

II. THE NELSON-DAVIS-ETZEL MODEL (NDE)

We have retained the basic integration scheme for spherical
stars and treatment of eccentric orbits, as originally stated by
Nelson and Davis. Their method of integration of the limb-darkened
stellar disks is based on a set of closed-form analytic formulae
for the areas of disks, rings, and sectors, rather than on ellip-
tic integrals, double Gaussian quadrature, or stellar surface grid
modeling. Proximity effects are not included in the actual treatment of eclipses, but are approximated as perturbations through
the whole light curve.

II A. Spherical Model

The basic spherical model parameters are given in Table I.

Table I

Basic Spherical Model Parameters

Principal	Associated
J_s=Central surface brightness of secondary component	$e \cos \omega$
	$e \sin \omega$
r_p=radius of primary	$L_3=1-L_1-L_2$=Third light
$k=r_s/r_p$ ratio of radii	$\Delta\theta$=ephemeris phase correction
i=inclination angle	mq=light curve normalization
u_p, u_s or u = limb darkening	$\Delta\gamma$= integration ring size

The value of J_p is defined as unity. The selection of the surface brightness ratio as a parameter, rather than the temperatures of the two components, is of fundamental importance. It is directly related to the ratio of eclipse depths for non-limb-darkened stellar disks for stars in circular orbits. Stellar temperatures are only indirectly related to the light curve through the many assumptions of the physical model. All synthesis computer programs calculate a monochromatic surface brightness ratio, but few of their users publish this information.

The unnormalized luminosity of each component is given by:

$$\mathcal{L}_i = J_i r_i^2 (1 - 1/3 \, u_i), \quad i = p \text{ or } i = s,$$

or in normalized units

$$L_i = \frac{\mathcal{L}_i}{\mathcal{L}_p + \mathcal{L}_s}.$$

The luminosity ratio is easily evaluated from the ratio of the radii and the surface brightness ratio with a correction for limb darkening.

The accuracy of the calculations of the eclipse functions is directly dependent on the integration ring size, $\Delta\gamma$, and the precision of the computer arithmetic. Tests were run on both the NDE and the Wood models for spherical stars using the seven-digit accuracy $\alpha(k, p)$ tables described by Davis (1963). The results are given in Table II.

Table II

NDE Model Integration Ring Size	(EBOP) Standard Error	Wood Model Integration Grid Point No.	(WINK) Standard Error
10°	.00037	4 x 4	.00190
5°	.00008	6 x 6	.00085
1°	.00001	16 x 16	.00011

Both computer programs were run with IBM 3033 standard precision, yielding about seven digits of computational accuracy. An integration ring size of 5° was used in all EBOP runs, although 10° would have been sufficient. A grid point number of 6 x 6 was normally used for WINK verification run.

II B. Perturbations for Proximity Effects

Wood's model is well-suited for the analysis of moderately distorted detached systems which exhibit proximity effects in their light curves. To modify the Nelson-Davis model to include the effects of oblateness in the eclipse calculations, or the addition of reflection, would essentially duplicate the efforts already made by others and increase the model complexity. Thus, to solve the light curves of spherical or mildly oblate systems, it was expected that a perturbation scheme for these effects would be adequate. Subsequent tests on real systems have verified this expectation by comparing the results with those from Wood's model. In fact, the NDE model can be used effectively in a complementary relationship with Wood's model. Reliable starting elements for distorted detached systems, and some semi-detached systems, can be obtained from the EBOP.

Development of the NDE perturbations was done by Etzel (1975) and by Popper and Etzel (1980). The treatment of oblateness is to convolve the phase-dependent stellar component luminosities with the eclipse calculations for a simple spherical limb-darkened star of equivalent volume. The evaluation of oblateness, ε, was taken from Binnendijk (1974) for a simple biaxial (rotational) ellipsoid. One computes, for specified values of mass ratio and equivalent spherical stellar volume, the triaxial dimensions a_3, b_3, c_3 for each star and obtains the dimensions of the equivalent biaxial ellipsoid, a_2, b_2 by imposing the conditions that:

$$a_2 \, b_2 \, b_2 = a_3 \, b_3 \, c_3, \text{ and}$$

$$\varepsilon_2(a_2, b_2) = \varepsilon_3(a_3, b_3) = 1 - b_3/a_3.$$

The unnormalized luminosity of each component at quadrature is then taken from Binnendijk (1960) to be:

$$\mathcal{L}_i^{ob} = \frac{\{(1 - u_i) + 2/3\ u_i\ (1 + 1/5\ \varepsilon_i)\}\ (1 + 3\ y_i\ \varepsilon_i)}{(1 - \varepsilon_i)}\ J_i\ b_i^2$$

where J_S is now the central surface brightness of the secondary at conjunction. Normalization proceeds as before. We then apply the phase-dependent variation in luminosity for biaxial ellipsoids from Binnendijk (1960):

$$L_i^{ob}(\phi) = L_i^{ob}\ [1 - \frac{(15 + u_i)}{(15 - 5u_i)}\ (1 + y_i)\ \varepsilon_i\ \cos^2 \phi],$$

to each component, where $\cos \phi = \sin i \cos \theta$.

To convolve this variation with the eclipses, we make the approximation that the fractional loss of light for the spherical model, call it α^{sp}, regardless of eclipse type, is identical to the fractional light loss, for the equivalent oblate component α^{ob}. This procedure would be strictly correct if both stars were similar ellipsoids with identical limb and gravity darkening terms. But since this perturbation model was made in the spirit of approximation, it is still useful within the limitations of its designed usage. A discussion of light loss for dissimilar ellipsoidal components may be found in Martynov (1973, pp. 156-161).

The treatment of reflection is essentially the same as in the simple uniformly illuminated hemisphere model as given by Binnendijk (1960). Some allowance is made for the eclipse of reflected light, however, which is again included in the spirit of approximation to illustrate the effect of reflection on the elements. Clearly, any system with a large amount of reflection present would violate the assumption of the NDE model that the isotopes of the components are only slightly distorted. The reflection model starts with the simple bolometric phase law:

$$f(\phi) = 0.2 + 0.4 \cos \phi + 0.2 \cos^2 \phi.$$

The luminosity of the primary will then vary by

$$L_p + \Delta L_p = L_p + L_s\ r_p^2\ f(\phi) = L_p + S_p\ (\tfrac{1}{2} + \cos \phi + \tfrac{1}{2} \cos^2 \phi),$$

where S_p is the contribution in luminosity to the heated side of the primary. A similar expression, with a negative $\cos \phi$ term, holds for the secondary component. As a first approximation in the case of monochromatic light, we can take

$$S_p = 0.4\ L_s\ r_p^2 \quad \text{and} \quad S_s = 0.4\ L_p\ r_s^2$$

if the two stars are of similar spectral types. One can refine this treatment further by incorporating a luminous efficiency factor and albedo as in Martynov (1973, pp. 161-176) or Kopal (1959, pp. 237-239).

The variation due to reflection is taken to be:

$$R = \Delta L_p + \Delta L_s - A\Delta L_e,$$

where ΔL_e is the reflected light from the eclipsed star and A is the fractional area of it that is eclipsed. This variation may also be approximated by:

$$R = (\Delta L_p + \Delta L_s)(1 - A),$$

since the contribution of the eclipsing star is very small during eclipses and increases only as A becomes very small.

Third light from a nearby optical companion can be added by using the definition:

$$L_p + L_s + L_3 = 1.$$

Thus, the general form of the light curve variation is:

$$\ell = \{L_p^{ob}(\phi) + L_s^{ob}(\phi) - \alpha^{ob} L_e(\phi) + \Delta L_p + \Delta L_s - A\Delta L_e\}$$
$$\cdot (1 - L_3) + L_3.$$

This in turn is normalized to reproduce the observed quadrature magnitude, or magnitude difference, through multiplication by an arbitrary numerical coefficient, which is usually included as a variable in least squares. For a system without reflection, this coefficient is identical to the luminosity equivalent of the quadrature magnitude.

II C. <u>Advantages, Limitations and Disadvantages of the Present NDE Model</u>

The advantages of EBOP, as compared to WINK, are that it is more accurate for systems with spherical components, that it is 15 to 40 times faster, and that repetitive runs on the same data to test the solution sensitivity are easily carried out. The accuracy difference is reflected in Table II. Two similar runs, including reflection, with both computer programs on V478 Cyg, including 910 observations and six variables, on the UCLA IBM 3033 computer, took 620 seconds (at a cost of $180) using WINK, and took 25 seconds (at a cost of $7) using EBOP. Thus, the majority of trials, finding initial solutions or testing sensitivity of k or u, for example, could be done at reduced cost with EBOP with WINK being used only for verification runs on the adopted solutions. As an example of the utility of repetitive solutions in one run, consider holding k fixed at several arbitrary values to determine the range of solutions, which represent the observations equally. Figure 1 graphically illustrates this technique for the V-light curve of BS Dra where an acceptable range of 0.95 to 1.05 in k exists, even though the optimal solution is 0.990 ± 0.015.

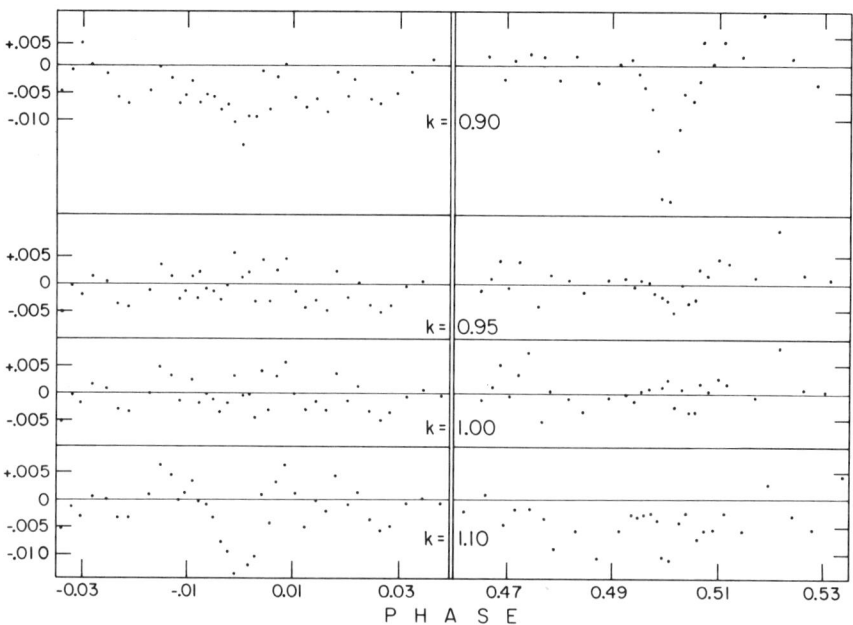

Figure 1. Residual plot for various solutions of the V-light curve of BS Dra showing the variation with k, the ratio of the radii. The plotted points are means of four successive phase residuals. Reprinted from Popper and Etzel (1980) by permission.

The NDE model does have limitations compared to those models of Wood and Wilson-Devinney due to the perturbation approximations to account for proximity effects. The greater the difference in the component oblatenesses, the greater the error in calculating α due to the errors in computing overlapping areas. Only the fractional areas of overlapping similar ellipses are the same as for the equivalent disks. Also, the greater the mean oblateness, the more error is introduced as the inclination decreases due to the biaxial model failing to represent off-centered isophotes. This effect was noticed to some extent by Popper and Etzel (1980) in the case of V478 Cyg where $<\varepsilon> \simeq 0.04$, giving an inclination value $0°5$ lower than the $78°3$ found by WINK. Lastly, there is the very rudimentary reflection model in EBOP. Reflection could be improved considerably even within the confines of the present model. This has not been done since Wood's program WINK is better suited for systems exhibiting moderate reflection and oblateness.

Since these disadvantages were acknowledged from the start, we simply consider them as limitations and use WINK exclusively for the final solutions for systems with moderately large proximity effects. The saving grace of the NDE model for slightly oblate stars is that, during eclipses, they will appear nearly circular in projection.

The reader is referred to the paper by Popper and Etzel (1980) for further details and results of applying the NDE model. There will also be found in that communication comments on such matters as the evaluation of realistic uncertainties of the parameters derived in the analysis, the determination of limb darkening, analysis of the observations between eclipses, evaluation of color indices, small orbital eccentricities, and the use of the surface brightness ratio as a fundamental parameter.

II D. Advantages of Synthesis Methods Over the Present Fourier Method in the Frequency Domain

The advantages of most synthesis computer methods over the Fourier-analysis frequency-domain method are several. First, the whole light curve is treated simultaneously, rather than evaluating the perturbation terms for proximity effects from the uneclipsed portion of the light curve. Second, the selection of luminosity scaling as an immutable constant is not necessary, thus preventing the introduction of any small systematic errors. Third, the value of the angle of external tangency need not be specified as a controlling parameter. In distorted systems this becomes difficult to evaluate accurately. Fourth, variables can be selected at will from a list of parameters in synthesis programs. The present Fourier analysis method demands that one must always include r_1, r_2, L_i and i as variables, and one cannot specify the ratio of radii or light ratio as constants if desired.

A philosophical point can also be raised about the use of the observations in the analysis. In the synthesis methods each observation enters individually into the solution (weighting optional). The Fourier analysis method uses an integral moment of the observations, thus masking the contribution of each point to the value of A_{2m}, particularly for small $\sin^{2m}\theta$ as m increases.

As of this date there have been few published applications of this new Fourier technique of Kopal (1979). Until many tests of application have been made, the advantages and limitations will remain largely unknown. It is doubtful that Kopal's Fourier technique will replace the widely-used synthesis methods, but it may take a place of its own as another tool in eclipsing binary light curve analysis. Ultimately, the utility of any such method will determine its acceptance by the astronomical community.

III. SUMMARY

The use of popular synthesis models is viewed as a viable method for the solution of eclipsing binary light curves. Fourier Analysis in the frequency-domain represents another method stressing analysis of the data through a highly developed mathematical model. Regardless of the model used, unique photometric solutions may not be possible due to the nature of some binary systems. As Professor Kopal <u>might</u> say, "...the indeterminacy lies with the binary star's message to us, not in our ability to receive it or decipher it." Constraints may need to be imposed from theory or spectrographic material.

The simple NDE synthesis model described here compares favorably with that of Wood for well-detached systems exhibiting only mild oblateness. The EBOP computer program is fast and flexible, and it is best used in a complimentary manner with Wood's program, WINK. The main advantage of EBOP is one of economy.

It is doubtful that the Fourier analysis method in the frequency domain will displace the synthesis models, but it does provide another independent method of analysis.

IV. ACKNOWLEDGMENTS

I would like to express my gratitude to Dr. Daniel M. Popper for his many suggested improvements to the NDE model and to the EBOP program. Because of these suggested improvements, the program became a useful utility for the study of eclipsing binaries.

This work was supported under NSF grant AST77-22672 Popper/Plavec. I would like to thank Professor Zdeněk Kopal for defraying my living expenses in Maratea, and my gratitude also goes to the NSF Division of International Programs for the travel grant to attend this NATO ASI.

REFERENCES

Binnendijk, L.: 1960, "Properties of Double Stars" (U. of Pennsylvania, Philadelphia), pp. 288-326.
Binnendijk, L.: 1974, Vistas in Astron. 16, pp. 61-74.
Davis, W.D.: 1963, Astron. J., 65, 277.
Etzel, P.B.: 1975, Masters Thesis. San Diego, CA: San Diego State Univ.
Kopal, Z.: 1959, "Close Binary Stars" (New York: John Wiley & Sons), pp. 237-239.
Kopal, Z.: 1979, "Language of the Stars" (Dordrecht, Holland: D. Reidel Publshing Co.).
Martynov, D.Ya.: 1973, Eclipsing Systems with Deformed Components. Fine Effects, in "Eclipsing Variable Stars," Tsesevich, V.P., ed. (New York: John Wiley & Sons), pp. 128-177.
Nelson, B., and Davis, W.: 1972, Astrophys. J., 174, 617.
Popper, D.M.: 1976, Astrophys. Space Sci., 45, 391.
Popper, D.M., and Etzel, P.B.: 1980, "Photometric Orbits of Seven Detached Eclipsing Binaries", submitted for publication, UCLA Astronomy and Astrophysics Preprint No. 93.
Wilson, R.E.: 1968, Astron. J., 73, S124.
Wilson, R.E., and Devinney, E.J.: 1971, Astrophys. J., 166, 605.
Wood, D.B.: 1971, Astron. J., 76, 701.
Wood, D.B.: 1972, Goddard Space Flight Center Report X-110-72-473.

Special thanks go to Mr. Robert L. O'Daniel for the fine job of editing and typing of this manuscript.

THE UNIQUENESS CRITERION: THE ANALYSIS OF ECLIPSING BINARY LIGHT CURVES

Burt Nelson

Mount Laguna Observatory, San Diego State University, San Diego, California, U. S. A.

Often, when scientific discussions are permitted to follow their natural course, fundamental questions at the heart of science; the philosophy of science, arise. The extensive and detailed discussions about binary stars at this meeting were no exception. The modeling of binary systems, based on the feeble signals from space, ultimately lead us to basic problems of epistemology and the theory of scientific knowledge.

During the sessions of this meeting, techniques of quantifying binary system models were repeatedly challenged in terms of uniqueness. This criterion itself is, of course, by no means unique. Plato raised it to the ultimate of sophistication in his rejection of empirical sources of knowledge. The Platonic fascination with the logical certainty of mathematics created the Realm of Ideas and the Doctrine of Recollection. True Platonic knowledge must be certain, unchanging, and eternal. Just as two plus two equals four--exactly four--in the past, present, and future, every word must have a unique and eternal referent. To this end, the ever changing physical world is to be totally ignored and true knowledge sought through pure deductive reason. Although the 17th Century witnessed a break from this rigid rationalist point of view, and hence the birth of modern science, the quest for certainty is still very much with us in one form or another. Even the most cautious and skeptical investigator may, at times, succumb to the lure of a "unique" solution to a problem.

In the course of discussions about the merits of analytic versus synthetic methods of evaluating the parameters for binary systems, it became clear that at times the unique evaluation of

the parameters was at issue and at other times the uniqueness
of the model itself became central--as though the model should
be logically deducible from the data! A tempting error of logic
indeed; to assume that the factual truth of a conclusion proves
the truth of the premises from which it was deduced. For-inst-
ance, recall the famous syllogism:

> Bread is made of stone,
> Stone is nourishing,
> Therefore, bread is nourishing.

A logically sound argument and the factual truth of the conclusion is beyond dispute but no one would subscribe to the factual
truth of the two premises from which it is deduced. However,
this type of mistake is often made when it is forgotten that a
model (of a binary system or an atom) is a set of premises.
The <u>usefulness</u> of the model is judged by the degree of correlation between the logical consequences of these premises and
relevant observational facts. But, no matter how close this
correlation may prove to be and no matter how "right" the model
(premises) may seem, the above example shows that an infinite
number of sets of premises can be fashioned which will logically
result in the same conclusion. The fact that some premises
would be classified as nonsense or absurd should not mislead
us--this is no test. Consider some aspects of quantum mechanics
or relativistic physics from a purely Newtonian point of view
and it becomes quite clear that <u>strangeness</u> cannot be used to
judge the validity of a premise. The fact that models are
modified by changing premises in order to achieve closer correlation with observed fact should, likewise, not deceive us as
the possible combination of premises are limitless--particularly
if it is remembered that today's "strangeness" can become tomor-
row's "common sense."

It is clear then that when uniqueness is used as a criterion, we should not--we cannot be referring to models. However,
given a model, that is accepted by all participants in the discussion, from which a scheme for the computation of parameters
is derived, it might then be asked; is there a set of parameter
values that is unique to a given set of data? To this question
a qualified yes is possible. Bearing in mind the fact that some
parameters for a binary system mimic each other and tend to mask
each other's effects, it is conceivable that a mathematical procedure, using something like <u>least squares</u> as a final arbiter,
can produce a unique set of parameters for the given data. But
when applied to a new set of observations of the same binary
system, the parameter values may or may not be the same--most
likely not.

The analytic solution of eclipsing binary light curves has

been developed to a very high degree of sophistication by Kopal and his collaborators. Further refinements of this technique are anticipated by investigators in this field including those who are presently developing methods of solution which involve comparison of synthetically generated curves with observed light curves.

No one at this time can predict which of the methods now before us will prove to be unique, that is, ultimately used by all investigators to the exclusion of all others. However, this is not the fundamental issue. The thesis is that whatever method is used, it must be remembered that certainty or uniqueness is a characteristic of mathematical logic and not of the physical world. Methods of scientific analysis, like scientific models, are judged, not on the basis of being true or false, but on the basis of their utility to the process of investigation.

Thus, whereas the sophisticated analytical methods of light curve solution will quickly close in to a precise unique result for a given set of data, the synthetic method by virtue of its faults keeps the investigator always mindful, as he watches the $\Sigma(o-c)^2$ curve walk through the valleys of false minima, that even his best models must always be tentative steps in a never ending process. He will be not afraid in the knowledge that God is a mathematician who <u>arranges</u> complicated puzzles to baffle the astronomer.

LIGHT CURVE ANALYSIS FOR THE BASIC SPHERICAL-MODEL OF ECLIPSING VARIABLES IN THE FREQUENCY DOMAIN

Osman Demircan

Department of Astronomy, University of Garyounis, Benghazi, Libya.

ABSTRACT

Recently a new approach has been developed chiefly by Kopal (cf. for a survey of the methods, Kopal, 1979) to the problem of an analysis of the light changes of eclipsing binary systems in the frequency domain. In Kopal's theory, a spherical model for the eclipsing binaries has been taken as basic and the solution for the elements of distorted systems has been reduced (cf., e.g., Kopal, 1976b) to one based on this spherical model. In the present communication the methods developed for the light curve analysis of basic spherical model of eclipsing variables have been reviewed and some important remarks given.

1. INTRODUCTION

Recently, Kopal, partly together with his colleagues (including the present author) developed Fourier techniques for an analysis of the light curves of eclipsing binaries. These techniques have been set forth in a series of papers starting in 1975 and culminated in a monograph (Kopal, 1979).

In Kopal's theory, a spherical model for the eclipsing binaries has been taken as basic and the solution for the elements of distorted systems has been reduced (cf., e.g., Kopal, 1975c; Kopal, 1976b, and Kopal and Demircan, 1978) to one based on this spherical model. For the basic spherical model:

1) The components are spherical, no distortion in shape.
2) No reflection effect.

3) Relative orbit is circular.
4) Distribution of brightness over the apparent discs is radially-symmetrical.

In this review we shall consider only the methods for this spherical model, for the reduction methods to the basic model, cf. the above-given references.

The fundamental data for the analysis of the basic model in the frequency domain are represented by a set of the quantities so called "the moments of the light curves" defined (cf., Kopal, 1975a, Eq. (3.1)) by

$$A_{2m} = \int_0^{\theta_1} (1-\ell) d(\sin^{2m}\theta) , \qquad (1.1)$$

where ℓ denotes the luminosity of the system normalized by the maximum value ℓ_1 and θ is the phase angle which is θ_1 at external contact of an eclipse. It is seen that, in the frequency domain, the light curve itself plays an auxiliary role for the fundamental quantities used as a basis for the solution of the eclipse elements. The quantities given by (1.1) are the simplest integral transforms of the light curves and can be related (cf. Section 2, Eq. (2.2)) algebraically with the Fourier coefficients a_n which may permit us to call the analysis "Fourier analysis of the light curves of eclipsing variables".

For the spherical model, it is known that the normalized luminosity at any time is the function of five unknowns, namely; the fractional radii $r_{1,2}$ of the components, inclination i of the relative orbit, uneclipsed luminosity L_1 and limb darkening coefficient u_1 of the undergoing star. If the limb darkening coefficient u_1 is taken to be known from the theory of the stellar atmospheres; this makes it necessary to utilize only four moments A_{2m} for four different values of m to derive the unknown eclipse elements $r_{1,2}$, i and L_1 from the equations of the form

$$(A_{2m})_{obs} - (A_{2m})_{theor} = 0 , \qquad (1.2)$$

where obs and theor stand for observational and theoretical, respectively, and

$$(A_{2m})_{obs} = A_{2m}(\ell,\theta), \quad (A_{2m})_{theor} = A_{2m}(L_1,r_{1,2},i).$$

Methods for the evaluation of the observational moments will be reviewed in Section 2. The theoretical moments in the case of total eclipses were obtained by Kopal (1975a,b) in a very simple closed form when the frequency m is 0, 1, 2 and 3. Thus, the eclipse elements for totally eclipsing systems can be solved algebraically which presents the power of the methods and the usefulness of the "moments of the light curves" A_{2m}. The theoretical moments for other (annular and partial) eclipses have been expressed in terms of the eclipse elements by Kopal (1976a, 1977b) and Demircan (1978a) for integer values of m. In Section 3 the author's results will be given. More general expressions for the theoretical moments were given by Kopal (1977b) and Demircan (1978a),c) in the forms of infinite series expansions which are valid for any type of eclipse, any degree of limb darkening and for any positive real value of m. Thus, when the observational values of the moments have been established (see Section 2), Equation (1.2) for different values of m constitutes relations between the unknown elements of the eclipses and the observed characteristics of the light curves and can be solved numerically for them. For the numerical solution of the four nonlinear equations of the form (1.2) a NAG library subroutine COSNAF which is based on a hybrid method due to Powell (1970) can be used. However, before the numerical solution the number of unknowns can be reduced from four to two (cf., Kopal and Demircan, 1978, and more generally Demircan, 1978a) by simple operations on the moments and will be reviewed in Section 4.

2. OBSERVATIONAL MOMENTS

The moments A_{2m} of the light curves as given by Eq. (1.1) define the areas subtended by the lines $\ell = 1$, $\sin^{2m}\theta = 0$ and the actual shape of the light curve in the $\ell - \sin^{2m}\theta$ coordinates. For $\theta > \theta_1$ by definition $\ell = 1$ for spherical stars, and thus any extension of the limits of integration on the right hand side of (1.1) would no longer affect the requisite areas.

As soon as we have the observation (ℓ_i, θ_i) for an eclipsing binary the empirical values of the moments A_{2m} can be readily ascertained by quadratures. In performing this planimetry, we should have enough observations and we may find it of advantage not to use the scattered individual observations themselves, but the artificial ones plotted on a smooth free hand curve drawn to represent the course of observed points. In doing so, however, we are not relying on any particular points of the light curve, but propose to give all a simultaneous representation which makes it difficult to estimate the uncertainty in the empirically determined moments. This uncertainty depends on the quality as well as number of observations and its probable value was given (cf. Demircan, 1977a, Equation (4.7)) by

$$\Delta A_{2m} = \Delta U \sin^{2m}\theta_1 . \qquad (2.1)$$

It is dominated mainly by the uncertainty in the unit of light U. Another error comes from the numerical integration of the unequally spaced data. The nature of integral (1.1) always increases this error even though the artificial points on the free hand curve are set initially with equal spaces. This error is not included in (2.1). The practical experience (cf., Demircan, 1978b) shows that the numerical values of the moments A_{2m} decrease rapidly by increasing values of m. Figures 1 and 2 show the m dependence of the moments A_{2m} for different parameters extracted from the above reference. Thus, there should be a restriction in use to the lowest terms of the sequence A_{2m}. The higher the value of m the less accurate are empirical values of the moment A_{2m}. The empirical values of the moments A_{2m} for $0 < m < 3$ may be obtained not more than three significant digits from the well-behaved photoelectric light curves.

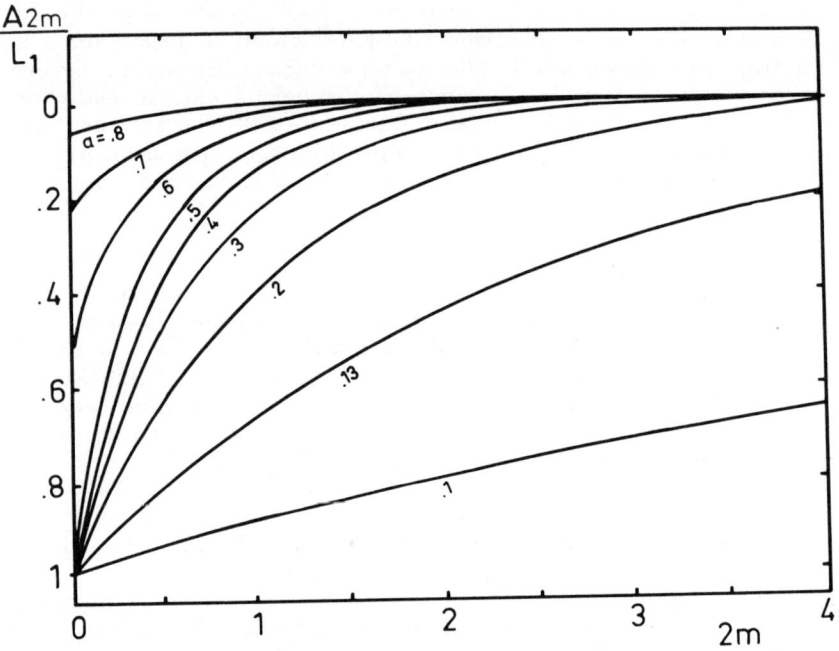

Figure 1 The m dependence of the moments A_{2m} for $r_1 = 0.1$, $i = 90°$ and fixed values of a (after Demircan, 1978b).

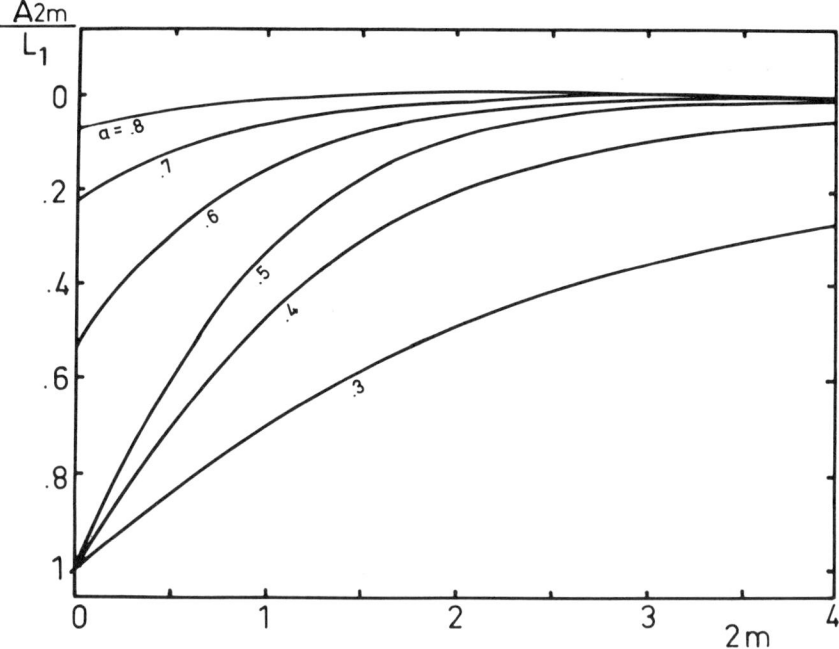

Figure 2 The m dependence of the moments A_{2m} for $r_1 = 0.3$, $i = 90°$ and fixed values of a (after Demircan, 1978b).

Kopal (1979) developed another method for a purely analytical determination of the areas A_{2m}. Kopal first expanded the observed luminosity ℓ into a finite Fourier cosine series. By writing this series for every observational point, the coefficients of the series can be obtained together with their probable errors by least-squares techniques. Then, the moments and their uncertainties can be readily obtained by making use of a relation (cf., Demircan, 1980, Eq. 3.23), given by

$$A_{2m} = \frac{\Gamma(2m+1)}{4^{m-1}} p^2 \sum_{j=0}^{\infty} \frac{(-1)^j (1+\frac{1}{2})^2 \sin[(2j+1)p]}{\Gamma(m+j+\frac{3}{2})\Gamma(m+\frac{1}{2}-j)} \times$$

$$\times \sum_{n=0}^{\infty} \frac{(-1)^n \varepsilon_n}{[(2j+1)p]^2 - (\pi n)^2} a_n \quad , \qquad (2.2)$$

between the moments A_{2m} and the Fourier cosine coefficients a_n. This relation is valid for any value of $m \geqslant 0$ - irrespective of whether it is integral or fractional - where p is a free parameter provided that $p \geqslant \theta$ (in radians) and $\varepsilon_0 = 1$ and $\varepsilon_n = 2$ for $n > 0$.

Note that the first infinite summation reduces to a finite one for half integral values of m. The second summation may have to be restricted to n = 0, 1, 2, ..., (N-1) if no more than N coefficients a_n of the Fourier cosine series can be determined significantly from the observed data. In practice, N happens to be not greater than 4 or 5. On the other hand, the first infinite summation for integral values of m has also to be approximated by N terms. This becomes possible only by the successful application of Lanczos' σ-factor method (cf., Lanczos, 1956, p.227) to the respective infinite Fourier series. Thus, in practice Equation (2.2) takes the form

$$A_{2m} = \frac{\Gamma(2m+1)}{4^{m-1}} P^2 \sum_{j=0}^{N-1} \frac{(-1)^j (j+1/2)^2 \sin(2j+1)\theta_1}{\Gamma(m+j+3/2)\Gamma(m-j+1/2)}$$

$$\frac{\sin[(2j+1)(\pi/(2N-1))]}{[(2j+1)(\pi/(2N+1))]} \sum_{n=0}^{N-1} \frac{(-1)^n \varepsilon_n}{[(2j+1)P]^2 - (\pi n)^2} a_n . \quad (2.3)$$

3. THEORETICAL MOMENTS

For total eclipses of the spherical model the moments A_2, A_4 and A_6 have been expressed (Kopal, 1975a,b) algebraically in terms of the eclipse elements L_1, $r_{1,2}$ and i but the same simplicity could not be obtained in the theoretical expressions of the moments for other eclipses. In the case of annular and partial eclipses, Kopal (1876a and 1977b) derived relatively complicated, but in closed form, expressions for the moments A_2 and A_4. Later, similar expressions for the moments A_2, A_4 and A_6 of the basic model valid for any type of eclipse have been obtained by the present author (Demircan, 1978a) independently, in terms of the new parameters $a = r_1/(r_1 + r_2)$, $b = 1 - a$ and $c_o = \cos i/(r_1 + r_2)$ which were introduced by Kopal (1977b) and vary between zero and unity. The author's results are:

$$A_2 = L_1 \left(\frac{r_1}{\sin i}\right)^2 \left\{ \frac{b^2 - c_o^2}{a^2} f_o + \left(\frac{b}{a}\right)^2 \sum_{\ell=0}^{\Lambda} \frac{2}{\ell+2} C^{(\ell)} S_1 \right\}, \quad (3.1)$$

$$A_4 = L_1 \left(\frac{r_1}{\sin i}\right)^4 \left\{ \left(\frac{b^2 - c_o^2}{a^2}\right)^2 f_o + \left(\frac{b}{a}\right)^2 \sum_{\ell=0}^{\Lambda} \frac{4}{\ell+2} C^{(\ell)} \alpha_\ell^o + \right.$$

$$\left. \left(\frac{b}{a}\right)^4 \sum_{\ell=0}^{\Lambda} \frac{2}{(\ell+2)(\ell+4)} C^{(\ell)} S_2 \right\}, \quad (3.2)$$

$$A_6 = L_1\left(\frac{r_1}{\sin i}\right)^6 \left\{ \left(\frac{b^2-c_o^2}{a^2}\right)^3 f_o + \left(\frac{b}{a}\right)^2 \left(\frac{b^2-c_o^2}{a^2}\right) \sum_{\ell=0}^{\Lambda} \frac{12}{\ell+4} C^{(\ell)} \alpha_\ell^o + \right.$$

(3.3)

$$\left. + \left(\frac{b}{a}\right)^2 \sum_{\ell=0}^{\Lambda} \frac{24}{(\ell+4)(\ell+6)} C^{(\ell)} \alpha_\ell^o + \left(\frac{b}{a}\right)^6 \sum_{\ell=0}^{\Lambda} \frac{2}{(\ell+2)(\ell+4)} C^{(\ell)} S_3 \right\},$$

where f_o comes from the zeroth moment

$$(A_o)_{obs} = 1 - \lambda, \quad (A_o)_{theo} = L_1 f_o$$

and

$$f_o = \sum_{\ell=0}^{\Lambda} C^{(\ell)} \alpha_\ell^o \quad . \tag{3.4}$$

$$S_1 = \left(\frac{b}{a}\right)^{\ell+2} Q_{\ell+2} , \tag{3.5}$$

$$S_2 = (\ell+8)\left(\frac{a}{b}\right)^2 J^o_{-1,\ell+2} + (\ell+4)\left(\frac{b}{a}\right)^\ell \frac{b^2-a^2-2c_o^2}{a^2} Q_{\ell+2} +$$

$$+ (\ell+8)\left(\frac{b}{a}\right)^{\ell+2} Q_{\ell+4} + 2(\ell+4)\left(\frac{b}{a}\right)^{\ell+2}\left(\frac{c_o}{b}\right)^3 I^1_{-1,\ell+2} , \tag{3.6}$$

$$S_3 = \frac{(\ell+10)(\ell+12)}{(\ell+6)}\left(\frac{a}{b}\right)^{\ell+2} Q_{\ell+6} + \left(\frac{b}{a}\right)^\ell\left[\frac{2(\ell+10)(b^2-a^2)-3(\ell+8)c_o^2}{a^2}\right] \cdot$$

$$Q_{\ell+4} + (\ell+4)\left(\frac{b}{a}\right)^{\ell-2}\left[\left(\frac{b^2-a^2}{a^2}\right)\left(\frac{b^2-a^2-3c_o^2}{a^2}\right) + 3\left(\frac{c_o}{a}\right)^4\right] Q_{\ell+2} +$$

$$\frac{1}{(\ell+6)}\left(\frac{a}{b}\right)^4\left[\frac{2(\ell+6)(\ell+10)b^2-3(\ell+6)(\ell+8)c_o^2}{a^2} - \ell(\ell+10)\right] \times J^o_{-1,\ell+2} +$$

$$\frac{(\ell+10)(\ell+12)}{(\ell+6)}\left(\frac{a}{b}\right)^4 J^o_{-1,\ell+4} + 2(\ell+12)\left(\frac{b}{a}\right)^{\ell+2}\left(\frac{c_o}{b}\right)^3 I^1_{-1,\ell+4} +$$

$$+ 2(\ell+4)\left(\frac{b}{a}\right)^\ell\left(\frac{c_o}{b}\right)^3 \left(\frac{b^2-a^2-2c_o^2}{a^2}\right) I^1_{-1,\ell+2} . \tag{3.7}$$

The quantities $J^o_{-1,\gamma}$ and Q_γ in the above expressions are the functions of well-known and best studied I-integrals of eclipsing binaries and defined by

$$J^o_{-1,\gamma} = \frac{r_2}{r_1}^{\gamma+2}\left[I^o_{-1,\gamma} - \frac{\delta_o}{r_2} I^1_{-1,\gamma}\right] \tag{3.8}$$

and

$$Q_\gamma = I^0_{-1,\gamma} + \frac{\delta_0}{r_2} I^1_{-1,\gamma}. \qquad (3.9)$$

For the I integrals the reader may refer chiefly to Kopal (1947) and Lanzano (1876a,b,c). The quantity λ stands for the light of the system at the moment of conjunction of the respective minimum; f_0 signifies the maximum obscuration of the star undergoing eclipse of luminosity L_1 (see Equation (3.4)), $C^{(\ell)}$'s are associated with the law of limb darkening of degree Λ and can be given (cf., e.g., Kopal, 1975b; Equations (2.4) and (2.5)) in terms of the coefficients of limb darkening u_1, u_2, ... u_Λ as

$$C^{(0)} = \frac{1-u_1-u_2- \cdots u_\Lambda}{1 - \sum_{\ell=0}^{\Lambda} \frac{u_\ell}{2+\ell}} \quad \text{and} \quad C^{(\ell)}(\ell > 0) = \frac{u_\ell}{\Lambda \ell u_\ell}{1 - \sum_{\ell=0}^{\Lambda} \frac{u_\ell}{2+\ell}}. \qquad (3.10)$$

The fractional loss of light α_ℓ^0 in Equation (3.4) was identified (Kopal, 1977a) as a Hankel transform of zero order and can be given in general, by

$$\alpha_\ell^0 = b^2(1-c^2)^{\nu+1} \Gamma(\nu) \sum_{n=0}^{\infty} \frac{n!(\nu+2n+2)}{(n+1)\Gamma(\nu+n+1)} [R_n^{(1,\nu)}(a)]^2 R_n^{(\nu+1,0)}(c^2), \qquad (3.11)$$

where $\nu = (\ell+2)/2$ and R's stand for the shifted Jacobi polynomials defining by

$$R_n^{(\alpha,\beta)}(x) = \frac{(-1)^n}{n!} \frac{\Gamma(n+\beta+1)}{\Gamma(\beta+1)} {}_2F_1\left(\begin{array}{c}-n,n+\alpha+\beta+1\\ \beta+1\end{array} \bigg| x\right) \qquad (3.12)$$

in terms of the ordinary hypergeometric functions ${}_2F_1$. For alternative expressions to (3.11) the reader may refer to Demircan (1977b and 1978c).

The expressions given by (3.1), (3.2) and (3.3) hold good for any type of eclipse and any degree Λ of the adopted law of limb darkening. It is known that for total eclipses all the I-integrals vanish, so do S_m's, since all the J's and Q's are zero there, thus, for total eclipses,

$$A_0 = L_1 \qquad (3.13)$$

$$A_2 = L_1 \left(\frac{r_1}{\sin i}\right)^2 \left(\frac{b^2-c_0^2}{a^2}\right) \qquad (3.14)$$

$$A_4 = L_1\left(\frac{r_1}{\sin i}\right)^4 \left\{ \left(\frac{b^2-c_o^2}{a^2}\right)^2 + \left(\frac{b}{a}\right)^2 \sum_{\ell=0}^{\Lambda} \frac{8}{(\ell+2)(\ell+4)} c^{(\ell)} \right\} \quad (3.15)$$

and

$$A_6 = L_1\left(\frac{r_1}{\sin i}\right)^6 \left\{ \left(\frac{b^2-c_o^2}{a^2}\right)^3 + \left(\frac{b}{a}\right)^2 \left(\frac{b^2-c_o^2}{a^2}\right) \times \right. \quad (3.16)$$

$$\left. \times \sum_{\ell=0}^{\Lambda} \frac{24}{(\ell+2)(\ell+4)} c^{(\ell)} + \left(\frac{b}{a}\right)^2 \sum_{\ell=0}^{\Lambda} \frac{48}{(\ell+2)(\ell+4)(\ell+6)} c^{(\ell)} \right\}$$

for any degree Λ of the law of limb-darkening, the summations on the right hand sides of (3.15) and (3.16) become

$$\frac{15-7u_1}{5(3-u_1)}, \quad \frac{3(15-7u_1)}{5(3-u_1)} \quad \text{and} \quad \frac{3(35-19u_1)}{35(3-u_1)}, \quad (3.17)$$

respectively, for the linear law of limb darkening; and

$$\frac{2(15-7u_1-10u_2)}{5(6-2u_1-3u_2)}, \quad \frac{6(15-7u_1-10u_2)}{5(6-2u_1-3u_2)} \quad \text{and} \quad \frac{3(140-76u_1-105u_2)}{70(6-2u_1-3u_2)} \quad (3.18)$$

for the quadratic law of limb-darkening. If the distribution of brightness over the disc of totally eclipsed component is uniform, then Equations (3.13) - (3.16) further reduce to

$$A_0 = L_1, \quad (3.18)$$

$$A_2 = L_1\left(\frac{r_1}{\sin i}\right)^2 \left(\frac{b^2-c_o^2}{a^2}\right), \quad (3.19)$$

$$A_4 = L_1\left(\frac{r_1}{\sin i}\right)^4 \left[\left(\frac{b^2-c_o^2}{a^2}\right)^2 + \left(\frac{b}{a}\right)^2\right] \quad (3.20)$$

and

$$A_6 = L_1\left(\frac{r_1}{\sin i}\right)^6 \left[\left(\frac{b^2-c_o^2}{a^2}\right)^3 + 3\left(\frac{b}{a}\right)^2\left(\frac{b^2-c_o^2}{a^2}\right)+\left(\frac{b}{a}\right)^2\right]. \quad (3.21)$$

The expressions (3.13) - (3.16) and (3.18) - (3.21) are equivalent to those given by Kopal (1975a,b) for total eclipses. Note from Equations (3.13) and (3.14) that A_0 and A_2 do not depend on limb darkening for total eclipses.

When the fractional loss of light α_ℓ^0 was identified with a Hankel transform of zero order and given in the form of general infinite series summation like (3.11), then it became possible to derive general expressions (Kopal, 1977b and Demircan, 1978a,c) for the moments A_{2m} in terms of the new parameters a, b and c_o (as defined before). For example, the Kopal's (1977b, Eq. (3.70)) formula is given in modified form, by

$$A_{2m} = L_1 b^2 \Gamma(m+1) \sin^{2m}\theta_1 \sum_{\ell=0}^{\Lambda} C^{(\ell)} \Gamma(\nu)(1-c_o^2)^{\nu+1} \quad (3.22)$$

$$\sum_{n=0}^{\infty} \frac{n!(n+\nu+1)(2n+\nu+2)}{(n+1)\Gamma(m+n+\nu+2)} [R_n^{(1,\nu)}(a)^2] R_n^{(m+\nu+1,-m)}(c_o^2) ,$$

where all the notations are as used before. This general expression and all its alternatives given in the above references hold good, equally well, for any type of eclipse, any degree Λ of the adopted law of limb darkening, and any positive real value of m. The reader may note that for m = 0, the foregoing Equation (3.22) reduces to the product $L_1 f_o(a,c_o)$, where $f_o(a,c_o)$ obtains for a combination of (3.11) and (3.4). The second summation on the right hand side of (3.22) should be approximated by the summation of first N terms. If the first 25 terms are considered, about 3 significant figures in the numerical approximations can be achieved. To achieve one more significant digit we have to increase N from 25 to about 80 and worse, summation of the terms beyond the eightieth term adds no more significance to the approximation and may falsify it by growing round-off error. It is also worth noting that three significant figures in the numerical values of the second summation on the right hand side of (3.22) will be enough for the analysis of the light curves of eclipsing variables for which the observational accuracy is not better than three figures.

4. SOLUTION FOR THE ELEMENTS

As has been pointed out already, in Kopal's theory the fundamental data for the analysis of basic spherical model in the frequency domain are represented by a set of the quantities so called "the moments A_{2m} of the light curves". They have been defined by Equation (1.1). The methods for evaluation of the observational moments and the theoretical moments in terms of the eclipse elements have been reviewed in Sections 2 and 3. Thus, for any eclipse of spherical model four unknown eclipse elements, namely, L_1, $r_{1,2}$ and i can be obtained by solving four non-linear equations of the form (1.2). If we write Equations (3.13) - (3.16) for total eclipses in terms of the parameters (cf., e.g., Kopal, 1977b, Equations (2.28) - (2.30)) $\bar{C}_{1,2,3}$ given by

$$\bar{C}_3 = (r_2^2 \csc^2 i - \cot^2 i) \; , \tag{4.1}$$

$$\bar{C}_2^2 = r_1^2 r_2^2 \csc^4 i \sum_{\ell=0}^{\Lambda} \frac{8}{(\ell+2)(\ell+4)} c^{(\ell)} \; , \tag{4.2}$$

and

$$\bar{C}_1 \bar{C}_2^2 = r_1^2 r_2^4 \csc^6 i \sum_{\ell=0}^{\Lambda} \frac{48}{(\ell+2)(\ell+4)(\ell+6)} c^{(\ell)} \; . \tag{4.3}$$

it has been shown (Kopal, 1975a,b) that those equations become algebraic and can easily be converted into the constants $\bar{C}_{1,2,3}$ and the eclipse elements then follow in terms of the deduced constants.

However, if the type of eclipse is not total, but annular or partial, then the equations of the form (1.2) for the moments A_{2m} are no longer reducible into a form such that they permit algebraic solution for the eclipse elements. In this case we have four non-linear equations from which the eclipse elements can be deduced only through the iterations. However, before the iterative solution the number of unknowns can be reduced from four to two. A method has been developed (Kopal and Demircan, 1978) and generalised (Demircan, 1978a) for this purpose which permits similar treatment of every type of eclipse (including those ending in totality), and can be made amenable to solution by automatic computers.

In order to recapitulate the essential steps of the generalized methods, let us return to the expressions given by (3.1) - (3.4) for the theoretical moments $A_{0,2,4,6}$ or more generally to (3.22) which can all be rewritten as

$$A_{2m} = L_1 \left(\frac{r_1}{\sin i}\right)^{2m} f_{2m}(a, c_0) \; . \tag{4.4}$$

If we now form the ratios

$$B \equiv \frac{A_{x_1}^{k_1} A_{x_2}^{k_2} \ldots A_{x_n}^{k_n}}{A_{y_1}^{j_1} A_{y_2}^{j_2} \ldots A_{y_m}^{j_m}} \; , \tag{4.5}$$

for the observational moments we obtain

$$B = \frac{f_{x_1}^{k_1} f_{x_2}^{k_2} \ldots f_{x_n}^{k_n}}{f_{y_1}^{j_1} f_{y_2}^{j_2} \ldots f_{y_m}^{j_m}} \equiv g(a, c_0) \; , \tag{4.6}$$

provided that

$$k_1 x_1 + k_2 x_2 + \ldots + k_n x_n = j_1 y_1 + j_2 y_2 + \ldots + j_m y_m ,$$ (4.7)

with real powers k_n's and j_m's, and the positive orders x_n's and y_m's (m,n = 1,2,3, ...). At least two values of y_m's should be different from x_n's. Thus, under condition (4.7) the right hand side of (4.6) becomes the function of two unknown parameters a and c_o only, and for two different sets of parameters k_n, x_n, j_m and y_m this equation constitutes two relations between the observed quantities and the unknown parameters a and c_o, since the quantities B given by (4.5) can be established from the observations as ratios of the respective powers of the respective moments of the light curves.

In particular, for example, if we suppose that n = m = 2; $j_1 = j_2 = k_1 = k_2 = 1$; $x_1 = x_2 = 2$ and $y_1 = 0$ and 2; $y_2 = 4$ and 6 respectively, we obtain

$$\frac{A_2^2}{A_0 A_4} = \frac{f_2^2}{f_0 f_4} \equiv g_2(a, c_o)$$ (4.8)

and

$$\frac{A_4^2}{A_2 A_6} = \frac{f_4^2}{f_2 f_6} \equiv g_4(a, c_o) ,$$ (4.9)

where the left hand sides are observational. These two simultaneous nonlinear equations can be solved numerically or otherwise for the parameters a and c_o. Once the value of the unknown parameters a and c_o have been determined we can use any one of the ratios A_{x_i}/A_{y_j} ($x_i \neq y_j$) to determine the fractional radius r_1 of the star undergoing eclipse. We can write with the aid of Equation (4.4) that

$$\frac{A_{x_i}}{A_{y_j}} = \left(\frac{r_1}{\sin i}\right)^{x_i - y_j} \frac{f_{x_i}}{f_{y_j}} = \left[\frac{(r_1 a)^2}{a - (r_1 c_o^2)}\right]^{\frac{x_i - y_j}{2}} \frac{f_{x_i}}{f_{y_j}} ,$$ (4.10)

thus it may be inverted to yield r_1. If, for example, $x_i = 2$ and $y_j = 0$, we obtain

$$r_1^2 = \frac{a^2 f_0 A_2}{c_o^2 f_0 A_2 + f_2 A_0} ,$$ (4.11)

which for known values of $A_0 \equiv 1 - \lambda$, A_2, a and c_o contains

r_1^2 as the only unknown. Once we have determined it, the values of r_2 and i follow from

$$r_2 = \frac{1-a}{a} r_1 \qquad (4.12)$$

and

$$\cos i = \frac{c_o}{a} r_1, \qquad (4.13)$$

while the fractional luminosity L_1 of the component undergoing eclipse follows, e.g., from the moment A_0 (see Equation (3.4)), as

$$L_1 = \frac{A_0}{f_0}. \qquad (4.14)$$

To make the methods reviewed in this section practicable, four digit numerical values of the functions $f_{2m}(a, c_o)$ and $g_{2m}(a, c_o)$ for different m's between zero and three have been calculated (Demircan, 1978d) with the aid of a CDC 7600 electronic computer of the University of Manchester. The work has been done for a quadratic law of limb-darkening, characterized by the coefficients $u_1 = 0.6500$ and $u_2 = -0.0226$, which reproduces the solution of the equation of radiative transfer in grey plane-parallel stellar atmospheres. The calculated values which are accurate to almost four significant digits were tabulated at intervals permitting linear interpolation between neighbouring entries within errors not in excess of 0.0004. The general behaviour of the functions f_0, g_2, g_4 and $g_{1/4}$ were illustrated (after the above reference) on the accompanying Figures 3-9. The graphically inverted almost two digit accurate a and c_o values are also given (after the above reference) in Tables 1 and 2. The entries for certain values of $g_{2,4}$ in these tables could not be completed because of the poor determinacy of the intersection points in certain regions. These regions in Figures 7-9 are the far left, right and top sides which all correspond to relatively shallow eclipses. In turn, the left side represents the data for an occultation of a small object by a large disc ($a \sim 0$, $r_1 \ll r_2$), while the right side represents the data for a transit eclipse by the same objects ($a \sim 1$, $r_1 \gg r_2$), the top side is for the grazing eclipses ($c_o \sim 1$). Thus, we can conclude that *the longer the distance from the point $P(\frac{1}{2}, 0)$ in a, c_o plane the poorer the determinacy could be attained for the eclipse parameters a and c_o.* The poor determinacy in a and c_o in fact becomes obvious if we consider the functional dependence between the g-functions for different values of m; for example,

$$g_2 = \frac{A_2 A_4}{A_0 A_6} / g_4.$$

For the similar functional behaviour of three different g-functions see Figures 7-9. The general behaviour of g-functions in three-dimensional space is illustrated (again after the above reference) in Figure 10.

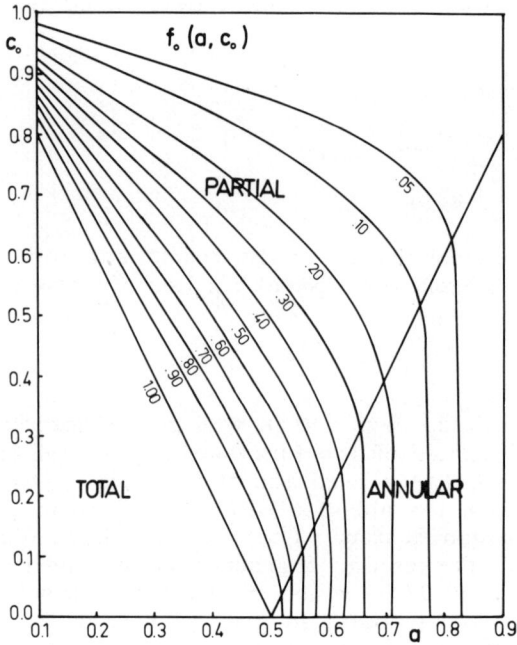

Figure 3. The functional behaviour of f_o as defined by Equation (3.4), in the (a, c_o) plane (after Demircan, 1978d).

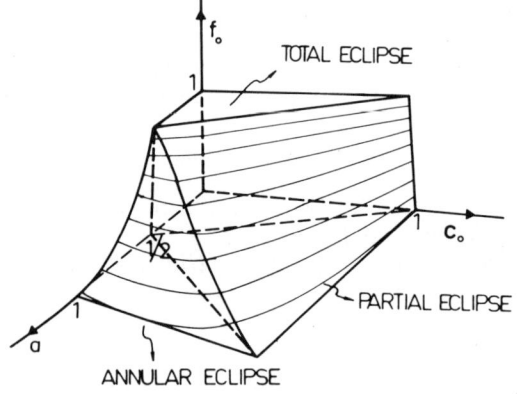

Figure 4. The functional behaviour of f_o as defined by Equation (3.4), in three dimensional space (after Kopal and Demircan, 1978).

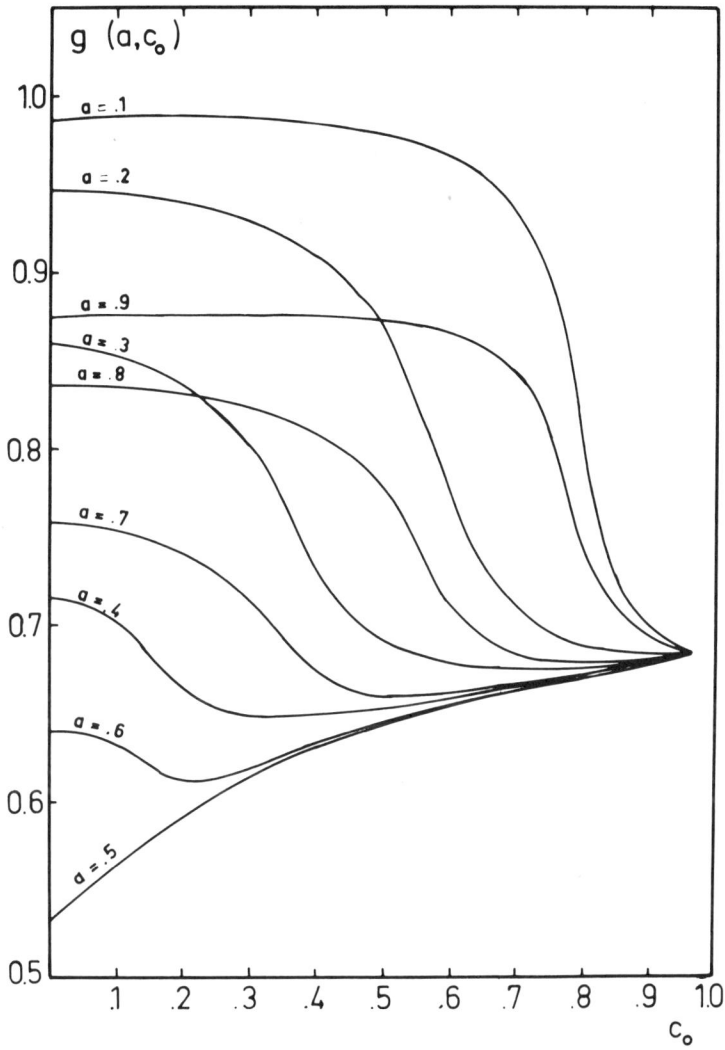

Figure 5. A plot of the function g_2 versus c_o for fixed a (after Kopal and Demircan, 1978).

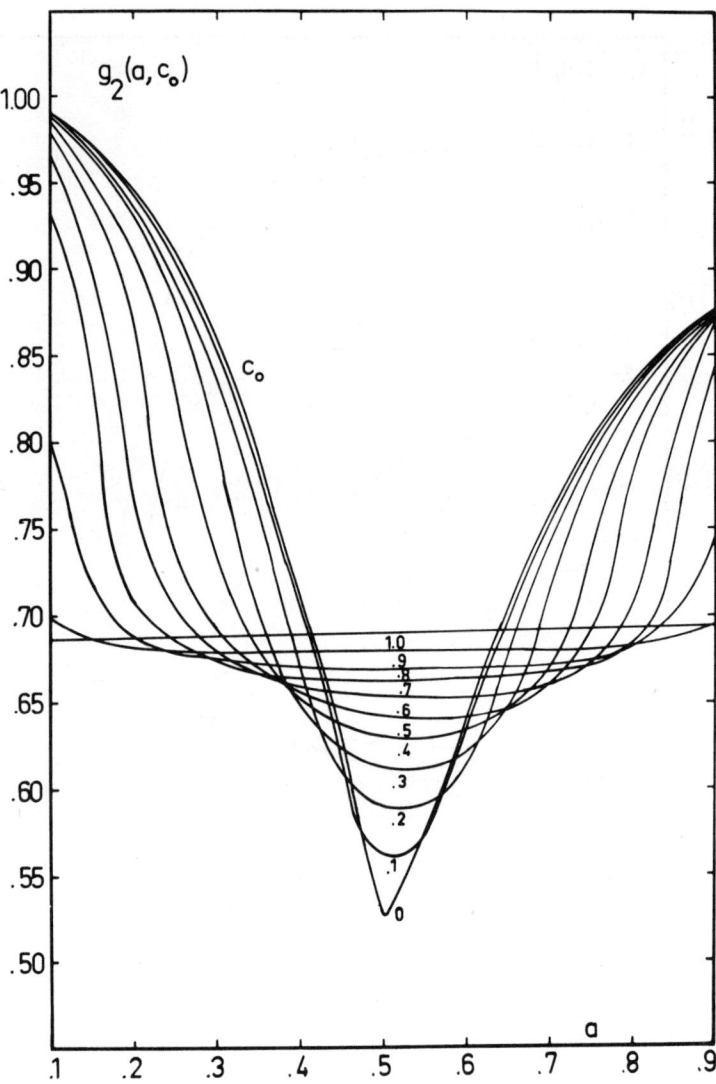

Figure 6. A plot of the function g_2 versus a for fixed values of c_o (after Demircan, 1978d)

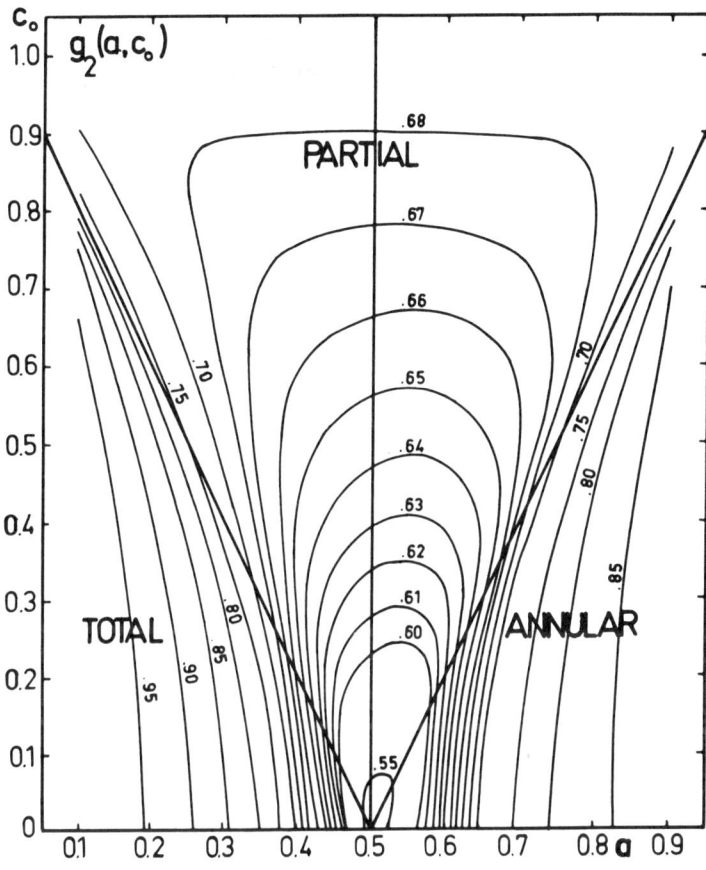

Figure 7. A diagrammatic representation of the function g_2 in the (a, c_o) plane for every type of eclipse (after Kopal and Demircan, 1978).

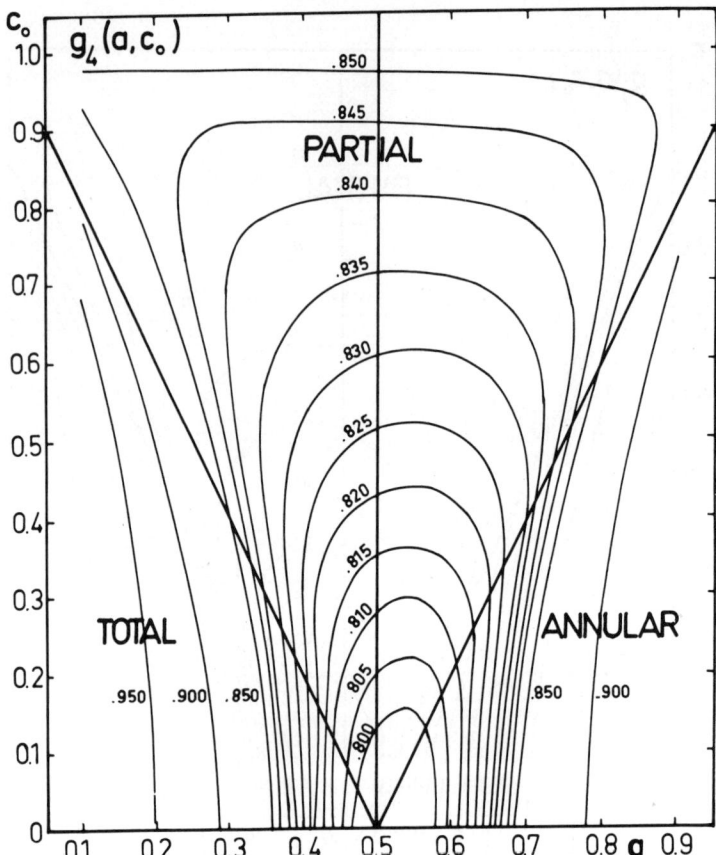

Figure 8. A diagrammatic representation of the function g_4 in the (a, c_o) plane for every type of eclipse (after Kopal and Demircan, 1978).

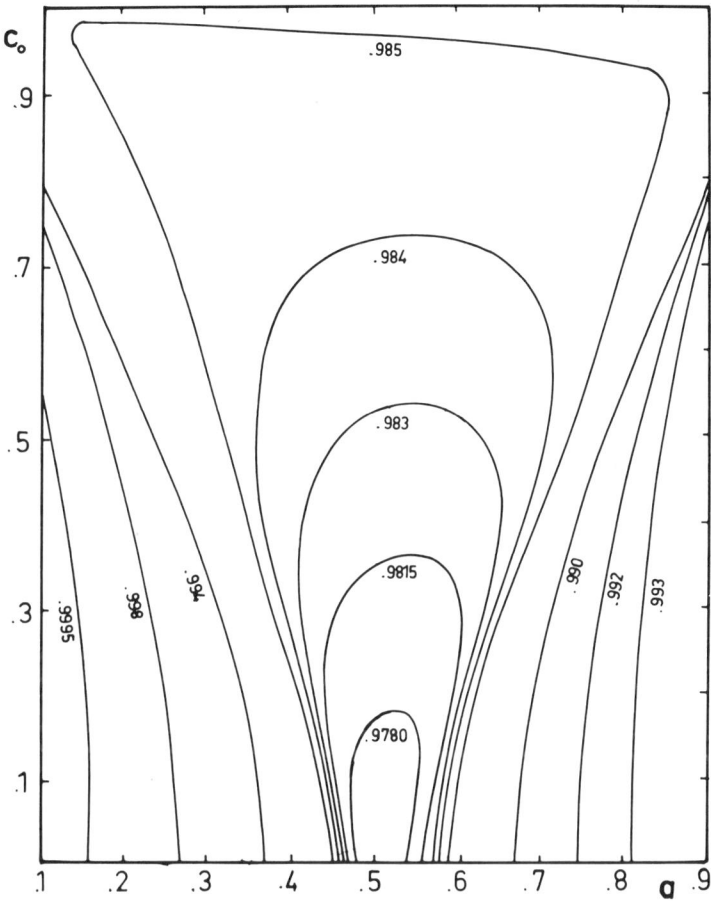

Figure 9. A diagrammatic representation of the function g_{1_4} in the (a, c_o) plane for every type of eclipse (after Demircan, 1978b).

Table 1a. $a(g_2, g_4)$ - functions for the occultation-type eclipses.

g_4 \ g_2	0.60	0.61	0.62	0.63	0.64	0.65	0.66	0.67	0.68	0.69	0.70	0.75
0.800	0.47											
0.805	0.46	0.46	0.46	0.45								
0.810		0.46	0.44	0.44	0.44	0.44						
0.815			0.44	0.43	0.43	0.43	0.43					
0.820					0.42	0.42	0.42	0.42				
0.825						0.40	0.41	0.41	0.41			
0.830							0.38	0.38	0.40	0.40		
0.835								0.34	0.37	0.38	0.39	
0.840									0.31	0.34	0.35	0.38
0.845												
0.850												

Table 1b. $c_o(g_2, g_4)$ - functions for the occultation-type eclipses.

g_4 \ g_2	0.60	0.61	0.62	0.63	0.64	0.65	0.66	0.67	0.68	0.69	0.70	0.75
0.800	0	0										
0.805	0.06	0.04	0.03	0.01								
0.810		0.23	0.12	0.09	0.02	0						
0.815			0.20	0.13	0.10	0.06	0					
0.820					0.24	0.12	0.07	0				
0.825						0.28	0.15	0.10	0			
0.830							0.43	0.23	0.13	0.05		
0.835								0.51	0.29	0.22	0.14	
0.840									0.56	0.39	0.30	0.05
0.845												
0.850												

Table 2a. $a(g_2, g_4)$ - functions for the transit-type eclipses.

g_4 \ g_2	0.60	0.61	0.62	0.63	0.64	0.65	0.66	0.67	0.68	0.69	0.70	0.75
0.800	0.58	0.58										
0.805	0.58	0.59	0.59	0.60								
0.810			0.61	0.62	0.61							
0.815				0.63	0.64	0.64	0.63					
0.820					0.65	0.66	0.65	0.64	0.64			
0.825						0.67	0.68	0.67	0.66	0.65	0.65	
0.830							0.70	0.70	0.70	0.69	0.67	
0.835								0.72	0.72	0.71	0.69	
0.840									0.75	0.73	0.72	
0.845									0.80	0.78	0.75	
0.850										0.84	0.82	0.70

Table 2b. $c_o(g_2, g_4)$ - functions for the transit-type eclipses.

g_4 \ g_2	0.60	0.61	0.62	0.63	0.64	0.65	0.66	0.67	0.68	0.69	0.70	0.75
0.800	0.10	0.03										
0.805	0.20	0.17	0.12	0.05								
0.810			0.23	0.20	0.18	0.07						
0.815				0.30	0.25	0.22	0.08					
0.820					0.36	0.30	0.25	0.22	0.05			
0.825						0.39	0.33	0.30	0.24	0.12	0	
0.830							0.48	0.41	0.39	0.32	0.21	
0.835								0.49	0.45	0.39	0.31	
0.840									0.56	0.47	0.41	
0.845										0.62	0.52	
0.850											0.68	0.20

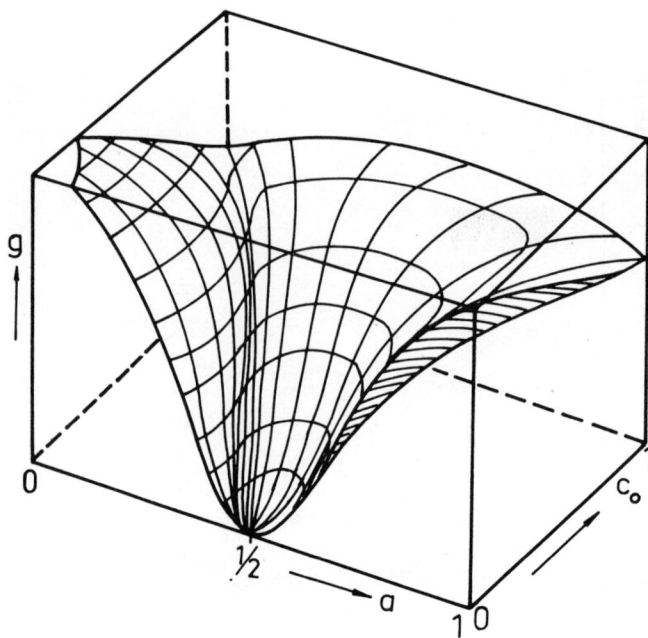

Figure 10. A schematic representation of the behaviour of g-functions (in general) in three-dimensional space (after Demircan, 1978d).

Thus, when we ascertained the observational values of two g-functions (for example, $g_{2,4}$) we can enter the above tables to establish the corresponding values of a and c_o. Note that if we do not know the type of eclipse under consideration, two different pairs of (a, c_o) will be established, one for the occultation, the other for the transit type of eclipse. The situation is illustrated geometrically in Figure 11. In this case we perform the solution of the elements with both pairs of (a, c_o). With the established values of a and c_o, the above-mentioned tables (which are available on request) make a determination of r_1 from (4.11) a matter of simple algebra and to obtain the other unknowns from Equations (4.12) - (4.14) is straightforward enough. The solution of the final elements for one of the pairs of (a, c_o) may turn out to be unrealistic, e.g., cos i > 1. This permits us to abandon that solution and accordingly the corresponding type of eclipse. Thus, the true solution can be established, but the procedure may not be applicable for certain cases, particularly in the case of partial eclipses for which both the solutions (under the assumption of occultation and transit eclipses) may turn out to be realistic. Then, the question arises, which solution is the true one? To answer the

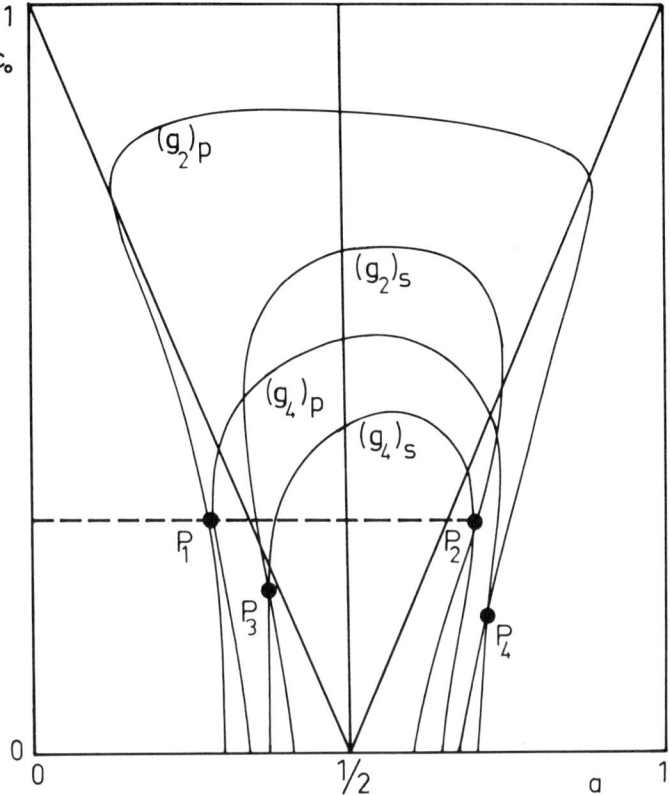

Figure 11. The geometrical illustration for the determination of the eclipse parameters a and c_o from the intersection points of g_2 and g_4-functions deduced from the primary (p) and secondary (s) eclipse minima of the same light curve. It is seen that four intersection points exist ($P_i(a_i, c_{oi})$, $i = 1,2,3,4$) which lead the analysis to two distinct solutions (one is true, the other is false) for the eclipse elements. If the primary minimum is an occultation eclipse, the intersection points P_1 and P_2 will lead us to the true solution, and in this case $c_{o1} = c_{o2} \neq c_{o3} \neq c_{o4}$ and $a_1 = 1-a_2$, $a_3 \neq 1-a_4$. If, on the contrary, the secondary minimum is an occultation eclipse, then the intersection points P_3 and P_4 will lead us to the true solution for the eclipse elements $r_{1,2}$, i and L_1, and this time $c_{o1} \neq c_{o2} \neq c_{o3} = c_{o4}$ and $a_1 \neq 1-a_2$, $a_3 = 1-a_4$.

question we must know prior to the analysis whether the minimum under consideration is due to an occultation or a transit eclipse. In the light curve of an eclipsing binary system, eclipses due to an occultation and a transit should occur alternatively, and if in the light curve observed during one orbital period one minimum is due to an occultation, the other minimum should be due to a transit. In the present methods it becomes possible to determine the type of eclipse minima prior to analysis in the following way: firstly, we can deduce from numerical data that, in general,

$$(g_m)_{occ} > (g_m)_{tra} \qquad (4.15)$$

for any value of m. For $a = b = \frac{1}{2}$ it is obvious that the relation is meaningless, and for grazing eclipses the inequality may not hold good. It can be observed from Figure 5, for example, that when c_o tends to unity (grazing eclipses) the difference between two quantities $(g_2)_{occ}$ and $(g_2)_{tra}$ approaches to zero and consequently in the case of grazing eclipses small observational errors in these quantities may alter the inequality sign in relation (4.15).

Thus, we first evaluate any g-function from the light variation in each primary and secondary minimum, then it will be easy with the aid of relation (4.15) (if the eclipses are not grazing type) to determine prior to analysis for the geometrical elements which one of the primary and secondary minima is due to an occultation or a transit by comparing the observational values of g-function deduced from the primary minimum with the corresponding one from the secondary. The larger numerical value of the quantity g will correspond to an occultation eclipse unless the eclipse minima are grazing type. As soon as we fix the type of eclipse minimum under consideration in this way the solution from that minimum becomes unified.

On the other hand, the unified true solution of the pair (a, c_o) may be ill-determined in certain domains because of the simulated functional dependence between the g-functions (see Figures 7-9). In such a case a combined information furnished from the alternate minima should be utilized. For the analysis of combined minima, the following relations between the observed quantities and the unknown parameters a and c_o have been given by Kopal (1979, p.180):

$$f_o(a,c_o) = (A_o)_p + \frac{(A_o)_s}{(a/b)^2 Y(a,c_o)} , \qquad (4.16)$$

$$\frac{(A_o)_s}{(A_o)_p} \frac{(A_{2m})_p}{(A_{2m})_s} = \left(\frac{a}{b}\right)^2 Y(a,c_o) \frac{f_{2m}(a,c_o)}{f_{2m}(b,c_o)} , \qquad (4.17)$$

where p and s refer to properties of the primary (deeper) and secondary (shallower) minima, respectively - regardless of which one happens to be an occultation or a transit. With

$$(A_o)_{p,s} \equiv 1 - \lambda_{p,s} \quad \text{and} \quad (A_{2m})_{p,s}$$

regarded as known, Equations (4.16) and (4.17) particularized for any value of m, can then be used to solve for a and c_o; and the rest of the solution for $r_{1,2}$, i and L_1 from known values of $(A_o)_{p,s}$ and $(A_{2m})_{p,s}$ carried out in the same way as before. The quantity Y in Equations (4.16) and (4.17) is a function of f_o, and can be given by

$$Y(a,c_o) = \left(\frac{b}{a}\right)^2 \frac{f_o(b,c_o)}{f_o(a,c_o)}, \qquad (4.18)$$

which is a slowly varying function of its parameters as well as of the coefficients of limb-darkening of both stars, whose theoretical values may be obtained by use of (3.11) in (3.4) (remember that b = 1-a). The solution of the nonlinear equations (4.16) and (4.17) for the unknowns a and c_o can be facilitated by a recourse to tables of the functions $f_{2m}(a,c_o)$ and $Y(a,c_o)$. The tables of function $Y(a,c_o)$ have also been constructed (Demircan, 1978d) by using the evaluated numerical values of $f_o(a,c_o)$ in (4.18) (note that $f_{2m}(1-a,c_o) = f_{2m}(b,c_o)$). The behaviour of function $Y(a,c_o)$ has been illustrated (after Demircan, 1978d) in Figure 12.

Alternatively, one could evaluate a and c_o from a pair of equations given (cf. Kopal and Demircan, 1978; Equations (2.26) and (2.27)) by

$$\frac{A_2^2}{A_o A_4}_p = \frac{f_2^2(a,c_o)}{f_o(a,c_o)f_4(a,c_o)} \equiv g_2(a,c_o), \qquad (4.19)$$

$$\frac{A_2^2}{A_o A_4}_s = \frac{f_2^2(b,c_o)}{f_o(b,c_o)f_4(b,c_o)} \equiv g_2(b,c_o), \qquad (4.20)$$

which are based on the same moments A_{2m} of the alternate minima. Such a procedure is generally applied to the analysis of the light curves of systems whose minima are due to partial eclipses of comparable depth; and tables of g_2-function continue to be available to facilitate this task. Note that Equations (4.19) and (4.20) specify two certain relations between the observed quantities on the left-hand sides and the parameters a and c_o on each half of Figure 7. Thus, if we mark each observational g_2 on a transparent paper, and overlay on each other flopped about the line a = 0.5; the co-ordinates of intersection of these two curves will indicate the res-

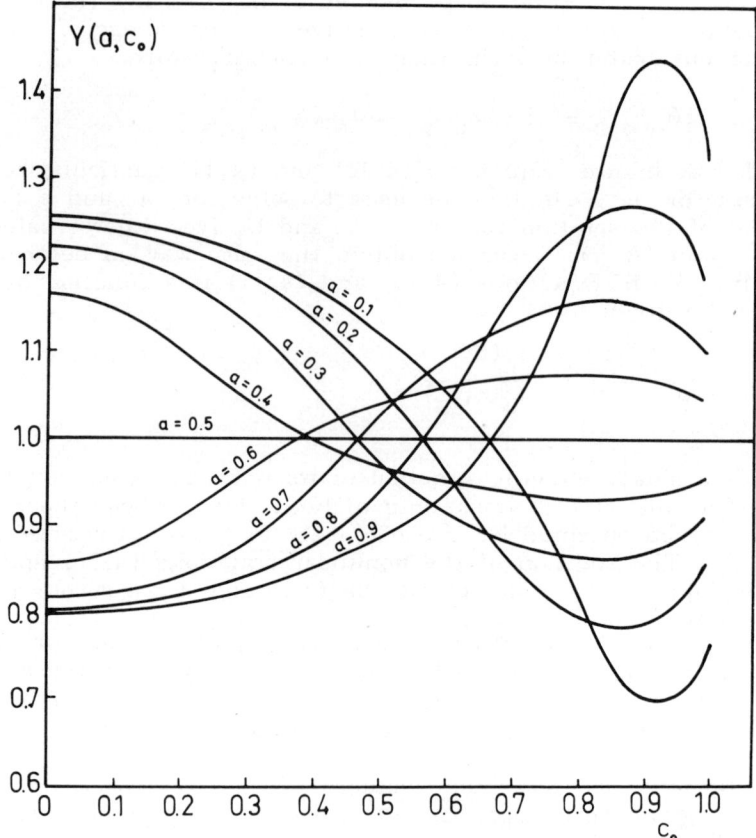

Figure 12. A plot of the function $Y(a,c_o)$ – as defined by Equation (4.18) – versus c_o for fixed values of a (after Demircan, 1978d).

pective values of a and c_o. Moreover, if $1 - \lambda_s$ is negligible, i.e., the component eclipsed at that time is effectively dark – this would imply that $L_2 \cong 0$ and accordingly $L_1 \cong 1$. In such a case the second independent relation to be adjoined to (4.19) would be (from (4.16)) as

$$f_o(a,c_o) = (A_0)_p \equiv 1 - \lambda_p \qquad (4.21)$$

which, together with (4.19) specifies the derived solution for the elements of the system $r_{1,2}$ and i consistent with $L_1 = 1$ and $L_2 = 0$.

CONCLUSIONS

The frequency domain methods discussed in the present paper for an analysis of the light changes of eclipsing variables are applicable to any type of eclipse of stars whose apparent discs are characterized by an arbitrary radially-symmetrical distribution of apparent brightness; but, let us remember that these techniques have been developed with an assumption that both components of the eclipsing system can be regarded as spherical. Moreover, no "reflection" effect has been considered. As is well known, this can be (approximately) the case only for relatively well-separated binaries. In any precise analysis of the light changes exhibited by close binaries, the mutual gravitational and radiative interaction between the components must be taken into account. It is already pointed out that in the frequency domain analysis of the problem a spherical model has been taken to be basic and a solution for the elements of distorted systems (by gravitational and radiative interaction between the components) have been reduced (cf., e.g., Kopal, 1975c, 1976b) to one based on this spherical model.

The principal advantages of the frequency-domain methodology are (i) that the quantities sought are related to integrals of the light curves which are thereby less sensitive to noise than would be particular points on the light curves as utilized in time-domain procedures; (ii) the formulation of the eclipsing binary light curve analysis problem is in general transcendental in the time domain, while in the frequency domain it becomes algebraic and most convenient for automatic computations; (iii) the possibility of a separation of the proximity effects (of any magnitude) from those caused by eclipses by purely algebraic means and (iv) the unification of the treatments for the eclipsing binary light curve analysis for all types of eclipse (partial, total and annular).

Together with all these advantages there are certain points to be considered with great care. Firstly, the A_{2m} - and consequently g_{2m}-functions for different values of m are found in practice (cf, Demircan, 1978b) to simulate roughly a functional dependence (see Figures 7, 8 and 9 for the similar behaviour of three different g-functions which do not make the analysis completely impossible, but weaken it considerably (cf, Demircan, 1978a, Section 5), particularly when we perform the analysis on one eclipse minimum alone. In this case, it is the author's conclusion that *the larger the separation between the m's for employed moments, the better the determinacy in four basic eclipse elements can be achieved*. Here, as was noted before, the restriction in use to the low index empirical moments should not be forgotten; it is well-known from the theory of approximation of functions by Fourier series that the higher the value of m, the more will an empirical determination of the moments A_{2m} be exposed to high-frequency noise arising from the dispersion of observations. It should also be noted that the determinacy of the eclipse elements is considerably improved by using the information from not only one but both eclipse minima.

The second point to be considered with great care arises in the light curve analysis of close eclipsing binary stars. In such a case, a process called "rectification" has been carried out to remove the proximity effects from the empirical moments A_{2m}. The reduction methods for the proximity effects may be perfect in its theoretical background, but in practice again the poor determinacy in the solution of certain quantities (by the analysis of observations between the minima) may occur (cf., e.g., Demircan, 1977a, Section 4.3) and in this case we become never sure whether the proximity effects were satisfactorily eliminated by the "rectification" and, consequently, whether the rectified light curve is correctly represented by the eclipse effect only.

ACKNOWLEDGMENT

The author is indebted to Professor Z. Kopal for his continuous encouragement and for initiating discussions about the subject matter of the present communication.

REFERENCES

Demircan, O.: 1977a, Astrophys. Space Sci., 47, p.459.
Demircan, O.: 1977b, Astrophys. Space Sci., 52, p.189.
Demircan, O.: 1978a, Astrophys. Space Sci., 56, p.389.
Demircan, O.: 1978b, Astrophys. Space Sci., 56, p.453.
Demircan, O.: 1978c, Astrophys. Space Sci., 59, p.313.
Demircan, O.: 1978d, Ph.D. Thesis, Univ. of Manchester (unpublished).
Demircan, O.: 1980, Astrophys. Space Sci., 67, p.375.
Kopal, Z.: 1947, Harvard Obs. Circ. No. 450.
Kopal, Z.: 1975a, Astrophys. Space Sci., 34, p.431.
Kopal, Z.: 1975b, Astrophys. Space Sci., 35, p.159.
Kopal, Z.: 1975c, Astrophys. Space Sci., 38, p.191.
Kopal, Z.: 1976a, Astrophys. Space Sci., 40, 461.
Kopal, Z.: 1976b, Astrophys. Space Sci., 45, p.269.
Kopal, Z.: 1977a, Astrophys. Space Sci., 50, p.225.
Kopal, Z.: 1977b, Astrophys. Space Sci., 51, p.439.
Kopal, Z.: 1979, *Language of the Stars*, D. Reidel Publ. Co., Dordrecht and Boston.
Kopal, Z. and Demircan, O.: 1978, Astrophys. Space Sci., 55, p.241.
Lanczos, C.: 1956, *Applied Analysis*, Prentice-Hall, Inc., Englewood Cliffs, N.J.
Lanzano, P.: 1976a, Astrophys. Space Sci., 42, p.425.
Lanzano, P.: 1976b, Astrophys. Space Sci., 45, p.419.
Lanzano, P.: 1976c, Astrophys. Space Sci., 45, p.483.
Powell, M. J. D.: 1970, in P. Rabinowitz (ed.) *Numerical Methods for non-Linear Algebraic Equations*, Gordon and Breach.

A COMPUTER PROGRAM FOR THE FREQUENCY-DOMAIN ANALYSIS OF THE LIGHT CURVES OF ECLIPSING VARIABLES

Alvaro Giménez and José M. García-Pelayo

Instituto de Astrofísica de Andalucía, Granada, Spain

ABSTRACT.

An algorithm for implementation of recently described Fourier transform techniques for eclipsing binary light curve analysis is presented and its use described. The algorithm has been applied by means of a desk top mini-computer in interactive mode. The required total amount of time for the whole analysis of a normal light curve (excluding light curve synthesis) is less than two hours of real time.

In a recent series of papers published in *Astrophysics and Space Science* and in this volume, a new method for the analysis of the light curves of eclipsing variables based on Fourier transforms has been presented by Z. Kopal and his collaborators. One of the main advantages of the method is that the fractional loss of light due to mutual eclipses of the components of the system can be expressed as a cross-correlation of two circular apertures in terms of Hankel transforms.

The aim of the present communication is to introduce the guidelines followed to write a computer program allowing any interested investigator to apply such a method to particular systems.

During this NATO Advanced Study Institute on Binary Stars, Drs. Jurkevich and Petty have presented an extensive investigation of the equations involved in the analysis of the light curves in the frequency-domain, specially concentrating in the determination of realistic errors for the final elements. The equations used by us, are essentially the same, being included in the recent book by Kopal (1979b) and therefore all the constraints and comments found by them are also valid for our approach that independently has confirmed their results.

The computer program that has been written for the complete analysis of the observations is mainly concerned with the practical procedure by means of which the raw data, i.e., photoelectric measurements, are handled towards the determination of the final elements: radii, inclination of the orbit and fractional luminosity, keeping throughout an empirical point of view.

The computational procedure is described by the flowchart given in Figure 1. The first task to be accomplished by the program is to read the observational data from the magnetic tape: phases and magnitudes. A codified number, heading the data file, indicates the characteristics of the input values (e.g., Julian dates instead of phases or luminosities instead of magnitudes). The light curve is then plotted and a preliminary inspection is made to see if eccentricity effects are present. If it is the case, corrections to the phases are applied according to the equations given by Kopal and Al-Naimiy (1978). After a new plot of the corrected light curve, the user is asked to select, by digitalization from the light curve, the magnitude outside eclipses m_o^d (ordinates) and the initial value for the external contact phase of the corresponding minimum, θ_1^d (abscissae). If another solution is available from previous works, the initial θ_1 may be computed using

$$\theta_1 = \arcsin((r_1 + r_2)^2 \cdot \csc^2 i - \ctg^2 i)^{\frac{1}{2}} .$$

The Fourier coefficients, a_i, are obtained together with their mean errors, δa_i, by a standard least-squares method according to the equation

$$1 - \ell(\theta) = \tfrac{1}{2} a_o + \sum_{i=1}^{\infty} a_i \cdot \cos(n\pi\theta/\theta_1) .$$

For the optimization of the θ_1 value, the vicinity of the initial θ_1^d is scanned for different values θ_1^c and the best one is found by the minimum sum of the squared residuals from the least-squares fitting. Once the a_i coefficients have been obtained together with their mean errors, statistical tests are applied for the significance of the fitting (F test) and the number of terms to be used avoiding the inclusion of noise by choosing the appropriate maximum value of i. As a further check, a plot of the observed points within the considered minimum is made together with the fitted curve.

The fundamental moments A_{2m} needed for the computation of the final elements are obtained by means of the fourier coefficients using equation 3.17 of Paper XXV by Kopal (1979a), except for

COMPUTER ANALYSIS OF THE LIGHT CURVES OF ECLIPSING VARIABLES

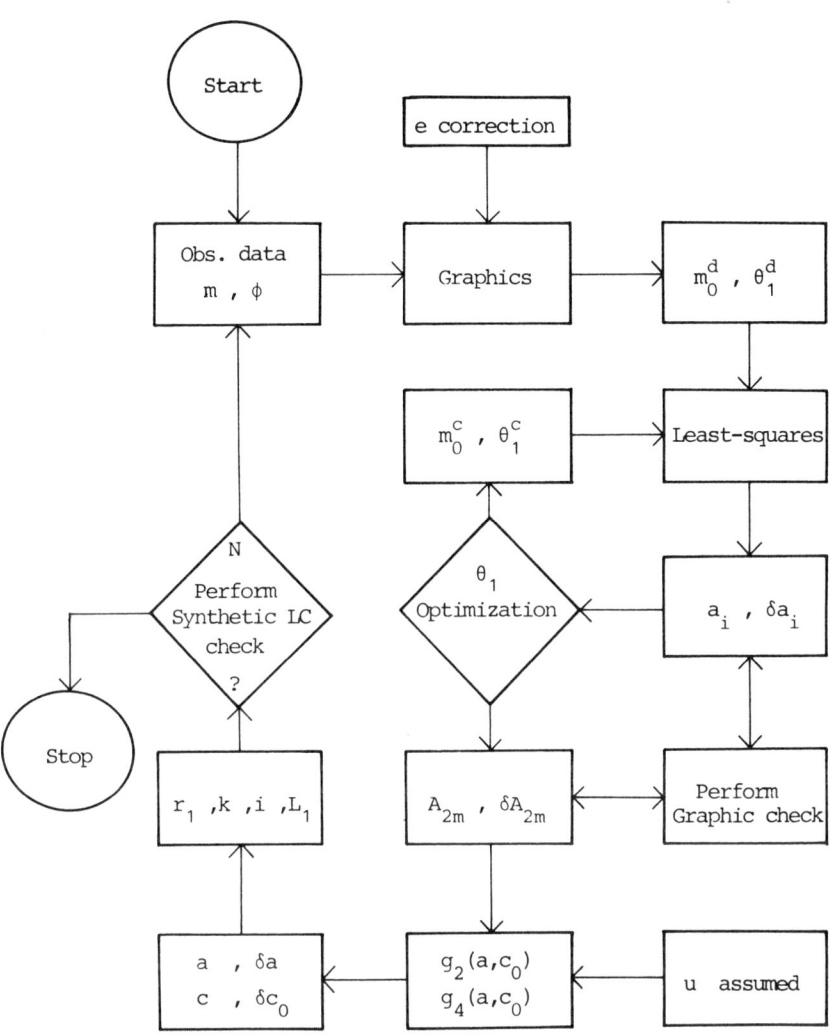

Figure 1. Flowchart of the program. See text.

the A_o moment that can be simply computed through

$$A_o = \tfrac{1}{2} a_0 + \sum_{i=1}^{n} a_i \quad .$$

In parallel to this procedure, the same moments are also obtained by numerical integration, trapezoidal rule, of the areas subtended by the light curve in $\ell - \sin^{2m}\theta$ coordinates as given by the expressions of Al-Naimiy (1977). The moments A_{2m} are the main observational parameters for the frequency-domain analysis of the light variations and with them, the g_{2m} functions are evaluated from

$$g_2(a, c_o) = \frac{A_2^2}{A_0 A_4} \quad \text{and} \quad g_4(a, c_o) = \frac{A_4^2}{A_2 A_6} \quad .$$

Using the method outlined by Kopal and Demircan (1978), by inversion of the above mentioned functions the values of the parameters a and c_o can be deduced as

$$a = \frac{r_1}{r_1 + r_2} \quad \text{and} \quad c_o = \frac{\cos i}{r_1 + r_2} \quad .$$

The solution is found numerically with a searching parameter space method and taking equations given in paper XII by Kopal (1977). The values for the limb-darkening coefficients, u, are assumed following the theory of stellar atmospheres and using the results tabulated by Al-Naimiy (1978).

With a and c_o, we can compute $f_2(a, c_o)$ and $f_4(a, c_o)$ by means of equations in paper XII, then together with the observed values of A_0 and A_2, it is straightforward to get the desired final elements: r_1, r_2, i and L_1, whose uncertainties are yielded by equations in paper XXV (Kopal, 1979a).

All computations are made for the secondary minimum as well as for the primary and the results are displayed together for comparison and also to disentangle indeterminacy in the case of partial eclipses. With the obtained final parameters, a synthetic light curve can be constructed and plotted with the original observed data as a final check of the whole analysis.

For the automatized processing we have used a small Hewlett-Packard 9845 A computer with 64K bytes of core and programs written in BASIC language. Another version divided in several subroutines and written in FORTRAN was used to check the numerical accuracy of the computations with a larger machine. As an example, for the a_i coefficients, the differences are within 2% of the mean errors δa_i. Programs also written in FORTRAN for the University of Manchester Regional Computer Centre CDC 7600, have been used as a basis for the main subroutines and were already checked by Z. Kopal and his collaborators. The mini-computer, obviously requires much more time for the calculations, but this is advantageously conterbalanced by its interaction working mode capability that allows the user to select during execution among all the possibilities of the method with different options. Moreover, the required total amount of time for the whole analysis of a normal light curve, not including the last step of the synthetic light curve check, is less than two hours real time, which is quite reasonable compared with other methods.

Printed outputs and plots of the results as well as the theoretical or the observational light curves are also included according to the special features of the equipment that has been used. Implementation of the programs as they are presently written should be no problem and fairly straightforward to modify them in order to suit other kinds of computers or individual requirements. It has to be noted that the whole procedure does not take into account the existence of perturbations or distortions of the components, being valid only for spherical stars. The extension of the procedure to include the existence of proximity effects will be part of subsequent improvements. It has also to be stressed that the last step of the analysis, namely the synthetic light curve check of the results, is not yet a part of the program itself and we used a separate program, WINK model given by Wood (1972), although translation of the prediction mode of the last version (Anderson et al, 1979) is being accomplished for the HP 9845 A mini-computer.

After the use of the program with several practical examples we have found a fairly fast convergence of the series for the required accuracy as well as a perfect agreement with previous analyses in the frequency-domain applying different methods. The maximum number of terms to consider for the a_i coefficients depends on the type of the light curve, but a good fit is provided by 8 terms. With respect to the computation of the A_{2m} moments, and therefore the g_{2m} finctions, probably the equations involved deserve a deeper insight since some amazing results have been obtained and a tendency to reach a constant 0.5 value of A_0 has been found, regardless of other parameters.

ACKNOWLEDGMENTS.

The authors should like to acknowledge Professor Z. Kopal for his very kind help and stimulation as well as for providing unpublished computer programs and discussion of the results.

REFERENCES.

Al-Naimiy, H. M.:   1977, Astrophys. Space Sci., 46, p.261.
Al-Naimiy, H. M.:   1978, Astrophys. Space Sci., 53, p.181.
Andersen, J., Clausen, J. V. and Nordström, B.:   1979, IAU Symp. No. 88, *Close Binary Stars: Observations and Interpretation*, Eds. M. J. Plavec, D. M. Popper and R. K. Ulrich, D. Reidel Publ. Co., Dordrecht, Holland.
Kopal, Z.:   1977, Astrophys. Space Sci., 51, p.439 (Paper XII).
Kopal, Z.:   1979a, Astrophys. Space Sci., 66, p.91 (Paper XXV).
Kopal, Z.:   1979b, *Language of the Stars*, D. Reidel Publ. Co., Dordrecht, Holland.
Kopal, Z. and Al-Naimiy, H. M.:   1978, Astrophys. Space Sci., 57, p.479 (Paper XIX).
Kopal, Z. and Demircan, O.:   1978, Astrophys. Space Sci., 55, p.241.
Wood, D. B.:   1972, "A Computer Program for Modeling Non-Spherical Eclipsing Binary Star Systems," X-110-72-473. GSFC, Greenbelt, Maryland.

A STUDY OF SPECIAL FUNCTIONS IN THE THEORY OF ECLIPSING BINARY SYSTEMS

Filaretti Zafiropoulos

Department of Astronomy,
University of Manchester, England.

ABSTRACT

The study summarized in the present paper deals with some special functions found by a process which employs Fourier's Theorem, in the theory of eclipsing binary systems.

In the first part the Fourier's Theorem is applied on three previously known expressions for α_ℓ^o so that a new generation of α_ℓ^o is created and studied numerically.

Subsequently, the new series obtained for α_ℓ^o are integrated so that three expansions giving $A_{2\mu}$ emerge. The results of numerical work are also included. On the series for α_ℓ^o and $A_{2\mu}$ the Lanczos's smoothing factor is applied.

Finally, certain expressions for α_ℓ^o are differentiated in an attempt to examine whether the Fourier Series for α_ℓ^o are differentiable.

I. SMOOTHING OF THE FOURIER SERIES

In the Fourier approach to a solution of the light curves of eclipsing binary systems already discussed in the preceding communications presented at this Conference, an important role is played by the Fourier Series for:

1. The fractional loss of light α_ℓ^o during eclipses of limb-darkened stars which are spherical in form;

2. The moments $A_{2\mu}$ of the light curves; and

3. The "boundary integrals" $I^m_{-1,n}$ and $J^o_{-1,n}$ expressing the modification of the loss of light which arises from distortion.

1. $\underline{\alpha}^o_\ell$.

In explicit terms these series can be expressed as:

$$\alpha^o_\ell = \frac{b^2 \sin \theta_1}{2\pi v} \sum_{n=0}^{\infty} \frac{v+2n+2}{(n+1)!} (v+1)_n \{G_n(v+2,v+1;a)\}^2 \times$$

$$\times \sum_{i=0}^{\infty} \frac{(n+v+2)_i}{i!} \binom{n}{i} \sum_{k=0}^{i} (-1)^{n+i+k} \binom{i}{k}(1-c_o^2)^{k+v+1} \times$$

$$\times B(\tfrac{1}{2}, k+v+2) \{{}_2F_1(\tfrac{1}{2}, \tfrac{1}{2}; k+v+\tfrac{5}{2}; \sin^2 \theta_1) +$$

$$+ 2 \sum_{m=1}^{\infty} {}_2F_1(\tfrac{1-m}{2}, \tfrac{1+m}{2}; k+v+\tfrac{5}{2}; \sin^2 \theta_1) \cos m\partial \} \quad (I.1)$$

or

$$\alpha^o_\ell = \frac{b^2 \sin \theta_1}{2\pi v} \sum_{n=0}^{\infty} (2n+\tfrac{1}{2}) P_{2n}(0) \times$$

$$\times F^{(4)}(-n+\tfrac{1}{2}, n+1; v+1, 2; a^2, b^2) \sum_{i=0}^{n} (-1)^i \frac{(n+\tfrac{1}{2})_i}{(\tfrac{1}{2})_i} \binom{n}{i} \times$$

$$\times (1-c_o^2)^i B(\tfrac{1}{2}, i+1) \{{}_2F_1(\tfrac{1}{2}, \tfrac{1}{2}; i+\tfrac{3}{2}; \sin^2 \theta_1) +$$

$$+ 2 \sum_{m=1}^{\infty} {}_2F_1(\tfrac{1-m}{2}, \tfrac{1+m}{2}; i+\tfrac{3}{2}; \sin^2 \theta_1) \cos m\theta \} \quad (I.2)$$

or

$$\alpha^o_\ell = \frac{b^2 \sin \theta_1}{2\pi v} \sum_{n=1}^{\infty} \frac{2}{(n-1)!} F^{(4)}(1-n, 1+n; 1+v, 2; a^2, b^2) \times$$

$$\times \sum_{i=0}^{\infty} \frac{(n+i-1)!}{i!} \binom{n}{i} \sum_{k=0}^{i} (-1)^{i+k} \binom{i}{k}(1-c_o^2)^k \times$$

$$\times \quad B(\tfrac{1}{2}, k+1) \{{}_2F_1(\tfrac{1}{2}, \tfrac{1}{2}; k+\tfrac{3}{2}; \sin^2\theta_1) +$$

$$+ 2 \sum_{m=1}^{\infty} {}_2F_1(\tfrac{1-m}{2}, \tfrac{1+m}{2}; k+\tfrac{3}{2}; \sin^2\theta_1) \cos m\theta \} \qquad (I.3)$$

where:

$$a = \frac{r_1}{r_1+r_2}, \quad b = \frac{r_2}{r_1+r_2}, \quad c_o = \frac{\delta_o}{r_1+r_2} = \frac{\cos i}{r_1+r_2}, \quad v = \frac{\ell+2}{2}, \qquad (I.4)$$

the $G_j(\alpha, \beta; \gamma)$ represent the Jacobi polynomials,

$$F^{(4)}(a,b;c,c';x,y) = \sum_{m=0}^{\infty} \sum_{n=0}^{\infty} \frac{(a)_{m+n}(b)_{m+n}}{m!n!(c)_m(c')_n} x^m y^n \qquad (I.5)$$

is the Appell's generalized hypergeometric series, and

$$P_{2n}(0) = (-1)^n \frac{(2n)!}{(n! 2^n)^2} \qquad (I.6)$$

stands for the Legendre polynomials of even order and argument zero.

The three series for α_ℓ^o, namely Eqs. (I.1), (I.2), (I.3), have been produced by having applied the Fourier's Theorem on the following three series:

$$\alpha_\ell^o = \frac{(1-c^2)^{v+1}}{v\Gamma(v+1)} \sum_{n=0}^{\infty} (-1)^n (v+2n+1) \frac{\Gamma(n+v+1)}{(n+1)!} \times$$

$$\times \{G_{n+1}(v, v+1, a)\}^2 G_n(v+2, 1, c^2) \qquad (I.7)$$

(Kopal, 1977c),

$$\alpha_\ell^o = \frac{b^2}{v} \sum_{n=0}^{\infty} (2n+\tfrac{1}{2}) F^{(4)}(-n+\tfrac{1}{2}, n+1; v+1, 2; a^2, b^2) \times$$

$$\times P_{2n}(\sqrt{1-c^2}) \qquad (I.8)$$

(Kopal, 1977c), and

$$\alpha_\ell^o = \frac{2b^2}{v} \sum_{n=1}^{\infty} F^{(4)}(1-n, 1+n; 1+v, 2; a^2, b^2) G_n(0, 1, c^2)$$

(I.9)

(Demircan, 1978d). At this point one must note that

$$c = \frac{\delta}{r_1 + r_2}$$

and

$$P_{2n}(\sqrt{1-c^2}) = \frac{(-1)^n (2n)!}{(n! 2^n)^2} {}_2F_1(-n, n+\tfrac{1}{2}; \tfrac{1}{2}; 1-c^2) =$$

$$= P_{2n}(0) G_n(\tfrac{1}{2}, \tfrac{1}{2}; 1-c^2) \qquad (I.10)$$

(cf. Kopal, 1979).

The three series (I.1), (I.2), (I.3) have been studied explicitly from the numerical point of view. The results were compared with values taken from the Tsesevich tables (Tsesevich, 1939, 1940). The convergence of the first two is as seen in Figures 1 and 2. Thorough numerical work has proved that the third series exhibits a very slow convergence and is soon accumulated by round-off errors, a fact which makes it unsuitable for future work.

2. $\underline{A}_{2\mu}$

Having studied Eqs. (I.1), (I.2) and (I.3) for α_ℓ^o, we can now produce the "moments" $A_{2\mu}$ by means of the well-known formula

$$A_{2\mu} = \int_0^{\theta_1} (1 - \ell(\theta)) d(\sin^{2\mu}\theta) \qquad (I.11)$$

where

$$1 - \ell(\theta) = L_1 \alpha = L_1 \sum_{\ell=0}^{\infty} c^{(\ell)} \alpha_\ell^o . \qquad (I.12)$$

The quantities L_1 and α represent the fractional luminosity of the eclipsed star and the instantaneous loss of light, respectively (cf. Kopal, 1975a); $c^{(\ell)}$ stand for the coefficients characterizing the limb-darkening law (cf. Kopal, 1975b). Therefore.

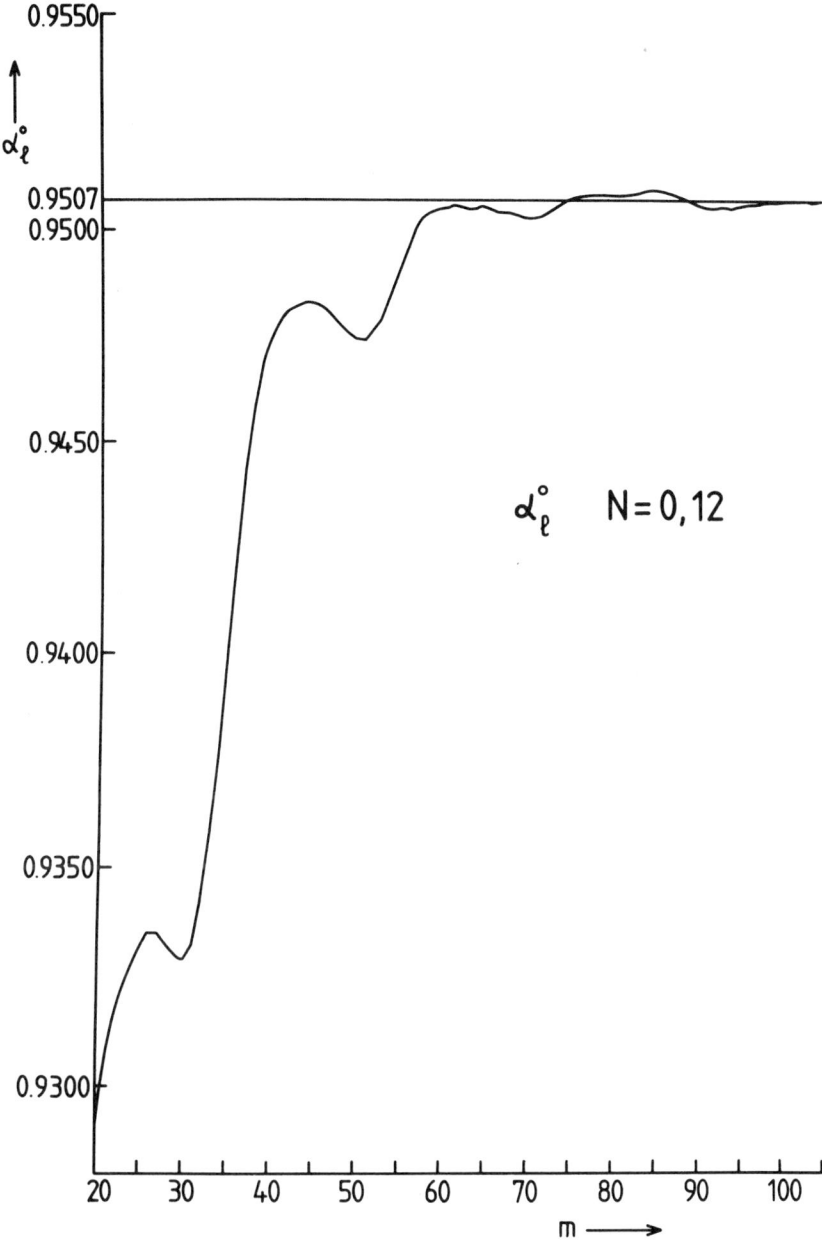

Figure 1. Convergence of Eq. (I.1) as a function of m .

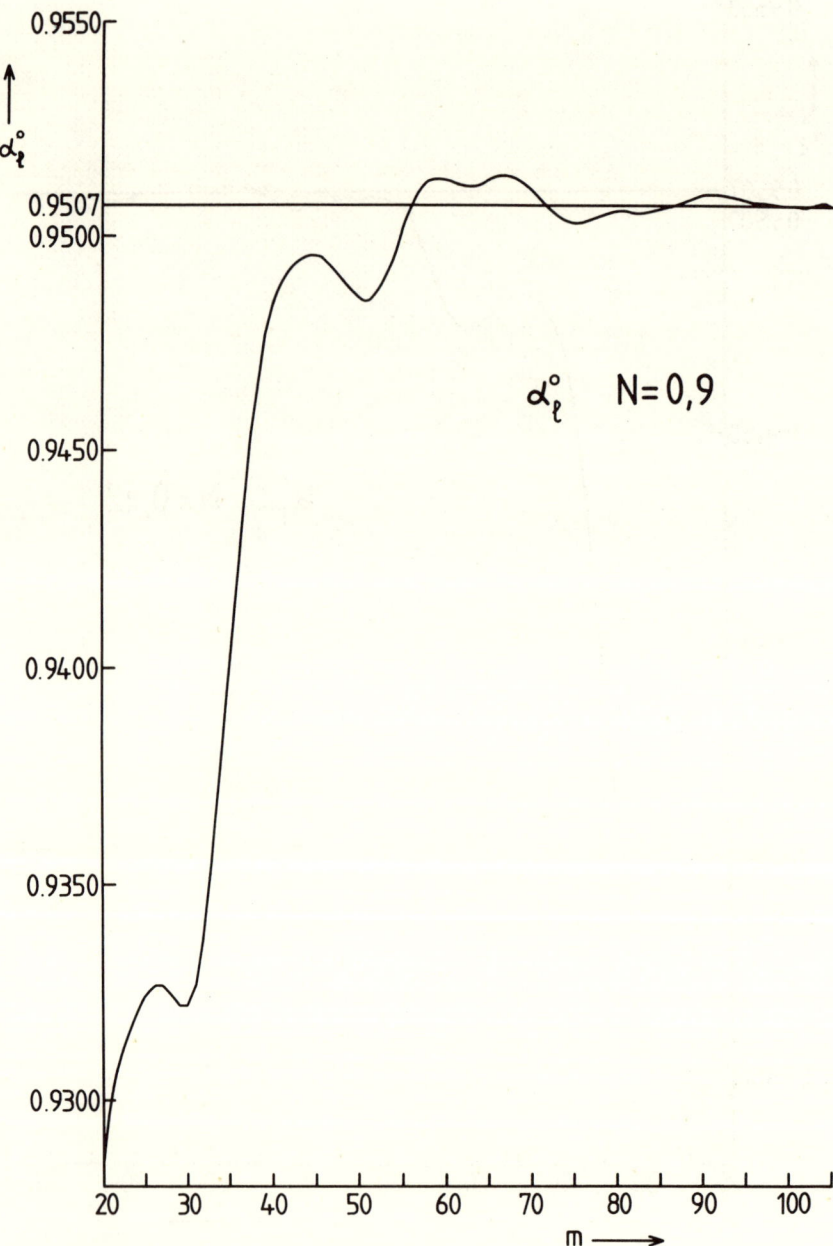

Figure 2. Convergence of Eq. (I.2) as a function of m.

$$A_{2\mu} = L_1 \sum_{\ell=0}^{\infty} c^{(\ell)} \int_o^{\theta_1} \alpha_\ell^o \, d(\sin^{2\mu}\theta) \tag{I.13}$$

where α_ℓ^o is given by one of Eqs. (I.1), (I.2) or (I.3).

To evaluate the integral on the right-hand side of Eq. (I.13), we make use of the transformation

$$u = \left(\frac{\sin \theta}{\sin \theta_1}\right)^2 \tag{I.14}$$

and the equation

$$\cos m\theta = {}_2F_1(\frac{-m}{2}, \frac{m}{2}; \frac{1}{2}; \sin^2\theta) \tag{I.15}$$

(Erdélyi et al, 1953), as a first step, and subsequently we employ the general identity

$$\int_0^1 (1-x)^{\lambda-1} x^{\nu-1} {}_pF_q(a_1, a_2, \ldots, a_p; b_1, b_2, \ldots, b_q; ax) dx =$$
$$= \frac{\Gamma(\lambda)\Gamma(\nu)}{\Gamma(\lambda+\nu)} {}_{p+1}F_{q+1}(\nu, a_1, \ldots, a_p; \lambda, b_1, \ldots, b_q; q) \tag{I.16}$$

which holds good under the following conditions:

(i) $\lambda > 0$,
(ii) $\nu > 0$,
(iii) $p < q + 1$, or
(iiia) if $p = q + 1$ then $|a| < 1$, satisfied in our case.

Consequently, the series for $A_{2\mu}$ emerging are of the form

$$A_{2\mu} = L_1 \frac{b^2(\sin \theta_1)^{2\mu+1}}{2\pi} \sum_{\ell=0}^{\infty} \frac{c^{(\ell)}}{\nu} \sum_{n=0}^{\infty} \frac{\nu+2n+2}{(n+1)!}(\nu+1)_n \times$$

$$\times \{G_n(\nu+2, \nu+1, a)\}^2 \sum_{i=0}^{n} \frac{(n+\nu+2)_i}{i!} \binom{n}{i} \sum_{k=0}^{i} (-1)^{n+i+k}\binom{i}{k} \times$$

$$\times (1-c_o^2)^{k+\nu+1} B(\frac{1}{2}, k+\nu+2) \{{}_2F_1(\frac{1}{2}, \frac{1}{2}; k+\nu+\frac{5}{2}; \sin^2\theta_1) +$$

$$+ 2 \sum_{m=1}^{\infty} {}_2F_1(\frac{1-m}{2}, \frac{1+m}{2}; k+\nu+\frac{5}{2}; \sin^2\theta_1) \times$$

$$_3F_2(\mu, \frac{-m}{2}, \frac{m}{2}; 1+\mu, \frac{1}{2}; \sin^2\theta_1)\} \qquad (I.17)$$

$$A_{2\mu} = L_1 \frac{b^2(\sin\theta_1)^{2\mu+1}}{2\pi} \sum_{\ell=0}^{\infty} \frac{c(\ell)}{v} \sum_{n=0}^{\infty} (2n+\frac{1}{2}) P_{2n}(0) \times$$

$$\times F^{(4)}(-n+\frac{1}{2}, n+1; v+1, 2; a^2, b^2) \sum_{i=0}^{n} (-1)^n \frac{(n+\frac{1}{2})_i}{(\frac{1}{2})_i}\binom{n}{i} \times$$

$$\times (1-c_o^2)^i B(\frac{1}{2}, i+1) \{_2F_1(\frac{1}{2}, \frac{1}{2}; i+\frac{3}{2}; \sin^2\theta_1) +$$

$$+ 2 \sum_{m=1}^{\infty} {}_2F_1(\frac{1-m}{2}, \frac{1+m}{2}; i+\frac{3}{2}; \sin^2\theta_1) \times {}_3F_2(\mu, \frac{-m}{2}, \frac{m}{2}; 1+\mu, \frac{1}{2};$$

$$\sin^2\theta_1)\} \qquad (I.18)$$

and

$$A_{2\mu} = L_1 \frac{b^2(\sin\theta_1)^{2\mu+1}}{2\pi} \sum_{\ell=0}^{\infty} \frac{c(\ell)}{v} \sum_{n=1}^{\infty} \frac{2}{(n-1)!} \times$$

$$\times F^{(4)}(1-n, 1+n; 1+v, 2; a^2, b^2) \times$$

$$\times \sum_{i=0}^{\infty} \frac{(n+i-1)!}{i!} \binom{n}{i} \sum_{k=0}^{i} (-1)^{i+k} \binom{i}{k} (1-c_o^2)^k \times$$

$$\times B(\frac{1}{2}, k+1) \times \{_2F_1(\frac{1}{2}, \frac{1}{2}; k+\frac{3}{2}; \sin^2\theta_1) +$$

$$+ 2 \sum_{m=1}^{\infty} {}_2F_1(\frac{1-m}{2}, \frac{1+m}{2}; k+\frac{3}{2}; \sin^2\theta_1) \times$$

$$\times {}_3F_2(\mu, \frac{-m}{2}, \frac{m}{2}; 1+\mu, \frac{1}{2}; \sin^2\theta_1)\} \ . \qquad (I.19)$$

The convergence of the first two series is given in Figures 3 and 4. The third one is as badly behaving as its parent, α_ℓ^o.

3. $\underline{I_{-1,n}^m, J_{-1,n}^o}$

Lastly, the Fourier expansions for the "boundary integrals" $J_{-1,n}^o$ and $I_{-1,n}^m$ are given as follows:

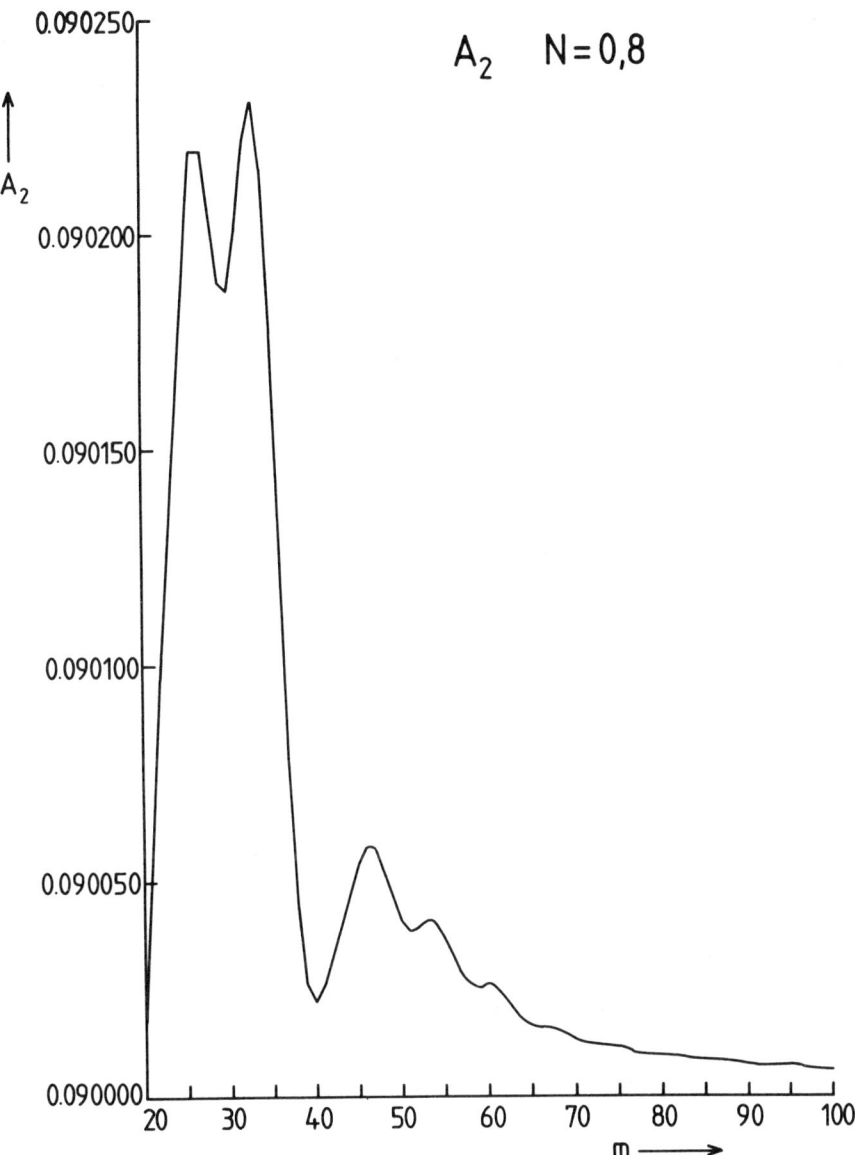

Figure 3a. Convergence of Eq. (I.17) as a function of m.

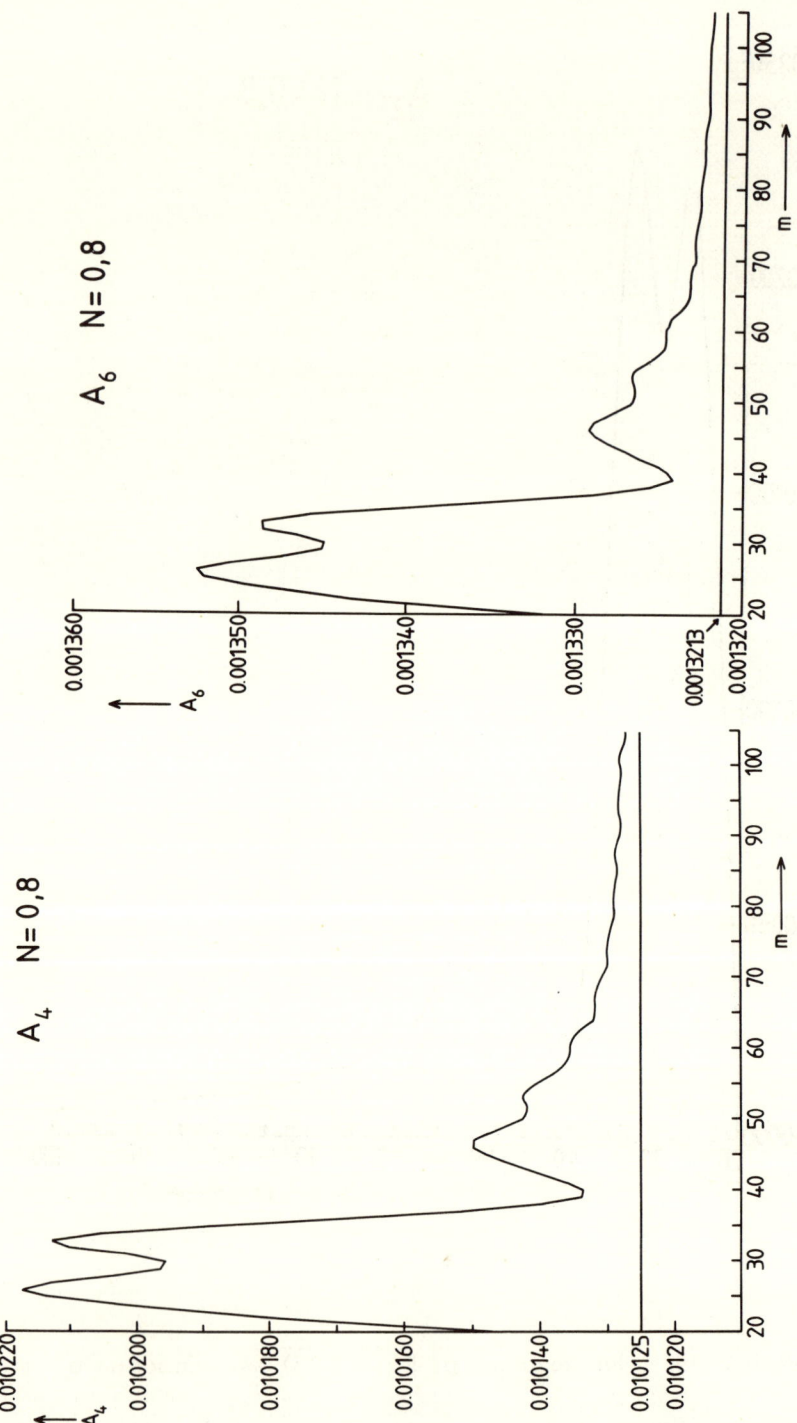

Figure 3b. Convergence of Eq. (I.17) as a function of m.

Figure 3c. Convergence of Eq. (I.17) as a function of m.

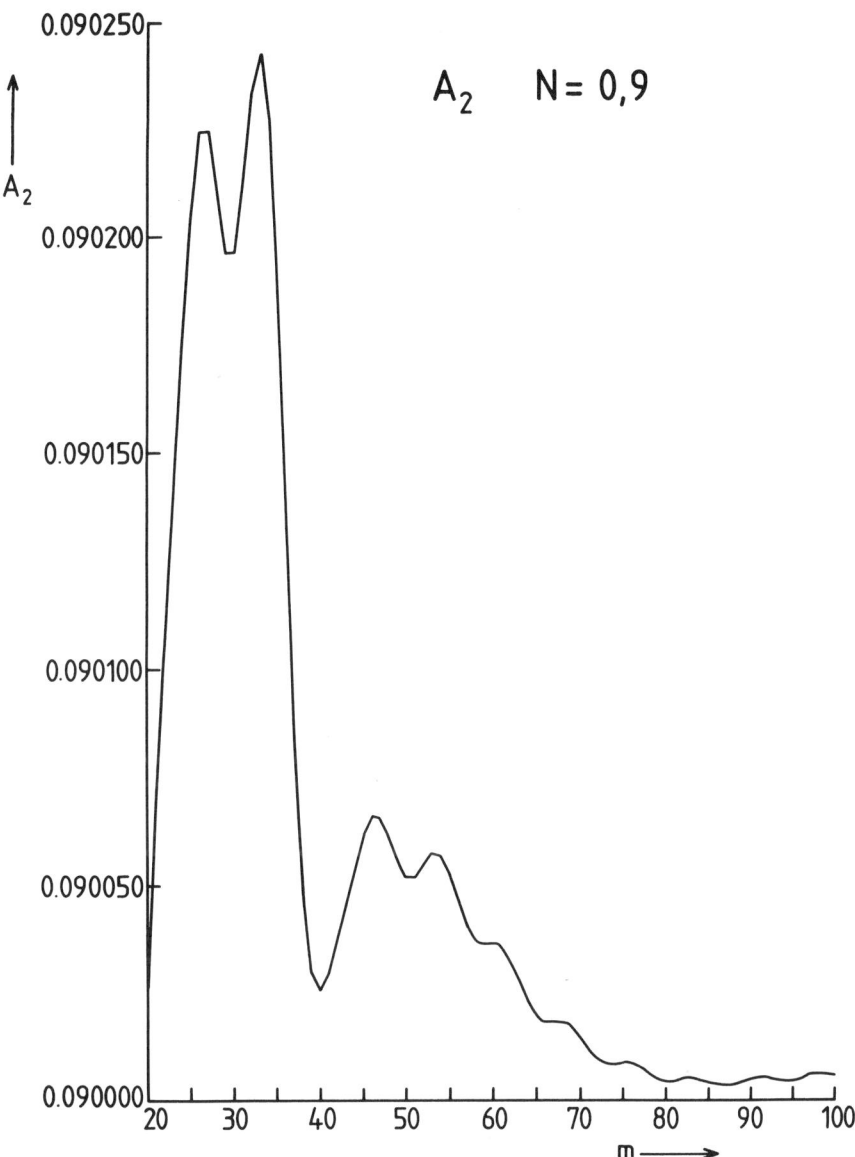

Figure 4a. Convergence of Eq. (I.18) as a function of m.

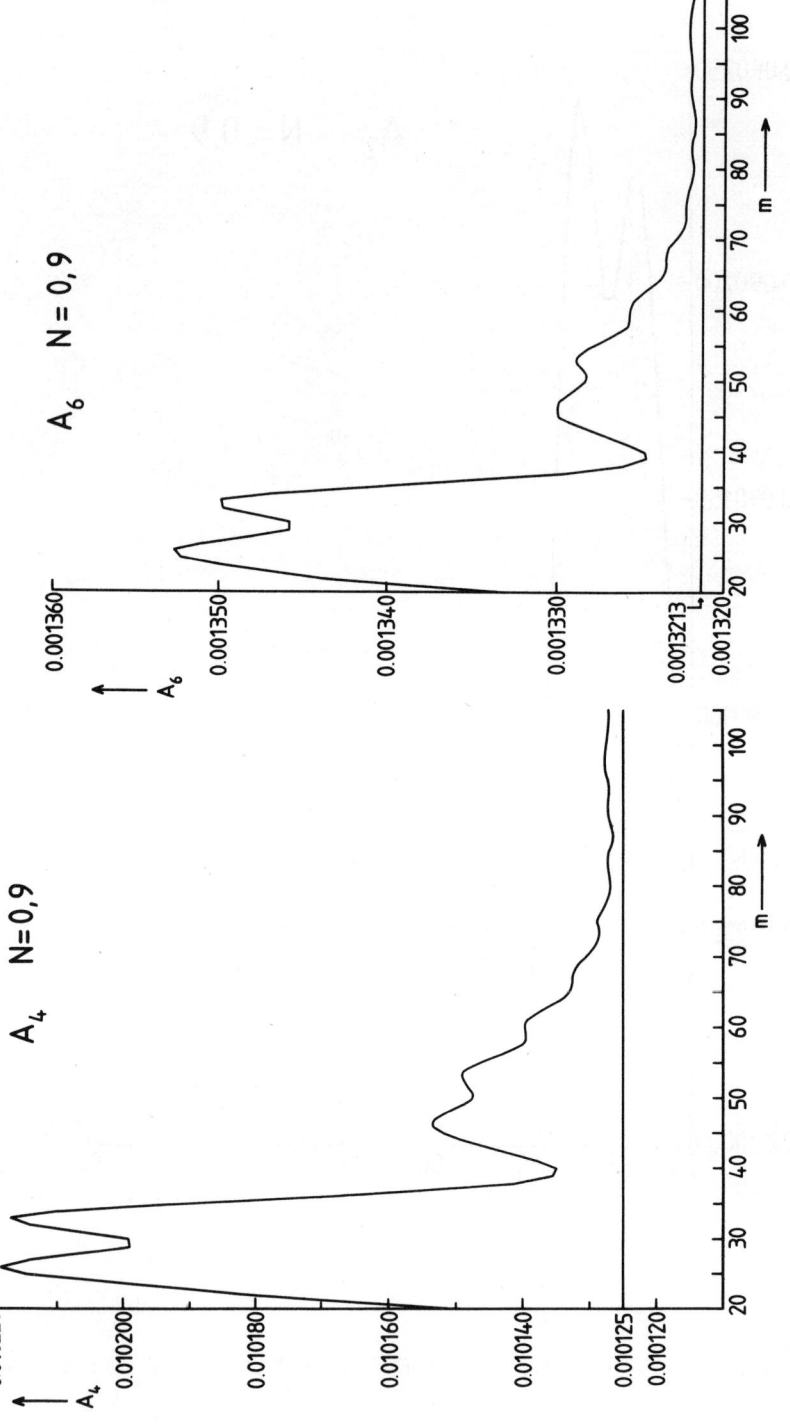

Figure 4b. Convergence of Eq. (I.18) as a function of m.

Figure 4c. Convergence of Eq. (I.18) as a function of m.

$$J^0_{-1,n} = \frac{a^2(1-c^2)^v}{v(v+1)\Gamma(v+2)} \sum_{j=0}^{\infty} \frac{(-1)^j}{j!} (v+2j+3)\Gamma(v+j+3) \times$$

$$\times \{G_{j+1}(v+1,v+2;a)\}^2 G_{j+1}(v+1,1;c^2) \qquad (I.20)$$

and

$$I^m_{-1,n} = \left(\frac{a}{b}\right)^{2v} \left(\frac{c}{b}\right)^m \frac{b^2(1-c^2)^v}{m!\,v\Gamma(v+1)} \sum_{j=0}^{\infty} \frac{(-1)^j}{j!} (m+v+2j+1) \times$$

$$\qquad (I.21)$$

$$\times \Gamma(m+v+j+1) \times \{G_{m+j}(v+1-m,v+1;a)\}^2 G_j(m+v+1,m+1;c^2)$$

for $m = 0$ or 1 only.

The derivation of these expressions can be located in Kopal's recent book on *The Language of the Stars* (D. Reidel Publ. Co., Dordrecht, 1979); for this reason, it need not be repeated in this place.

These two equations (I.20), (I.21) will be used in the second part of this work.

As could have been anticipated - and as is typical of the approximations represented by the Fourier Series - the convergence of the expressions of the form (I.1) - (I.2) and (I.17) - (I.18), as seen in Figures 1, 2, 3 and 4, is generally non-uniform; that is, the errors ε_n, ε_m of the respective approximations do not diminish monotonously with increasing n, m, but do so in an oscillatory manner (so that ε_{n+1} or ε_{m+1} may happen to be actually larger than ε_n or ε_m, respectively.

In order to lessen this oscillatory nature of the errors ε_n, ε_m, clearly manifested in Figures 1-4, Demircan (1980), recently proposed to apply the Lanczos's "smoothing factor" to Eqs. (I.1) - (I.2) and (I.17) - (I.18). In doing so he multiplies by the factor

$$s_j = \frac{\sin(\pi j/J)}{(\pi j/J)} \quad , \quad j = 0, 1, 2, \ldots, J \qquad (I.22)$$

It is clear that for $j = 0$, $s_0 = 1$, whereas $s_J = 0$ (cf. Lanczos, 1956, 1961).

In order to test the quality of the improvement brought about by an application of the Lanczos's smoothing factor, we have evaluated the partial sums of the series on the right-hand sides of Eqs. (I.1) - (I.2) and (I.17) - (I.18) as functions of m; the results are shown on the accompanying Figures 5, 6, 7 and 8. Their comparison with Figures 1-4 discloses that while Demircan's resort to the Lanczos's smoothing factor did, in fact, smooth the convergence to render $\varepsilon_{m+1} < \varepsilon_m$, it caused a dramatic shift and affected the accuracy.

II. DIFFERENTIABILITY OF THE FOURIER SERIES FOR α_ℓ^o

Let us consider the Eq. (I.7):

$$\alpha_\ell^o = \frac{(1-c^2)^{v+1}}{v\Gamma(v+1)} \sum_{j=0}^{\infty} (-1)^j (v+2j+2) \frac{\Gamma(v+j+1)}{(j+1)!} \times$$

$$\times \{G_j(v, v+1; a)\}^2 \, G_j(v+2, 1; c^2) \,.$$

This can be readily differentiated with respect to the elements r_1, r_2 and δ. For example,

$$r_1 \frac{\partial \alpha_\ell^o}{\partial r_1} = \frac{-2a(1-c^2)^v}{v\Gamma(v+2)} \sum_{j=0}^{\infty} \frac{(-1)^j}{j!} (v+2j+2) \Gamma(v+j+2) \times$$

$$\times \{(1-a)G_j(v+2,v+2;a)G_{j+1}(v,v+1;a)(1-c^2)G_j(v+2,1;c^2) -$$

$$- (v+1)[G_{j+1}(v,v+1;a)]^2 \, c^2 G_j(v+2,2;c^2) \} \tag{II.1}$$

In accordance with the equation:

$$r_1 \frac{\partial \alpha_\ell^o}{\partial r_1} = \frac{2}{r_2} \left(\frac{r_2}{r_1}\right)^{\ell+2} \{\delta I^1_{-1,\ell} - r_2 I^o_{-1,\ell}\} \equiv -2J^o_{-1,\ell} \tag{II.2}$$

where:

$$2\pi I^{-m}_{-1,\ell} = \left(\frac{\delta}{r^2}\right)^{\ell/2} B\left(\frac{1}{2}, \frac{\ell+2}{2}\right) (2K)^{\ell+1} \times$$

$$\times {}_2F_1\left(\frac{1}{2}-m, \frac{1}{2}+m; \frac{\ell+3}{2}; K^2\right) \tag{II.3}$$

if the eclipse is partial, and

THEORY OF ECLIPSING BINARY SYSTEMS

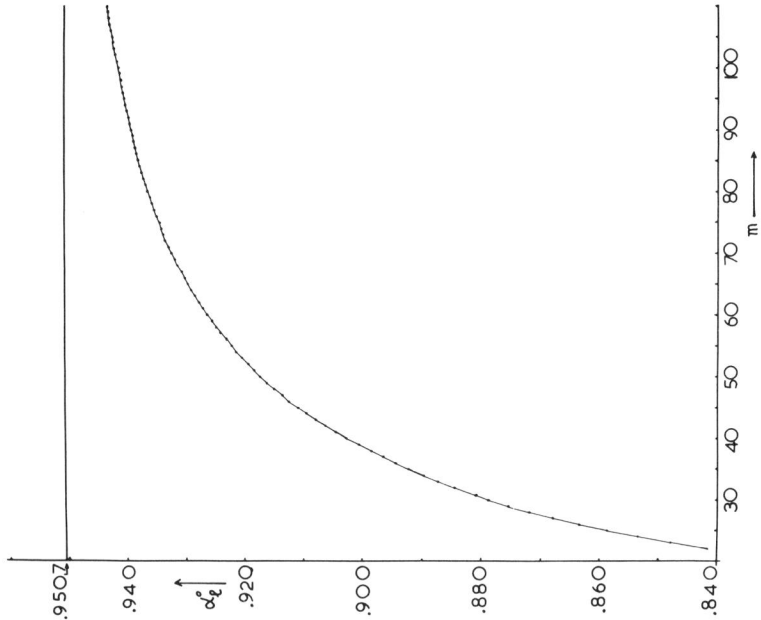

Figure 5. Convergence of Eq. (1.1) as a function of m, after the Lanczos's smoothing factor has been applied.

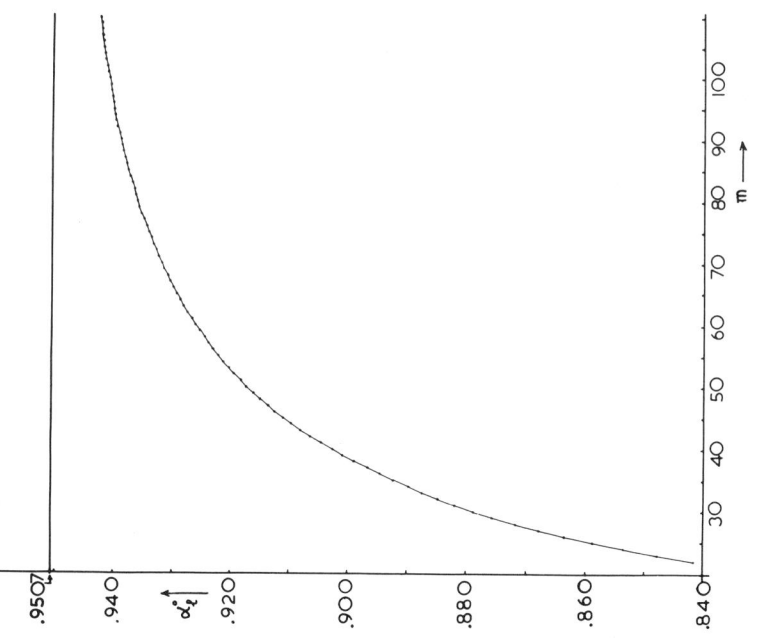

Figure 6. Convergence of Eq. (1.2) as a function of m, after the Lanczos's smoothing factor has been applied.

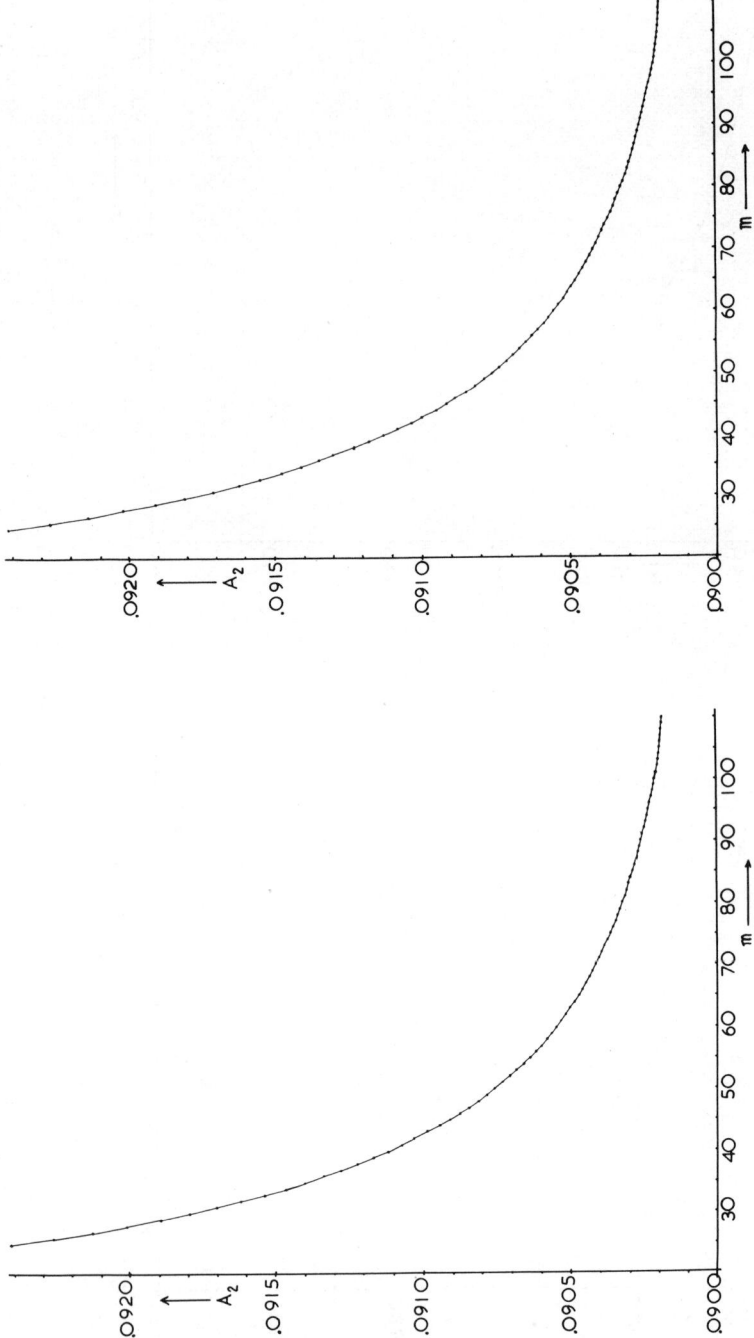

Figure 7. Convergence of Eq. (I.17) as a function of m, after the Lanczos's smoothing factor has been applied.

Figure 8. Convergence of Eq. (I.18) as a function of m, after the Lanczos's smoothing factor has been applied.

$$I^m_{-1,\ell} = \left(\frac{\ell}{4K^2}\right)^m \left(\frac{\delta}{r_2}\right)^{\ell/2} (2K)^\ell {}_2F_1(m-\tfrac{\ell}{2}, m+\tfrac{1}{2}; 2m+1; \tfrac{1}{K^2}) \quad (II.4)$$

if it is annular, with

$$K^2 = \frac{r_1^2 - (\delta - r_2)^2}{4\delta r_2} \quad (II.5)$$

for both types of eclipse, and m = 0 or 1 only, we can obtain a closed form for $r_1 \frac{\partial \alpha_\ell^o}{\partial r}$ (cf. Kopal, 1959).

On the other hand, if we use the Fourier approximations to $J^o_{-1,\ell}$, $I^m_{-1,\ell}$, namely Eqs. (I.20), (I.21), we get other expressions for $r_1 \frac{\partial \alpha_\ell^o}{\partial r_1}$ which are Fourier expansions (Kopal, 1979).

Another aspect of the problem which we set out to investigate is the extent to which the series for α_ℓ^o is differentiable with respect to the parameters. All the series for α_ℓ^o quoted in this communication are Fourier Expansions; and as such they are known to be integrable (thus making the evaluation of the moments $A_{2\mu}$, on the basis of the Fourier series for α_ℓ^o, mathematically legitimate); but need not be differentiable.

The numerical study gave the following results: all the Fourier approximations for $r_1 \frac{\partial \alpha_\ell^o}{\partial r_1}$ give very similar results. But these are very much different from what the closed expression gives, a fact which indicates that the Fourier expansion for α_ℓ^o is not differentiable. The results are summarized on the graphs 9 and 10 which represent plots of Eq. (II.1) and the one obtained by a combination of Eq. (II.2) and Eq. (I.20), for $J^o_{-1,\ell}$: namely the equation

$$r_1 \frac{\partial \alpha_\ell^o}{\partial r_1} = \frac{-2a^2(1-c^2)^v}{v(v+1)\Gamma(v+2)} \sum_{j=0}^{\infty} \frac{(-1)^j}{j!} (v+2j+3)\Gamma(v+j+3) \times$$

$$\times \{G_{j+1}(v+1, v+2; a)\}^2 G_{j+1}(v+1, 1; c^2). \quad (II.6)$$

At this stage, having in mind that the Lanczos's smoothing factor not only smooths the approximation, but it may transform a divergent Fourier Series into a convergent one and may make it differentiable, we applied it on Eqs. (II.1) and (II.6). The result was a smoother approximation and a slightly shifted curve, but it did very little to render the difference between the result of the closed form and this of the Fourier approximation. This is

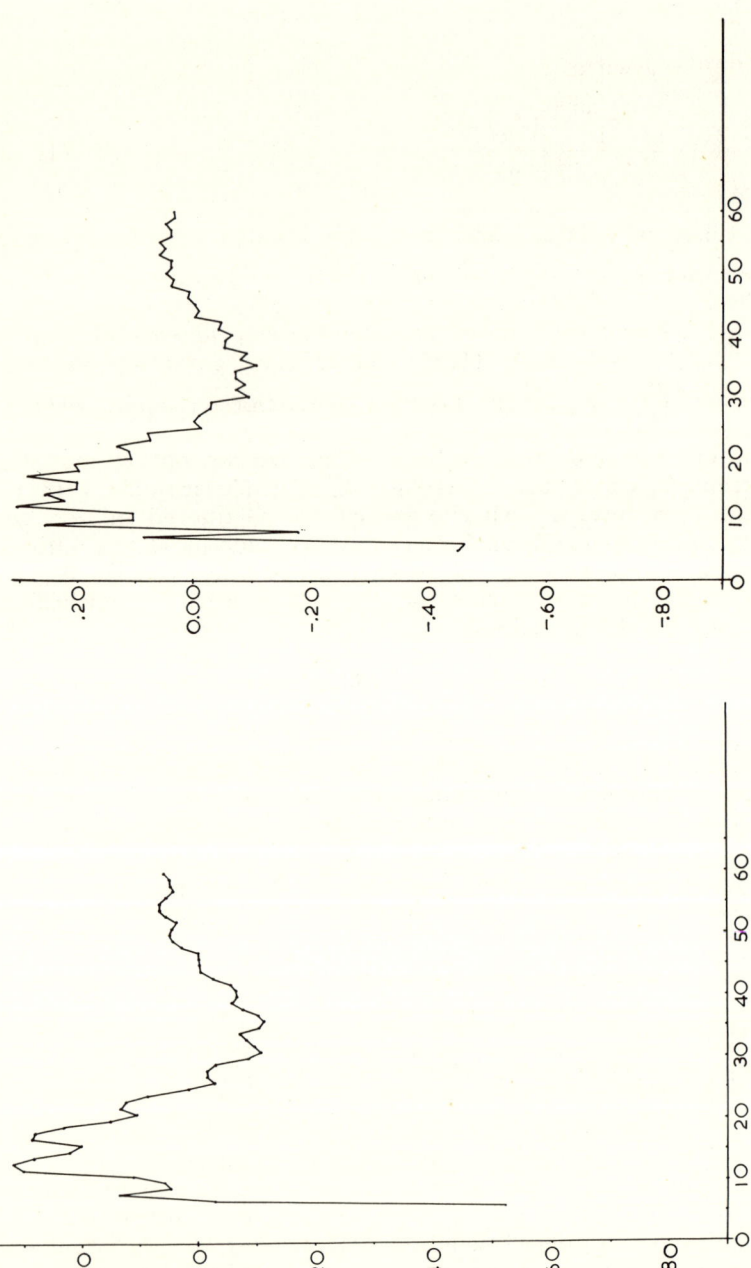

Figure 10. Convergence of Eq. (II.6) as a function of j.

Figure 9. Convergence of Eq. (II.1) as a function of j.

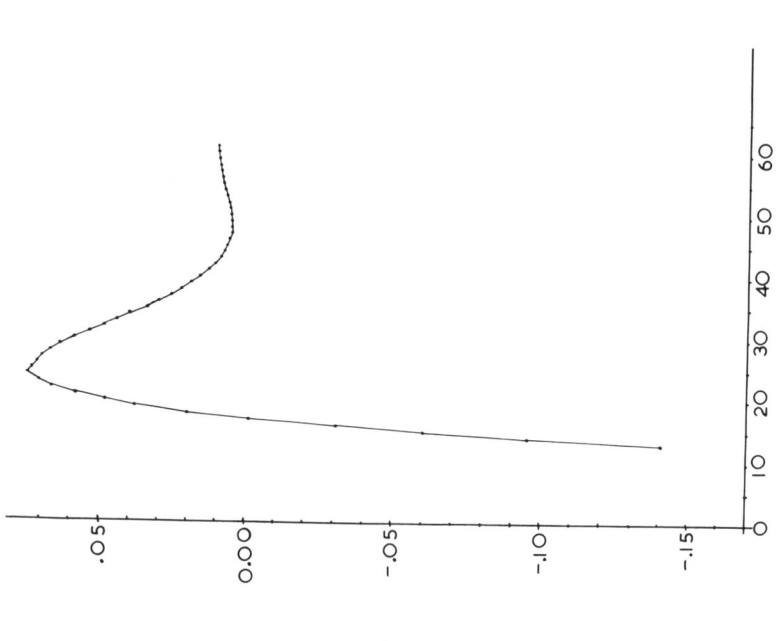

Figure 12. Convergence of Eq. (II.6) as a function of j, after the Lanczos's smoothing factor has been applied.

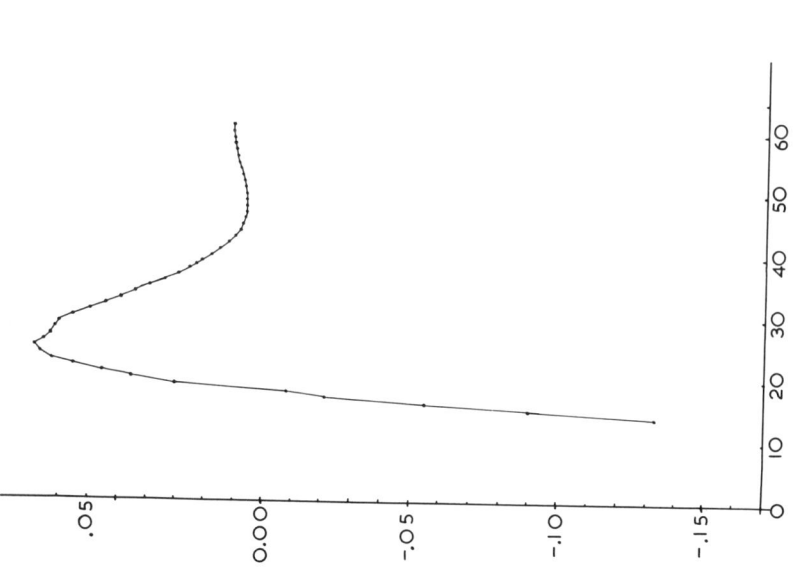

Figure 11. Convergence of Eq. (II.1) as a function of j, after the Lanczos's smoothing factor has been applied.

clearly manifested in the accompanying Figures 11 and 12.

All the numerical work, prepared for this communication, has been carried out by use of the CDC 7600 computing machine of the University of Manchester's Regional Computer Centre.

REFERENCES

Demircan, O.: 1978d, Astrophys. Space Sci., 59, p.313.
Demircan, O.: 1980, Astrophys. Space Sci., 67, p.375.
Erdélyi, A., Magnus, W., Oberhettinger, F. and Tricomi, F. G.: 1953, *Higher Transcendental Functions*, McGraw Hill Publ. Co., New York, vol.1.
Kopal, Z.: 1959, *Close Binary Systems*, Chapman-Hall and John Wiley, London and New York.
Kopal, Z.: 1975a, Astrophys. Space Sci., 34, p.431.
Kopal, Z.: 1975b, Astrophys. Space Sci., 35, p.159.
Kopal, Z.: 1977c, Astrophys. Space Sci., 51, p.439.
Kopal, Z.: 1979, *Language of the Stars*, D. Reidel Publ. Co., Dordrecht.
Lanczos, C.: 1956, *Applied Analysis*, Prentice-Hall, Inc., Englewood Cliffs, N.J.
Lanczos, C.: 1961, *Linear Differential Operators*, Van Nostrand, New York.
Tsesevich, V. P.: 1939, Bull. Astr. Inst. USSR Acad. Sci., No.45.
Tsesevich, V. P.: 1940, Bull. Astr. Inst. USSR Acad. Sci., No.50.

PHOTOMETRIC PERTURBATIONS OF CLOSE BINARIES IN THE FREQUENCY-DOMAIN

Helen Rovithis - Livaniou

Department of Astronomy, University of Athens,
Athens, Greece.

ABSTRACT

In the present work the way with which the photometric perturbations of close binary systems, whose components are distorted by axial rotation and mutual tidal action, can be found in the frequency-domain is represented and discussed.

1. INTRODUCTION

In the last years, a new method of analysis of the light changes of binary stars in the frequency-domain has been developed by Kopal and his collaborators, known as Fourier analysis. Not only all types of eclipses of spherical stars have been examined (Kopal, 1975 a, b, c, d) but also the light variations exhibited by close binaries whose components are distorted by axial rotation and mutual tidal action have been studied (Kopal, 1975 e; 1976b). In this latter case the observed light changes occur not only at the times of minima, but, extend over the whole cycle. Light changes arising either from tidal and rotational distortion of the two components, or from the "proximity" effects are independent of, and supplementary to, the light changes arising from eclipses, if our binary happens to be an eclipsing variable; moreover, their magnitudes increase with increasing proximity of the components. We want to separate one from the other and to find the elements of the system free from any systematic error. This can be done easily enough in the frequency-domain. Because, we have to find the moments A_{2m} (Kopal, 1975a), which in the case of distorted binary systems will consist of three parts (Kopal, 1976b)

$$A_{2m} = -m! \sum_{j=1}^{n} \frac{\Gamma(\frac{1}{2}j + 1)}{\Gamma(\frac{1}{2}j + m + 1)} c_j + \bar{A}_{2m} + \mathfrak{P}_{2m}, \qquad (1.1)$$

where A_{2m} are the observed moments and $\widehat{A}_{2m}$ the so-called "rectified" moments; that is the light changes which our system would exhibit if it was not distorted. The first and the third term of the right-hand side of Eq. (1.1) represent the "corrections" we must apply to the moments A_{2m} to find the $\bar{A}_{2m}$'s. The first term arises from the "proximity" effects, while the third represents the "photometric perturbations". Our task will be to see how one can find the $\mathfrak{P}_{2m}$'s. This will be done in the following two sections in which we shall only give the very general equations from which the photometric perturbations can be evaluated; because, their explicit forms for total, transit and partial eclipses have been given in three previous communications (Livaniou, 1977; Rovithis-Livaniou, 1977 and 1978; hereafter referred to as Paper I, II and III, respectively). Evaluating the $\mathfrak{P}_{2m}$'s we have limited ourselves to first order approximation. If we want to obtain only the three geometrical elements r_1, r_2 and i then three values of m - namely m = 1,2 and 3 - are enough. But, if we want to solve for more elements - e.g. for the limb-darkening coefficients - then more values of m are needed (as many as the unknowns). In such a case besides the complicated form of the photometric perturbations we shall have difficulties in the evaluation of the moments A_{2m} since, as it is known, the greater being the m the smaller becomes the areas A_{2m}.

2. ECLIPSE PERTURBATIONS IN THE FREQUENCY-DOMAIN

In the frequency-domain, if we want to find the terms which arise from mutual eclipses of rotationally and tidally distorted components of close binary systems, we have to evaluate (Kopal, 1975e) integrals of the form

$$\mathscr{L}_{2m} = mL_1 \csc^{2m} i \sum_{h=1}^{n+1} C^{(h)} \int_{\delta_0^2}^{\delta_1^2} (\delta^2 - \delta_0^2)^{m-1} \times \qquad (2.1)$$
$$\times \{\alpha_{n-1}^0 + f_*^{(h)} + f_1^{(h)} + f_2^{(h)}\} d\delta^2,$$

where, L_1 stands for the fractional loss of light of the star undergoing eclipse, i is the inclination of the eclipsing system to the celestial sphere, δ is the apparent separation of the centres of the two components, the α_n^m's denote the associated α-functions and the coefficients $C^{(h)}$ are due to the limb-darkening (h = 1 for uniformly bright discs, while h > 1 for limb-darkened ones). The limits of integration are two values of $\delta = \delta_1 = r_1 + r_2$ - where r_1, r_2 are the fractional radii of the two stars - and δ_0 = cosi.

The foregoing Eq. (2.1) can be rewritten as

$$U_{2m} = mL_1 \csc^{2m} i \sum_{h=1}^{n+1} C^{(h)} \int_{\delta_o}^{\delta_1^2} (\mathcal{E}^2 - \delta_o^2)^{m-1} \alpha_{n-1}^{o^2} d\delta^2 + \mathcal{P}_{2m}$$

(2.2)

where, the first part on the right-hand side of the foregoing Eq. (2.2) depends only on the α-functions and constitutes the photometric effects of the eclipses of spherical stars arbitrarily darkened at the limb - which have been studied by Kopal (1975a,b,c,d) - while the second stands for the photometric perturbations and it is obviously given by

$$\mathfrak{P}_{2m} = mL_1 \csc^{2m} i \sum_{h=1}^{n+1} C^{(h)} \int_{\delta_0^2}^{\delta_1^2} (\delta^2 - \delta_0^2)^{m-1} \{f_*^{(h)} + f_1^{(h)} + f_2^{(h)}\} d\delta^2$$

(2.3)

or

$$\mathcal{P}_{2m} = mL_1 \csc^{2m} i \sum_{h=1}^{n+1} C^{(h)} \sum_{l=0}^{m-1} \binom{m-1}{l} (-\delta_0^2)^{m-l-1} \mathfrak{P}_h^{(l)}$$

(2.4)

if we put

$$\mathfrak{P}_h^{(l)} = \int_{\delta_0^2}^{\delta_1^2} \delta^{2l} \{f_*^{(h)} + f_1^{(h)} + f_2^{(h)}\} d\delta^2.$$

(2.5)

From the foregoing f-functions, the $f_*^{(h)}$'s depend on the direction cosines n_o, n_1, n_2 and ℓ_o, ℓ_1, ℓ_2 as well as on α-functions and constitute the contributions of distortion and gravity-darkening over the circular portion of the eclipsed discs. On the other hand, the $f_1^{(h)}$ and $f_2^{(h)}$ depend on the $\mathcal{I}_{\pm 1, \gamma}^m$ and $I_{\pm 1, \gamma}^m$ integrals as well as on the direction cosines and represent the photometric contributions of the "boundary corrections" arising from the distortion of the primary and secondary component, respectively. The expressions of these functions being well known (cf., e.g. Kopal, 1959; 1975e) are not repeated here. We shall only add that everyone of their terms is multiplied either by $w_i^{(j)}$ or by $v_i^{(2)}$, $i = 1,2$; where both $w_i^{(j)}$ and $v_i^{(2)}$ are small quantities of first order and specify the amount of distortion: the $v_i^{(2)}$'s denote the polar flattening of the respective configuration, while the $w_i^{(j)}$'s are proportional to its equatorial ellipticity produced by the j-th partial tide.

Thus, if we limit ourselves to terms of first order approximation in f's, the equilibrium theory of tides signifies a restriction to the three first partial tides of seconds-, third- and fourth- harmonic symmetry. Their amplitudes are proportional to r_i^{j+1}, i = 1,2 where j denotes again the order of the respective harmonic. The apparent separation δ of the centres of the two components will be a quantity of the order of magnitude of r_1, r_2 within eclipses. Therefore, the products $\delta^m w_i^{(j)}$ should be of the order of $w_i^{(m+j)}$ and ignorable within the scheme of first order approximation for m + j > 4. As to the rotational distortion of the two components, which affects the light variation within eclipses only, in order to evaluate the photometric perturbations $\eta_{h,rot}^{(\ell)}$ causing by it we assume that the inclination i is not deviating much from 90°. Then, $v_i^{(2)}$, $n_0^2 \cos^2 i$ are small enough and can be regarded as negligible; while for the direction cosines n_0, n_1 and n_2 we have: $n_0^2 = n_2^2 = 0$; $n_1^2 = 1$.

Under these conditions, the photometric perturbations $\eta_{h,tid}^{(\ell)}$ coming from the three first tidal harmonics - and the $\eta_{h,rot}^{(\ell)}$ - coming from rotational distortion - can be expressed in terms of the $a_n^{(\ell)}$, $b_n^{(\ell)}$ and $I_{-1,0}^0$ integrals, of $\alpha_n^0(\delta_0)$ and of $(\partial \alpha_n^0 / \partial \delta^2) \delta_0$ for every type of eclipses as follows:

A. Uniformly-Bright Discs (h = n + 1 = 1)

A_1. Tidal Distortion.

Second-harmonic:

$$\mathfrak{P}_{1,t\alpha}^{(l)} = -\frac{1}{(l+1)} w_1^{(2)} \{6\tau_1 [a_2^{(l+1)} + b_2^{(l+1)} - \delta_0^{2(l+1)} \alpha_2^0(\delta_0)] -$$

$$- (2\tau_1 - l - 2)[a_0^{(l+1)} + b_0^{(l+1)}] + (2\tau_1 - 1)\delta_0^{2(l+1)} \alpha_0^0(\delta_0)\} +$$

$$+ \left(\frac{r_2}{r_1}\right)^2 [w_1^{(2)} - w_2^{(2)}] \int_{\delta_0^2}^{\delta_1^2} \delta^{2l} I_{-1,0}^0 \, d\delta^2. \qquad (2.6)$$

Third-harmonic:

$$\mathfrak{P}_{1,t\alpha}^{(l)} = -8\tau_1 w_1^{(3)} [a_3^{(l+1)} + b_3^{(l+1)}] - \frac{25}{2} \frac{\tau_1}{(l+1)} w_1^{(3)} [a_3^{(l+1)} +$$

$$+ b_3^{(l+1)} - \delta_0^{2(l+1)} \alpha_3^0(\delta_0)] +$$

$$+ \frac{15}{2} \frac{\tau_1}{(l+1)} w_1^{(3)} [a_1^{(l+1)} + b_1^{(l+1)} - \delta_0^{2(l+1)} \alpha_1^0(\delta_0)]. \qquad (2.7)$$

Fourth-harmonic:

$$\mathfrak{P}_{1,tid}^{(l)} = \frac{3}{2(l+2)} w_1^{(2)} \{6\tau_1 [a_2^{(l+2)} + b_2^{(l+2)} - \delta_0^{2(l+2)} \alpha_2^0(\delta_0)] -$$

$$- (2\tau_1 - 1)[a_0^{(l+2)} + b_0^{(l+2)} - \delta_0^{2(l+2)}\alpha_0^0(\delta_0)]\} -$$
$$- \tfrac{3}{8}w_1^{(4)}[4\tau_1(l + 2) + 50\tau_1][a_4^{(l+1)} + b_4^{(l+1)}] +$$
$$+ \tfrac{3}{2}w_1^{(4)}(2l + 3 + 5\tau_1)[a_2^{(l+1)} + b_2^{(l+1)}] -$$
$$- \tfrac{9}{4}w_1^{(4)}[a_0^{(l+1)} + b_0^{(l+1)}] + \frac{3}{4(l + 1)}w_1^{(4)}[a_0^{(l+1)} + b_0^{(l+1)} -$$
$$- \delta_0^{2(l+1)}\alpha_0^0(\delta_0)] - \frac{3\tau_1}{4(l + 1)}w_1^{(4)}\{35[a_4^{(l+1)} + b_4^{(l+1)} -$$
$$- \delta_0^{2(l+1)}\alpha_4^0(\delta_0)] - 30[a_2^{(l+1)} + b_2^{(l+1)} - \delta_0^{2(l+1)}\alpha_2^0(\delta_0)] +$$
$$+ 3[a_0^{(l+1)} + b_0^{(l+1)} - \delta_0^{2(l+1)}\alpha_0^0(\delta_0)]\} +$$
$$+ 3w_2^{(2)}[a_0^{(l+2)} + b_0^{(l+2)}] - \tfrac{3}{2}(2l + 3)r_1^2 w_2^{(2)}[a_2^{(l+1)} + b_2^{(l+1)}] -$$
$$- 3w_2^{(4)}[a_0^{(l+1)} + b_0^{(l+1)}] + \tfrac{3}{2}\tau_1 w_1^{(4)}\delta_0^{2(l+2)}\left(\frac{\partial \alpha_4^0}{\partial \delta^2}\right)_{\delta_0} -$$
$$- 3w_1^{(4)}\delta_0^{2(l+2)}\left(\frac{\partial \alpha_2^0}{\partial \delta^2}\right)_{\delta_0} + 3r_1^2 w_2^{(2)}\delta_0^{2(l+2)}\left(\frac{\partial \alpha_2^0}{\partial \delta^2}\right)_{\delta_0} -$$
$$- \frac{3}{4}\left(\frac{r_2}{r_1}\right)^2 [w_1^{(4)} - w_2^{(4)}] \int_{\delta_0^2}^{\delta_1^2} \delta^{2l} I_{-1,0}^0 \, d\delta^2, \tag{2.8}$$

where τ_1 stands for the gravity-darkening coefficient of the star undergoing eclipse.

A_2. Rotational Distortion

$$\mathfrak{P}_{1,\text{rot}}^{(l)} = -\frac{(\tau_1 - 1)}{(l + 1)} v_1^{(2)}[a_2^{(l+1)} + b_2^{(l+1)} - \delta_0^{2(l+1)}\alpha_2^0(\delta_0)] +$$
$$+ \frac{(\tau_1 - 2)}{3(l + 1)} v_1^{(2)}[a_0^{(l+1)} + b_0^{(l+1)} - \delta_0^{2(l+1)}\alpha_0^0(\delta_0)] -$$
$$- \tfrac{1}{2}(l + 1)(\tau_1 - 1)r_1^2 v_1^{(2)}[a_4^{(l)} + b_4^{(l)}] +$$
$$+ \tfrac{1}{2}(\tau_1 - 1)r_1^2 \delta_0^{2(l+1)} v_1^{(2)}\left(\frac{\partial \alpha_4^0}{\partial \delta^2}\right)_{\delta_0} +$$
$$+ \frac{1}{3}\left(\frac{r_2}{r_1}\right)^2 \left(\frac{3r_2^2 - r_1^2}{r_1^2}\right) v_1^{(2)} \int_{\delta_0^2}^{\delta_1^2} \delta^{2l} I_{-1,0}^0 \, d\delta^2 -$$
$$- \frac{1}{r_1^2}\left(\frac{r_2}{r_1}\right)^2 v_1^{(2)} \int_{\delta_0^2}^{\delta_1^2} \delta^{2(l+1)} I_{-1,0}^0 \, d\delta^2 -$$

$$-\frac{1}{6}\frac{(6r_2^2+7r_1^2)}{r_1^2}v_1^{(2)}[a_0^{(l+1)}+b_0^{(l+1)}]+\frac{1}{r_1^2}v_1^{(2)}[a_0^{(l+2)}+b_0^{(l+2)}]-$$

$$-\tfrac{1}{2}(r_2^2-r_1^2)v_1^{(2)}[a_0^{(l)}+b_0^{(l)}]+\tfrac{1}{4}(3r_2^2-5r_1^2)v_1^{(2)}[a_2^{(l)}+b_2^{(l)}]+$$

$$+\tfrac{9}{4}v_1^{(2)}[a_2^{(l+1)}+b_2^{(l+1)}]+\tfrac{5}{8}r_1^2v_1^{(2)}[a_4^{(l)}+b_4^{(l)}]-$$

$$-\frac{2}{3}\left(\frac{r_2}{r_1}\right)^2 v_2^{(2)}\int_{\delta_0^2}^{\delta_1^2}\delta^{2l}I_{-1,0}^0\,d\delta^2+v_2^{(2)}\Big\{[a_0^{(l+1)}+b_0^{(l+1)}]-$$

$$-\frac{(2l+1)}{2}r_1^2[a_2^{(l)}+b_2^{(l)}]+r_1^2\delta_0^{2(l+1)}\left(\frac{\partial\alpha_2^0}{\partial\delta^2}\right)_{\delta_0}\Big\}. \qquad (2.9)$$

B. Limb-Darkened Discs (h > 1)
B_1. Tidal Distortion

Second-harmonic:

$$\mathfrak{P}_{h,tid}^{(l)}=-\frac{1}{2(l+1)}w_1^{(2)}\{3(4\tau_1-h+1)[a_{h+1}^{(l+1)}+b_{h+1}^{(l+1)}-\delta_0^{2(l+1)}\times$$

$$\times\alpha_{h+1}^0(\delta_0)]-2(2\tau_1-h+1)[a_{h-1}^{(l+1)}+b_{h-1}^{(l+1)}-$$

$$-\delta_0^{2(l+1)}\alpha_{h-1}^0(\delta_0)]+(h-1)[a_{h-3}^{(l+1)}+b_{h-3}^{(l+1)}-\delta_0^{2(l+1)}\alpha_{h-3}^0(\delta_0)]\}-$$

$$-\frac{2}{(h+1)}w_2^{(2)}\Big\{\frac{(h+1)}{2}[a_{h-1}^{(l+1)}+b_{h-1}^{(l+1)}]-lr_1^2[a_{h+1}^{(l)}+b_{h+1}^{(l)}]+$$

$$+r_1^2\delta_0^{2(l+1)}\left(\frac{\partial\alpha_{h+1}^0}{\partial\delta^2}\right)_{\delta_0}\Big\}. \qquad (2.10)$$

Third-harmonic:

$$\mathfrak{P}_{h,tid}^{(l)}=-\frac{6(4\tau_1-h+1)}{(h+2)}w_1^{(3)}[a_{h+2}^{(l+1)}+b_{h+2}^{(l+1)}]-$$

$$-\frac{5(5\tau_1-2h+2)}{2(l+1)}w_1^{(3)}[a_{h+2}^{(l+1)}+b_{h+2}^{(l+1)}-\delta_0^{2(l+1)}\alpha_{h+2}^0(\delta_0)]+$$

$$+\frac{5(3\tau_1-2h+2)}{2(l+1)}w_1^{(3)}[a_h^{(l+1)}+b_h^{(l+1)}-\delta_0^{2(l+1)}\alpha_h^0(\delta_0)]. \qquad (2.11)$$

Fourth-harmonic:

$$\mathfrak{P}_{h,tid}^{(l)}=\frac{3}{4(l+2)}w_1^{(2)}\{3(4\tau_1-h+1)[a_{h+1}^{(l+2)}+b_{h+1}^{(l+2)}-\delta_0^{2(l+1)}\alpha_{h+1}^0(\delta_0)]-$$

$$- 2(2\tau_1 - h + 1)[a_{h-1}^{(l+2)} + b_{h-1}^{(l+2)} - \delta_0^{2(l+1)}\alpha_{h-1}^0(\delta_0)] +$$
$$+ (h - 1)[a_{h-3}^{(l+2)} + b_{h-3}^{(l+2)} - \delta_0^{2(l+2)}\alpha_{h-3}^0(\delta_0)]\} -$$
$$- \frac{3}{(h+1)(h+3)}[(l+2)(4\tau_1 - h + 1) + 5(h+1) \times$$
$$\times (5\tau_1 - 2h + 2)][a_{h+3}^{(l+1)} + b_{h+3}^{(l+1)}]w_1^{(4)} +$$
$$+ \frac{3}{(h+1)}[(l+2) + 5(\tau_1 - h + 1)]w_1^{(4)}[a_{h+1}^{(l+1)} + b_{h+1}^{(l+1)}] -$$
$$- 3w_1^{(4)}[a_{h-1}^{(l+1)} + b_{h-1}^{(l+1)}] - \frac{3}{8(l+1)}w_1^{(4)}\{35(2\tau_1 - h + 1) \times$$
$$\times [a_{h+3}^{(l+1)} + b_{h+3}^{(l+1)} - \delta_0^{2(l+1)}\alpha_{h+3}^0(\delta_0)] -$$
$$- 15(4\tau_1 - 3h + 3)[a_{h+1}^{(l+1)} + b_{h+1}^{(l+1)} - \delta_0^{2(l+1)}\alpha_{h+1}^0(\delta_0)] +$$
$$+ (4\tau_1 - 9h + 9)[a_{h-1}^{(l+1)} + b_{h-1}^{(l+1)} - \delta_0^{2(l+1)}\alpha_{h-1}^0(\delta_0)] -$$
$$- (h - 1)[a_{h-3}^{(l+1)} + b_{h-3}^{(l+1)} - \delta_0^{2(l+1)}\alpha_{h-3}^0(\delta_0)]\} +$$
$$+ 3w_2^{(2)}[a_{h-1}^{(l+2)} + b_{h-1}^{(l+2)}] - \frac{3(2l+3)}{(h+1)}r_1^2 w_2^{(2)}[a_{h+1}^{(l+1)} + b_{h+1}^{(l+1)}] -$$
$$- \tfrac{9}{4}w_2^{(4)}[a_{h-1}^{(l+1)} + b_{h-1}^{(l+1)}] - \frac{3l}{2(h+1)}r_1^2 w_2^{(4)}[a_{h+1}^{(l)} + b_{h+1}^{(l)}] +$$
$$+ \frac{3(4\tau_1 - h + 1)}{(h+1)(h+3)}w_1^{(4)}\delta_0^{2(l+2)}\left(\frac{\partial \alpha_{h+3}^0}{\partial \delta^2}\right)_{\delta_0} - \frac{3}{(h+1)}w_1^{(4)}\delta_0^{2(l+2)} \times$$
$$\times \left(\frac{\partial \alpha_{h+1}^0}{\partial \delta^2}\right)_{\delta_0} + \frac{6}{(h+1)}r_1^2 w_2^{(2)}\delta_0^{2(l+2)}\left(\frac{\partial \alpha_{h+1}^0}{\partial \delta^2}\right)_{\delta_0} +$$
$$+ \frac{3}{2}\frac{1}{(h+1)}r_1^2 w_2^{(4)}\delta_0^{2(l+1)}\left(\frac{\partial \alpha_{h+1}^0}{\partial \delta^2}\right)_{\delta_0}. \qquad (2.12)$$

B_2. Rotational Distortion

$$\mathfrak{P}_{h,\text{rot}}^{(l)} = -\frac{(2\tau_1 - h - 1)}{2(l+1)}v_1^{(2)}[a_{h+1}^{(l+1)} + b_{h+1}^{(l+1)} - \delta_0^{2(l+1)}\alpha_{h+1}^0(\delta_0)] +$$
$$+ \frac{(2\tau_1 - 3h - 1)}{6(l+1)}v_1^{(2)}[a_{h-1}^{(l+1)} + b_{h-1}^{(l+1)} - \delta_0^{2(l+1)}\alpha_{h-1}^0(\delta_0)] -$$
$$- \frac{2(2\tau_1 - h - 1)}{(h+1)(h+3)}r_1^2(l+1)v_1^{(2)}[a_{h+3}^{(l)} + b_{h+3}^{(l)}] +$$
$$+ \frac{2(2\tau_1 - h - 1)}{(h+1)(h+3)}r_1^2 v_1^{(2)}\delta_0^{2(l+1)}\left(\frac{\partial \alpha_{h+3}^0}{\partial \delta^2}\right)_{\delta_0} +$$
$$+ \frac{(h+4)}{2(h+3)}r_1^2 v_1^{(2)}[a_{h+3}^{(l)} + b_{h+3}^{(l)}] - \frac{1}{2(h+1)} \times$$
$$\times [(2h+3)r_1^2 - (h+2)r_2^2]v_1^{(2)}[a_{h+1}^{(l)} + b_{h+1}^{(l)}] +$$

$$+ \frac{3(h+2)}{2(h+1)} v_1^{(2)} [a_{h+1}^{(l+1)} + b_{h+1}^{(l+1)}] + \frac{1}{2}(r_1^2 - r_2^2) v_1^{(2)} \times$$

$$\times [a_{h-1}^{(l)} + b_{h-1}^{(l)}] - \frac{3}{2} v_1^{(2)} [a_{h-1}^{(l+1)} + b_{h-1}^{(l+1)}] - \frac{(h+1)}{6(l+1)} \times$$

$$\times \frac{(r_1^2 - 3r_2^2)}{r_1^2} v_1^{(2)} [a_{h-1}^{(l+1)} + b_{h-1}^{(l+1)} - \delta_0^{2(l+1)} \alpha_{h-1}^0(\delta_0)] -$$

$$- \frac{(h+1)}{2(l+2)} \frac{1}{r_1^2} v_1^{(2)} [a_{h-1}^{(l+2)} + b_{h-1}^{(l+2)} - \delta_0^{2(l+2)} \alpha_{h-1}^0(\delta_0)] +$$

$$+ \frac{(h-1)}{6(l+1)} \frac{(r_1^2 - 3r_2^2)}{r_1^2} v_1^{(2)} [a_{h-3}^{(l+1)} + b_{h-3}^{(l+1)} - \delta_0^{2(l+1)} \alpha_{h-3}^0(\delta_0)] +$$

$$+ \frac{(h-1)}{2(l+2)} \frac{1}{r_1^2} v_1^{(2)} [a_{h-3}^{(l+2)} + b_{h-3}^{(l+2)} - \delta_0^{2(l+2)} \alpha_{h-3}^0(\delta_0)] +$$

$$- \frac{1}{3(h+1)} v_2^{(2)} \Big\{ (h+1)[a_{h-1}^{(l+1)} + b_{h-1}^{(l+1)}] - (2l+3)r_1^2 \times$$

$$\times [a_{h+1}^{(l)} + b_{h+1}^{(l)}] + 2r_1^2 \delta_0^{2(l+1)} \left(\frac{\partial \alpha_{h+1}^0}{\partial \delta^2} \right)_{\delta_0} \Big\}. \qquad (2.13)$$

Of course, this is a very general way of expressing the photometric perturbations due to both tidal and rotational distortion and it stands for every type of eclipses; but, as we know the $\alpha_n^0(\delta_0)$'s, the $(\partial \alpha_n^0/\partial \delta^2)_{\delta_0}$'s and the $a_n^{(\ell)}$, $b_n^{(\ell)}$ and $I_{1\,0}^0$ integrals have different expressions for every kind of eclipses: thus, we have to examine everyone of them separately. This will be done in the following Section.

3. EVALUATION OF THE PHOTOMETRIC PERTURBATIONS

As was shown before, the photometric perturbations arising from tidal, $\mathcal{P}_{h,tid}^{(\ell)}$, and rotational, $\mathcal{P}_{h,rot}^{(\ell)}$, distortion of the two components of a close eclipsing binary are expressed by Eqs. (2.6) - (2.13) of Section 2 in a very general way. From the $\mathcal{P}_h^{(\ell)}$'s the perturbations $\mathcal{P}_{2m} = \mathcal{P}_{2m,tid} + \mathcal{P}_{2m,rot}$ can be found, where

$$\mathcal{P}_{2m,tid} = mL_1 \csc^{2m} i \sum_{h=1}^{n+1} C^{(h)} \sum_{l=0}^{m+1} \binom{m-1}{l} (-\delta_0^2)^{m-l-1} \mathcal{P}_{h,tid}^{(l)}$$

and
$$\qquad (3.1)$$

$$\mathcal{P}_{2m,rot} = mL_1 \sum_{h=1}^{n+1} C^{(h)} \mathcal{P}_{h,rot}^{(m-1)}.$$
$$\qquad (3.2)$$

Moreover, for uniformly bright discs $n = 0$ and $C^{(1)} = 1$. Thus, the foregoing Eqs. (3.1) and (3.2) simply become

$$^{U}\mathfrak{P}_{2m,\text{tid}} = mL_1 \csc^{2m} iC^{(1)} \sum_{l=0}^{m-1} \binom{m-1}{l}(-\delta_0^2)^{m-l-1}\mathfrak{P}_{1,\text{tid}}^{(l)}, \qquad (3.3)$$

and

$$^{U}\mathfrak{P}_{2m,\text{rot}} = mL_1 \mathfrak{P}_{1,\text{rot}}^{(l)}, \qquad (3.4)$$

respectively.

For linear limb-darkening n = 1 and the coefficients $C^{(1)}$ and $C^{(2)}$ are related to the limb-darkening coefficient u_1 of the star undergoing eclipse by

$$C^{(1)} = \frac{3(1-u_1)}{3-u_1}, \qquad C^{(2)} = \frac{3u_1}{3-u_1}. \qquad (3.5)$$

Thus, Eqs. (3.1) and (3.2) by virtue of Eq. (3.5) become, respectively

$$^{L}\mathfrak{P}_{2m,\text{tid}} = \frac{3(1-u_1)}{3-u_1}{}^{U}\mathfrak{P}_{2m,\text{tid}} +$$

$$+ \frac{3u_1}{3-u_1} mL_1 \csc^{2m} i \sum_{l=0}^{m-1}\binom{m-1}{l}(-\delta_0^2)^{m-l-1}\mathfrak{P}_{2,\text{tid}}^{(l)}; \qquad (3.6)$$

and

$$^{L}\mathfrak{P}_{2m,\text{rot}} = \frac{3(1-u_1)}{3-u_1}{}^{U}\mathfrak{P}_{2m,\text{rot}} + \frac{3mL_1 u_1}{3-u_1}\mathfrak{P}_{2,\text{rot}}^{(m-1)}, \qquad (3.7)$$

Finally, for quadratic limb-darkening n = 2 and the coefficients $C^{(1)}$, $C^{(2)}$ and $C^{(3)}$ are related to the limb-darkening coefficients u_1, u_2 of the star undergoing eclipse by

$$C^{(1)} = \frac{6(1-u_1-u_2)}{6-2u_1-3u_2},$$

$$C^{(2)} = \frac{6u_1}{6-2u_1-3u_2},$$

$$C^{(3)} = \frac{6u_2}{6-2u_1-3u_2}. \qquad (3.8)$$

Thus, Eqs. (3.1) and (3.2) by virtue of (3.8) become, respectively

$$^Q\mathfrak{P}_{2m,\text{tid}} = \frac{2(3 - u_1)}{6 - 2u_1 - 3u_2} {}^L\mathfrak{P}_{2m,\text{tid}} - \frac{6mu_2}{6 - 2u_1 - 3u_2} L_1 \csc^{2m} i \times$$

$$\times \left[\sum_{l=0}^{m-1} \binom{m-1}{l} (-\delta_0^2)^{m-l-1} \mathfrak{P}_{1,\text{tid}}^{(l)} - \right.$$

$$\left. - \sum_{l=0}^{m-1} \binom{m-1}{l} (-\delta_0^2)^{m-l-1} \mathfrak{P}_{3,\text{tid}}^{(l)} \right] \tag{3.9}$$

and

$$^Q\mathfrak{P}_{2m,\text{rot}} = \frac{2(3 - u_1)}{6 - 2u_1 - 3u_2} {}^L\mathfrak{P}_{2m,\text{rot}}$$

$$- \frac{6m\, u_2}{6 - 2u_1 - 3u_2} L_1 \{\mathfrak{P}_{1,\text{rot}}^{(m-1)} - \mathfrak{P}_{3,\text{rot}}^{(m-3)}\}$$

$$\tag{3.10}$$

Thus, we can evaluate the photometric perturbations for linear limb-darkening in terms of those for uniformly bright discs - which are rather simple - and those of higher order for limb-darkening in terms of linear, quadratic and so on.

A. OCCULTATION ECLIPSES

In the case of occultation eclipses terminating in totality the $b_n^{(l)}$ integrals and the partial derivatives $(\partial \alpha_n^0 / \partial \delta^2)_{\delta^0}$ are equal to zero. The $a_n^{(l)}$ integrals can be expressed (Kopal, 1975b) in a closed form as

$$a_n^{(l)} = - \int_{r_2-r_1}^{r_2+r_1} \delta^{2l} \left(\frac{\partial \alpha_n^0}{\partial \delta} \right)_p d\delta = r_2^{2l} a_n^{(0)} F\left(-l, 1-l, \frac{n+4}{2}; k^2 \right), \tag{3.11}$$

where

$$a_n^{(0)} = \frac{2}{n+2} \quad \text{and} \quad k = \frac{r_1}{r_2}; \tag{3.12}$$

while the integration of $I_{-1,0}^0$ integrals between δ_0^2 and gives δ_1^2 gives (Kopal, 1975e)

$$\int_{\delta_0^2}^{\delta_1^2} \delta^{2l} I_{-1,0}^0 \, d\delta^2 = r_2^{2(l+1)} k^2 F(-l, -l, 2; k^2). \tag{3.13}$$

Moreover, the hypergeometric series on the right-hand side of Eqs. (3.11) and (3.13) reduce to polynomials, since l is an integer. Thus, in this case, both $_{oc.}\mathcal{P}_{h,tid}^{(l)}$ and $_{oc.}\mathcal{P}_{h,rot}^{(l)}$ have a simple and accurate algebraic expression for every value of h and l (cf. Eqs. (3.4) - (3.13) and (3.15) - (3.23) of Paper I).

B. TRANSIT ECLIPSES

In the case of transit eclipses terminating in an annular phase the $a_n^{(l)}$ and $b_n^{(l)}$ integrals are defined as

$$a_n^{(l)} = -\int_{r_1-r_2}^{r_1+r_2} \delta^{2l} \left(\frac{\partial \alpha_n^0}{\partial \delta}\right)_p d\delta, \quad b_n^{(l)} = -\int_{\delta_0}^{r_1-r_2} \delta^{2l} \left(\frac{\partial \alpha_n^0}{\partial \delta}\right)_{an} d\delta. \tag{3.14}$$

and their sum is generally expressed (Kopal, 1976a; Kurutac, 1976) by

$$a_n^{(l)} + b_n^{(l)} = [a_n^{(l)}]_{oc} - \frac{n}{2}(kq)^2 \cos^{2l} i (1-k^2)^{(n-2)/2} \times$$
$$\times S[(g)_j, j!(j+l+1)], \tag{3.15}$$

where we have abbreviated

$$S[(g)_j, j!(j+l+1)] \equiv$$
$$\equiv \sum_{j=0}^{\infty} (g)_j \frac{1}{j!(j+l+1)} \left(\frac{q}{1-k^2}\right)^{2j} F\left(-j, \frac{n+2}{2} - j, 2; k^2\right) \tag{3.16}$$

and

$$g = \frac{2-n}{2}, \quad (g)_j = \frac{\Gamma(g+j)}{\Gamma(g)}; \tag{3.17}$$

and where, for transit eclipses,

$$k = \frac{r_2}{r_1}, \quad q = \frac{\cos i}{r_1} \quad \text{and} \quad k + q \le 1. \tag{3.18}$$

Moreover,

$$[a_n^{(l)}]_{oc} = \begin{cases} k^2 F\left(-\dfrac{n}{2}, 1, 2; k^2\right) & \text{for } l = 0 \\[2ex] \dfrac{2}{n+2} r_2^{2l} F\left(-l, 1-l, \dfrac{n+4}{2}; k^{-2}\right) & \text{for } l > 0 \end{cases} \quad (3.19)$$

Thus, the sums of $\{a_n^{(l)} + b_n^{(l)}\}_{tr}$ and $\{a_n^{(l)} + b_n^{(l)} - \delta_o^{2l} \alpha_n^0(\delta_o)\}_{tr}$ for transit eclipses can be generally expressed in terms of the $\{a_n^{(l)} + b_n^{(l)}\}_{oc}$ and $\{a_n^{(l)} + b_n^{(l)} - \delta_o^{2l} \alpha_n^0(\delta_o)\}_{oc}$ for occultation eclipses. Furthermore, if n is an even integer, the infinite series on the right-hand side of Eq. (3.16) reduce to polynomials and thus the sums $\{a_n^{(l)} + b_n^{(l)}\}_{tr}$ and $\{a_n^{(l)} + b_n^{(l)} - \delta_o^{2l} \alpha_n^0(\delta_o)\}_{tr}$ can be expressed in a closed form; if, moreover, $l = 0$ they become even simpler. So, for uniformly bright discs (n = 0) as well as for quadratic limb-darkening (n = 2) the terms arising from rotational distortion and from the second and fourth tidal harmonics are expressed in a closed form and only the terms coming from the third tidal harmonic contain infinite series. The opposite is true for linear limb-darkening (n = 1).

Furthermore, integration of the $I_{-1,0}^0$ integrals gives (Kopal, 1975e)

$$\int_{\delta_0^2}^{\delta_1^2} \delta^{2l} I_{-1,0}^0 \, d\delta^2 = \frac{1}{(l+1)} r_1^{2(l+1)} \{F(-l, -l-1, 1; k^2) - \cos^{2(l+1)} i\} \quad (3.20)$$

while for the partial derivatives $(\partial \alpha_n^0 / \partial \delta^2) \delta_0$ we have that for any value of $n \geq 1$

$$\left(\frac{\partial \alpha_n^0}{\partial \delta^2}\right)_{\delta_0} = -\frac{n}{2} \frac{1}{r_1^2} k^2 [1 - (q-k)^2]^{(n-2)/2} F\left(\frac{2-n}{2}, \frac{3}{2}, 3; \kappa_0^2\right) \quad (3.21)$$

where the modulus κ_0^2 is given by

$$\kappa_0^2 = \frac{r_1^2 - (\delta_0 - r_2)^2}{4 \delta_0 r_1} \quad (3.22)$$

We notice that the hypergeometric series on the right-hand side of Eq. (3.21) reduce to polynomials for even values of n.

This being the situation, then for transit eclipses terminating in an annular phase we manage to express both $\text{tr} \mathcal{L}_{h,\text{tid}}^{(\ell)}$ and $\text{tr} \mathcal{L}_{h,\text{rot}}^{(\ell)}$ in terms of $\{\mathcal{L}_{h,\text{tid}}^{(\ell)}\}_{oc}$ and $\{\mathcal{L}_{h,\text{rot}}^{(\ell)}\}_{oc}$, respectively and some additional terms. (Eqs. (4.23), (4.25), (4.27), (4.29), (4.31), (4.33) and (4.35) - (4.37) of Paper I for the $\text{tr}\mathcal{L}_{h,\text{tid}}^{(\ell)}$ and Eqs. (4.38) - (4.41) and (4.43) - (4.47) for the $\text{tr}\mathcal{L}_{h,\text{rot}}^{(\ell)}$). In the additional terms there are 28 different infinite series of the form of Eq. (3.16) depending on two variables - namely, $\kappa = \frac{r_2}{r_1}$ and $q = \frac{\cos i}{r_1}$ - and 5 hypergeometric series of argument κ^2. The 28 infinite series have been computed (Livaniou, 1976) for 10 values of κ, 0.0(0.1)0.9 and 100 values of q, 0.01(0.01)1.0 having in mind that for transit eclipses $\kappa+q \leq 1$. Moreover, in Paper I the 28 infinite series are represented graphically for all values of κ but for only three values of q (0.1, 0.2, 0.3), while the 5 hypergeometric series have been computed for 11 values of the argument κ^2, 0.0(0.1)1.0 and are given in Table I of the same paper.

C. PARTIAL ECLIPSES

In the case of partially eclipsing systems the $b_n^{(\ell)}$ integrals are zero, while the $a_n^{(\ell)}$'s are defined as

$$a_n^{(l)} = -\int_{\delta_0}^{\delta_1} \delta^{2l} \left(\frac{\partial \alpha_n^0}{\partial \delta}\right)_p d\delta \tag{3.23}$$

where the apparent separation δ of the two stars centres obeys the inequality

$$\delta_1 = r_1 + r_2 \geq \delta \geq \delta_0 = \cos i \geq \delta_2 = |r_1 - r_2|. \tag{3.24}$$

Two values of $\ell > 0$ are sufficient for the determination of the three geometrical elements r_1, r_2 and i, if two functions of the family of $a_n^{(\ell)}$'s - say $a_0^{(0)} = \alpha_0^0$ and $a_1^{(6)} = \alpha_1^0$ - can be regarded as known, since they are available in tabular form (Tsesevich, 1939; 1940). This is true if the two components can be regarded as spheres, but it is not so in the case of those binary systems whose components are distorted. In this case, as is immediately obvious from Eqs. (2.6) - (2.13) of Section 2, it is necessary to know the $a_n^{(\ell)}$ integrals for values of ℓ up to 4 and for $-1 \leq n \leq 6$. This can be done easily enough (e.g. Paper II) since for even values of $n \geq$ we have that

$$a_n^{(0)} = \frac{2}{n+2} \{\alpha_0^0 - [2(kq)^2 \mathfrak{J}_{1,0}^0 + 4(kq)^3 \mathfrak{J}_{1,2}^0 + \cdots + n(kq)^{(n+2)/2} \mathfrak{J}_{1,n-2}^0] +$$
$$+ k^2[(kq)\mathfrak{J}_{-1,2}^0 + (kq)^2 \mathfrak{J}_{-1,4}^0 + \cdots + (kq)^{n/2} \mathfrak{J}_{-1,n}^0]\};$$
(3.25)

while for odd values of $n \geq 3$

$$a_n^{(0)} = \frac{1}{n+2} \{3\alpha_1^0 - 2[3(kq)^{5/2} \mathfrak{J}_{1,1}^0 + 5(kq)^{7/2} \mathfrak{J}_{1,3}^0 + \cdots + n(kq)^{(n+2)/2} \mathfrak{J}_{1,n-2}^0] +$$
$$+ 2k^2[(kq)^{3/2} \mathfrak{J}_{-1,3}^0 + (kq)^{5/2} \mathfrak{J}_{-1,5}^0 + \cdots + (kq)^{n/2} \mathfrak{J}_{-1,n}^0]\}$$
(3.26)

where

$$k = \frac{r_2}{r_1}, \quad q = \frac{\cos i}{r_1} = \frac{\delta_0}{r_1};$$
(3.27)

regardless of whether $r_1 \gtrless r_2$. Thus, the ratio of the two radii can in this case assume any value between 0 and ∞.

Moreover, Kopal (1975d) has shown that the $a_n^{(\ell)}$ integrals for partial eclipses can be found, for every value of $n \geq -1$ and $\ell \geq 0$, with the aid of the recursion relation

$$(n + 4l + 2)a_n^{(l)} - n[r_1^2 a_{n-2}^{(l+1)} + (1 - k^2)a_{n-2}^{(l)}] =$$
$$= 2\delta_0^{2l}[(n+2)a_n^{(0)} - na_{n-2}^{(0)} - 2k^{(n+4)/2}q^{n/2}\mathfrak{J}_{-1,n}^0(\delta_0)];$$
(3.28)

while the α_n^0 functions, for $n > 0$, obey the recursion relation (Kopal, 1959; Section IV.5)

$$(n+2)\alpha_n^o - n\alpha_{n-2}^o = \delta\left(\frac{\partial \alpha_n^o}{\partial \delta}\right)_p + 2\left(\frac{r_2}{r_1}\right)^{n+2}\left(\frac{\delta}{r_2}\right)^{n/2} \mathfrak{J}_{-1,n} \cdot$$
(3.29)

The $\mathfrak{J}_{\pm 1,n}^o$'s belong to a family of integrals $\mathfrak{J}_{\beta,\gamma}^m$, which has been extensively investigated by Kopal (1947) and are available in tabular form (Kopal, 1947; Demircan, 1976); so they can be regarded as known.

Furthermore, the $\partial \alpha_n^o / \partial \delta^2$ δo for partial eclipses take the form

$$\left(\frac{\partial \alpha_n^0}{\partial \delta^2}\right)_{\delta_0} = -\frac{1}{2\pi r_1^2} \frac{k^{1/2}}{q^{3/2}} B\left(\frac{1}{2}, \frac{n+2}{2}\right)[1-(q-k)^2]^{(n+1)/2} \times$$

$$\times F\left(-\frac{1}{2}, \frac{3}{2}, \frac{n+3}{2}; \kappa_0^2\right),$$

(3.30)

where the modulus κ^2 continues to be given by Eq. (3.22); while the integration of the $I_{-1,0}^0$ integrals gives

$$I_{part}^{(l)} \equiv \int_{\delta_0^2}^{\delta_1^2} \delta^{2l} I_{-1,0}^0 \, d\delta^2 =$$

$$= \frac{r_1}{(l+1)\pi} \int_{\varphi_0}^{\pi} (r_1^2 - 2r_1 r_2 \cos\varphi + r_2^2)^l (r_1 - r_2 \cos\varphi) \, d\varphi -$$

$$- \frac{\delta_0^{2(l+1)}}{(l+1)\pi} \varphi_2,$$

(3.31)

where the angles φ_0 and φ_2 are

$$\varphi_0 = \cos^{-1} \frac{r_1^2 + r_2^2 - \delta_0^2}{2r_1 r_2},$$

(3.32)

$$\varphi_2 = \cos^{-1} \frac{r_2^2 - r_1^2 + \delta_0^2}{2\delta_0 r_2},$$

(3.33)

such that

$$\frac{\sin\varphi_0}{\sin\varphi_2} = \frac{\delta_0}{r_1} = \frac{\cos i}{r_1}$$

(3.34)

Under these conditions, we can express the $\mathcal{L}_{h,tid}^{(\ell)}$ in terms of $\mathcal{J}_{\pm 1,n}^{0}$ and $I_{part}^{(\ell)}$ integrals, of α_0^0 and α_1^0 of the associated α_n^m functions and of hypergeometric series of the form $F(-\frac{1}{2}, \frac{3}{2}, \frac{n+3}{2}; \kappa_0^2)$ (cf. Papers II and III). Evaluating the $\mathcal{L}_{2m,tid}^{(\ell)}$ from the $\mathcal{L}_{h,tid}^{(\ell)}$ — we notice that for m = 2 and 3 and independently of the limb-darkening low all the foregoing hypergeometric series add up to zero. Unfortunately this is not so for m = 1 as well as for the photometric perturbations arising from the rotational distortion of the two components.

4. DISCUSSION

As we show in the previous Sections the calculation of the photometric perturbations in the frequency-domain is not too difficult but their explicit forms have a simple algebraic expression only in the case of occultation eclipses terminating in totality. In the case of transit or partial eclipses their expressions become complicated and, moreover, we have to use Tables for their evaluation. Tables which give the 5 hypergeometric and 28 infinite series in the first case or the α_o^o and α_1^o functions as well as the $J_{\pm1,n}^o$ integrals in the latter.

In spite of these, applications have been made to many particular systems = cf. Tsouroplis (1977) to RW Tam and U Sge, Caracatsanis (1977) to Algol taking into account only third harmonic terms for tidal distortion and Niarchos (1978) to 14 W Ursae Majoris-type stars.

In order to do things easier for practical use, Edalati and Budding (1978) made numerical evaluation of Eq. (2.3) by trapezoidal quadrature and set up tables of values of the photometric perturbations for adopted values of the mass-ratio $\frac{m_2}{m_1} = 1$), limg-darkening (n = 0.6) and gravity-darkening ($\tau = 1$).

Meanwhile, Kopal (1977a,b) had expressed the basic α_o^o and α_1^o of the associated α^m functions in terms of Haykel transforms and later (Kopal, 1978) he did so for the integrals of the form $J_{\pm1,\gamma}^m$ — which constitute the "boundary corrections" to the light curves — as well as for the expressions which are used in the evaluation of the respective terms of the perturbations ℓ_{2m} for any type of eclipses. Alkan and Edalati (1978) extended this work and evaluated the ℓ_{2m}'s in terms of convergent series very similar to those in which the A_{2m}'s had been expressed and what is more important for non-integral values of m.

Finally, it should be added that although we can express the photometric perturbations for linear limb-darkening in terms of those for uniformly bright discs — Eqs. (3.6) and (3.7) — and these for quadratic in terms of those for linear — Eqs. (3.9) and (3.10) — we have used this property only in the case of partially eclipsing systems. Then, as we show, evaluating the $\ell_{2m,tid}$, coming from tidal distortion, for $m = 2$ and $m = 3$ all the hypergeometric series of the form $F(-\frac{1}{2}, \frac{3}{2}, \frac{n+3}{2}; \kappa_o^2)$ were added up to zero. Thus, it looks worthwhile to do so for transit eclipses; then, it seems possible some of the series to concel each other. As to the case of occultation eclipses terminating in totality both the $\ell_{oc\,h,tid}^{(\ell)}$ and $\ell_{oc\,h,rot}^{(\ell)}$ have a very simple and accurate algebraic expression in all cases and the use of the foregoing property will be good only for practical use.

REFERENCES

Alkan, H. and Edalati, M. T.: 1978, Astrophys. Space Sci. 59, p. 431.
Caracatsanis, V. A.: 1977, M.Sc. Thesis, Manchester University.
Demircan, O.: 1977, Astrophys. Space Sci. 47, p. 459.
Edalati, M. T. and Budding, E.: 1978, Astrophys. Space Sci. 57, p. 181.
Kopal, Z.: 1947, Harvard Obs. Circ., No. 450.
Kopal, Z.: 1959, *Close Binary Systems*, Chapman-Hill and John Wiley, London and New York.
Kopal, Z.: 1975a, Astrophys. Space Sci. 34, p. 431.
Kopal, Z.: 1975b, Astrophys. Space Sci. 35, p. 159.
Kopal, Z.: 1975c, Astrophys. Space Sci. 35, p. 171.
Kopal, Z.: 1975d, Astrophys. Space Sci. 36, p. 227.
Kopal, Z.: 1975e, Astrophys. Space Sci. 38, p. 191.
Kopal, Z.: 1976a, Astrophys. Space Sci. 40, p. 461.
Kopal, Z.: 1976b, Astrophys. Space Sci. 45, p. 269.
Kopal, Z.: 1977a, Astrophys. Space Sci. 50, p. 225.
Kopal, Z.: 1977b, Astrophys. Space Sci. 51, p. 439.
Kopal, Z.: 1978, Astrophys. Space Sci. 57, p. 439.
Kurutac, M.: 1976, Ph.D. Thesis, Manchester University.
Livaniou, H.: 1976, Ph.D. Thesis, Manchester University.
Livaniou, H.: 1977, Astrophys. Space Sci. 51, p. 77 (Paper I).
Niarchos, P.: 1978, Astrophys. Space Sci. 58, p. 301.
Rovithis-Livaniou, H.: 1977, Astrophys. Space Sci. 52, p. 271 (Paper II).
Rovithis-Livaniou, H.: 1978, Astrophys. Space Sci. 59, p. 463 (Paper III).
Tsesevich, V. P.: 1939, Bull. Astro. Inst. U.S.S.R. Acad. Sci., No. 45.
Tsesevich, V. P.: 1940, Bull. Astro. Inst. U.S.S.R. Acad. Sci., No. 50.
Tsouroplis, A. G.: 1977, Astrophys. Space Sci. 47, p. 361.

AN ANALYSIS OF THE LIGHT CHANGES OF W UMa-TYPE SYSTEMS IN THE FREQUENCY DOMAIN

P. G. Niarchos

Department of Astronomy, University of Athens, Greece

ABSTRACT

The analysis of the light curves of 30 well observed W UMa-type systems by Kopal's method, using frequency domain techniques, is presented. New geometric and photometric elements are derived and the problem, according to our results, of whether these systems are contact or not is considered.

1. INTRODUCTION

The close eclipsing binary systems of W UMa-type have been the subject of many investigations in the past and continue to pose a challenge in the general framework of stellar evolution. They represent the commonest type of binary systems in the space around us and they are among the most interesting objects to test empirically the theories of the internal structure of stars of low mass. The close vicinity of the two stars causes a remarkable deformation of both components, due to tidal interaction and to rotational flattening as a consequence of the orbital motion around the common centre of gravity. It is not clear whether the components of W UMa-type stars are in physical contact with the photospheres touching each others.

The contributions of two authors, within the last years, were especially important for the understanding of W UMa-type stars. Eggen (1967) was able to show the existence of a period-colour relation which is a proof of the close vicinity of the two components, independent of the derivation of photometric elements. There is no need, however, for physical contact of the two stars. Lucy

(1968a,b) succeeded in demonstrating that contact systems of the two zero-age Main Sequence stars with different masses are theoretically possible and, in certain cases, hydrodynamically stable. Lucy's ideas were subsequently developed further by several investigators (Hazlehurst, 1970; Moss and Whelan, 1970; Moss, 1971, 1972, and others); but a satisfactory agreement between theory and all aspects of observational evidence has not yet been attained. In any case the structure and evolution of contact binary systems is a major unsolved problem in close double star theory.

2. PHOTOELECTRIC LIGHT CURVES USED

The light curves of 30 well observed systems, listed in Table I, are analysed and studied. Most of the information about these systems has been taken from the papers of Eggen (1967), Batten (1967) and Binnendijk (1970). The references in the last column of Table I contain the light curves of the systems which have been used in the present analysis.

TABLE I

W UMa-type Systems

System	Spectra	Period	References
AB And	G5	0.3319	Binnendijk (1959)
V 535 Ara			
(≡BV 419)	A3	0.6293	Chambliss (1967)
OO Aql	G5	0.5068	Binnendijk (1968)
AC Boo	F0	0.3524	Binnendijk (1965)
TY Boo	G3-G7	0.3171	Carr (1972)
VW Boo	G5	0.3422	Binnendijk (1973)
XY Boo	F8	0.3705	Binnendijk (1971)
CW Cas	G5	0.3189	Broglia (1964)
RR Cen	F2	0.6057	Knipe (1961)
VW Cep	G5-K1	0.2783	Kwee (1966)
CC Com	K5	0.2206	Rucinski (1976)
RZ Com	K0	0.3385	Broglia (1960)
DK Cyg	A2-F2	0.4707	Binnendijk (1964)
1073 Cyg			
(≡BV 342)	A3-F0	0.7859	Kondo (1966)
UX Eri	G0	0.4453	Binnendijk (1967)
YY Eri	G5	0.3215	Purgathofer (1960)
AK Her	F2-F8	0.4215	Binnendijk (1961)
EM Lac	G5	0.3891	Broglia and Conconi (1974)
SW Lac	G3	0.3207	Bookmyer (1965)
AM Leo	F8	0.3658	Binnendijk (1969b)

Table I (continued)

UZ Leo	A7	0.6180	Binnendijk (1972)
V502 Oph	G2	0.4534	Binnendijk (1969a)
V566 Oph	F4-F5	0.4096	Bookyer (1969)
V839 Oph	G0	0.4090	Binnendijk)1960a)
ER Ori	G2	0.4234	Binnendijk (1962)
U Peg	F3	0.3748	Binnendijk (1960b)
AE Pho	G0	0.3623	Williamon (1975)
RZ Tau	F0	0.4157	Binnendijk (1963)
W UMa	F8	0.3336	Binnendijk (1966)
AH Vir	K0	0.4075	Binnendijk (1960c)

3. EQUATIONS OF THE PROBLEM

The first step for the determination of the elements of distorted systems in the frequency-domain is to evaluate the 'empirical moments' (cf. Figure 1);

$$A_{2m} = \int_0^{\pi/2} \{l(\tfrac{1}{2}\pi) - l(\psi)\} \, d(\sin^{2m} \psi), \qquad (3.1)$$

where $l(\tfrac{1}{2}\pi)$ denotes the light of the system at quadratures ($\psi = 90°$ or $270°$), and $l(\psi)$ stands for the light of the system at any phase angle ψ.

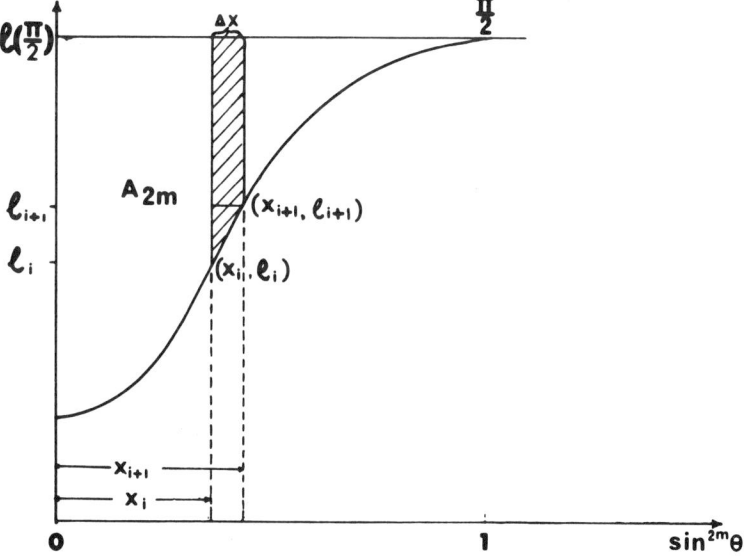

Figure 1. Empirical moments A_{2m} for close eclipsing systems.

The observed light of a close eclipsing system can be expressed as

$$l = \mathscr{L}_1 + \mathscr{L}_2 - L_1 \Delta \mathscr{L}, \tag{3.2}$$

where $\mathscr{L}_{1,2}$ denote the luminosities of the two components mutually distorted by rotation and tides, as given by Equations (2.28) - (2.29) of Kopal (1975e); L_1 is the fractional luminosity of the eclipsed star; and $\Delta \mathscr{L}$ stands for the loss of light arising from the eclipse of the primary component and given by Equations (3.30) - (3.34) of Kopal (1975e).

The integrand in Equation (3.1) can be expressed by means of Equation (3.2) as (Kopal, 1975e)

$$l(\tfrac{1}{2}\pi) - l(\psi) = -\sum_{j=1}^{n} c_j \cos^j \psi + L_1 \sum_{h=1}^{K} C^{(h)} \{\alpha_{h-1}^0 + f_*^{(h)} + f_1^{(h)} + f_2^{(h)}\}, \tag{3.3}$$

where n in the first summation denotes the order of the highest harmonics retained and, in the second summation, K = n + 1 stands for the degree of limb darkening.

Integrating Equation (3.3) with limits of integration from 0 to $\tfrac{1}{2}\pi$, we get

$$A_{2m} = \int_0^{\pi/2} \{l(\tfrac{1}{2}\pi) - l(\psi)\} \, d(\sin^{2m} \psi) = \\ = \tilde{A}_{2m} + \bar{A}_{2m} + \mathfrak{P}_{2m}, \tag{3.4}$$

where

$$\tilde{A}_{2m} = -\sum_{j=1}^{n} c_j \int_0^{\pi/2} \cos^j \psi \, d(\sin^{2m} \psi) = \\ = -m! \sum_{j=1}^{n} \frac{\Gamma(1 + \tfrac{1}{2}j)}{\Gamma(1 + m + \tfrac{1}{2}j)} c_j, \tag{3.5}$$

$$\bar{A}_2 = L_1 \sum_{h=1}^{K} C^{(h)} \int_0^{\theta'} \alpha_{h-1}^0 \, d(\sin^{2m} \psi), \tag{3.6}$$

$$\mathfrak{P}_{2m} = L_1 \sum_{h=1}^{K} C^{(h)} \int_0^{\theta'} \{f_*^{(h)} + f_1^{(h)} + f_2^{(h)}\} \, d(\sin^{2m} \psi), \tag{3.7}$$

where θ' denotes the phase angle of the first contact; α_{h-1}^0 expresses the well-known 'associated α-functions' (of zero order); $f_*^{(h)}$, $f_{1,2}^{(h)}$, stand for photometric corrections due to distortion, the explicit forms of which have been given by Kopal (cf. Kopal, 1942,

1959, 1975e); and the $C^{(h)}$ are functions of the coefficient of limb darkening, given by Equation (2.30) of Kopal (1975e).

The term A_{2m} on the left-hand side of Equation (3.4) denotes the empirical moments which can be evaluated from the observations by quadratures or otherwise.

The first term $\widetilde{A}_{2m}$ on the right-hand side of the same equation represents the proximity effects outside eclipses. Their evaluation can be obtained by a suitable modulation of the part of the light curve which is unaffected by eclipses. The method for the evaluation of the proximity effects has been fully described by Kopal (1976b) and an application of that method has been carried out by Niarchos (1977).

The second term $\widetilde{A}_{2m}$ in the same equation stands for the theoretical expression of the eclipse moments in terms of the elements of the system. It assumes different expressions for different types of the eclipse. All these expressions have been given by Kopal (cf. Kopal, 1975a,b,c,d,e) and can be regarded as known.

The last term $\mathcal{B}_{2m}$ expresses the photometric perturbations arising from the distortion of both components. The explicit forms of $\mathcal{B}_{2m}$ terms for any type of eclipse have been given by Kopal (1975e) and Livaniou (1977a,b).

4. METHOD OF ANALYSIS

A. Evaluation of the Empirical Moments

In order to evaluate the empirical moments A_{2m}, the magnitudes of the observational points are expressed in terms of the m_0 (m_0 being mean value of the magnitude at quadratures, when $\sin^{2m}\theta = 1$). The new magnitudes are converted into light intensities and the phases into phase angles. Then the intensities for each minimum are plotted against the powers $\sin^{2m}\theta$ (m = 1, 2, 3) in a large piece (50 x 50 cm) of graph paper reflecting the observations in the range 180°-360° back over those of the 0°-180° range. After careful inspection to judge the extent and influence of observational scatter a smooth free-hand curve is drawn through the points. In such a case (Figure 1) the area confined by the lines $l = l(\frac{1}{2}\pi)$, $\sin^{2m}\theta = 0$ and the drawn light curve is obviously given by

$$A_{2m} = \sum_{i=1}^{n}(\Delta x)\{l(\tfrac{1}{2}\pi) - \tfrac{1}{2}(l_i + l_{i+1})\}, \qquad (4.1)$$

where

$$(\Delta x) \equiv x_{i+1} - x_i, \qquad x_i \equiv \sin^{2m} \theta_i, \qquad x_{i+1} \equiv \sin^{2m} \theta_{i+1}. \qquad (4.2)$$

The evaluation of the areas A_{2m} (m = 1, 2, 3) defined by Equation (4.1) is carried out by a computer program in which the points $(\sin^{2m} \theta_i, l_i)$ are read straight from the curve.

The right-hand side of Equation (4.1) represents nothing but the integral

$$A_{2m} = \int_0^{\pi/2} \{l(\tfrac{1}{2}\pi) - l(\theta)\} \, d(\sin^{2m} \theta), \qquad (4.3)$$

which expresses the empirical 'moments' A_{2m}, defined by Equation (3.4) (Figure 2).

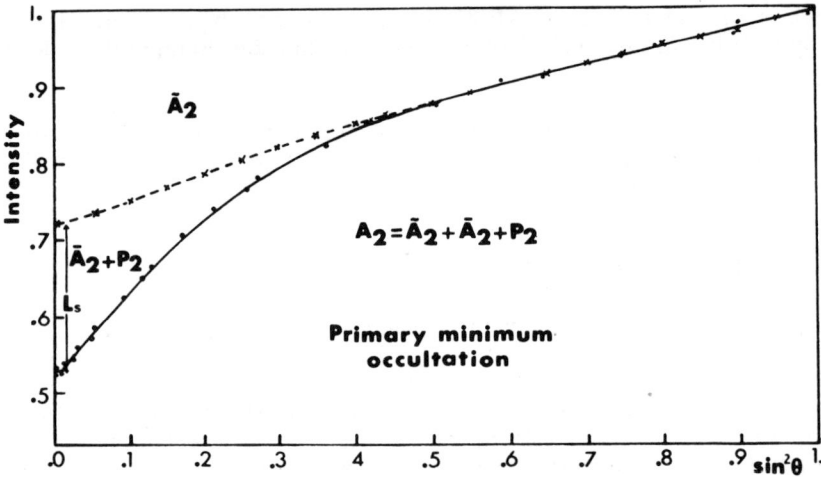

Figure 2. The empirical moment A_2 of W UMa. Normal points of Yellow Observations.

B. Evaluation of the Proximity Effects

The next step in our analysis is the evaluation of the proximity terms $\tilde{A}_{2m}$ defined by Equation (3.5)

$$\tilde{I}_{2m} = -\sum_{j=1}^{n} c_j \int_0^{\pi/2} \cos^j \psi \, d(\sin^{2m} \psi) =$$

$$= -m! \sum_{j=1}^{n} \frac{\Gamma(1 + \tfrac{1}{2}j)}{\Gamma(1 + m + \tfrac{1}{2}j)} c_j, \qquad (3.5)$$

The only unknown in Equation (3.5) is the coefficient c_j in terms of which the moments $\tilde{A}_{2m}$ can be expressed. The evaluation of the coefficients c_j (j = 1, 2, 3, 4) has been carried out by a method explicitly described by Niarchos (1977). Then the proximity effects can be evaluated by means of Equation (3.5) which becomes

for m = 1

$$\tilde{A}_2 = -2 \sum_{j=1}^{4} \frac{c_j}{j+2} \qquad (4.4)$$

for m = 2

$$\tilde{A}_4 = -8 \sum_{j=1}^{4} \frac{c_j}{(j+2)(j+4)} \qquad (4.5)$$

and for m = 3

$$\tilde{A}_6 = -48 \sum_{j=1}^{4} \frac{c_j}{(j+2)(j+4)(j+6)} \qquad (4.6)$$

Having evaluated the 'proximity effects' we may now turn back to the empirical moments A_{2m} to free them from any influence of the proximity effects by subtraction. So, Equation (3.4) can be written as

for m = 1: $A_2 - \tilde{A}_2 = \bar{A}_2 + \mathfrak{P}_2,$ (4.7)

for m = 2: $A_4 - \tilde{A}_4 = \bar{A}_4 + \mathfrak{P}_4,$ (4.8)

for m = 3: $A_6 - \tilde{A}_6 = \bar{A}_6 + \mathfrak{P}_6;$ (4.9)

or

$$A'_2 = \bar{A}_2 + \mathfrak{P}_2, \qquad (4.10)$$

$$A'_4 = \bar{A}_4 + \mathfrak{P}_4, \qquad (4.11)$$

$$A'_6 = \bar{A}_6 + \mathfrak{P}_6, \qquad (4.12)$$

where A'_2, A'_4, A'_6 represent the 'rectified' empirical moments.

C. Determination of the Elements

In order to evaluate the geometric elements we first neglect the perturbation terms $\mathscr{B}_2$, $\mathscr{B}_4$ and $\mathscr{B}_6$ in Equations (4.10) - (4.12) and solve for the elements using the equations

$$A'_2 = \overline{A}_2, \qquad (4.13)$$

$$A'_4 = \overline{A}_4, \qquad (4.14)$$

$$A'_6 = \overline{A}_6, \qquad (4.15)$$

where the right-hand side of the above equations assumes particular expressions for different types of eclipse. The explicit forms of equations used to determine the geometric elements are given below.

(i) Systems with complete eclipses

Two methods have been used for the evaluation of the elements of totally eclipsing systems. The first makes use of one minimum alone (that corresponding to occultation) and will be hereafter referred to as Method I. The equations used (cf. Kopal, 1975b, Equations (3.13)-(3.23)) are

$$r_{1,2}^2 = \frac{C_{1,2}^2}{(1 - C_3)C_1 + C_2^2}, \qquad (4.16)$$

$$\sin^2 i = \frac{C_1}{(1 - C_3)C_1 + C_2^2}, \qquad (4.17)$$

where the constants C_1, C_2, C_3 are intermediate constants related to the empirical moments A'_{2m} through the equations

$$A'_2 = L_1 \overline{C}_3, \qquad (4.18)$$

$$A'_4 = L_1(\overline{C}_3^2 + \overline{C}_2^2), \qquad (4.19)$$

$$A'_6 = L_1(\overline{C}_3^3 + 3\overline{C}_2^2 \overline{C}_3 + \overline{C}_1 \overline{C}_2^2), \qquad (4.20)$$

where

$$\overline{C}_1 = \left\{ \frac{3(35 - 19u_1)}{7(15 - 7u_1)} \right\} C_1, \qquad (4.21)$$

$$\overline{C}_2 = \left\{ \frac{15 - 7u_1}{5(3 - u_1)} \right\}^{1/2} C_2, \qquad (4.22)$$

$$\overline{C}_3 = C_3, \qquad (4.23)$$

where u_1 denotes the linear coefficient of limb darkening of the star undergoing eclipse.

The second method (which hereafter will be referred to as Method II) makes use of all the information contained in the light curve: namely, it uses both minima simultaneously. The necessary steps of this method have been described by Budding (1977). The equations used are

$$K = \tfrac{1}{6}\{[12(3 - u_g)(a_2'/a_2) + (a_4 - a_2)/Y_2^{(2)}a_2^2 u_g]^{1/2} + \qquad (4.24)$$
$$+ [(a_4 - a_2^2)/Y_2^{(2)}a_2^2]^{1/2}u_g\},$$

$$K_n = \{a_2'/a_2[1 - f_1(K_{n-1}, u_g, u_s)]\}^{1/2}, \quad n = 1, 2, 3 \qquad (4.25)$$

$$f_1(K, u_g, u_s) = \frac{2u_g}{(3 - u_g)K^2}\left\{\left[\frac{a_4 - a_2^2}{K^2 a_2^2 Y_2^{(2)}}\right]^{1/2} - 1\right\} \times \qquad (4.26)$$
$$\times (1 - K^2)[1 - (1 - K^2)^{1/2}],$$

$$\cot^2 i = \frac{C_2}{K} - C_3, \qquad (4.27)$$

$$r_g^2 = \frac{C_2}{C_2 + K(1 - C_3)}, \qquad (4.28)$$

where

$$a_2 = A_2'(\mathrm{oc})/L_s, \qquad a_2' = A_2'(\mathrm{tr})/L_g, \qquad a_4 = A_4'(\mathrm{oc})/L_s,$$

$$K = r_s/r_g, \qquad Y_2^{(2)} = \frac{15 - 7u_s}{5(3 - u_s)}, \qquad K_0 = (a_2'/a_2)^{1/2}; \qquad (4.29)$$

and u_s, u_g denote the coefficients of limb darkening of the smaller and larger component, respectively; L_s and L_g stand for the fractional luminosities of the two components.

(ii) Systems with partial eclipses

In the case of partially eclipsing systems the method described by Kopal (1975d, 1976a) has been used. Following an analysis made by Demircan (1977), the expressions for the geometric elements used in our analysis can be obtained.

For occultations,

$$r_g^2 \equiv r_2^2 = A_2'(\mathrm{oc})/[L_1\alpha_0(1 - q^2) + L_1 S_1^{(\mathrm{oc})} + A_2'(\mathrm{oc})q^2], \qquad (4.30)$$

$$i = \cos^{-1}(qr_2), \qquad (4.31)$$

$$r_s^2 \equiv r_1^2 = [A_4'(\mathrm{oc})\sin^4 i/L_1 - \alpha_0(r_2^2 - \cos^2 i)^2 - r_2^4 S_2^{(\mathrm{oc})}]/r_2^2(\alpha_0 - \tfrac{1}{2}x_s\alpha_1^0), \qquad (4.32)$$

where

$$k \equiv r_1/r_2, \quad q \equiv (\cos i)/r_2 = 1 + kp_0, \tag{4.33}$$

and

$$\alpha_0 \equiv (1 - x_s)\alpha_0^0(k, p_0) + x_s\alpha_1'^0(k, p_0), \quad x_s \equiv \frac{2u_s}{3 - u_s}. \tag{4.34}$$

For transits, on the other hand,

$$r_g^2 \equiv r_1^2 = A_2'(\text{tr})/[L_1\alpha_0(k^2 - q^2) + L_1k^2 S_1^{(\text{tr})} + A_2'(\text{tr})q^2], \tag{4.35}$$

$$i = \cos^{-1}(qr_1), \tag{4.36}$$

$$r_s^2 \equiv r_2^4 = [A_4'(\text{tr}) \sin^4 i/L_1 - r_1^2 r_2^2(\alpha_0 + \tfrac{1}{5}x_g\alpha_1^{0\prime\prime}) -$$
$$- (r_2^2 - \cos^2 i)^2 \alpha_0]/S_2^{(\text{tr})}, \tag{4.37}$$

where

$$k \equiv r_2/r_1, \quad q \equiv (\cos i)/r_1 = 1 + kp_0, \quad x_g = \frac{2u_g}{3 - u_g}; \tag{4.38}$$

$$\alpha_0 = \frac{(1 - x_g)\alpha_0^0(k, p_0) + \tfrac{3}{2}x_g\Phi(k)\alpha_1^{0\prime\prime}(k, p_0)}{1 - x_g + \tfrac{3}{2}x_g\Phi(k)}, \tag{4.39}$$

$$\Phi(k) = \frac{4}{3\pi k^2}\{\sin^{-1}\sqrt{k} + \tfrac{1}{3}(4k - 3)(2k + 1)\sqrt{k(1 - k)}\}. \tag{4.40}$$

The α-functions for both occultation and transit eclipses can be simply evaluated by means of the available Tsesevich tables (Tsesevich, 1939, 1940). The $S_m(k, q, u) \equiv f(J_{\beta,\gamma}^0)$ functions can be evaluated by using the $J_{\beta,\gamma}^0$ tables prepared by Kopal (1947). In the course of our analysis we re-evaluated numerically the required $J_{\beta,\gamma}^0$ integrals using a different method in order to increase the accuracy of the S_m functions. The explicit forms of the S_m functions have been given by Demircan (1977). The whole procedure for the determination of the elements of systems which undergo complete as well as partial eclipses has been fully automated in a computer program.

Having determined the preliminary elements, the actual magnitude of the perturbation terms $\mathcal{B}_{2m}$ occurring on the right-hand side of Equation (3.4) can be evaluated in terms of the preliminary elements and other physical parameters of the system (such as its mass ratio or limb and gravity darkening of the component undergoing eclipse). The outcome is transposed to the left-hand side of Equations (4.10)-(4.12) and the solution is iterated until the resulting elements no longer differ from those used to compute the perturbations by any significant amount. In practical cases one iteration turns out to be enough for a satisfactory agreement between the preliminary elements and those found after using the perturbation terms.

For the evaluation of the $\mathcal{B}_{2m}$ terms the mass ratio from spectroscopic observations, the assumed values of the coefficient of limb darkening and computed values of gravity-darkening coefficients have been used. The latter have been evaluated by using Kopal's (1968) method. For systems without spectroscopic observations, estimated values of the mass ratio according to the Roche model have been obtained by using Kopal's (1959) or Plavec and Kratochvíl's (1964) tables. The final elements for totally eclipsing systems derived after applying the perturbation terms $\mathcal{B}_{2m}$ are given in Tables II and III, while Tables IV and V contain the elements of partially eclipsing systems derived without the $\mathcal{B}_{2m}$ terms. In tables containing the results of our analysis, all symbols have the usual meaning.

Table II. Final Elements. Yellow Observations.

System	r_s	r_g	i	k	u^a	L_s	L_g
BV 419	0.213	0.426	81°6	0.50	0.6	0.107	0.893
AC Boo	0.216	0.496	76.7	0.44	0.6	0.167	0.833
RR Cen	0.269	0.450	78.5	0.60	0.6	0.093	0.907
CC Com	0.226	0.405	85.1	0.56	0.6	0.198	0.802
RZ Com	0.235	0.446	81.9	0.53	0.6	0.200	0.800
DK Cyg	0.227	0.452	80.4	0.50	0.6	0.127	0.873
AK Her	0.227	0.521	78.9	0.44	0.6	0.119	0.881
AM Leo	0.213	0.446	80.9	0.48	0.6	0.152	0.848
UZ Leo	0.200	0.476	80.3	0.42	0.6	0.088	0.912
V 566 Oph	0.216	0.486	82.2	0.44	0.6	0.108	0.892
AE Pho	0.192	0.458	81.1	0.42	0.6	0.157	0.843
RZ Tau	0.191	0.400	86.1	0.48	0.6	0.114	0.886
W UMa	0.226	0.456	78.2	0.50	0.6	0.186	0.814
AH Vir	0.223	0.476	77.9	0.47	0.6	0.149	0.851

Table III. Final Elements. Blue Observations.

System	r_s	r_g	i	k	u^a	L_s	L_g
BV 419	0.206	0.396	86°5	0.52	0.7	0.075	0.925
AC Boo	0.223	0.495	78.5	0.45	0.7	0.173	0.827
CC Com	0.234	0.438	82.3	0.53	0.7	0.210	0.790
DK Cyg	0.288	0.492	77.8	0.58	0.7	0.151	0.849
AK Her	0.230	0.506	81.4	0.45	0.7	0.126	0.874
AM Leo	0.219	0.437	83.4	0.50	0.7	0.150	0.850
UZ Leo	0.204	0.496	75.1	0.41	0.7	0.108	0.892
V 566 Oph	0.211	0.454	79.6	0.46	0.7	0.094	0.906
AE Pho	0.186	0.457	83.1	0.41	0.7	0.141	0.859
RZ Tau	0.216	0.415	83.1	0.52	0.7	0.101	0.899
W UMa	0.230	0.438	79.2	0.52	0.7	0.181	0.819
AH Vir	0.221	0.464	77.2	0.48	0.7	0.169	0.831

Table IV. Final Elements. Yellow Observations.

System	r_s	r_g	i	k	u^a	L_s	L_g
AB And	0.258	0.456	73°.7	0.56	0.6	0.297	0.703
OO Aql	0.304	0.377	77.2	0.81	0.6	0.310	0.690
TY Boo	0.232	0.483	69.6	0.48	0.6	0.213	0.787
VW Boo	0.254	0.480	66.2	0.53	0.6	0.219	0.781
CW Cas	0.255	0.445	69.8	0.57	0.6	0.228	0.772
VW Cep	0.183	0.556	56.9	0.33	0.6	0.129	0.871
1073 Cyg	0.276	0.503	63.9	0.55	0.6	0.115	0.885
UX Eri	0.223	0.448	70.6	0.50	0.6	0.140	0.860
YY Eri	0.228	0.450	73.3	0.51	0.6	0.171	0.829
EM Lac	0.223	0.497	66.9	0.45	0.6	0.149	0.851
SW Lac	0.212	0.436	72.1	0.49	0.6	0.199	0.801
V 502 Oph	0.195	0.516	64.9	0.38	0.6	0.103	0.897
V 839 Oph	0.208	0.530	60.3	0.39	0.6	0.152	0.848
ER Ori	0.327	0.399	72.6	0.82	0.6	0.348	0.652
U Peg	0.212	0.477	67.8	0.44	0.6	0.172	0.828

Table V. Final Elements. Blue Observations.

System	r_s	r_g	i	k	u^a	L_s	L_g
AB And	0.255	0.456	73°.5	0.56	0.7	0.301	0.699
OO Aql	0.316	0.393	77.9	0.80	0.7	0.321	0.679
TY Boo	0.228	0.478	71.0	0.48	0.7	0.202	0.798
VW Boo	0.258	0.489	65.4	0.53	0.7	0.230	0.770
XY Boo	0.209	0.607	55.1	0.34	0.7	0.076	0.924
VW Cep	0.189	0.560	57.8	0.34	0.7	0.132	0.868
1073 Cyg	0.286	0.528	63.1	0.54	0.7	0.077	0.923
UX Eri	0.328	0.468	68.7	0.51	0.7	0.132	0.868
YYEri	0.234	0.466	73.7	0.50	0.7	0.187	0.813
EM Lac	0.248	0.503	67.5	0.49	0.7	0.184	0.816
SW Lac	0.227	0.436	72.3	0.52	0.7	0.203	0.797
V 502 Oph	0.175	0.514	63.3	0.34	0.7	0.101	0.899
V 839 Oph	0.197	0.517	62.8	0.38	0.7	0.139	0.861
ER Ori	0.339	0.389	72.7	0.87	0.7	0.352	0.648
U Peg	0.218	0.482	68.9	0.45	0.7	0.194	0.806

Note: In Tables II-V, a = assumed.

5. ROCHE EQUIPOTENTIALS

The derivation of the analytical expressions of Roche equipotentials is based on the following assumptions: (1) the mass of each component is concentrated at its centre; (2) the components revolve in circular orbits; (3) the axes of rotation are perpendicular to the orbital plane; and (4) the period of axial rotation and orbital revolution are identical. The geometry of the Roche model has been thoroughly investigated by Kopal (1959, Chapter III) and extensive tables listing the sizes of Roche equipotentials as functions of the mass-ratio (q) have been published by Kopal (1959) or Plavec and Kratochvíl (1964). An inspection of these tables reveals that the sum $r_s + r_g$ of the fractional radii of both components in contact binary systems is very nearly constant and equal to 0.75 ± 0.01 for a very wide range of mass ratio q ; whereas the ratio $K = r_s/r_g$ decreases monotonically with diminishing value of q.

The inner surfaces of Roche equipotentials, called Roche limits or critical lobes, are most important here because they set an upper limit to the possible sizes of the components of W UMa-type systems.

The dimensions of the systems determined by the present analysis of their light changes can now be compared with the sizes of their lobes. In Table VI the radii r_s, r_g (r_s, r_g are the radii of spheres capable of containing the same volume as the respective distorted configurations) are given along with the sizes (a_s, a_g) of their lobes. Only those systems are included for which spectroscopic mass ratios are available. The mean values of the radii r_s and r_g from yellow and blue observations are used. Table VII contains the fractional sizes of systems without known spectroscopic mass ratio. The mean values from yellow and blue observations are used. For partially eclipsing systems the elements without $2m$ terms have been considered for reasons explained below.

Table VI. Comparison between Sizes and Lobes.

System	q	r_s	a_s	r_g	a_g	Ecl.
AB And	0.62	0.246	0.316	0.483	0.433	P
VW Cep	0.32	0.183	0.277	0.556	0.484	P
RZ Com	0.48	0.235	0.310	0.446	0.444	C
1073 Cyg	0.35	0.276	0.282	0.503	0.478	P
YY Eri	0.65	0.228	0.334	0.474	0.416	P
SW Lac	0.86	0.220	0.360	0.465	0.388	P
V 502 Oph	0.40	0.214	0.294	0.561	0.462	P
V 566 Oph	0.34	0.213	0.282	0.470	0.478	C
ER Ori	0.61	0.268	0.328	0.468	0.422	P
U Peg	0.80	0.218	0.354	0.509	0.395	P
RZ Tau	0.54	0.204	0.319	0.408	0.433	C
W UMa	0.50	0.228	0.313	0.446	0.440	C
AH Vir	0.42	0.222	0.299	0.470	0.457	C

C = complete eclipses (total and annular)
P = partial eclipses.

Table VII. Fractional Sizes of Systems without known Spectroscopic Mass Ratios

System	r_s	r_g	$r_s + r_g$	Ecl.
BV 419	0.210	0.411	0.621	C
AC Boo	0.220	0.495	0.715	C
RR Cen	0.269	0.450	0.719	C
CC Com	0.230	0.421	0.651	C
DK Cyg	0.257	0.472	0.729	C
AK Her	0.228	0.513	0.741	C
AM Leo	0.216	0.442	0.658	C
UZ Leo	0.202	0.486	0.688	C
AE Pho	0.189	0.458	0.647	C
OO Aql	0.310	0.385	0.695	P
TY Boo	0.230	0.481	0.711	P
VW Boo	0.256	0.485	0.741	P
XY Boo	0.209	0.607	0.816	P
CW Cas	0.255	0.445	0.700	P
UX Eri	0.231	0.458	0.689	P
EM Lac	0.236	0.500	0.736	P
V 389 Oph	0.204	0.523	0.727	P

C = complete eclipses (total and annular)
P = partial eclipses.

6. DISCUSSION

It is well known that an analysis of the light changes of distorted eclipsing systems is a very difficult task. Such an analysis becomes more complicated if the system under consideration happens to be one with partial eclipses. In order to check our results, the derived elements were used to construct theoretical light curves and compare them with observations. Two well-behaved systems have been chosen: W UMa itself, which undergoes complete eclipses, and the partially eclipsing system YY Eri. The theoretical light curves were computed by a program provided by Al-Naimiy (1977). Keeping the geometric and photometric elements within the limits of their uncertainties and varying only the coefficient τ of gravity darkening, a satisfactory agreement between the observed and theoretical light curves was obtained. It should be noted that large values of τ (~ 2.0) gave the best fit. The results are shown in Figures 3 and 4.

The elements in the present analysis are derived from the rectifiable model and the non-contact hypothesis is used. The results show that the totally eclipsing systems are "undersize" ($r_s + r_g < 0.75$) and in most cases the greater (more massive) star fills its critical lobe, while the smaller does not. An application of the perturbation terms $\mathcal{B}_{2m}$ causes a decrease of the fractional sizes of the components

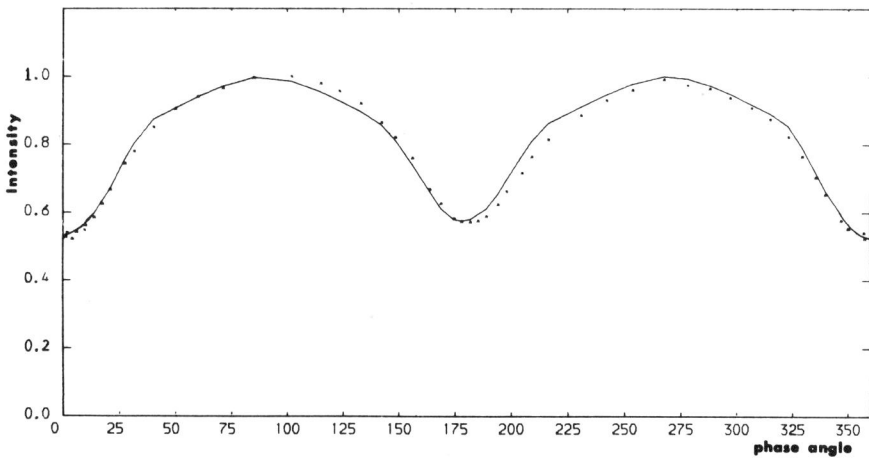

Figure 3. Observed and theoretical light curve of W UMa.
Yellow observations

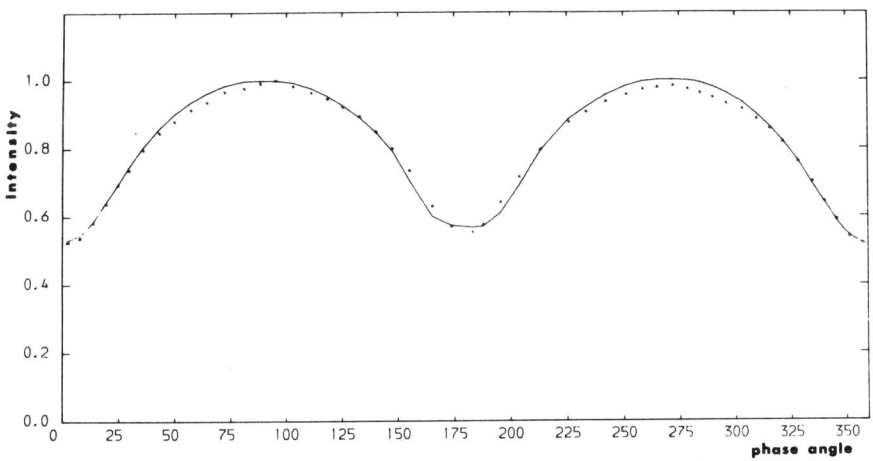

Figure 4. Observed and theoretical light curve of YY Eridani.
Yellow observations.

and an increase of the inclination. However, the situation is different in the case of partially eclipsing systems. The fractional dimensions of these systems become, after applying the perturbations, larger than those derived without φ_{2m} terms. It should be noted that for some systems (YY Eri, SW Lac, AB And) undergoing grazing eclipses, the perturbations converge and give reasonable elements, while for deeper partial eclipses the determined elements cannot as yet be regarded as reliable. Therefore the question arises whether the infinite series giving the α-functions, involved in the formulae giving the perturbation terms, converge for deep partial eclipses.

In a theoretical investigation Lucy (1967) concluded that for W UMa-type systems the gravity-darkening coefficient should be approximately equal to 0.32. According to our calculations (τ 2.0) and those of Kopal (1968), a discrepancy between theory and observations is evident. The origin of this discrepancy remains so far unexplained; a good discussion of its significance has been given by Kopal (1968).

While for totally eclipsing systems all aspects of an analysis of the light curves (including photometric perturbations within eclipses arising from the mutual distortion of the two stars) are reasonably well in hand as a result of previous work by Kopal (1975e) and Livaniou (1977a), this is not yet the case, for reasons we mentioned above, when the eclipses become partial. Difficulties in establishing appropriate numerical values of the photometric perturbations for partially eclipsing systems from the analytical expressions established previously by Livaniou (1977b) still make it difficult for us to arrive at more definite conclusions concerning the contact nature of partially eclipsing systems of W UMa-type, or the amount of gravity darkening of their components.

REFERENCES

Al-Naimiy, H.: 1977, Ph.D. Thesis, University of Manchester (unpublished).
Binnendijk, L.: 1959, Astron. J., 64, p.65.
Binnendijk, L.: 1960a, Astron. J., 65, p.79.
Binnendijk, L.: 1960b, Astron. J., 65, p.88..
Binnendijk, L.: 1960c, Astron. J., 65, p.358
Binnendijk, L.: 1961, Astron. J., 66, p.27.
Binnendijk, L.: 1962, Astron. J., 67, p.86.
Binnendijk, L.: 1963, Astron. J., 68, p.22.
Binnendijk, L.: 1964, Astron. J., 69, p.157.
Binnendijk, L.: 1965, Astron. J., 70, p.201.
Binnendijk, L.: 1966, Astron. J., 71, p.340.
Binnendijk, L.: 1967, Astron. J., 72, p.82.
Binnendijk, L.: 1968, Astron. J., 73, p.32.
Binnendijk, L.: 1969a, Astron. J., 74, p.218.

Binnendijk, L.: 1969b, Astron. J., 74, p.1031.
Binnendijk, L.: 1970, *Vistas in Astronomy*, vol. 12, p.217.
Binnendijk, L.: 1971, Astron. J., 76, p.923.
Binnendijk, L.: 1972, Astron. J., 77, p.246.
Binnendijk, L.: 1973, Astron. J., 78, p.103.
Bookmyer, B. B.: 1965, Astron. J., 70, p.415.
Bookmyer, B. B.: 1969, Astron. J., 74, p.1197.
Broglia, P.: 1960, Contr. Milano-Merate, 165.
Broglia, P.: 1964, Contr. Milano-Merate, 226.
Broglia, P. and Conconi, P.: 1974, Contr. Milano-Merate, 361.
Budding, E.: 1977, Astrophys. Space Sci., 46, p.407.
Carr, R.: 1972, Astron. J., 77, p.155.
Chambliss, C. R.: 1967, Astron. J., 72, 512.
Demircan, O.: 1977, Astrophys. Space Sci., 47, p.459.
Eggen, O. J.: 1967, Mem. Roy. Astron. Soc., 70, p.111.
Hazlehurst, J.: 1 1970, Mon. Not. Roy. Astron. Soc., 149, p.129.
Knipe, G. F. G.: 1961, Johannesburg Circ. No. 120.
Kondo, Y.: 1966, Astron. J., 71, p.54.
Kopal, Z.: 1942, Proc. Amer. Phil. Soc., 85, p.399.
Kopal, Z.: 1947, Harvard Obs. Circ. No. 450.
Kopal, Z.: 1959, *Close Binary Systems*, Chapman Hall and John
    Wiley, London and New York.
Kopal, Z.: 1968, Astrophys. Space Sci., 2, p.23.
Kopal, Z.: 1975a, Astrophys. Space Sci., 34, p.431 (Paper I).
Kopal, Z.: 1975b, Astrophys. Space Sci., 35, p.159 (Paper II).
Kopal, Z.: 1975c, Astrophys. Space Sci., 35, p.171 (Paper III).
Kopal, Z.: 1975d, Astrophys. Space Sci., 36, p.277 (Paper IV).
Kopal, Z.: 1975e, Astrophys. Space Sci., 38, p.191 (Paper V).
Kopal, Z.: 1976a, Astrophys. Space Sci., 40, p.461 (Paper VIII).
Kopal, Z.: 1976b, Astrophys. Space Sci., 45, p.269 (Paper IX).
Kwee, K. K.: 1966, Bull. Astron. Inst. Netherlands, Suppl. 1,
    No. 6, p.265.
Livaniou, H.: 1977a, Astrophys. SpaceSci., 51, p.77.
Livaniou, H.: 1977b, Astrophys. Space Sci., 52, p.271.
Lucy, L. B.: 1967, Z. Astrophys., 65, p.89.
Lucy, L. B.: 1968a, Astrophys. J., 151, p.1123.
Lucy, L. B.: 1968b, Astrophys. J., 153, p.877.
Moss, D. L.: 1971, Mon. Not. Roy. Astron. Soc., 153, p.41.
Moss, D. L.: 1972, Mon. Not. Roy. Astron. Soc., 157, p.433.
Moss, D. L. and Whelan, J. A. J.: 1970, Mon. Not. Roy. Astron.
    Soc., 149, p.147.
Niarchos, P.: 1977, Astrophys. Space Sci., 47, p.79.
Plavec, M. and Kratochvíl, P.: 1964, Bull. Astron. Inst. Czech.,
    15, p.165.
Purgathofer, A.: 1960, Wien Mitt. 10, p.211.
Rucinski, S. M.: 1976, Publ. Astron. Soc. Pacific, 88, p.777.
Tsesevich, V. P.: 1939, Bull. Astron. Inst. USSR Acad. Sci.,
    No. 45.
Tsesevich, V. P.: 1940, Bull. Astron. Inst. USSR Acad. Sci.,
    No. 50.
Williamon, M. R.: 1975, Astron. J., 80, p.140.

DATA ACQUISITION IN ASTRONOMICAL PHOTOELECTRIC PHOTOMETRY

Giorgio Sedmak

Osservatorio Astronomico di Trieste
Via G.B.Tiepolo, 11
34131 Trieste, Italia

Abstract:
The data acquisition in astronomical photoelectric photometry is analyzed with reference to the observational environment and to the technology and methodology of operation of the measuring system for the optical band from UV to near-IR up to 1µ wavelength.
The model of the measurement is detailed and an expression for the overall accuracy is given that allows to evaluate the various operating modes.
The measuring system is analyzed and described in terms of the transducer and the data processor. The structure of the transducer is detailed for the analog and digital photon counting operation on the basis of the photomultiplier physics. The photon counting mode is recommended for higher accuracy photometry. The data management is discussed with reference to currently available computer systems.
The integration of the analysis in the design of optimum photometers is finally presented through working examples together with some information on future systems for astronomical applications.

1. INTRODUCTION.

The point source photometry is one major task in observational astronomy. The photoelectric photometry is the most effective technique for the measurement of optical sources in the ultraviolet to near infrared band.
The applications of photoelectric photometers are limited by the technology of the detector and the informa-

tion processing system available.
The purpose of this review is to analyze the observational environment, the technological realization, and the methodology of operation of state of the art photoelectric photometers in the ultraviolet to near infrared band.
The model of the measurement, the measuring system, the transducer, the detector and the data management are reported and discussed separately in the following sections. The information reported can be used to select the configuration of a photoelectric photometer optimized for an assigned observational task.

2. THE MODEL OF THE MEASUREMENT.

The model of the measurement in astronomical point source photometry is shown in Fig.2.1. The observer inputs the signal from the target source projected on the sky background through a channel of propagation by means of a measuring system of finite performance.

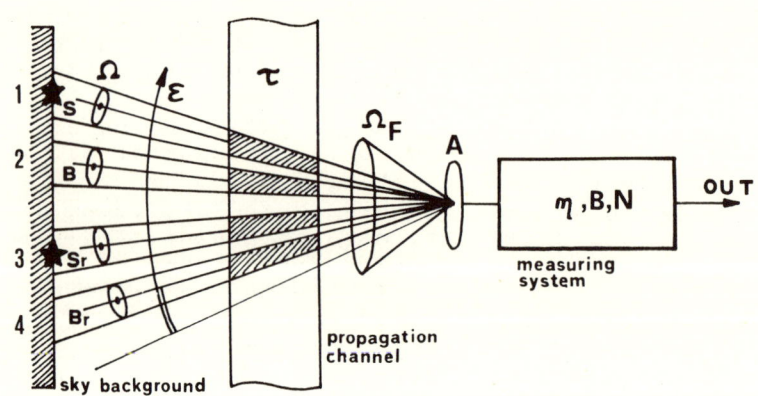

Figure 2.1   The model of the measurement.

If the transfer function of the propagation channel is to be calibrated, like for ground-based photometry, then an astronomical reference source is also needed to estimate correctly the target signal.
Four basic measurements are therefore necessary in the general case, two on the target source and background and two on the reference source and background, for any

of the selected bands. The basic measurements can be reduced to the two concerned to the target if the propagation channel does not need to be calibrated, like in space-based observations.

The signal from the target source is estimated by processing the measures obtained from the basic measurements. It is assumed that the measuring system is stable during the measurements. If not, a suitable calibration sequence must be applied to the measuring system and the calibration data introduced in the data processing. The model can be expressed by the following equations of the four basic measurements:

$$m_{1j} = ((S+B\Omega)\tau\eta b_j A + N)T_{1j} \quad (j=1,n) \quad (2.1)$$

$$m_{2j} = (B\Omega\tau\eta b_j A + N)T_{2j}$$

$$m_{3j} = ((S_r + B_r\Omega)\tau\eta b_j A + N)T_{3j}$$

$$m_{4j} = (B_r\Omega\tau\eta b_j A + N)T_{4j}$$

where:

S, S_r (photons $m^{-2} s^{-1} Hz^{-1}$) are the target and reference signals.

B, B_r (photons $m^{-2} s^{-1} Hz^{-1} sterad^{-1}$) are the target and reference backgrounds.

Ω (sterad) is the solid angle of vision.
τ is the transmittance of the propagation channel.
η (counts/photons) is the digital responsivity of the transducer.
b_j (Hz) is the bandwidth of the measurement.
A (m^2) is the collecting area.
N (counts s^{-1}) is the noise of the measuring system.
T_{kj} (s) is the measuring time of the k-th basic measurement.
n is the total number of bands.

Notice that the separation ε_{ik}, $i,k=1,4$, of the axes of the four basic measurements cannot be smaller than the aperture of Ω. Notice also that the total field of view Ω_F cannot be smaller than 4Ω if the basic measurements are to be performed contemporarily.

The data array $(m_{kj}, k=1,4, j=1,n)$ is used to estimate the target signal against the reference signal as follows:

$$(S/S_r)_j = (m_{1j} - m_{2j} \cdot T_{1j}/T_{2j})/(m_{3j} - m_{4j} \cdot T_{3j}/T_{4j}) \cdot T_{3j}/T_{1j}$$

$(j=1,n)$ \hfill (2.2)

The overall accuracy of measurement depends through equation (2.2) on the statistics of (a) the signal and background, (b) the propagation transfer function, and (c) the transducer responsivity and noise.

(a) Signal and background statistics.
The statistics of the signal and background of the target and reference sources approximates the Poisson statistics. The Bose Einstein correction should be applied only in the far infrared.

(b) Propagation channel statistics.
The statistics of the propagation channel is of fundamental importance for ground-based observations through the earth atmosphere. In this case two effects must be considered, the multiplicative transmittance noise (scintillation, variability) and the blurring spatial noise (seeing).
The transmittance noise can be characterized by the frequency spectra of the ergodic and non-ergodic components. The typical transmittance noise spectra are shown in Fig. 2.2 (Paternò,1976).

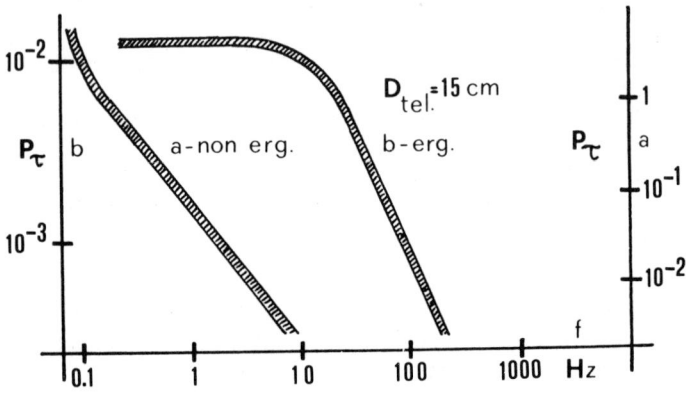

Figure 2.2 Frequency spectrum of the (a) non ergodic and (b) ergodic transmittance noise.

The seeing noise is characterized by a spatial and a temporal spectrum. The temporal spectrum shows a frequency behaviour similar to the transmittance noise spectrum. The spatial noise can be characterized by the average point spread function (PSF) of the seeing. The typical PSF of the seeing is rotationally symmetrical. The average cross-profile of the average seeing PSF is shown in Fig. 2.3. Notice that the seeing PSF represents also the correlation function of the transmittance noise with respect to the angular separation.

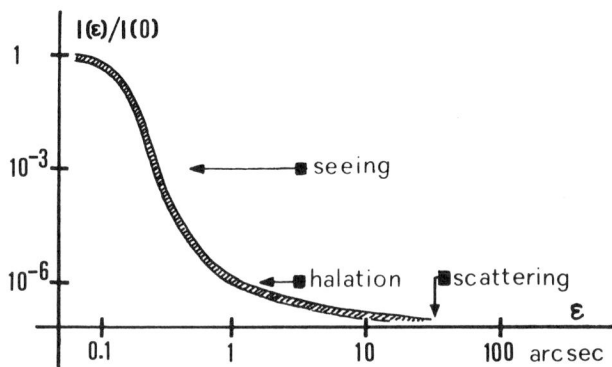

Figure 2.3 PSF of average seeing.

The values of the cut-off frequencies of the transmittance noise spectra and the width of the PSF of the average seeing are very important parameters for the configuration and operation of ground-operated astronomical photometers.
It must be stressed that the transfer function of the propagation channel can be assumed to show unity transmittance and zero noise in space-based observations.

(c) Transducer statistics.
The statistics of the responsivity and noise of the measuring system depend on the technology of the transducer used in the system.
For a digital photon counting transducer the responsivity is substantially constant and free of multiplicative noise. The dark noise approximates the Poisson

statistics with excess factors up to 1.2 at lower levels of noise.
For an analog transducer the responsivity shows a multiplicative plus an additive noise characterized by a 1/f shaped frequency spectrum. The effective noise variance depends on the technology of the detector and the front end amplifier.

The overall relative error of measurement can be estimated by computing the variance of S/S_r in equation (2.2) taking into account the statistics and the time variability of the various terms.
The overall relative error for one band results as follows (Sedmak, 1973):

$$RE_j = (\sigma/\mu)_j = (\rho^2_{S_j} + \rho^2_{S_{rj}} + \rho^2_{b_j})^{1/2} + \qquad (2.3)$$

$$+ (\Delta_{S_j} + \Delta_{S_{rj}} + \Delta_{SS_{rj}} + \Delta_{b_j}) \quad (j=1,n)$$

where:

$$\rho^2_{S_j} = (S+(B\Omega+N/\tau n b_j A)(1+T_{1j}/T_{2j}))/\tau n b_j A S^2 T_{1j} +$$

$$+ (1+B\Omega/S)^2 \sigma^2_\tau/\tau^2 \big|_{T_{1j}} + (B\Omega/S)^2 \sigma^2_\tau/\tau^2 \big|_{T_{2j}}$$

$\rho^2_{S_{rj}}$ = same for S_r, B_r

$$\rho^2_{b_j} = \sigma^2_\tau/\tau^2 \big|_{T_{1j}}$$

$$\sigma^2_\tau \big|_{T_{kj}} = \int_0^{1/2T_{kj}} P_\tau(f) df \approx P_\tau(0)/2T_{kj} \quad (k=1,4, j=1,n)$$

$$P_\tau(f) = <|FT(\tau(t))|^2> \quad (FT = \text{Fourier Transform})$$

and where:

$$\Delta_{S_j} = (B\Omega/S)\Delta\tau_{j12}/\bar{\tau}_{j12}$$

$$\Delta_{S_{rj}} = (B_r\Omega/S_r)\Delta\tau_{j34}/\overline{\tau}_{j34}$$

$$\Delta_{SS_{rj}} = \Delta\tau_{j13}/\overline{\tau}_{j13}$$

$$\Delta_{bj} = \Delta\tau_{1j1}/\overline{\tau}_{1j1}$$

where:

$$\Delta\tau_{jik}/\overline{\tau}_{jik} = 2(\tau_{ji}-\tau_{jk})/(\tau_{ji}+\tau_{jk}) \quad (i,k=1,4, j=1,n)$$

$$\tau_{ji} = (1/T_{ji}) \cdot \int_{T_{ji}} \tau(t)dt$$

The physical meaning of the terms of the overall relative error RE is the following: the RE shows a random and a systematic component with several time-dependent sub-components.

The random component includes contributions from the source and the reference source and associated backgrounds plus an inter-band distortion term (conventionally referred to band j=1). The random terms are ρ_{S_j}, $\rho_{S_{rj}}$, and ρ_{bj}.

The first two include one sub-component contributed by the signal and background statistics and two sub-components contributed by the transmittance noise, each consisting of an ergodic and a non-ergodic part.

The systematic component includes contributions by the source and the reference source and the associated backgrounds plus an inter-source and an inter-band distortion error term (referred conventionally to band j=1). The systematic terms are Δ_{S_j}, $\Delta_{S_{rj}}$, $\Delta_{SS_{rj}}$, and Δ_{bj}.

It must be stressed that the overall RE depends on the time subdivision within the total measuring time. Consequently it is possible to define optimum operating modes and optimum configurations of the measuring system by the minimization of the overall RE against the terms that are time and/or configuration dependent. The analysis of the optimization is given in next section concerning the operating modes. The analysis of the optimization of the configuration of the measuring system is given in Section 6.

2.1 OPTIMUM OPERATING MODES.

The optimum operating mode can be identified by the analysis of the random and systematic components of the overall relative error as given by equation (2.3) after the definition of the observational environment (sources, telescope, photometer, ground or space operation, accuracy, time resolution, total measuring time available).

Optimization of the random component of overall RE.

The contributions due to sources and transducer statistics cannot be optimized in general by means of the operating mode because they depend only on the cumulative measuring time.
For analog transducers, however, the transducer noise contribution can be minimized by operating the photometer in switched mode between the sources and the associated backgrounds. The switching frequency must be higher than the cut-off frequency of the noise spectrum of the transducer/s used.
The contributions due to the transmittance noise can be optimized only in regard of the non-ergodic component by using a photometer in multi-transducer configuration or by using a switched photometer for switching between the target and the reference source. The switching rate must be higher than the cut-off frequency of the non-ergodic component of the transmittance noise (typically 0.3 Hz). The ergodic component of the transmittance noise can be only averaged down because it is impossible to use the two-sources comparison to compare out the scintillation noise. This limitation follows from the finite size of the average seeing PSF, that allows to compare scintillation correlated sources only within the angular separation set by the seeing disc. In practice, this implies intensity correlation of the two sources concerned if an average source tracking is used or a high speed dynamic tracking with subarcsecond accuracy if the intensity correlation must be avoided as it is necessary in point source photometry. Such tracking system are not currently available at the present state of the art for astronomical instrumentation.

Optimization of the systematic component of overall RE.

The contributions to the systematic component are due in general to the variability of the transmittance between two of the concerned measurements.

The optimization can be made by using multi-transducer configurations or operating a switched single transducer photometer by switching between the concerned channels at a rate higher than the cut-off frequency of the transmittance variations spectrum (typically 0.1 to 0.3 Hz depending on the observing height).

The discussion made above emphasizes that the use of parallel operating multitransducers photometers or of high switching rate multiplexing single transducer photometers allows to minimize the non-ergodic component of the transmittance noise and the effects of the transmittance variability. This is very important because it implies the capability of achieving an assigned accuracy of measurement by increasing the overall observing time only when using parallel or high switching rate photometers. Other operating modes cannot ensure the convergence of the measures by averaging an increasing number of them.

3. THE MEASURING SYSTEM.

The measuring system is responsible for the execution of the basic measurements required to estimate the signal in the selected bands.
The configuration and operating mode of the measuring system must be defined taking into account the physics of the various contributions to the overall relative error RE as given by equation (2.3).
The configuration and operating mode of the measuring system can be optimized for a given observational environment only with respect to an assigned criterion of optimum.
Two basic criteria of optimum can be given:

(a) Assigned Time Mode (ATM) : to achieve the maximum precision of measurement in the assigned total measuring time.
(b) Assigned Precision Mode (APM) : to achieve the assigned precision of measurement in the minimum total measuring time.

The ATM and APM criteria do not affect directly the design of the measuring system but only its operating mode. The measuring times concerned to the basic measurements as well as the subdivision of these times must be determined against a given optimization criterion by computing the constrained minimum of the over-

all error RE as given by equation (2.3).
The general architecture of the measuring system is shown in Fig.3.1. The measuring system is organized in two functional blocks, the transducer and the processor (De Biase and Sedmak,1974).

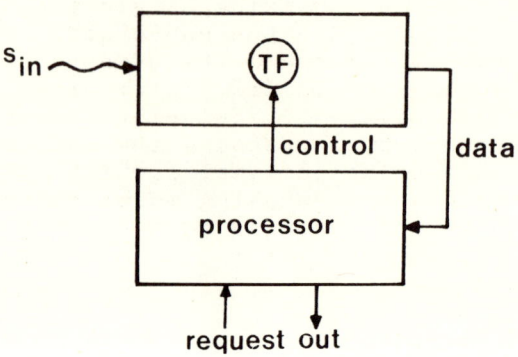

Figure 3.1 Functional structure of the measuring system.

The transducer (a)inputs the signal and (b)outputs the concerned parameters of the input signal. The transducer is controlled through the transfer function control port. The status of the transducer is communicated through the output port.
The processor (a)inputs the commands of the user,(b)controls the transducer so as to execute the wanted measurement, (c)inputs the data output from the transducer and (d)processes these data so as to generate the final output in the form requested by the user who commands the measuring system.
Two basic operational configurations can be defined, the multiplexing photometer and the parallel photometer. Any practical configuration can be realized by means of a suitable combination of the basic ones by taking into account the particular technical or observational constraints of the case considered.

3.1 MULTIPLEXING PHOTOMETERS.

The multiplexing photometers are based on multiplexing one single transducer over the measuring channels concerned to the measurement.
The technique which is commonly used is the time multiplexing implemented by sharing in time one single transducer over the various channels. Frequency multiplexing is also possible (Sedmak,1973).

The advantages of this approach are the optimum self-consistency of the measurements, the optimum calibration capability, and the optimum performance to complexity and cost ratios.

The disadvantages are the minimum efficiency, the additivity of the individual measuring times involved in the measurement, and the limited time resolution capability. The performance of the multiplexing photometer is inversely proportional to the total number of bands. The functional structure of the multiplexing photometer is shown in Fig. 3.1.1. The band multiplexing is commonly implemented as shown in Fig. 3.1.2.

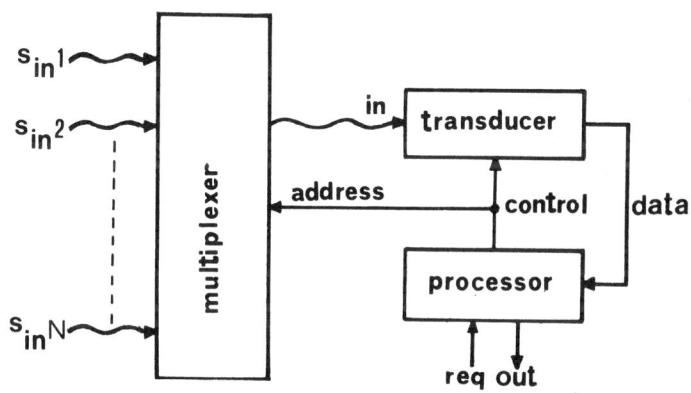

Figure 3.1.1 Functional structure of the multiplexing photometer.

The source and background switching is commonly implemented by moving the telescope or better by moving mirrors and/or fiber optics switching subassemblies.

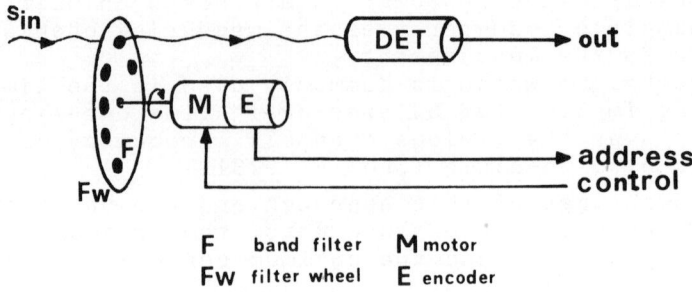

Figure 3.1.2 Band multiplexing by means of a motor-filter wheel-encoder set.

3.2 PARALLEL PHOTOMETERS.

The parallel photometers are based on using one single transducer for each measuring channel, that is 4n transducers for the general ground-based case with n bands. The advantages of the parallel approach are the optimum efficiency and time resolution, the minimum total measuring time, and the optimum logical simplicity.
The disadvantages are the minimum self-consistency of the various basic measurements, the difficulty of calibration, the difficulty of interfacing the transducers to the optical system even for moderate numbers of bands, and the minimum performance to complexity and cost ratios.
It must be emphasized that the disadvantages of the parallel photometers depend strictly on the technology available for the detectors used in the transducer.
From a methodological point of view the parallel photometer is substantially superior to any other solution. The functional structure of the parallel photometer is shown in Fig.3.2.1. The bands interfacing is commonly implemented as shown in Fig. 3.2.2.
Notice that the parallel photometer is defined to be parallel only against the input channels. The output to the processor may be multiplexed without affecting

the parallel nature of operation of the photometer.

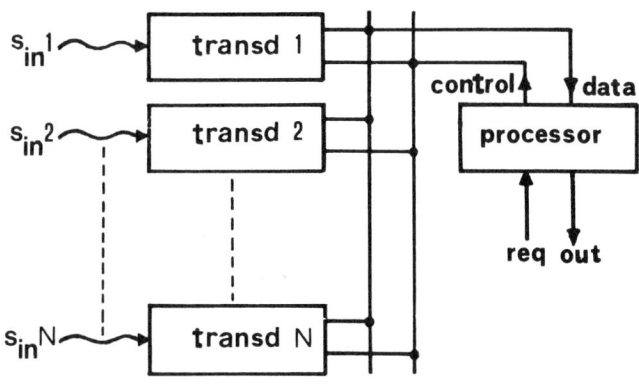

Figure 3.2.1 Functional structure of the parallel photometer.

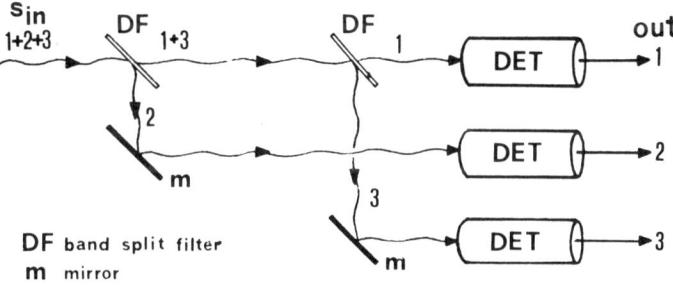

Figure 3.2.2 Band interfacing for a parallel photometer implemented by band split filters and mirrors set.

3.3 OPTIMUM CONFIGURATION SELECTION.

The selection of the optimum configuration of the measuring system must be made against the observational task to be performed.
The multiplexing and parallel approaches must be analyzed with respect to the sources levels, the total number of bands, the overall accuracy of measurement and the time resolution wanted. Also the ground or space based capability for performing the observations must be included in the analysis.
Special care must be given to the evaluation of the technology available for the transducer/s and/or the multiplexing, and to the problem of the system calibration.
A quantitative aid to the decision can be obtained by computing the overall error RE as given by equation (2.3) for the cases to be compared. One typical set of conditions is compared in this mode in the plots shown in Fig. 6.1.
The general scenary that results from the analysis is reported in the following table.

OBSERVATIONAL TASK	OPTIMUM CONFIGURATION
Strong stationary source High accuracy measurements	Fully multiplexed single transducer photometer
Variable source Good accuracy measurements Good time resolution	Two transducers multiplexed over the channels of the source concerned (Reduces to one transducer for space-based observations)
Faint rapidly varying source Reasonable accuracy High time resolution Maximum efficiency	One transducer per band per source multiplexed only between the source and the background or One fully parallel image photometer

Table 3.3.1 Task to configuration match.

4. THE TRANSDUCER.

The transducer inputs the signal from the concerned signal channel and outputs the measure of this signal integrated over the measuring time.
The transfer function of the transducer must be calibrated against a given unity of measurement. The transducer output is a number representable by a digital data word of finite length. Also the transfer function can be controlled by a digital data word of finite length.
The transducer is characterized by two basic parameters, the responsivity and the noise.
The responsivity is defined as the ratio of the output to input quantities and is generally wavelength and signal level dependent.
The noise is defined as the output due to zero input.
The responsivity and the noise of any physical transducer are generally dependent on the operating environment, particularly on the temperature.
The functional structure of the transducer is shown in Fig.4.1. The transducer includes three main blocks, the detector, the integrator, and the digital converter.

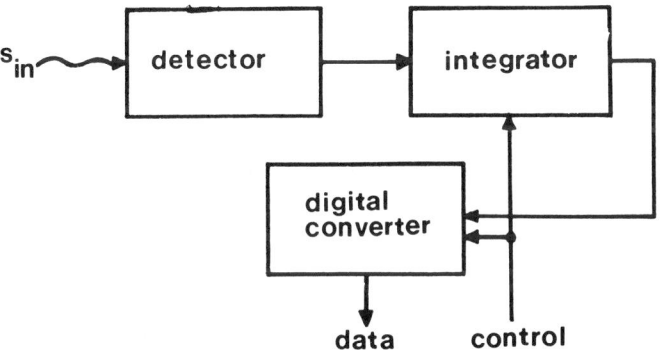

Figure 4.1 Functional structure of the transducer.

The detector inputs photons from the source and converts the photons carried information into an easy to use

physical quantity (generally an electrical quantity). The integrator sums the detected information over the assigned measuring time. The digital converter converts the integrated information into a number represented by a digital word of finite length. Notice that this last block follows from the need of mathematically processing the measures and exchanging the information to other users in a conventionally codified manner.
The three basic functions may be resident in a different number of physical blocks according to the technology used in the realization of the transducer.
The transducer can be realized following one of two main approaches, the analog and the digital or photon counting technique. Both techniques approximate the optimum detection mode within the limits of the readout noise (Young,1969).
The analog and digital techniques are substantially equivalent for medium level signals (say from m_V=+5 to m_V=+15 on 1m^2) from UV to near-IR up to 1 micron, while the analog technique is the only presently available for far-IR photometry due to limitations in available IR detectors. The best detector for astronomical applications below 1 micron is the photomultiplier. Since the present review is not dedicated to infrared photometry, the analog and digital transducers will be detailed in the following only with reference to the photomultiplier detector, which can be used successfully in both techniques.

4.1 THE PHOTOMULTIPLIER DETECTOR.

The photomultiplier (PM) is a well known and proven photons detector. The PM approximates the physical limits of the photons detection in the optical band from 0.1 micron to 1.1 microns.
Full details on the PM detector are given by Morton (1868) and Young (1969) and Sareyan and Ischi (1973). Technical details are reported in the technical data catalogues by the manufacturers (EMI,EMR,RCA).
The PM physics which is relevant to astronomical applications is shown in Fig.4.1.1 to 4.1.7. The structure and operation of the PM are shown in Fig.4.1.1. The detected photon ejects one photoelectron from the photocathode. The photoelectrons are focussed on the first of a series of electrodes (dynodes) coated with a material of high secondary emission. The electron packet from the last dynode is focussec on the anode. The average current at the anofde is approximated by:

$$I_a = \eta e^- M^m n \quad (A) \tag{4.1.1}$$

where η is the quantum efficiency of the photocathode, e^- the electron charge, M the secondary emission gain, m the number of dynodes, and n the input photons rate. The overall gain increases exponentially with the overall anode to cathode voltage. Gains of 10^7 are common with tubes of 13 dynodes operated at 1.5 kV supply. The exponential behaviour of the gain against the supply voltage allows to accommodate a very large dynamic range of the input signal by simply varying the anode voltage. Dynamic ranges of up to seven decades can be obtained on currently available tubes. The limit to the dynamic range is set by anode and/or cathode dissipation and fatigue effects. The sensitivity of the gain to the anode voltage implies a careful regulation of the anode voltage supply for accurate and stable operations. A regulation to 10^{-4} is recommended.

The typical single photon anode current pulse is shown

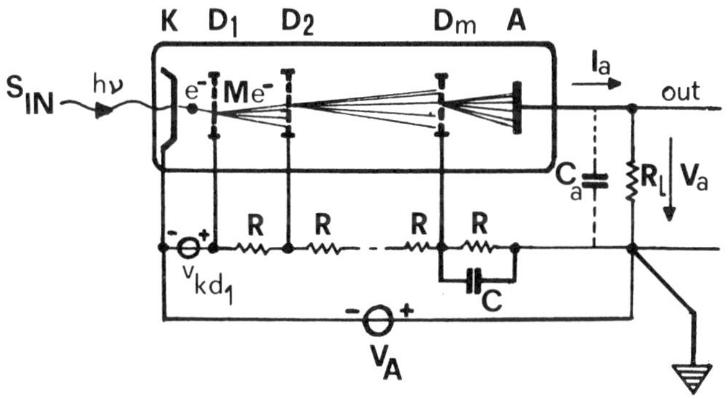

Figure 4.1.1 Structure of the photomultiplier.

The typical single photon anode current pulse is shown in Fig.4.1.2. The pulse shape approximates that of a pulsed charge generator into an R-C load equivalent to the PM load resistance in parallel to the parasitic anode capacitance. The pulse width depends on the PM

technology. Typical astronomical tubes shows values of
1 to 30 ns, corresponding to peak anode currents of
about 0.1 mA for 1 pC output charge.
The anode pulses occurr randomly in time with randomly
distributed peak amplitudes. The time distribution of
the occurrences approximates a Poisson statistics as a
consequence of the Poisson statistics of the input photons.

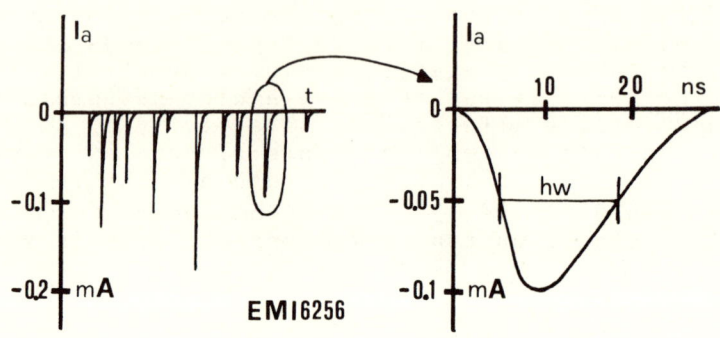

Figure 4.1.2 Anode current pulses in a PM.

The amplitude distribution of the anode pulses is characterized by the signal and noise amplitude spectra
shown in Fig. 4.1.3 and 4.1.4.
The average dark noise current is temperature dependent
as shown in Fig. 4.1.5.
The quantum efficiency spectral sensitivity depends on
the composition of the photocathode. The typical response of a blue S20 and a red S1 photocathodes is shown
in Fig. 4.1.6. The effective spectral responsivity is
temperature dependent and the thermal behaviour is also
dependent on the composition of the photocathode.
The typical behaviour of the spectral responsivity with
respect to the temperature for the blue and red bands
is shown in Fig.4.1.7. Notice that for red operations
cooling the tube may not result in an improvement of
the S/N ratio due to the decrease of the responsivity
in parallel to the decrease of the thermal noise.

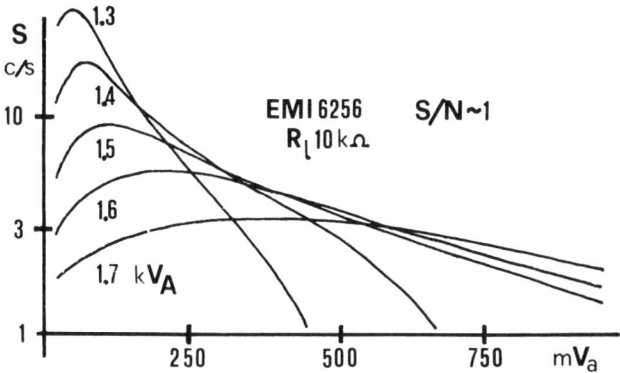

Figure 4.1.3　Pulse amplitude spectrum of the signal for typical photomultiplier operated at S/N ~1.

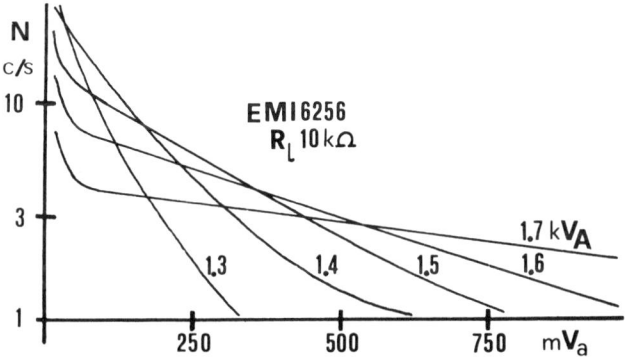

Figure 4.1.4　Pulse amplitude spectrum of the dark noise for typical photomultiplier.

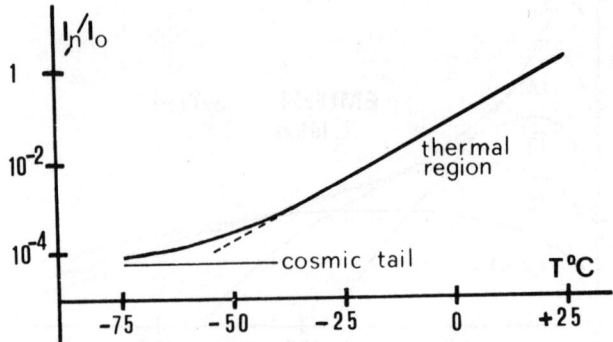

Figure 4.1.5   Thermal behaviour of the average dark noise current for typical photomultiplier.

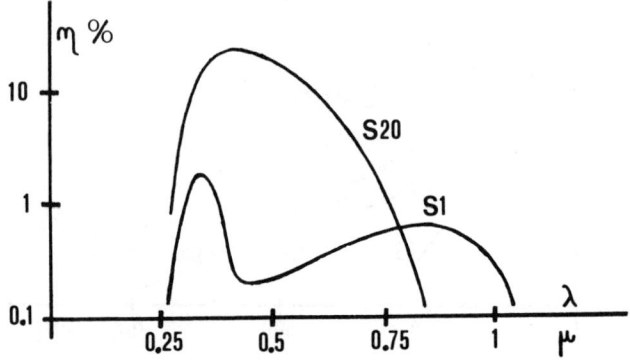

Figure 4.1.6   Spectral quantum efficiency of typical blu and red photocathodes.

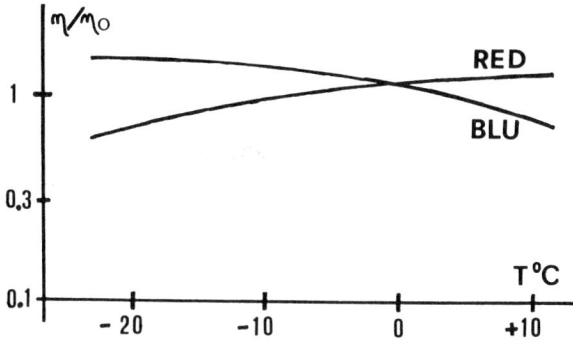

Figure 4.1.7 Thermal behaviour of the relative spectral efficiency in blu and in red bands.

The temperature dependence of the main parameters of current PM detectors implies the need for a careful temperature regulation of the PM tube. Also cooling the PM tube is important for decreasing the dark noise, but the temperature stability is more important for high accuracy operations.

4.2 THE ANALOG TRANSDUCER.

The analog transducer is based on operating the PM detector in analog mode. The analog operation is implemented by measuring the input signal through the integration of the PM output current over the assigned measuring time (This technique is used in IR band by replacing the PM with a suitable IR detector and using a fron end electronics consistent to the high read out noise condition typical of available IR detectors and front end amplifiers).
The state of the art analog transducer as shown in Fig. 4.2.1 uses a high impedance, low noise FET operational amplifier to convert the PM current output into a low source impedance voltage. The equivalent voltage source is then connected to a voltage to frequency converter

which generates a train of standard pulses of average rate linearly proportional to the input voltage.
The integration over the assigned measuring time and the final conversion to the number which constitutes the final measure are implemented contemporarily by means of a counter which counts the pulses output from the voltage to frequency converter. Th RC time constant of the current to voltage amplifier must be matched to the period resolution of the voltage to frequency converter to ensure a proper integration mode. In practice the time constant must be higher than the reciprocal of the average frequency rate at nominal operating point.

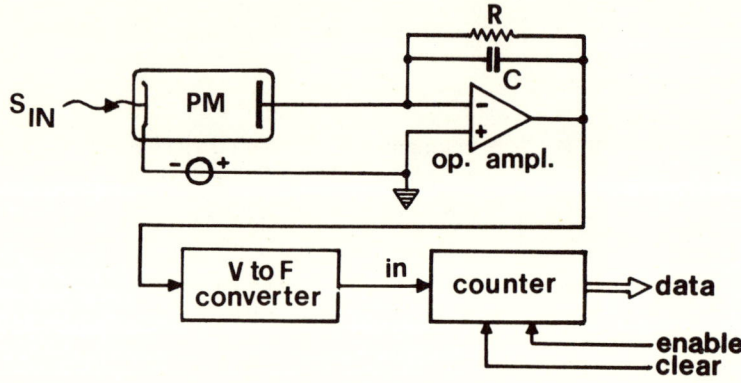

Figure 4.2.1   Structure of the PM based analog transducer.

The linear dynamic range of the electronics is typically of three decades. The electronics dynamics can be matched to the much larger dynamics of the PM by varying the feedback resistor of the current to voltage amplifier. Smaller variations can be accommodated by using an amplifier in non-inverting variable gain configuration. Very large gain variations can be accommodated exclusively by varying the PM anode voltage. Care must be given to the stabilization time required by the PM detector after a voltage supply change.
The transfer function of the analog transducer as shown

is controlled by the two digital lines that enable and clear the counter (that define the integration start and duration times) and by an analog quantity, the feedback resistor R. It is easy to digitize the value of R and to select one of the values through a digital address word. The same solution can also be used to set the gain of the input amplifier if a variable gain configuration is used. Consequently, the transfer function of the analog transducer can be in any case controlled by means of a digital data word. This is important for easy matching to a digital processor.

The present technology (See ANALOG DEVICES and Teledyne Philbrick Technical Data Catalogues) allows to build analog transducers based on PM detectors that are suitable for astronomical operations up to $m_V=+12$ on a telescope of 1 m² aperture.

The limitation of analog transducers to the above level follows from the considerations on the S/N ratio at the anode of available PM detectors rather than on limitations of input amplifiers. The amplifier noise figures limit the analog operations only if the PM dark noise current is substantially reduced by cooling the PM tube to very low temperatures. However, analog operations at fainter source levels require high values of the feedback resistor R and extremely high stability figures for the temperature and the voltage supplies and a very careful and effective shielding of the transducer with respect to external electrical and/or magnetic noise. All this (and the availability of photon counting as an ideal faint source measuring mode) limits in practice the analog operation to the figure reported above.

This discussion is summarized in the data reported in the following table. The values used in computing the magnitude scale origin were taken frm Allen (1964).

m_V	I_a (A)	S/N (PM)	S/N (Amplifier)
0	10^{-3}	10^6	—
+5	10^{-5}	10^4	—
+10	10^{-7}	10^2	—
+15	10^{-9}	1	10^4
+20	10^{-11}	10^{-2}	10^2

The data reported in the above table are computed by taking a PM gain of 10^7, a PM noise current of 1 nA, a PM efficiency of 17%, a collecting area of $1m^2$, and a total transmittance of 0.3. The data on the S/N ratio for the amplifier only are computed by taking an amplifier noise current of 0.1 pA and an average level of output voltage of 1V corresponding to a feedback resistor of 10^{10} ohms. It must be understood that the S/N figures reported are limiting figures computed in the absence of any stability problem. In practice the overall instability will limit the operations from one to two orders of magnitude below the theoretical limits. A very important fact to be emphasized (Sareyan and Ischi,1973) is that analog PM transducers suffer of a variety of small, concurrent non-linear effects that are very difficult to be understood and calibrated out. The overall limiting accuracy for analog transducers seems to be limited to 0.3 to 1% in best cases. Notice that this is not exceedingly important in practical astronomical applications if a two sources comparison operating mode is used that allows to operate the analog transducer in a differential mode.

4.3 THE DIGITAL PHOTON COUNTING TRANSDUCER.

The digital photon counting transducer is based on operating the PM detector in digital mode by counting each detected photon as an individual event.
The photon counting transducer is shown in Fig. 4.3.1. The photon counting is implemented by passing all the PM anode pulses that shows amplitude within an assigned window to a counter for the integration over the assigned measuring time. The selection of the pulses is made by means of a two-levels discriminator circuit. The discriminator is connected to the PM through a pulse amplifier if the amplitude range of the PM output and the discriminator input are not directly matched together.
The fundamental advantage of the photon counting transducer consists in showing a practically zero read out noise and an intrinsic linearity of operation. The linearity of operation depends however on the statistics of the signal in presence of a finite transducer bandwidth, as discussed in the following. The photon counting transducer also shows a superior stability of operation due to the soft dependence of the total count on the discriminator window stability. In practice, the photon counting transducer approximates the optimum

detection mode. The operational limits are therefore set by the signal statistics and by the observational environment rather than the transducer noise. This recommends the photon counting transducer for any low level application.

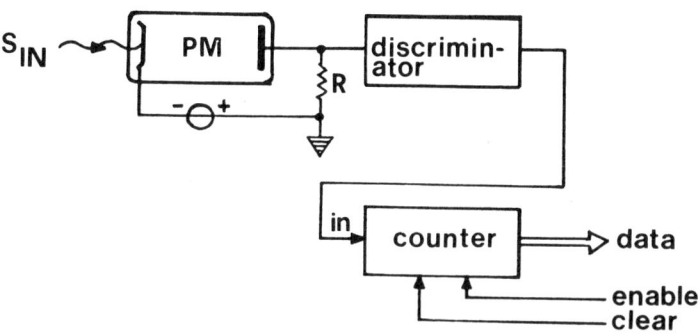

Figure 4.3.1 Structure of the PM based digital photon counting transducer.

The discriminator window must be set according to the PM signal and noise amplitude spectra so as to optimize the overall accuracy of the measurements. A rough setting of the counting window can be made by the direct comparison of the signal and noise amplitude spectra of the PM at low S/N input signal, as shown in Fig.4.3.2. The accurate setting of the counting window can be made by computing the precision of measurement at the given S/N input ratio against the lower and upper thresholds of the discrminator (Sedmak,1972). Typical results are shown in Fig.4.3.3.
Notice that for astronomical applications a single threshold discriminator is generally sufficient due to the soft dependence of the resulting accuracy on the upper threshold and to the low ratio of the noise of available PM detectors to the sky background counts. However, a two thresholds discriminator is clearly recommended if a good rejection of the multipulsing induced by cosmic generated giant pulses is wanted.

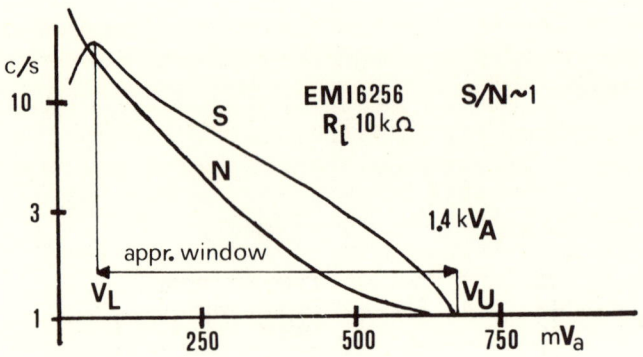

Figure 4.3.2 Approximate location of the counting window by direct comparison of the signal and noise pulse amlitude spectra of the photomultiplier detector.

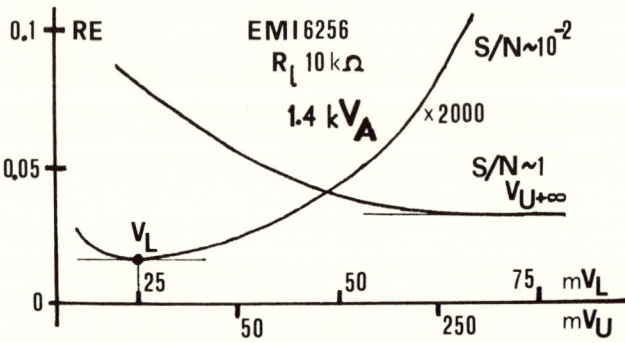

Figure 4.3.3 Final precision of measurement versus the lower and upper discrimination thresholds for a typical photomultiplier detector.

ASTRONOMICAL PHOTOELECTRIC PHOTOMETERS

The photon counting transducer is limited to relatively low signal levels due to the effect of the transducer dead time on the counting statistics. The dead time is defined as the average time elapsing from one count that is sufficient to allow the next count. The dead time is limited by the PM technology and by the technology of the front end amplifier and discriminator. The observed count can be corrected for the dead time in the case of an optical stationary source by means of the following expression:

$$n_c = n_o/(1-n_o t_d) \quad \text{(counts)} \quad (4.3.1)$$

This equation can be used to correct the observed count n_o into the corrected count n_c up to 75% of the limiting count set by the reciprocal of the dead time t_d. The present technology (See SSRI and EMR Technical Data Catalogues) allows to realize photon counting transducers that can be directly used down to $m_V = +5$ on a telescope of $1m^2$ aperture, as shown in the following table. Notice that the photon counting transducer can be easily interfaced to strong sources simply by means of a neutral density filter in front of the PM tube. The data used to compute the table are the same used for the table of performance of the analog transducer reported in the preceding section.

m_V	$S(\text{counts.s}^{-1})$	Dead Time Correction $t_d = 10$ ns, %	Dead Time Correction $t_d = 100$ ns, %
0	5.10^8	———	———
+5	5.10^6	5.3	100
+10	5.10^4	5.10^{-2}	0.5
+15	5.10^2	5.10^{-4}	5.10^{-3}
+20	50	5.10^{-5}	5.10^{-4}

It should be understood that the linearity corrections are not equivalent to the linearity errors, that result typically ignorable under common operating conditions.

4.4 SYSTEM CALIBRATION.

The transducer transfer function must be calibrated before operating the measuring system. The calibration must be monitored in time for granting the accuracy of the measures obtained. This follows from the unavoidable variation of the transducer transfer function due to aging and fatigue effects.
The calibration may be absolute or relative. An absolute calibration is generally not necessary for astronomical applications concerned to an astronomical reference source. However, a rough absolute calibration can be useful to make sure that the transducer is operated within the proper dynamic range.
The calibration must apply to the overall system. Any component which is not included in the calibration must be calibrated against an astronomical reference (atmospheric transmittance, telescope loss, etc.).
The calibration of the transducer must define the transducer responsivity and noise. The calibration stability must be validated by a suitable control of the transducer environment and by a careful operation of the transducer according to the main settling times involved. Temperature regulations of 0.1 ^{o}C for the PM tube and 0.5 ^{o}C for the electronics should be in any case guaranteed for high accuracy operations. Voltage regulation of 10^{-4} is also recommended, particularly for analog transducers. A careful MUMETAL shielding of the PM tube is fundamental for avoiding position-dependent PM gains as a function of the relative position of the PM tube against the Earth magnetic field vector.
The aging effects can be taken into account exclusively by periodical re-calibration of the transducer. The critical components are the PM tube and narrow band-pass filters.
The fatigue effects are much more delicate to be taken into account. Non-reversible fatigue effects are generally very small and can be included in the aging if anomalous operations are avoided (PM burn-out, supply voltage shocks, etc.). The reversible fatigue effects depend on the particular measuring situation, and must be evaluated on a case per case basis. The analog transducers are partcularly sensitive to fatigue effects. The recovery time of the transducer transfer function when switching between sources of levels different by more than one order of magnitude may be longer than one minute. This must be taken very seriously into account for the design of high switching rate multiplexed photometers using analog transducers.

The photon counting photometers suffer considerablly less of the fatigue effects on rapidly varying sources if the maximum source level does not approximate the photocathode saturation level. Care must be given in this case to the fatigue effects in the discriminator electronics, which may be more important than the PM effects even if much lower than in analog photometers. Finally, it must be emphasized that the calibration of a high accuracy photometer requires a calibration system of consistent accuracy. The most critical parameters are the photometric and the spectral responses.
The photometric response must be calibrated over a dynamic range of 10 decades for general astronomical applications. Such a broad range can be accommodated by maintaining a high accuracy by means of the fractional aperture technique in the realization of the signal attenuator (Pitz,1979). The spectral response can be calibrated by means of a set of neutral and band filters integrated into a star simulator (Sedmak et al.,1977). An important facility of the transducer for field astronomical applications is the availability within the transducer of a standard calibration source which can be addressed before the execution of the measurements to make sure that the system transfer function maintains the nominal calibration, or to identify the need for a new calibration.

5. THE PROCESSOR AND THE DATA MANAGEMENT.

The processor inputs the commands of the user, controls the transducer so as to execute the wanted measurement, inputs the data output from the transducer, and outputs the measure in the form requested by the user who commands the system. The output includes the documentation of the measuring operations and the archive.
These tasks constitute the data management. The control of the transducer and the data acquisition are on-line tasks that must be executed in real time. In particular the control of the transducer includes the fundamental task of the timing. The timing is implemented by means of one or more real time clocks that are used to maintain the local time base and to assign the start times and the integration times for the various measurements. The documentation and the archive tasks may be executed off-line, or on-line but not in real time.
The data management is presently implemented by means of a digital computer loaded with the necessary peripherals.

The transducer is interfaced to the computer, which is programmed so as to execute the measurements and to archive the data on the assigned supports.
The operations can be optimized against a given criterion of optimization by implementing a closed feedback control loop through the transducer transfer function control port (De Biase and Sedmak,1974).
The basic configuration of the computer integrated data management system is shown in Fig. 5.1. The user interacts to the system through the computer consolle. The current data are monitored on the graphic video display and on the hard copy monitor (usually a strip chart recorder). The measures are stored on the standard magnetic tape unit. The system disc serves as the support for the operating system software and as a buffer memory for better data handling in high speed data acquisition operations. The graphic terminal and the magnetic tape unit can be deleted if low speed operations only are foreseen and if the disc is a removable carttridge or floppy disc unit, in which case the data exchange can be made directly on the disc support.

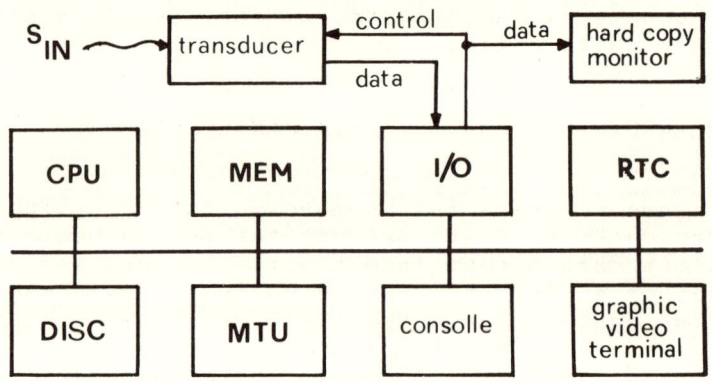

Figure 5.1 Configuration of the computer integrated data management system.

The computer can also be used for dedicated on-line data processing, like direct two-sources comparisons, Fourier spectral analysis, synchronous photometry,etc. (Santin and Sedmak,1976;Furlani and Sedmak,1977;

Rayner and Watson,1974).
The basic configuration can be implemented by means of
a standard minicomputer system, or it can be realized
by means of a dedicated network of microprocessors, or
any combination of these two approaches.
The current technology recommends using a dedicated array of microprocessors as the optimum performance to
flexibility and cost ratios solution. Care must be given to evaluate correctly the cost and effort of the
software required to operate the hardware. It is recommended to use a consolidated microprocessor or low-end
minicomputer for the system pole so as to be able to
interface standard peripherals through standard software (an array of INTEL 8080 family micros bus-connected to a DEC PDP11/23 microcomputer may be a nice 1980
compromise of all parameters).
The importance of the software for both dedicated tasks
and general system management must be emphasized with
special reference to the possibility of writing the
software in high level languages (bit-oriented FORTRAN
is an example of available possibilities). Details on
the control software cannot be given here, but it may
be useful to make reference to the documentation on the
system hardware and software of state of the art computer manufacturers (See for example INTEL and DEC System Data Catalogues).

6. CONFIGURATION AND OPERATION OF OPTIMUM PHOTOMETERS.

The configuration and operation of optimum photometers
follows from matching the observational environment to
the available technology.
The optimum solutions may be very different according
to the assigned observational tasks, but some general
considerations can be made that in practice imply some
standardization of the configuration of the photoelectric photometers for astronomical applications and the
associated operating modes.
The considerations are the following:

(a) The atmospheric transmittance is the major source
    of errors on ground-based observations of sources
    brighter than m_v=+10 on a telescope of 1m aperture.
(b) Commonly available telescopes show a total field of
    view that is sufficient to observe contemporarily
    two sources and their backgrounds.
(c) Parallel photometers are the most efficient systems
    but the difficulty of calibration and of maintenan-

ce of the calibration is exceedingly high unless using an imaging photometer.
(d) Photon counting is the most effective detection technique.
(e) Computer integrated data management systems are the most effective for optimum transducer control and data acquisition and archive tasks.

Moreover, some quantitative considerations are possible for any given configuration by computing the overall relative error of measurement as given by equation (2.3). One sample calculation is reported in Fig.6.1. Notice the strong equalization effect of the multiplicative transmittance noise on sources brighter than $m_V = +10$.

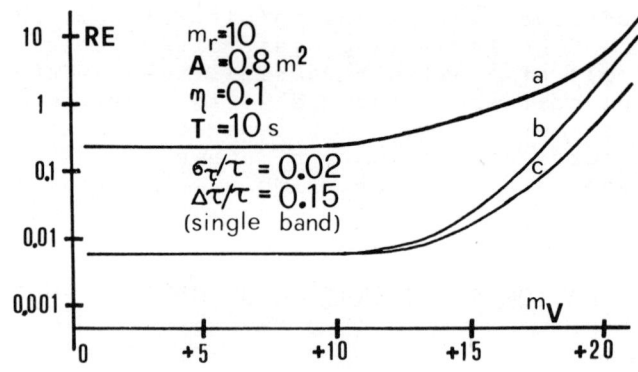

Figure 6.1   Sample overall relative error of measurement for single band measures obtained by (a) fully multiplexed, (b) two transducers, and (c) full parallel four transducers photometers.

The considerations (a) to (e) and the plots of Fig.6.1 allow to identify two main fields of applications: strong sources and low time resolutions, and faint sources and high time resolutions. The critical parameter is the source level to data acquisition rate ratio. On this basis the following classification can be made:

For strong to medium level sources at high time resolution or for faint sources at medium to low time resolution the optimum practical configuration for ground-based photoelectric photometry is the two transducers, band switched, photon counting computer controlled system.
For faint sources at high time resolution the optimum practical configuration is the one transducer per band photon counting computer controlled system.
In both cases the transducers are switched between the sources and the associated backgrounds. Notice that for space-based observations only one transducer per source per band is needed.
Presently available imaging photometers are not competitive to the solutions quoted above at high accuracy operating conditions. However, it must be understood that the fully parallel imaging photometer represents the limiting solution as soon as the quality of the available two-dimensional detectors approximates the quality figures of available photomultipliers.
One example of the first approach is the URSULA photometer (DeBiase et al.,1978). One example of the second approach is the photometer designed for the Space Telescope. The Wide Field Camera and the Faint Object Camera designed for the Space Telescope also constitute two examples of the best machines presently available for imaging astronomical photometry.

7. FUTURE SYSTEMS AND CONCLUDING REMARKS.

The photoelectric photometers discussed in this review are the consolidated result of the application of field proven technology to astronomical observations.
It is clearly possible to design and build non-standard photoelectric photometers for astronomical applications that can yield superior performance in normal cases or that allow to execute measurements otherwise impossible or extremely difficult or exceedingly unaccurate.
One example of a standard photometer based on a consolidated but not astronomy-standard technology is the high speed photopolarimeter designed for the Space Telescope. This instrument shows a conventional multibeam optical design integrated to an array of photon counting image dissector photomultipliers. The advantages of this solution consist in the capability of the image dissector to be operated as an array of photomultipliers with equal photocathode and non-mechanical addressing.
One example of non-standard photometer based on a proven

but actually developing technology is the imaging two-dimensional photon counting photometer realized by means of a two-dimensional multielement detector. The typical working machine of this type is the Faint Object Camera designed and built for the Space Telescope on the basis of the Boxenberg original UCL photon counter. The use of a multielement imaging detector allows to measure contemporarily several sources on several bands with the substantial advantages of the optimum estimation of the sky background, the optimum efficiency, and the optimum detection capability following from the oversampling of the source image.

The application of new technologies and methodologies will undoubtedly change the scenary of astronomical photoelectric photometry in the near future.
The major improvements in the UV to near-IR band are foreseen with respect to the accuracy to thruput ratio. This follows from the current availability of photometers that approximate the limits of detection of ideal photometers on a limited number of channels. The future machine will probably be a multielement photon counting two-dimensional imaging detector with high speed readout for high speed operation.
A substantial improvement is also foreseen in the far-IR band with the introduction of more efficient IR detectors, like solid state IR arrays and broad-band IR heterodyne systems.
Some improvements are also foreseen in the data management, particularly in regard of the optimum control of not fully parallel systems and the general data processing.

REFERENCES

Allen C.W., Astrophysical quantities,1964,The Athlone Press, London.
Analog Devices,Data Acquisition Products Catalogue, Route One,Industrial Park,Norwood MA 02026,USA.
De Biase G.A., Sedmak G., Astron.Astrophys.33,1974,1.
De Biase G.A.,Paternò L.,Pucillo M.,Sedmak G.,Appl.Opt. 17,1978,435.
DEC Computer Systems Data Catalogue, Digital Equipment Corporation, 146 Main Street, Maynard MA 01754,USA.
EMI Photomultiplier Data Catalogue, EMI Ltd,243 Blyth Road,Hayes,Middlesex UB3 1HJ,UK.
EMR Photomultiplier Data Catalogue and PAD631 Data Sheets,EMR Photoelectric, Princeton NJ 08540,USA.

Furlani S., Sedmak G., Astrophys.Space Sci. 48,1977,65.
Young A.T., Appl.Opt. 8,1969,2431.
INTEL System Data Catalogue, INTEL,3065 Bowers Avenue, Santa Clara CA 95051,USA.
Morton G.A., Appl.Opt. 7,1968,1
Paternò L., Astron.Astrophys. 47,1976,437.
Pitz E., Appl.Opt. 18,1979,1360.
Rayner P.T.,Watson R.D., ASA Proceed. 2,1974,274.
RCA Photomultiplier Manual, RCA Electronic Components, Harrison NJ 07029, USA.
Santin P.,Sedmak G., Astrophys.Space Sci. 48,1977,57.
Sareyan J.P.,Ischi E., Astron.Astrophys. 27,1973,183.
Sedmak G., Astron.Astrophys. 18,1972,232.
Sedmak G., Astron.Astrophys. 25,1973,379.
Sedmak G., Astron.Astrophys. 25,1973,41.
SSR Instruments Co., 1120 Amplifier Discriminator Data Sheets, SSRI, 1001 Colorado AV. CA 90404, USA.
Teledyne Philbrick Linear Circuit Modules, Teledyne Philbrick, Allied Drive Route 128, Dedham MA 02026,USA.

DETECTION OF CLOSE BINARY SYSTEMS BY MEANS OF LUNAR OCCULTATIONS

G. Longo and M. Rigutti

Osservatorio Astronomico di Capodimonte,
Naples, Italy.

ABSTRACT

Applications of the lunar occultation technique to the detection of double systems with angular separations less than 0.01" are thoroughly discussed. Particular emphasis is given to the analysis of the various factors limiting the resolution power of the method. A complete list (up to June 1980) of all the double systems revealed or measured with this technique is also given.

1. INTRODUCTION

Although the suggestion of considering lunar occultations of stars to detect close binary systems is older than one hundred years, some first clear indications on the use of this method appeared only in 1909. MacMahon (1909) recognizes the potentialities of this technique and Eddington (1909) describes the presence of diffraction effects during an occultation. However, the theory was developed thirty years later by Williams (1939) and, with the exception of an attempt performed by Whitford in 1938 (Whitford, 1939), actual measurements with the occultation method had to wait for the new techniques of fast photometry, because of the extreme short duration of the phenomenon. In 1950 and 1952 O'Keefe and Anderson (1952) observed photoelectrically the star 228 B Aur for which they measured an angular separation of 0.05". But only in the last ten years a systematic and extensive use of this method was made to contribute in a significant way to the discovery and resoltuion of very close star systems.

2. BINARY SYSTEMS: SOME GENERAL RESULTS

First of all it will be convenient to give a glance at the general situation to secure a sufficiently clear and precise picture of it, in order to know the frame where the method we are interested in has to be placed, and what we reasonably can expect to get from it.

Down to the 10th (apparent) magnitude, there are about 75000 visual catalogued double stars, but for only 1% (about 600) of them the observations are sufficiently complete for computing orbits. Further, about 4000 photometric binary stars are known, but for only 10% (about 400) of them are the light curves good enough to obtain orbits. The majority of these systems are spectroscopic binaries. Finally, about 2000 spectroscopic pairs are listed. Some 50% have known orbits.

We then know, more or less accurately, 3000 binary system orbits in all. This is not a very large number, although not too small to allow some statistics to be based on it. For a given instrumentation (for example, a given telescope) the parameters which determine the possibility of discovering and observing a binary system are somewhat different for different types of pairs.

For visual systems these parameters are:

i) the visual total (combined) magnitude of the system : m_T ;
ii) the magnitude difference between the two components A and B:
$$\Delta m = m_B - m_A \quad \text{with} \quad m_B > m_A \;;$$
iii) the angular separation ρ (in arc sec) between the components A and B.

Factors like the ones here mentioned, which clearly affect the measurements and lead to selection effects, have to be allowed for, when we want to make statistics. This means that we have to take into account, in some way, that the most luminous systems may also be observed at large distances and that large separations between the two components of a visual double system help in the observations. A "measure of difficulty" C to observe visual pairs, considering the influence of Δm and ρ, was given by Opik (1924) and later by Heintz (1978), who gives for C the following definition:

$$C = 0.22 \, \Delta m - \log \rho$$

and one can say that down to the magnitude 9.5, binaries with $C < 0.5$ may be considered well known, whereas binaries with $C > 1.0$ are essentially unknown.

As for the spectroscopic systems the most luminous are obviously more easily detected (they can be observed at large distances while the Doppler effect does not depend on the distance). Also, large Doppler shifts help the discovery and the observation of binary systems.

Instead, photometric pairs are best observed when the two components are little different in magnitude. This is easily seen observing that

$$m_A - m_T = 2.5 \log(1 + 10^{-0.4 \Delta m}).$$

For $\Delta m = 0$ one gets $m_A - m_T = 0.75$, but
for $\Delta m = 6$, $m_A - m_T = 0.00$,
which means that for $\Delta m = 6$ the secondary minimum in the lightcurve is missing. Anyway, also for $\Delta m = 5$, the situation is not much better. In fact, for this value of Δm,

$$m_A - m_T = 0.01.$$

Let us now consider angular separations. Visual binaries are still observable when the angular separation is about 0.2", but below this limit observations are extremely difficult because of seeing effects. For angular separations less than 0.2", the fact that the observed object is a binary system may be deduced from the periodic variations of the radial component of the orbital velocity. Of course, this kind of observation is made easier either by large values of the orbital velocity and/or large values of the inclination of the orbital plane to the plane tangent to the celestial sphere.

It is also possible to make observations photographically. With measurements obtained from one hundred images, it is possible to reach internal errors as small as 0.01", but external errors are much larger than this value because, generally speaking, the limits in the total accuracy come from seeing effects. Good results have been obtained also for angular separations of 1", but seeing influences are to be seriously taken into consideration for angular separations $\lesssim 2"$. Special photographic techniques have also been used but here we can disregard them because they do not substantially change the view, and no systematic work has been done with them.

Interferometric measurements are also possible. However, this technique has not been used extensively because of its severe limitations. Only brilliant objects with a not too large value Δm between the two components may be observed. The fringe system is in fact more indistinct for larger Δm's. In any case, as an example, we can quote an angular separation of 0.09" measured with the 67cm Johannesburg refractor (Finsen, 1954, 1964). Hanbury Brown (1974) used an intensity interferometer which works like a radio interferometer on the correlation of photo-currents of two reflectors. But also in this kind of interferometry, in the optical range an increase of the length of the baseline makes the signal-to-noise ratio worse, and the instrument can only be used for brilliant objects. As an example, we may quote the observations of α Vir (a pair with a period of 4 days). Angular separations of 0.0015" have been measured using baselines up to 90m, but it seems reasonable that 0.0001" should be reached with larger baselines.

It must always be remembered that one of the weakest aspects of all the kinds of measurement we mentioned are the sources of errors. This is, of course, almost obvious. Measurements are, in fact, made near the limits of the instruments' possibilities.

Let us conclude with some statistical results which will complete the general view of the field of research where the lunar occultation technique may be usefully applied. According to Couteau (1960), there are at least 5.10^5 binary systems in the magnitude range from 6 to 11. Only 7% of these show angular separations within the values of $0.4" < \rho < 1.0"$; most of the remaining pairs are expected to be much closer; for larger values of the magnitude the ratio should be still worse.

According to Heintz (1969), 100 objects contain 8 double systems with semi-axis from 0.01 to 0.1 AU and 12 double systems with semi-axis from 0.1 to 1, etc., as shown in Figure 1. Further, 100 objects should include 205 components as follows:

30 single stars :	30 components,
47 double stars :	94 components,
23 multiple stars :	81 components,
100 objects :	205 components.

In other words, 85% of the stars should be components of double or multiple systems.

Finally, according to Batten, Worley and Baize, double systems represent, respectively, 66.7% (Batten, 1967), 78.1% (Worley, 1967), and 86.9% (Baize, 1975) of all the stars.

3. THE LUNAR OCCULTATION METHOD, INTRODUCTION

The practical possibility of measuring the angular separation between the two components of a binary system using lunar occultations is based on high time resolution photometry. Nowadays photoelectric photometers are common devices and also pulse-counting equipment is within the means of small observatories. The observations consist of very accurate recording of the fluctuations in brightness of the observed object during the very short period of time of occultation. It is not very difficult to prepare predic-

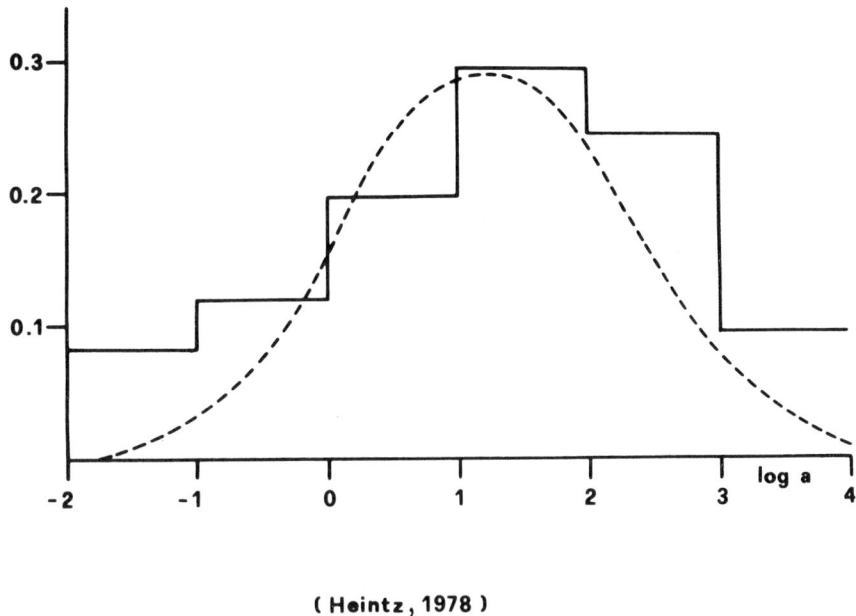

(Heintz, 1978)

Figure 1. Double star semi-axes major distribution. Ordinate is the frequency per star and per unit of log a (Heintz, 1978).

tions of occultations for particular sites, but, in any case, the Naval Observatory provides them on request. Procedures to reduce data of observations have also received sp,e s@ecoa; attention (von Hoerner, 1964).

By means of this method, angular separations of the order of 0.01" are safely achievable, and this means that the occultation method is a very powerful technique for discovering close binary systems. Indeed, in some cases a double system for which no spectroscopic evidence was available, was resolved by this technique. Furthermore, very close pairs with i = 90° (or near 90°) may also be discovered. We will consider later the main aspects of this method, but it may be useful to state that it lies on the fact that a point-like source of light gives rise to a diffraction fringe system when it is occulted by a straight-edge.

Of course, the occultation of a particular object is a rather rare event. According to Herr (1969), in a typical month 160 SAO (Smithsonian Astrophysical Observatory Star Catalog, 1966) stars may be occulted at a given site. But some factors affecting the observations have to be determined experimentally (sky brightness, altitude of the star above the horizon, instrumental set up, etc.), others may be evaluated. Taylor (1966), for example, considers several telescope apertures and gives for them limiting resolutions as a function of apparent visual magnitude.

The main weak point of the method is, of course, that one has to wait for the occultation of a given object, but during observations the discovery of new pairs is a common experience and this aspect of the matter deserves some attention. Evans (1977), who also used the occultation method to measure many stellar diameters, reports that some three or four times per month it may happen that double or multiple systems will be discovered and that even pairs separated by less than 0.01" are recognizable from the diffraction pattern.

Because of the inclination of the lunar orbit to the ecliptic and the motion of the nodes, about one sixth of the whole sky is scanned by the Moon. In fact, the average value of inclination of the lunar orbit to the ecliptic and the motion of the nodes, about one sixth of the whole sky is scanned by the Moon. In fact, the average value of inclination of the lunar orbit to the ecliptic is about 5°09', with a main oscillation of about ±9', which takes place in a period of 173 days. Moreover, the nodes move westward describing the ecliptic in 18.61 years. Taking also into account the size of the Earth, it is easy to see that a band 13° in total width along the ecliptic is the region of the sky where lunar occultation may occur.

4. THE MEASUREMENTS

To get an accurate determination of the angular separation and the corresponding position angle, it is necessary to combine observations performed from several (at least two) sites sufficiently apart in longitude. Programs for this kind of observation are not heavy (good occultations are not very frequent, the observations are not time consuming, and moonless nights are never used) and might be considered at least by all those Observatories which are already equipped for this work.

One of the first pairs discovered through lunar occultations was 27 Tauri at the McDonald Observatory, Texas. For this object a component of the angular separation, perpendicular to the lunar limb of 0.006" was measured (Nather and Wild, 1973).

The geometry of the occultation is shown in Figure 2. As is seen, one does not measure the angular separation of the two components but its projection on the direction of the lunar velocity at the position angle determined by the occultation point on the lunar limb. Generally speaking, the integration time per record during the intensity fluctuation observations lasts one or two milliseconds. For illustrative purposes, we may assume that the velocity of the lunar limb is 0.4" s^{-1}. Let us also assume that the angular separation of a binary system in the direction of the lunar motion is 0.01". In this case the time interval between the occultation of the two stars is 0.025^S. Such a time interval is easily measurable and the corresponding angular separation is one order of magnitude smaller than the values we can visually measure.

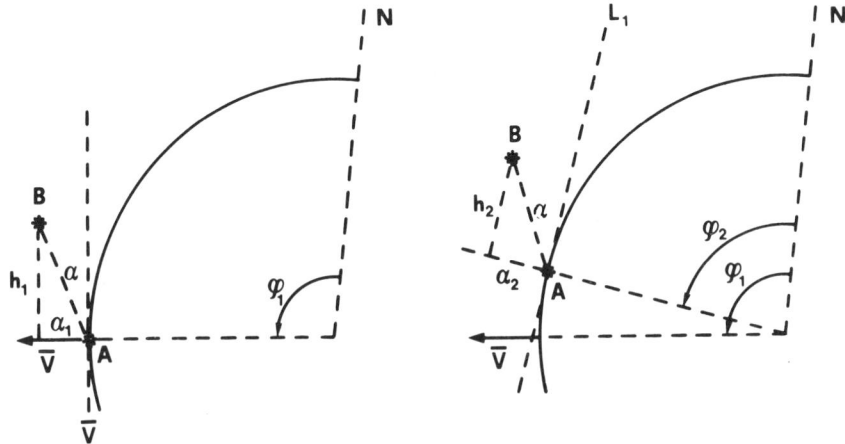

Figure 2. Geometry of an occultation.

Figure 3 is a nice example of the kind of record one can get from fast photometry of lunar occultation (Africano et al, 1975). The occulted star is τ Capricorni. From top to bottom are shown the observed pattern, the model, the fit, and the residuals. The integration time was 2 ms/reading. The angular separation obtained from this observation was 0.352" at the position angle 113.9°.

Another example of lunar occultation is shown in Figure 4 (White, 1977). The whole record lasts 0.15^S. It was performed using a Johnson V passband. It is very interesting to note that the binary nature of the star in Figure 4 came out from this kind of observation although the star had already been observed spectroscopically (Cowley et al, 1969). Through the occultation record it was possible to get $\Delta V = 0.64^m \pm 0.04^m$ for the two components, and a value of 0.052" ± 0.0005" for the projection of the angular separation in the position angle 123.4°.

A third example is shown in Figure 5. Here the star is σ Sco (Nather et al, 1974). This star was simultaneously observed from

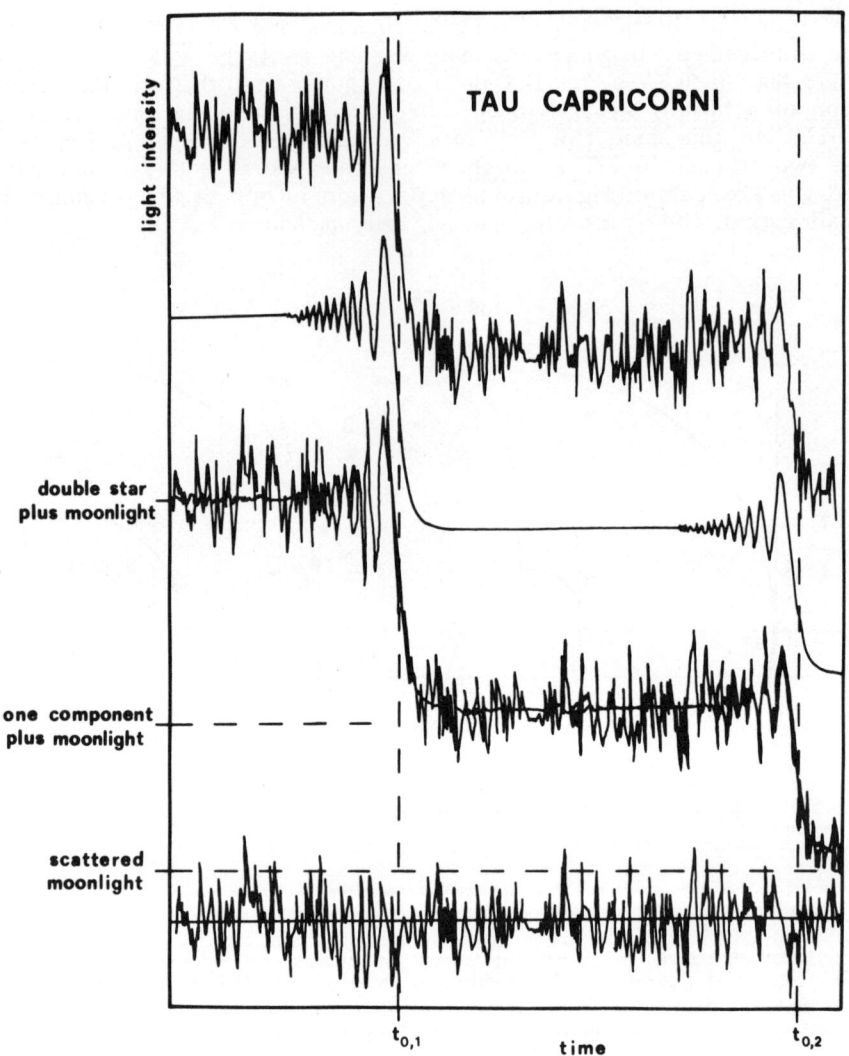

Figure 3. Occultation trace of τ Cap.
From top to bottom: observed trace,
model,
fit,
residuals,
(adapted from White, 1977).

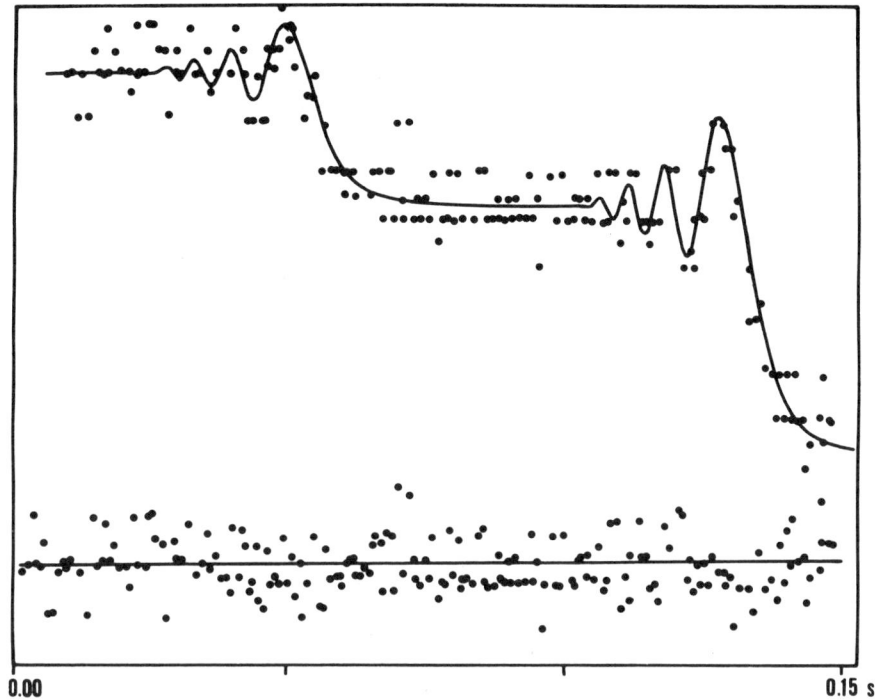

Figure 4. Occultation trace of BS 2304 (adapted from White, 1977).

the South African Astronomical Observatory at Cape Town and the S.A.A.O. outstation at Sutherland. Figure 5 is a reproduction of the trace recorded at Sutherland with a 20-inch reflector and a conventional photometer with a B filter. The integration time was 4 ms/reading. The presence of the companion is shown by the small drop in the first part of the trace, while the larger drop corresponds to the occultation of the primary.

As already stated, an actual determination of the angular separation between the two components of a double system, in the position angle determined by the occultation point on the lunar edge, is possible only when observations are performed at least from two different sites.

Let us assume (see Figure 2 and Figure 6) that

α = angular separation = $\overline{AB}$,
α_1 = projected separation for site 1 = $\overline{AD}$,
α_2 = projected separation for site 2 = $\overline{AC}$,
ϕ_1 = position angle for site 1,

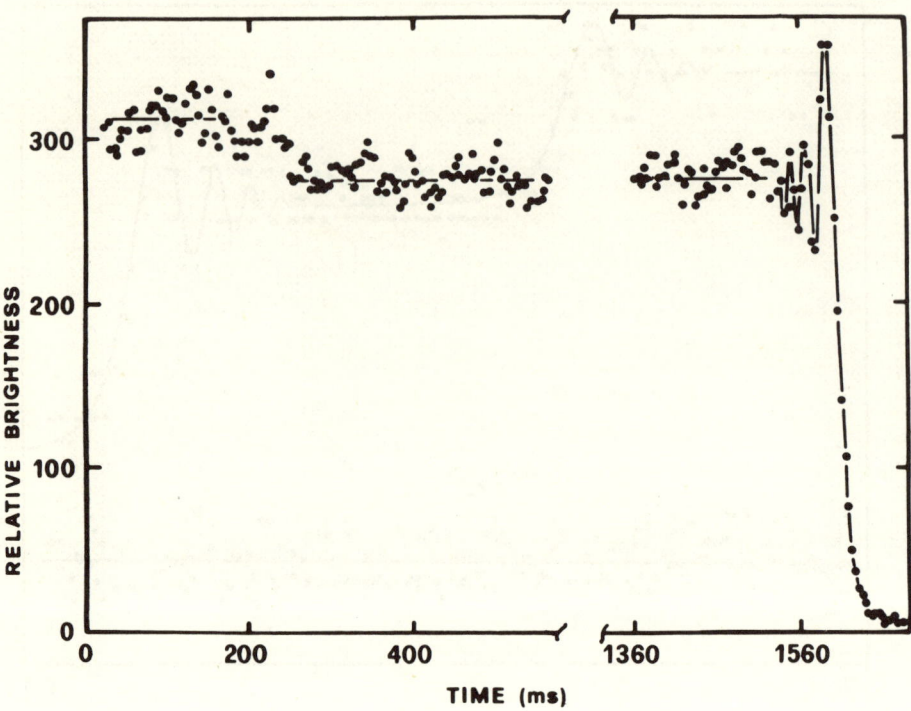

Figure 5. Occultation trace of σ Sco (Evans, 1970)

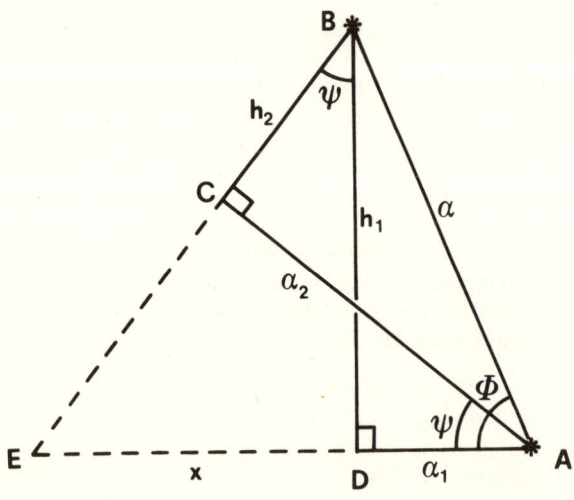

Figure 6. For computing α and φ (see text)

ϕ_2 = position angle for site 2,
$\psi = \phi_1 - \phi_2$,
$\widehat{ACE} = 90°$;

then

$$\cos \psi = \alpha_2/(\alpha + x)$$

or

$$x = (\alpha_2 - \alpha_1 \cos \Delta)/\cos \psi .$$

But $\widehat{CBD} = \widehat{CAE}$ so that, from the triangle EBD, one gets

$$\overline{BD} = h_1 = x/\tan \psi .$$

From this we obtain

$$\alpha = (h_1^2 + \alpha_1^2)^{\frac{1}{2}} \quad \text{and} \quad \phi = \tan^{-1} \frac{h_1}{\alpha_1}$$

or

$$\alpha = \frac{(\alpha_1^2 + \alpha_2^2 - 2\alpha_1 \alpha_2 \cos \psi)^{\frac{1}{2}}}{\sin \psi}$$

and

$$\phi = \tan^{-1} \frac{\alpha_2 - \alpha_1 \cos \psi}{\alpha_1 \sin \psi} ,$$

which give, respectively, the angular separation and the position angle.

5. THE DIFFRACTION FRINGE PATTERN

In order to better understand the meaning of the observational data we get from a lunar occultation observed with a fast photometry system, it is useful to start from a theoretical model containing some simplifying assumptions.

Let us consider an ideal monochromatic receiver out of the Earth's atmosphere and suppose it to observe a lunar occultation of a star sufficiently far away to be considered a point source.

Disregarding the lunar limb curvature we may treat the occultation as one does with a typical phenomenon of Fresnel diffraction by a straight edge. With reference to Figure 7, classic theory shows that the intensity measured by the receiver O at the time t is given by:

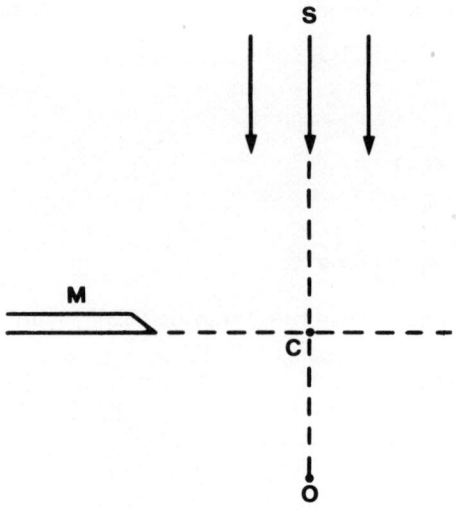

Figure 7. Scheme of a lunar occultat on of a star:
S = point-like star
M = lunar edge
O = observer.

$$f[\omega(t),\lambda] = I_\lambda p(\omega) \quad (1)$$

where I_λ is the monochromatic flux density from the star at the wavelength λ outside the Earth's atmosphere, and $p(\omega)$ is the so-called Fresnel function (see Born and Wolf, 1959)

$$p(\omega) = \tfrac{1}{2}[\tfrac{1}{2} + C(\omega)]^2 + \tfrac{1}{2}[\tfrac{1}{2} + S(\omega)]^2 ,$$

where

$$C(\omega) = \int_{-\infty}^{\omega} \sin\left(\frac{\pi\tau^2}{2}\right) d\tau \quad \text{and} \quad S(\omega) = \int_{-\infty}^{\omega} \cos\left(\frac{\pi\tau^2}{d}\right) d\tau ,$$

the shape of the function in $p(\omega)$ is given in Figure 8.

Particularly important is the shape of the Fresnel adimensional variable ω, which in the occultation problems may be written in the form (see de Vegt and Gehlich, 1976)

$$\omega = v_r(t - t_o) \frac{2D}{\lambda}^{1/2} ,$$

where v_r (measured in arcsec s^{-1}) is the angular velocity of the lunar limb in the direction of the occultation, D is the distance of the lunar edge from the observer and t_o is the time of the geo-

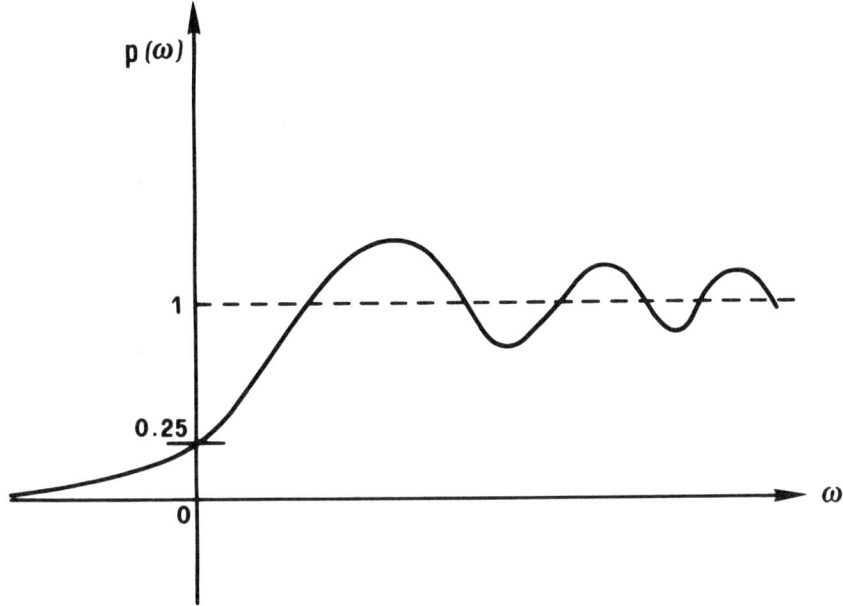

Figure 8. The Fresnel function $p(\omega)$

metrical occultation.

Taking into account that, when $t = t_o$, $\omega = 0$ and $p(0) = 0.25$, it comes out that when the geometrical occultation takes place the occultation curve reaches a quarter of its initial value. However, it must be noted that this is true only when Equation (1) is valid, and, particularly, when we can consider the star to be a point source of light.

Considering two close point sources, the occultation curve is the linear sum of those one would separately get for the two single occultation events. However, the two events are clearly shifted by $\Delta t = (t_{o,1} - t_{o,2})$ where $t_{o,1}$ and $t_{o,2}$ are respectively the moments of geometrical occultations of the two stars (Figure 3). The analytical form of the occultation curve is then (see the theoretical pattern in Figure 3)

$$f[\omega(t)] = I_{\lambda,1} \, p(\omega_1) + I_{\lambda,2} \, p(\omega_2), \qquad (2)$$

with

$$\omega_i = v_{r,i}(t - t_{o,i})\left(\frac{2D}{\lambda}\right)^{\frac{1}{2}}, \qquad i = 1,2 \ ;$$

where $I_{\lambda,1}$ and $I_{\lambda,2}$ are the values of the monochromatic flux densities of the two single sources.

For double stars with a large value of the angular separation it is not possible to assume that $v_{r,1} = v_{r,2}$ because $v_{r,i}$ is a function both of the geometry of the occultation and of possible irregularities of the lunar edge at the occultation point.

Finally, it must be taken into account that, with the exception of double systems with ρ very large, $I_{\lambda,1}$ and $I_{\lambda,2}$ are unknown, because the only measurable quantity is the total monochromatic flux density

$$I_{\lambda,tot} = I_{\lambda,1} + I_{\lambda,2} \ .$$

6. FACTORS AFFECTING MEASUREMENTS

Before writing some notes about the way usually adopted to get the quantities $t_{o,1}$, $t_{o,2}$, $v_{r,1}$, $v_{r,2}$, $I_{\lambda,1}$ and $I_{\lambda,2}$ from the occultation trace, it is necessary to consider several factors which disturb the observations.

First we must note that the receiver is not ideal; therefore its size, passband, integration and observation times are finite. Besides, there is instrumental noise. Moreover,

i) the lunar edge is not straight and its shape is not regular;
ii) the observations are subject to the influence of seeing and scintillation; and
iii) the light sources may not be point sources.

Finally, we must recall, as a limiting factor, that the stars which can be occulted are only those in the zodiacal band.

We shall discuss now, in some detail, all the listed disturbing factors.

6.1 Finite Size of the Receiver

To write Eqs. (1) and (2) we suppose that the receiver has a constant gain over all angles. However, the receiver gives different weights to the contributions of different parts of the wavefront. If we assume that the instrumental gain is constant within a solid angle 2Ω and zero outside of it, it is possible to show that one gets no information on the Fourier components of the entrance signal with spatial frequency $\nu > \Omega D/\lambda$ (Reboul, 1979). Therefore, the maximum angular resolution Ω_m may be about

$$\Omega_m \simeq \frac{\lambda}{D\Omega} \quad \text{which, with } \Omega \simeq \frac{\lambda}{d} \ , \text{ gives}$$

$\Omega_m \sim \frac{d}{D}$, where d is the telescope aperture.

With d = 100 cm, D = 4.10^{10} cm,

$\Omega_m \sim 5.2 \times 10^{-4}$ arcsec.

6.2 Finite Passband

Another source of errors lies in the fact that observations are never strictly monochromatic. Particularly with very weak sources, wide passband receivers are necessary.

Therefore, the occultation curve (1) must be re-written as follows:

$$f[\omega(t)] = \int_{-\infty}^{+\infty} R(\lambda_o + \Delta\lambda)\, p_\lambda(\omega)\, d(\Delta\lambda), \qquad (3)$$

where $R(\lambda) = T(\lambda)\, S(\lambda)\, I(\lambda)$, with

$T(\lambda)$: transmission coefficient of Earth's atmosphere plus optics,
$S(\lambda)$: chromatic sensitiviety of the receiver,
$I(\lambda)$: monochromatic flux density from the star at the wavelength λ, outside the Earth's atmosphere.

The Fourier transform od Eq. (3) (see Hazard, 1976; and Krishnon, 1971) is

$$F(a) = \int_{-\infty}^{+\infty} R(\lambda_o + \Delta\lambda)\, P_\lambda(a)\, d(\Delta\lambda),$$

but

$$P_\lambda(a) = P_{\lambda_o}(a)\, \exp(i\pi a^2\, \Delta\lambda/D\, \text{sng } a).$$

Therefore,

$$F(a) = P_{\lambda_o}(a) \int_{-\infty}^{+\infty} R(\lambda_o + \Delta\lambda)\, \exp(i\pi a^2 \Delta\lambda/D\, \text{sgn } a)\, d(\Delta\lambda) =$$

$$= P_{\lambda_o}(a)\, B(a),$$

where

$$B(a) \equiv \int^{+\infty} R(\lambda_o + \Delta\lambda)\, \exp(i\pi a^2 \Delta\lambda/D\, \text{sgn } a)\, d(\Delta\lambda).$$

We then get

$$f[\omega(t)] = p(\lambda_o,\omega) * b(\omega), \qquad (4)$$

where

$$b(\omega) = \int_{-\infty}^{+\infty} B(a) \exp(2\pi i a \omega) \, da \;.$$

Taking into account the smoothing introduced by a convolution process, Eq. (4) permits us to estimate the loss of resolving power due to the finite passband.

Assuming that $R(\lambda_o + \Delta\lambda)$ is gaussian, we may write

$$R(\lambda_o + \Delta\lambda) = \exp(-\Delta\lambda^2/2\sigma^2),$$

which has a halfwidth $\Delta\lambda_o = \sigma(8 \ln 2)^{\frac{1}{2}}$. From this we get (see Hazard, 1976)

$$B(a) = \exp(-\gamma a^4),$$

with

$$\gamma = \pi^2 \Delta\lambda_o^2/16 D^2 \ln 2.$$

Then

$$b(\omega) = \int_{-\infty}^{+\infty} \exp(-\gamma a^4) \cos(2\pi a \omega) \, da \;. \qquad (5)$$

Sheuer (1965) has shown that Eq. (5) may be approximated very well by a gaussian function having a halfwidth $\beta_{\Delta f}$ given by

$$\beta_{\Delta f} \simeq 6.5 \, \lambda^{\frac{1}{2}} \left(\frac{\Delta f}{f}\right)^{\frac{1}{2}}$$

which, using the value $\Delta f/f = 0.13$, corresponding to the Johnson V passband, gives

$$\beta_{\Delta f} \simeq 0.02.$$

6.3 Finite Times of Integration and Observation

Both the finite time of integration and the finite time of observation originate a loss in the ability of reconstructing the Fourier components of higher frequency in the occultation curve.

As for the integration time τ_o, it is possible to neglect the loss of resolving power, is

$$\tau_o < 0.002"/v_r \ .$$

For example, with $v_r = 0.35"\ s^{-1}$ one gets $\tau_o \gtrsim 5.10^{-3} s$.

As for the observation time, the problem is more complex. In fact, from one hand, the modern data acquisition techniques gives the possibility of making observations as long as we need them, in order to get a good resolving power, but, from the other hand, the practical need of limiting the duration of the observations (presence of field diaphragms, etc.) prevents the measurements on double stars showing an angular separation larger than about 10".

6.4 Irregularities of the Lunar Limb

The many irregularities of the lunar edge cause distortion in the fringe pattern and, therefore, affect the values of the parameters one can get from the occultation trace.

Moreover, when we consider double star systems separated by more than 0.1", it is necessary to take into account also the lunar limb curvature. In fact, the velocity of the occultation v_r is the radial component of the lunar angular velocity, at the occultation point, and, from Figure 9a one gets (Evans, 1970)

$$v_{r,i} = |\bar{v}| \cos(\phi_1 - \phi_{2,i}), \quad i = 1, 2, \tag{6}$$

where ϕ_1 is the position angle of the lunar velocity vector $\bar{v}$, and $\phi_{2,1}$ and $\phi_{2,2}$ are, respectively, the position angles of the occultation positions of the two components.

Therefore, if the separation between the two components is large (>0.1"), the position angles and the occultation velocity are different.

So far, we have assumed that the lunar edge is smooth and perpendicular to radial direction. Instead, we must also consider that it may form an angle different from 90° to the radial direction. The consequence is a change in the occultation velocity and therefore an expansion or a compression of the occultation trace (see de Vegt and Gehlich, 1975; Evans, 1970).

Let us consider a double system with $\rho < 0.1"$. According to Figure 9b it may be seen that for a limb slope angle $\theta \neq 0$ (Evans, 1970),

$$v'_r = |\bar{v}| \cos(\phi_1 - \phi_2 - \theta). \tag{7}$$

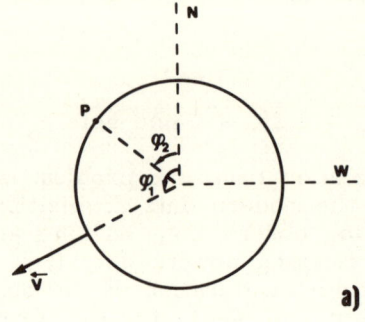

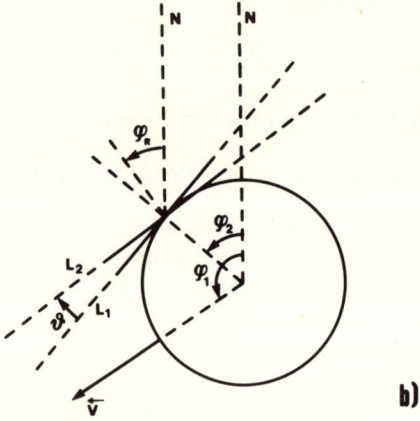

Figure 9. Geometry of an occultation:
   a) without anomalous limb slopes
   b) with anomalous limb slopes.

From this equation θ may be obtained because v'_r comes from the occultation curve by the fitting method (see §7) and the other parameters are theoretically computed. The only problem is to fix the slope corresponding to the value θ in a non-ambiguous way.

By putting $\psi = \phi_1 - \phi_2$ one gets from Eq. (7)

$$\theta = \psi \pm \arccos \frac{v'_r}{|\overline{v}|} . \qquad (8)$$

By definition, the positive solution of Eq. (8) is taken if $\phi_R > \phi_2$ (Figure 9). When the equation gives two positive solutions one takes the smaller value. This procedure is justified by several theoretical works (Fujinami, 1952) which showed that the lunar limb is prevailingly smooth and minor slopes are preferred; so that

$$\theta = \begin{cases} \psi - \text{sgn}(\psi) \arccos(v'_r \cos \psi / v_r) & \psi \neq 0, \\ |\arccos(v'^3_r / v_r)| & \psi = 0. \end{cases}$$

Another source of distortion of the fringe pattern is to be cound in the irregularities present along the lunar limb. Diercks and Hunger (1952) showed that this type of distortion reaches its maximum value for irregularities with size comparable to that of the first Fresnel zone at the distance of the lunar limb.

The only way to analyze the phenomenon is to divide it in two parts (Evans, 1970):

i) for a given irregularity of a given shape, one tries to compute how large the distortion introduced into the fringe pattern may be,

ii) one tries to find statistically how large is the probability of getting a given type of irregularity.

As for the first problem, one has to point out that a general treatment for an irregularity arbitrarily shaped is very difficult. Evans (1970) considered rectangular irregularities of the types: $h \times 2H$ and $h \times h$, centred at the occultation point, for several values of h (*).

For $h < 0.3$ one gets an effect of smoothing of the fringes. It may simulate an effect due to a non point-like source. For $0.4 < h < 1.0$, the distortion becomes larger and simulates a duplicity effect (Figure 10) like the one observed in the case of very close binary stars ($\rho \sim 0.1''$) with very small Δm ($\Delta m \sim 0$). For $h \gg 1.0$ the distortion has a very peculiar aspect and may easily be recognized and eliminated (see the fitting method; §7).

There are two ways to overcome these difficulties. First (Diercks and Hunger, 1952), one can observe the same event from two sites very far from each other. On the other hand, this kind of observation is necessary to get the separation vector.

The second way (Evans, 1970) makes use of observations from the same place but in two different colours. In fact, the distortions are a strong function of h, which is a function of λ. Therefore,

---

(*) h is given in natural units: that is, $(D\lambda/2)^{\frac{1}{2}}$.

a given irregularity causes distortions which are functions of λ, too. Then if we observe a different trend of the occultation trace in different colours, we also can discriminate limb effects from duplicity effects.

Note that in the case of well separated double systems ($\rho >$ 0.01") all these problems are practically non-existent. In this case, in fact, the only limb effect to be taken into account in separation measurements is the already mentioned expansion of the occultation curve.

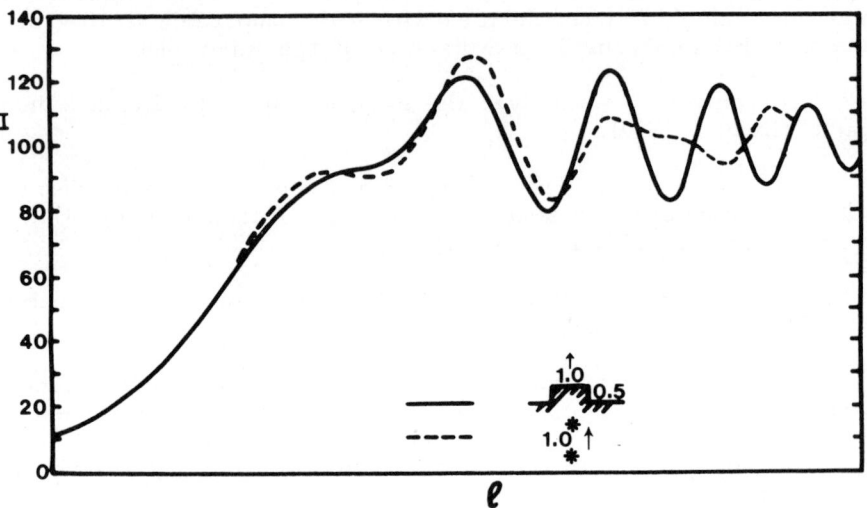

Figure 10 Theoretical occultation curve of a star in the presence of an irregularity, compared with the theoretical occultation curve of a double star (Evans, 1970).

6.5 Finite Diameter of the Star

For a number of the nearest stars the assumption that the source of light is a point-like source may be incorrect. Then, if $b(\omega)$ is the brightness distribution of the source, the theoretical occultation curve will be given by (Evans, 1970)

$$f(\omega) = b(\omega) * p(\omega) .$$

This means that a smoothing of the fringe pattern will take place, and the effect is more and more important when the diameter of the source becomes larger and larger.

A finite diameter of the source introduces further ambiguities in the identification and determination of double system parameters. First, for the geometrical occultation instant one cannot any longer take the one corresponding to 25% of the maximum value of the occultation trace. Second (White, 1977), it is very difficult to discriminate between an occultation curve for a star with a small (2-3m arc sec) angular diameter and an occultation curve of a very close ($\rho \sim$ 2-3m arc sec) binary system.

However, it is not particularly difficult to overcome this last difficulty, if we know the spectral type, the luminosity class, and the apparent magnitude of the occulted star. In fact, in this case it is possible (White, 1977) to estimate values for the angular diameter in order to make tests to find which of the two mechanisms may be the cause of the fringe pattern distortion.

6.6 Noise Effects

Instrumental noise usually is the most prominent source of error. Its presence prevents the identification and reconstruction of the highest order fringes in the occultation trace.

Hazard (1976) showed that if $q_o = S^{(*)}/N$ is the signal-to-noise ratio in the occultation curve, the resolving power p_r is given by

$$P_r \geq 45/q^2_o \ .$$

For $q_o = 100$, $P_r = 4.5 \cdot 10^{-3}$.

6.7 Scintillation and Seeing Effects

The occultation curve is also affected by a blurring due to seeing and scintillation effects. However, taking into account that the duration of a typical occultation curve is about 0.1s, one could show that both effects are statistically negligible (Young, 1971).

In any case, it is easy to pick up observations affected by the atmospheric turbulence because these are characterized by a clear modulation of the occultation curve.

7. THE METHOD OF FITTING

The best method to get from an occultation curve the parameters we need for the observed object is the method of fitting, developed by Nather and McCants (1970).

$^{(*)}$ $q_o = S/N$ is the ratio rms obtained through an integration of the occultation curve over a $\Delta t = 1"/v'_r$.

Assuming that for very close point-like stars $v_{r,1} = v_{r,2} = v$, and neglecting seeing and scintillation effects as well as the finiteness of the source size, they have, taking also into account the finiteness of the passband,

$$f[\omega(t_i)] = \delta_1 \int_{\lambda_o - \Delta\lambda}^{\lambda_o + \Delta\lambda} T(\lambda)S(\lambda)I_1(\lambda)p(\omega_{1,i})d\lambda +$$

$$+ \delta_2 \int_{\lambda_o - \Delta\lambda}^{\lambda_o + \Delta\lambda} T(\lambda)S(\lambda)I_2(\lambda)p(\omega_{2,i})d\lambda + B_o ,$$

where B_o is the sky background intensity,

$$\omega_{j,i} = v_r(t_j - t_{o,i})(2\frac{D}{\lambda})^{\frac{1}{2}} \qquad i = 1, 2; \quad j = 1, 2, 3, \ldots$$

($\omega_{j,i}$ is measured in arc sec s^{-1}).

$I_1(\lambda)$, $I_2(\lambda)$ are normalized monochromatic flux densities, and δ_1 and δ_2 are scale factors, proportional to the intensities of the two stars.

Usually it is impossible to know separately $I_1(\lambda)$ and $I_2(\lambda)$. For this reason one has to assume that

$$I_1(\lambda) = I_2(\lambda) = I_{tot}(\lambda) ,$$

where $I_{tot}(\lambda)$ is the normalized monochromatic flux density of the unresolved double star.

The fitting operation goes on working on six free parameters: δ_1, δ_2, v, B_o, $t_{o,1}$ and $t_{o,2}$.

In the case of a binary system with components very far from each other, one must take into consideration that $v_{r,1} \neq v_{r,2}$, and therefore introduce a seventh parameter ($v_{r,2}$) in the fitting process.

DETECTION OF CLOSE BINARY SYSTEMS 275

TABLE I

SAO	BD/ZC	M	Sp	Date	α		Δα	Φ	Δm	εm	Ref
75635	+20°0475	8.7	F2	691220	0.010			204°5	0.37	0.05	45
75673	+20°0484	4.6	A2	691220	T2.06	?		279.8	0.6	0.2	45
				721120	0.458		.001	285.5			45
				730114	0.045		.001	158.0			45
				730114	0.7		.1				45
75881	+22°0474	8.9	A0	720221	0.050		.004	253	3.3	0.6	45
75945	+22°0495	6.1	G5	721218	0.10		.002	120	1.4	0.4	45
75987	+22°0504	8.3	A3	721218	0.114		.001	238.7	1.7	0.3	45
76032	+24°0520	8.6	A2	721218	0.069		.002	238.7	2.2	0.4	45
76103	+23°0496	8.1	A5	720125	0.046		.001	200	1.3	0.3	45
76126	+23°0505	5.5	B7	730211	0.0231			116.3	0.95		45
76131	+23°0507	3.7	B6	690323	0.006		.001	247			45
76140	+24°0547	4.4	B5	691220	0.0025			108		2.02	45
76152	+23°0512	7.2	A1	690806							45
76155	+23°0516	3.9	B7	710910	0.255		.005	184	2.3	0.8	45
76159	+24°0553	5.8	B8	720319	0.155		.005	48.7			45
				720319	0.03	? ?					45
76172	+23°0522	4.2	B6	690930							45
76183	+24°0562	6.8	B9	691211							45
76184	+24°0563	8.2	A2	720319	0.299		.002	173.0	3.0	1.4	45
76185	+23°0528	8.4	A2	710910	0.326		.001	249.0	0.8	0.2	45
76188	+23°0531	8.7	F5	710910		? ? ?					45
76192	+23°0536	6.3	A0	730211							45
				720319	0.0019			28.5	1.11		45
				690323	0.031	?	.001	207.1	1.57	0.25	45
76199	+23°0541	2.8	B5	711104							45
76200	+23°0540	6.8	A0	720319							45
76214	+23°0556	8.4	K0	740105	0.512		.002	270.1	0.3	0.4	1
76225		6.3	F0	730211	0.0093			94.9	2.87		45

SAO	BD/ZC	M	Sp	Date	α	Δα	Φ	Δm	εm	Ref
76228	+23°0557	3.6	B8	681231	0.006	.001	56	2.0	0.4	45
				720319	0.0074		124	1.63		45
				730211	0.0040		48.6	1.5	0.1	45
76254		8.0	F5	740105	0.051	.002	220	0.7	0.3	1
76425	+23°0609	6.4	F5	730115	0.032	.001	156.3	0.04	0.06	45
76955	+26°0783	6.6	B5	711230	0.076	.002	327	1.4	0.4	45
77111		8.8	A2	750320	0.116	.004	303.7	2.2	0.5	40
77221	+26°0829	8.5	F5	720223	0.063	.001	66	0.6	0.2	45
77423			A0	750221	0.0159	.001	36.1	0.8	0.1	1
77588	+28°0874	8.1	B8	710305	0.017	.001	157	1.0	0.4	45
77606		8.7	A2	740330	0.027	.001	59.4	1.2	0.4	1
77625	+27°0888	5.6	G7	700315	0.010			1.3		45
77638	+28°0918	8.0	A0	700315	0.010					45
77646	+25°1008	9.1	K0	720418						45
77776		7.8	B8	730213	0.175	.002	145	0.9	0.4	1
77819	+27°0943	6.8	K0	710206	0.040	.002	338	0.7	0.2	45
77926		8.3	F0	750125	0.039	.001	272.2	1.3	0.3	1
					?			*0.8	0.4	1
78349	+23°1356	6.0	A0	730409	0.067	.001	124	0.5	0.2	45
78378		8.7	A2	730117	0.068	.001	337.6			45
78417	+27°1122	6.5	F5	700922	0.053		255.9	0.5	0.1	45
78440	+28°1138	6.8	A2	500423	0.023	.001	128	0.0	0.2	45
				690326	0.125	.001	283	1.5	0.4	45
78484		8.4	A0	720224	0.019	.002	241.5	0.6		1
78507		8.0	A3	740204	≲0.2					1
78762		8.8	K0	740331	0.015	.002	100	1.6	0.4	1
79797	+23°1843	9.2	A2	710501	0.0874			0.73		45
79804	+24°1805	7.5	G0	700413	0.131	.003	222.4	0.9	0.5	45
92106		8.8	K0	740102	0.078	.003	25	0.9	0.5	1
92120		8.8		740102						1

DETECTION OF CLOSE BINARY SYSTEMS

SAO	BD/ZC	M	Sp	Date	α	Δα	Φ	Δm	εm	Ref
92369		8.1	F5	750120	0.024	.001	252.3	0.5	0.1	1
92429		9.1	G5	740130	0.036			*0.5	0.3	
92801	+17°0315	6.5	A3	730113	0.030	.002	22.8	0.95	0.14	1
92645	+15°0268	7.9	F0	711102		.002	287	2.4	0.4	45
92919		9.0	F7	740131	0.037					45
93022		5.6	A0	770930	0.021	.002	101	1.1	0.5	1
93062	+190403	5.7	F5	691219	0.013	.001	265.7	0.1	0.1	42
93070		9.1	F5	770930	0.006			0.81		45
93085		7.8	F5	770224	0.3799	.002	327.7	0.6	0.8	42
					?	.003	55.7	3.1	0.5	42
								*2.3	0.6	42
93414		8.3	G5	760307	5.814	.001	101.7	0.6	0.3	41
93415		9.1	G5	760307	0.035	.003	13.0	0.8	0.6	41
93484		7.0	F5	770225	0.010	.003	30.7	3.5	0.8	42
93778		7.9	A2	760113	0.018	.002	154.4	2.7	0.5	40
93835		8.8	G0	760308	0.395	.001	102.0	1.2	0.06	41
93870		6.9		780315				1.61	0.09	42
								*1.16	0.1	42
94031		7.8	A2	760210	0.079	.001	290.2	0.0	0.3	40
								*0.3	1.1	40
94422		8.8	F5	770227	0.036	.007	281.1	0.9		42
94431		7.5	B3	770227	3.07		159.6	0.6		42
								*0.65		42
94554		5.3	B3	780217	0.099	.001	98.1	1.13	0.06	42
								0.90	0.06	42
95166		5.1	B8	760310	0.054	.001	50.3	1.15	0.04	41
								1.0	0.3	41
95166		5.1	B8	760115	0.041	.001	122.7	1.1	0.2	40

SAO	BD/ZC	M	Sp	Date	α		Δα	Φ	Δm	ε_m	Ref
95183		8.8	A5	760310					0.7	0.8	40
95229		8.3	B9	770228	0.038		.002	273.8	0.9	0.4	41
95265		8.8	A0	770228	0.016		.006	235.6	0.1	1.2	42
95456		6.8	K0	770228	0.014	?	.002	64.4	1.3	0.5	42
95794		7.8	F8	771001	8.51			277.5	0.44		42
									*0.7		42
95866		8.3	G0	780122	0.019		.004	61.1	1.8	0.8	42
95988		8.9	A2	750222	0.021	?	.002	162.7	0.9	0.5	1
96090		8.3	A2	760407							41
96515		9.0		770301	0.057		.005	261.8	0.6	0.8	42
96561		8.7	G5	050126	0.078		.006	317.4	2.4	0.9	1
96634		9.0		780219	0.238		.006	107.6	1.9	0.9	42
96634				770908	0.308		.002	111.5	1.2	0.3	42
									*1.3	0.4	42
96646		9.0	A5	750126	0.071		.004	262.0	1.9	0.5	1
96687		8.3	G0	750126	0.045		.001	83.6	0.6	0.3	1
96977		9.0		770329	8.24			75.2	0.14		42
									*0.23		42
96991		8.8	F2	770329	0.006		.001	85.2	1.2	1.0	42
97609		8.6	F5	770426	0.039		.006	87.4	0.6	1.1	42
97687		9.2		740526	0.245		.003	135.6	1.0	1.2	42
									0.6	0.5	1
97813	+18°1942	9.1	G5	720518	0.288		.002	333.3	0.6	0.8	1
97919		8.3		730315	0.057		.007	39	0.6	0.3	45
97953		8.4	A0	750224	0.046		.001	318.8	0.06	0.15	1
									1.3	0.2	1
98007		7.8	K2	730315	0.0132		.007	243.2	*1.4	1.1	1
98132	+19°2105	9.3	F2	710502	0.317		.001	256	1.10	0.15	1
									0.1	0.2	45

DETECTION OF CLOSE BINARY SYSTEMS

SAO	BD/ZC	M	Sp	Date	α	Δα	Φ	Δm	ε_m	Ref
98161	+18°2057	6.7	G5	710502	0.086	.001	119	0.08	0.07	45
98627		5.1	G5	730316	0.007	.001	254.1	1.91	0.12	1
98696	+13°2131	6.9	F5	710503	0.006					45
98709		3.5	A2	780222						42
98767		6.3	A3	710627	0.031					44
109244		9.0	F2	771121	0.034					42
109262		5.6	A0	751115	TRIPLA					
109269	+8°0069	8.7	G0	721215	0.013	.005	111	0.2	0.5	45
109596		8.1	F0	751213	0.043	.002	16			40
109719		8.6	F8	770222	0.026	.002	282	1.1	0.4	40
109790		8.8	G5	771122	0.060	.002	21.6	0.04	0.3	40
109923		8.6	F8	761105	0.034	.007	285.8	*0.14	0.8	42
117767		8.5	A3	770525	0.041			0.7	0.8	42
117837		8.7	A2	750225	0.029	.001	272.8	0.7	0.7	42
117991		9.2	F2	730510	1.74			0.6	0.3	42
				771105	1.83	.005	201.5	0.6	0.4	41
118224		8.6	F0	770526	0.195	.008	89.0	*1.0	1.5	42
						.003	248.7	2.4	1.1	
						.05	39	1.2	0.6	
							227.7	0.19	0.07	
						.003	57.0	0.3		
118289		8.6	K2	730317	TRIPLA			2.4	0.8	42
				770526	0.470			*0.2	0.8	42
118355	+10°2126	3.8	B0	691229	0.0036	.003	75.6	3.2	1.0	1
118425	+5°2374	8.4	K0	720327	0.206			0.40		42
118397	+5°2359	8.6	G5	720327				0.2	0.2	45
118443		6.5	F5	770429	0.118	.001	264.1			45
								1.6	0.2	45
						.008	110.0	1.5	0.2	42
118473	+3°2408	6.6	K2	730511	0.1			0.0		42
118577		6.9	K0	740529	0.012	.002	135.2	1.3	0.6	1

SAO	BD/ZC	M	Sp	Date	α		Δα	Φ	Δm	ε_m	Ref
118786	+0°2769	8.5	F8	720328	0.022		.001	310	0.2	0.1	45
128212		9.2	F8	731010	0.322		.006	64.0	0.1	0.6	1
128368	+0°5042	8.9	F0	711031	0.337		.005	320	1.8	1.0	45
				761130		?					41
128467	+2°4730	8.2	K0	711031	0.010		.003	115	0.6	0.6	45
138462	-3°3198	8.8	G5	720425	0.365		.004	105.8	0.1	0.3	45
138692		9.0	G5	750521	0.318		.002	124.4	0.6	0.3	41
138777		8.0	F5	690527	0.021		.001	141	0.6	0.2	45
138911		8.0	M	750715	0.025		.002	277.7	2.1	0.3	40
138925		9.0	K0	750715	0.055		.003	108.1	1.8	0.8	40
145502		8.0	K0	751111		?					40
145635		7.0	K0	761225	0.062		.008	260.3	2.5	0.8	41
145973		8.1	G5	751112	0.228		.001	353.9	1.1	0.2	40
146067		6.7	A0	741102	0.098		.001	221	0.20	0.00	1
146239		6.4	G5	731009	0.129		.002	233.5	2.7	0.3	1
146239				740929	0.065		.001	254.1	1.11	0.10	1
									*1.0	0.3	
146307		7.5	F5	751210	0.009		.003	72.3	2.0	1.1	40
146419	-6°6112	8.7	K5	701012	0.040		.002	69	0.2	0.4	45
157584		6.0	A0	740504	0.100		.003	1.7	0.0	0.6	1
157613		7.4	K5	740504	0.118		.003	339.0	1.2	0.3	1
									*2.5	0.4	
157836	-10°364	5.6	A2	690528	0.010						1
157998	-14°3739	6.2	K0	701223	TRIPLA	?					45
158147		2.8	A0	740505	0.01						45
158840	-15°3966	6.8	A3	660531							1
159085			K0	770725	0.054		.001	338.7	2.6	0.1	45
									2.6	0.1	
159786		8.3	K0	770526	0.103		.001	238.0	*2.5	0.2	42
160399		8.8	A0	770727	0.030		.003	302.3	1.3	0.5	42

DETECTION OF CLOSE BINARY SYSTEMS

SAO	BD/ZC	M	Sp	Date	α	Δα	Φ	Δm	εm	Ref
161153		6.3	A2					*0.9	0.9	42
161399		8.8	A0	771018	0.090	.001	65.9	1.4	0.2	41
161463		8.8	B8	771018	0.077	.004	60.8	*0.4	0.3	42
161848		6.5	A0	770507	0.0120	.002	78.4	2.7	0.8	42
								0.4	0.6	42
								0.1	0.3	42
161852		7.5	A2	771213	9.49		73.1	*1.7	0.4	42
163481		3.0	G0	751207	0.023	.001	60.3	0.02		42
				761126	0.015	.001	293.4	1.1	0.2	40
								0.0	0.1	41
				761126	0.015	.001	113.9	*2.6	0.4	41
				770315	0.019	.001	110.0	1.8	0.1	42
				771020	0.007	.001	24.0	1.1	0.1	42
163563		9.1	G5	750916	0.088	.002	190.5	1.02	0.07	42
163563				771020	0.088		190	*2.2	0.2	42
163666	−17°6045	7.8	F0	730909	0.061	.001	194	0.5	0.7	40
163760		8.2	K5	721113	0.133	.001	199.6	0.09	0.09	42
163771		6.3	B5	740830	0.257	.001	64.7	0.4	0.3	1
								0.42	0.05	45
164213		8.5		750917	0.003	.003	302.3	0.4	0.2	1
164222		8.6	G0	750917	0.045	.004	250.8	0.5	0.7	40
164222				761127	0.318	.001	302.6	1.15	0.08	40
								1.1	0.2	41
164231	−17°6262	8.8	K0	751208	0.016	.001	228.1	*0.8	0.3	41
164376	−17°6251	9.0	A2	701204	0.010			1.2	0.3	40
164520	−20°6251	4.6	B3	680904	0.005	.001		1.3		45
164717	−10°5785	6.5	B9	721018	0.033			1.4	0.1	45

SAO	BD/ZC	M	Sp	Date	α	Δα	Φ	Δm	ε_m	Ref
164750	$-14°6163$	7.8	M2	701011	0.010					45
164935	$-12°6209$	7.1	G5	701107	?					45
164971	$-11°5792$	8.8	K0	701205	0.013	.002	27	0.7	0.4	45
183333	$-23°12173$	7.2	A2	720623	0.015	.002	164	0.4	0.3	45
183445	$-22°10975$	8.1	F0	720720	0.054		308.3	0.3		45
				730613	0.349	.001	329.1	0.8	0.2	1
183565	$-24°12155$	7.1	A3	720527	0.159	.001	274.8	0.0	0.2	45
184141		7.9	G5	750525	0.052	.001	104.4	1.00	0.08	40
								*1.4	0.3	40
184336	$-25°11485$	3.1	B1	720721	0.296		317.4	2.2		45
184415	$-26°11358$	1.2	M0		3.16		274.6	3.9	0.1	45
186152		6.9	B3	730906	0.174	.003	203.0	2.2	0.4	1
186579	$-23°14161$	9.4	B8	731003	0.030	.003	277.7	0.15		45
186917	$-25°13207$	8.6	F8	720724	0.115	.001	243	0.3	0.2	45
188452	$-23°15652$	8.6	A0	710929	0.202	.002	222	2.3	1.3	45
188470	$-25°14257$	6.6	A5	701105	0.010					45
189609	$-21°5805$	8.8	F8	701106	0.117	.001	76	0.6	0.3	45
189663	$-21°5820$	8.9	F8	701106	0.022	.002	104	0.4	0.4	45

Explanation of Table Headings

Column 1: Number in the SAO Catalogue
Column 2: Number in the BD/ZC Catalogue
Column 3: Visual magnitude
Column 4: Spectral type
Column 5: Date
Column 6: Angular separation (see text)
Column 7: Error in α
Column 8: Position angle (see text)
Column 9: Magnitude difference with B filter asterisque indicates that observations were performed through an R filter)
Column 10: Error in Δm
Column 11: Bibliographic reference

REFERENCES

Africano, J. L., Cobb, C. C., Dunham, D. W., Evans, D. S., Fekel, F. C. and Vogt, S. S.: 1975, Astron. J., 80, p.689.
Baize, P.: 1975, L'Astronomie, 89, p.159.
Bartholdi, P.: 1975, Ap. J., 80, p.445.
Batten, A. H.: 1967, Comm. Obs. R. Belgique, 17, p.221.
Born, M. and Wolf, E.: 1959, *Principles of Optics*, Pergamon Press, New York.
Couteau, P.: 1960, J. Obs., 43, p.41.
Cowley, A., Cowley, C., Jaschek, M. and Jaschek, C.: 1969, Astron. J., 74, p.375.
de Vegt, C. and Gehlich, U. K.: 1976, Astron. Astrophys., 48, p.245.
Dierks, H. and Hunger, K.: 1952, Z. f. Astrophys., 31, p.182.
Eddington, A. E.: 1909, Mon. Not. Roy. Astron. Soc., 69, p.178.
Evans, D. S.: 1970, Astron. J., 75, p.575.
Evans, D. S.; 1977, Sky and Telescope, p.229.
Finsen, W. S.: 1954, Un. Obs. Circ., 4, p.225.
Finsen, W. S.: 1964, Astron. J., 69, p.317.
Fujinami, S.: 1952, Publ. Astron. Soc. Japan, 4, p.1-5.
Hanbury Brown, R.; 1974, *The Intensity Interferometer*, Taylor and Francis, London.
Hazard, C.: 1976, *Methods of Experimental Physics*, vol.12C, p.92, Academic Press.
Heintz, W. D.: 1969, Journ. Roy. Astron. Soc. Canada, 63, p.275.
Heintz, W. D.: 1978, *Double Stars*, D. Reidel Publ. Co., Dordrecht.
Herr, R. B.: 1969, Publ. Astron. Soc. Pacific, 81, p.105.
Hoerner, S. von: 1964, Ap. J., 140, p.65.
Kopal, Z.: 1969: *The Moon*, D. Reidel Publ. Co., Dordrecht.
Krishnan, T.: 1971, Highlights of Astronomy, 2, p.646.
MacMahan, P. A.: 1909, Mon Not. Roy. Astron. Soc., 69, p.126.
McGraw, J. T., Dunham, D. W., Evans, D. W. and Moffett, T. J.: 1974, Astron. J., 79, p.1299.
Nather, R. E. and McCants, M. M.: 1970, Astron. J., 75, p.963.
Nather, R. E. and Evans, D. S.: 1971, Astrophys. Space Sci., 11, p.28.
Nather, R. E. and Wild, P. A. T.: 1973, Astron. J., 78, p.628.
Nather, R. E., Churms, J. and Wild, P. A. T.; 1974, Publ. Astron. Soc. Pacific, 86, p.116.
O'Keefe, J. A. and Anderson, J. P.: 1952.
Öpik, E.: 1924, Publ. Obs. Tartu, 25, Part 6.
Reboul, H.: 1979, *Introduction a la Theorie de l'Observation en Astrophysique*, Masson, Paris.
Scheuer,P. G.: 1962, Australian Journ. Physics, 15, p.333.
Smithsonian Astrophysical Observatory Staff: 1966, *Star Catalogue*, Smithsonian (special) Publ. No.4652.
Taylor, J. H.: 1966, Nature, 210, p.1105.
White, N. M.: 1977, Revista Mexicana de Astron. y Astrof., 3,p.43.

Whitford, A. E.: 1939, Ap. J., 89, p.472.
Williams, J. D.: 1939, Ap. J., 89, p.467.
Young, A. T.: 1971, Highlights of Astronomy, 2, p.622.

Bibliography Quoted only in Table I

Africano, J. L., Evans, D. S., Fekel, F. C. and Ferland, G. J.:
    1976, Astron. J., 81, p.650.
Africano, J. L., Evans, D. S., Fekel, F. C. and Montemayor, T.:
    1977, Astron. J., 82, p.631.
Africano, J. L., Evans, D. S., Fekel, F. C., Smith, B. W. and
    Morgan, C. A.: 1978, Astron. J., 83, p.1100.
Beivers, W. I. and Eitter, J. J.: 1971, Astron. J., 76, p.1131.
Beivers, W. I. and Eitter, J. J.: 1974, Ap. J. Suppl. Series, 28,
    p.405.
Böhme, D.: 1978, Astron. Nachr., 299, p.243.
Morgan, C. A.: 1978, Astron. J., 83, p.1100.
Radick, B. B.: 1979, Astron. J., 84, p.257.
Ridgway, S. T., Wells, D. C., Joyce, R. R. and Allen, R. G.:
    1979, Astron. J., 84, p.247.
Sandmann, W.: 1977, Astron. J., 82, p.503.

Other References

Born, E.: 1979, Astrophys. Space Sci., 63, p.439.
Born, E. and Debrunner, H.: 1979, Astrophys. Space Sci., 63,
    p.457.
Davis, J.: 1971, Highlights of Astronomy, 2, p.713.
Deeming, J. T.: 1871, Highlights of Astronomy, 2, p.662.
Evans, D. S.: 1965, Astron. J., 60, p.432.
Evans, D. S. and Nather, R. E.: 1970, Astron. J., 76, p.1107.
Hoerner, S. von: 1965, Ap. J., 142, p.1264.
Kopal, Z. and Carder, R. W.: 1974, *Mapping of the Moon*, D.Reidel
    Publ. Co., Dordrecht.
Lang, K. R.: 1969, Ap. Jl, 158, p.1189.
Poss, H.: 1971, Highlights of Astronomy, 2, p.626.
Rakos, K. D.: 1971, Highlights of Astronomy, 2, p.675.

PROGRESS AND PROBLEMS IN RS CVn STAR RESEARCH

Marcello Rodonò

Osservatorio Astrofisico di Catania and
Istituto di Matematica, Università di Messina, Italia

Abstract.

Recent observational results obtained from the ground and with instruments on board of satellites are presented. Particular emphasis is given to the photometric observations carried out at Catania Observatory. A brief outline of the models, which have been proposed to interpret the RS CVn star variability, and of the numerous still unresolved problems deserving future observational and theoretical efforts follows. The importance of studying the intrinsic variability of RS CVn stars, in the contest of stellar activity, is stressed.

1. INTRODUCTION

It was my great pleasure to accept Professor Kopal's invitation to attend the 1980 Advanced Study Institute on "Binary Stars" and give a lecture on RS Canum Venaticorum Binaries. In fact, the 1980 marks the 15th year after the Catania observations of RS CVn itself led us to disclose the most important photometric feature of this interesting type of binaries: the outside eclipse wave-like distortion and its systematic migration on the light curve. Actually, I had planned the 1965 observations of RS CVn in order to ascertain the cause of some inconsistencies between the 1963 and 1964 light curves obtained at Catania by Chisari and Lacona (1965): these observations together with older ones, were suggestive of a systematic variation of the light curves as, indeed, turned out to be the case (Fracastoro 1965, Catalano and Rodonò 1967). This fundamental discovery is best illustrated by the series of light-curves shown in Figure 1, which covers a full 9.7

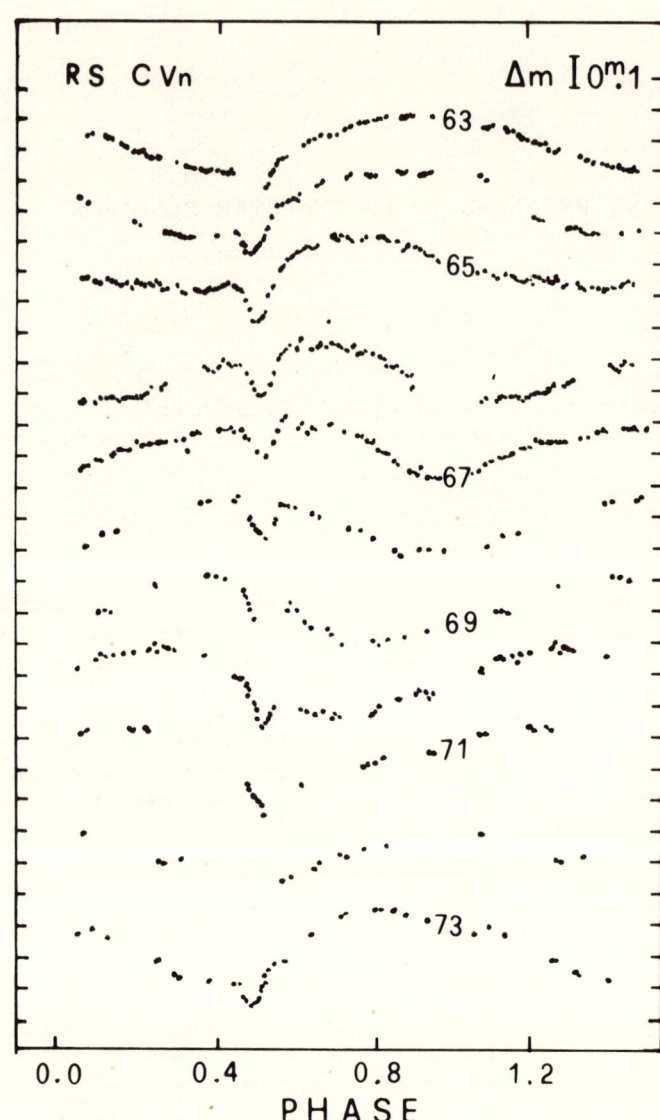

Figure 1. The outside eclipse RS CVn light curves from 1963 to 1973 showing the migrating wave-like distortion (Catalano et al. 1980).

-year cycle of variability. From the residual light at totality in successive years, the hotter component being completely eclipsed at primary minimum, it was possible to ascertain that the distortion wave on RS CVn light-curve and its slow migration is produced by the periodic variability of the cooler and larger component: its period of variability is slightly shorter than the orbital one, so that a given phase of its light variation is anticipated with respect to the orbital one, as defined by eclipses. Actually, these results were obtained because, in the eclipsing binary research carried out at Catania Observatory, rather than to the mutual eclipses of the star discs, particular attention was being paid to the intrinsic variability of the stars, which gives rise to distorted light-curves.

In my talk I will be reviewing the most significant progress made in the optical, radio, UV and X-ray regions; then a brief outline of the proposed models will follow. I will conclude presenting the several problems on RS CVn binaries which are still open and deserve to be investigated. Therefore, my presentation of RS CVn binaries should not be regarded as a comprehensive review, but as a stimulating discussion on the most intersting aspects of the systems, with special emphasis on the observations. More specific data on the individual members of the group can be found in the reviews by Hall (1976, 1981) and in the December 1978 issue of the Astronomical Journal. Also the papers given at the Joint Meeting of the I.A.U. Commissions 10, 27, 40, 42 and 48 (Larsson-Leander 1980) constitute a useful background.

2. HOW TO DEFINE RS CVn BINARIES

No obvious and unique definition of a class of variable stars can be given: it can be too broad or too restrictive. Both give rise to some problems which can be misleading when one is asked to present the general characteristics of any class. The criticism by Eggen (1978) of the dangerous tendency of astronomers in establishing new rigid classes of variable stars seems to me quite appropriate. Actually, a too restrictive definition can leave out important borderline members, whose study is often essential in understanding the variability phenomenon involved. It is also true that any definition is guided by a particular aim, so that we must be aware that the adopted definition can be perfectly valid and useful only within a confined area of application.

Of the several characteristic features listed by Hall (1976) in presenting a working definition of RS CVn binaries, I would retain only those few which have their roots in the lists of binaries with variable spectroscopic and photometric features, such as those compiled by Struve (1946) and Catalano and Rodonò (1968):

a) detached and semi-detached eclipsing and non-eclipsing binaries;

b) spectral type of the hotter (h) and cooler (c) component $(F-G\ V-IV)_h$ and $(G-K\ IV)_c$, respectively, i.e. the cooler component has sizable surface convective zones;

c) variable emission in the Ca II H and K lines;

d) outside eclipse low-amplitude light variation with period almost equal to the orbital one, so that a slow migrating wave-like distortion on the light curve is observed;

e) occasional radio bursts correlating with the intensification of chromospheric lines; and,

f) according to recent UV and X-ray observations with the International Ultraviolet Explorer (IUE) and the HEAO-2 (EINSTEIN) satellites, respectively, abnormally high emission in UV chromospheric and transition region lines and soft X-ray continuum, also in the quiescent phase.

It is apparent that the rather broad definition given above is guided by the stress laid on the intrinsic variability of the stars rather than by their binary nature. Actually, the latter characteristic is a fundamental one because, as will become apparent later on, not only does it help in discovering and studying the variability of the individual components, but most probably it has a key role in favouring and maintaining their intrinsic variability.

I would also attribute marginal importance to the evolutionary status of RS CVn binaries as a characteristic feature of the class. Of course, this does not mean that the evolutionary status should not be considered an important parameter, but that it would be more meaningful to concentrate on the physical characteristics of the stars themselves. On the other hand, there is no final agreement yet, whether the cooler components of RS CVn systems are pre- or post-main-sequence stars. The conclusion by Popper and Ulrich (1977) that they are in the post-MS seems quite convincing: with a few exceptions, as Z Her, the cooler and larger component is definitely the more massive and more luminous one, therefore it should be more evolved than its MS companion. Moreover, WW Dra and V 711 Tau have an additional low-mass MS companion of low luminosity which, being the less evolved, requires the subgiant more massive component to be in the post-MS phase. Morgan and Eggleton (1979) have presented additional evidence supporting the Popper and Ulrich (1977) conclusion.

However, since the mass ratio is very close to unity and, in a few systems, the hotter component is definitely the more massive one, Popper and Ulrich's (1977) suggestion needs additional corroboration. Actually, Vivekananda Rao and Sarma (elsewhere in this volume) have reached the opposite conclusion. Moreover, the kinematic characteristics, the emission line strengths and the λ 6707 lithium line in II Peg are suggestive of young age.

3. RECENT OBSERVATIONAL PROGRESS

Thanks to the promotive activity of the I.A.U. Working Group on RS CVn stars (1) of Commission 42, a significant amount of observational material has become available in the past few years.

3.1 Photometry and Optical Spectroscopy

For several systems seasonal light curves are now available allowing a better description of the distortion wave and of its migration rate on the light-curves. However, much work remains to be done and the individual groups of observers are in the process of "adopting" selected systems with the purpose of obtaining systematic and homogeneous seasonal ligh-curves, as the RS CVn ones obtained at Catania Observatory.

Typically, the outside-eclipse light-curve of RS CVn systems can be approximately represented by the following equation:

$$L = L_o + L_A \cos(\phi - \phi_{min}) \\ = L_o + A \cos\phi + B \sin\phi \quad (1)$$

where $A = L_A \cos\phi_{min}$ and $B = L_A \sin\phi_{min}$. The meaning of the symbols used are indicated graphically in Figure 2. A least squares fit to the data by Equation (1) allows us to determine L_o, A and B, and then $\phi_{min} = \tan^{-1}(A/B)$ and $L_A = (A^2 + B^2)^{\frac{1}{2}}$. This is of course only an ideal situation which applies only to a restricted number of systems, such as RS CVn. The distortion wave is sometimes asymmetric or complicated by classical photometric effects in close binaries, such as the ellipticity or reflection effect. In this case, the following truncated Fourier expansion is required:

$$L = L_o + A \cos\phi + B \sin\phi + A_2 \cos 2\phi \quad (2)$$

---

(1) Those interested both in theory and observations of RS CVn stars can contact D.S. Hall (Dyer Observatory, Vanderbilt University, Nashville, Tenn. 37235, USA) if in the Eastern Hemisphere, or M. Rodonò (Catania Astrophysical Observatory, 95125 Catania, Italy) if in the Wetsern Hemisphere.

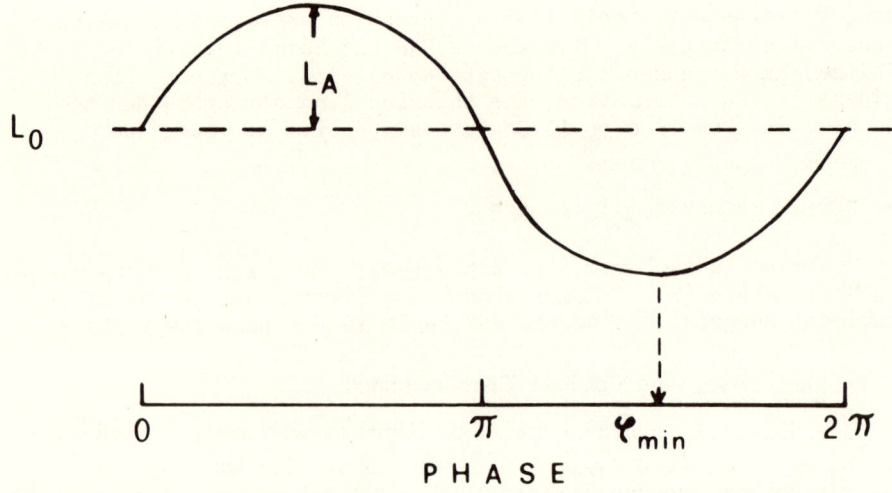

Figure 2. A graphical definition of the various parameters entering the analytical equation (1) of the distortion wave.

and a solution for L_o, A and B, i.e. for L_o, L_A and ϕ_{min}, can be sought after an initial guess of A_2 is made.

The simple method here indicated can be applied to most of the RS CVn binary light-curves. Usually L_A and ϕ_{min} are not constant, i.e. the wave has a variable amplitude and shape, and migrates on the light curve. Any general conclusion about the behaviour of the mean luminosity level, L_o, is hampered by the inhomogeneity of the presently available data.

Representative examples of the kind of photometric data which are of interest in the study of RS CVn systems are shown in Figure 3. These are recent unpublished light-curves obtained at Catania by Blanco, Catalano, Marilli and myself showing the sinusoidal wave with variable amplitude in V 711 Tau and WW Dra and the asymmetric wave in SS Boo light curves. Also the B-V and U-B colour behaviour of V 711 Tau is shown. These colours show almost no variation, as is typical for most RS CVn systems.
In Figure 4 (Blanco et al. 1980) two typical light-curves (CQ Aur and RU Cnc), distorted both by a wave and other photometric effects, are presented. In this case Equation (2) must be used in order to extract the wave characteristics. An extreme example of highly variable light-curve is that of VV Mon (Cerruti-Sola

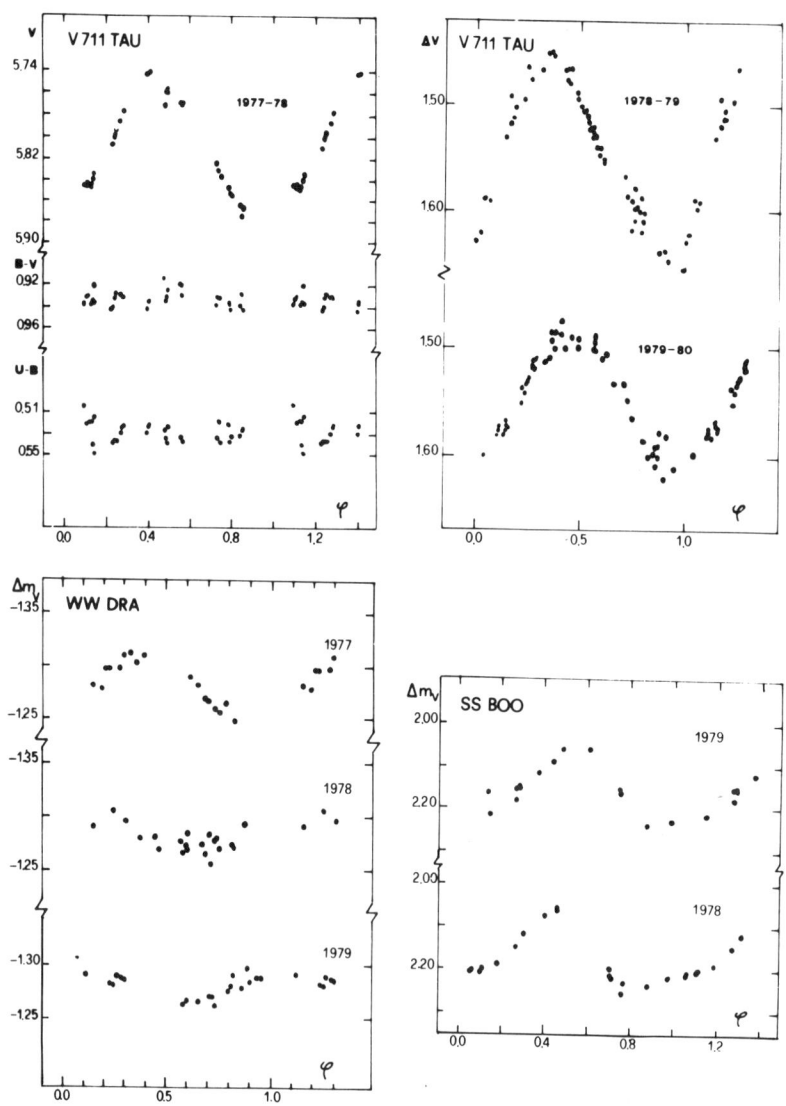

Figure 3. Representative light curves of RS CVn systems with quasi sinusoidal (V 711 Tau and WW Dra) and asymmetric waves (SS Boo). The almost constant U-B and B-V colours of V 711 Tau are typical to the majority of RS CVn systems (Blanco et al. 1980a).

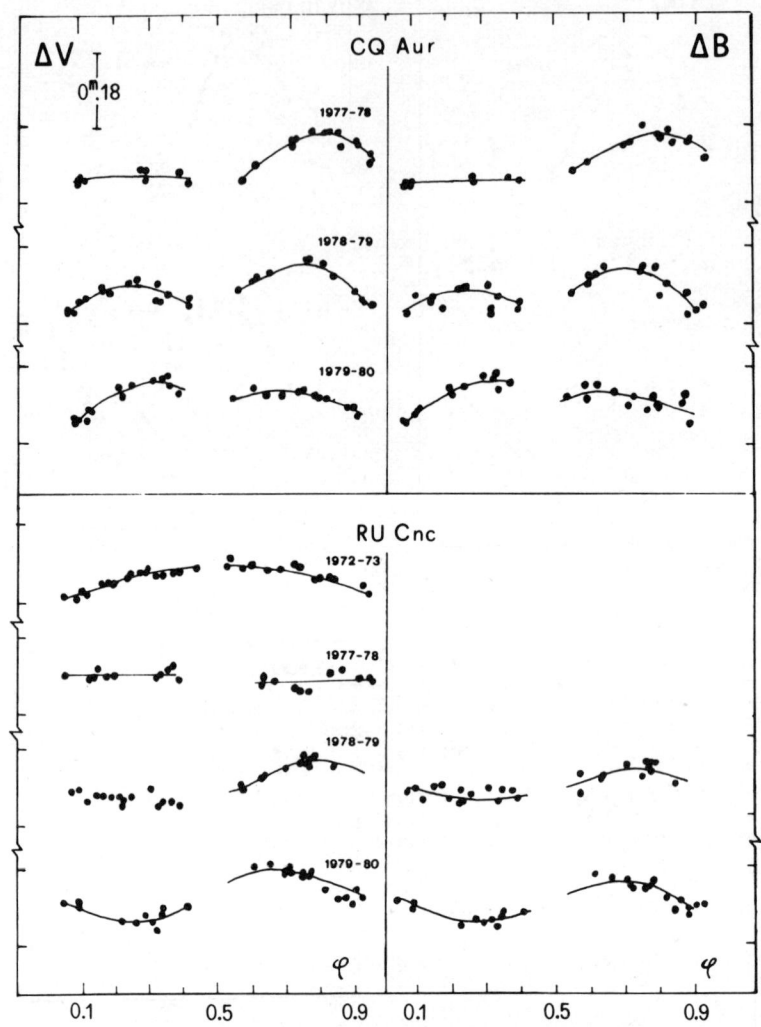

Figure 4. Representative light curves of RS CVn systems distorted by a "wave" and other photometric effects. The wave migration produces striking light curve changes (from Blanco et al. 1980a).

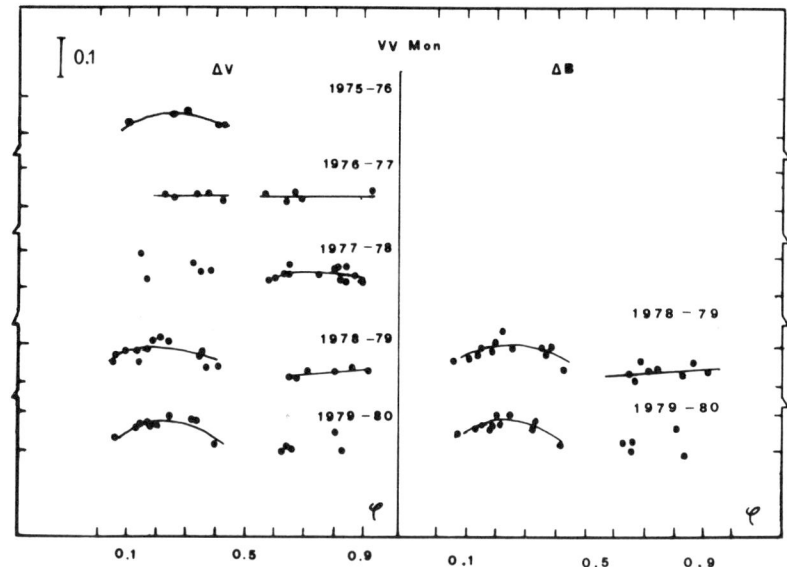

Figure 5. The variable light curve of VV Mon showing no clear evidence of a wave-like distortion (Cerruti-Sola et al. 1980).

et al. 1980): no wave is apparent in any of the seasonal light curves between 1975 and 1980 and the maxima are flat in 1976-77 (Figure 5).

The light curves presented in Figures 1, 3, 4, and 5 do not exaust the whole scenario made up by RS CVn stars, however they are representative enough of the phenomenology involved.

Contrary to initial evidence, both direct and retrograde migration of the distortion wave is observed (Figure 6) and in four systems - SS Boo, CG Cyg, AR Lac and V711 Tau - changes in the migration direction has been observed (Figure 7). Cyclically variable directions seem to occur in SS Boo and possibly in V 711 Tau: only sufficiently extended future observations can confirm or deny this indication and tell us whether it is a general behaviour pattern of RS CVn systems.

With the possible exception of SS Boo, the migration direction does not appear to correlate with the wave amplitude. In the system studied best, namely RS CVn, while the wave has shown a continuous almost linear retrograde migration, at least in the past 30 years, the wave amplitude varies cyclically in about 5 years (Figure 8).

As mentioned in the INTRODUCTION, the wave migrates on the

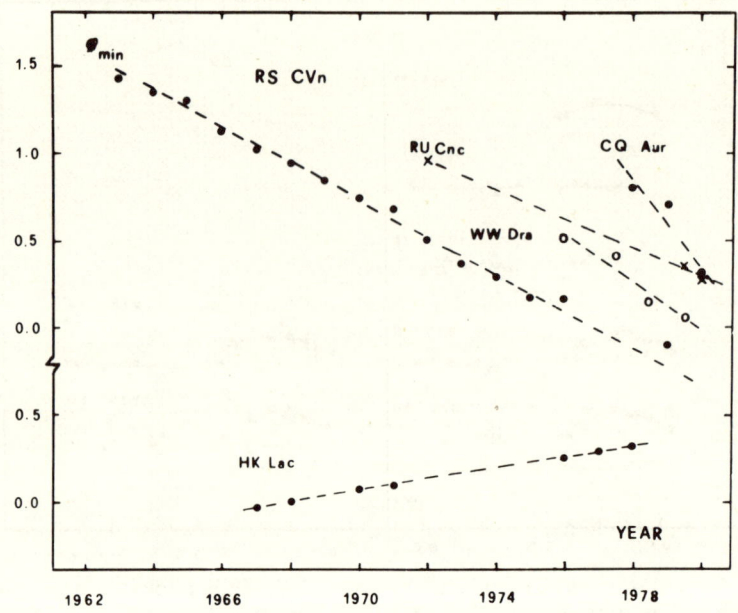

Figure 6. Retrograde and direct migrations of the distortion wave on the light curves of some RS CVn systems studied at Catania Observatory. φ_{min} is the phase of the wave light-minimum.

light curves of RS CVn systems because of the difference between the orbital period and the photometric period of variability of the variable component. Therefore it is of interest to investigate any possible correlation between the wave migration period, P_w, as defined by the time required for the wave to travel over the complete light curve, and the orbital period. Figure 9 shows that a linear negative correlation exists between $\log P_w$ and $\log P_{orb}$, when P_w is expressed in unity of P_{orb} (dots), while no correlation is apparent when P_w is expressed in years (crosses). The latter values show only a large scatter around the mean value of about 9 years. Therefore, it appears that the distortion wave travels faster on the light curve as the orbital period increases, in such a way that the time required for the wave to travel once on the light curve turns out to be similar, within a factor of two, for all systems. This is an important piece of information to be taken into account in modelling the activity phenomena observed in RS CVn stars.

From the data presented above, it is apparent that significant photometric works on RS CVn stars require decades of systematic and homogeneous collection of data: it is not fortuitous that the most

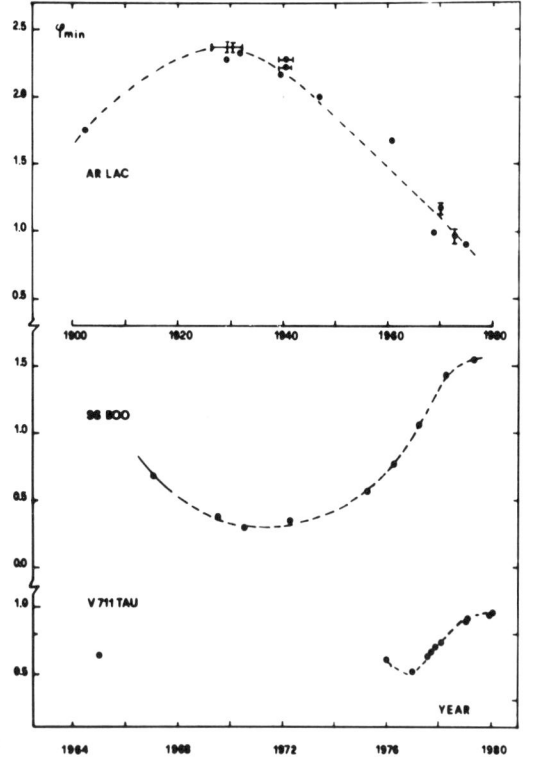

Figure 7. Changes in the migration direction of the distortion wave in three RS CVn binaries (Blanco et al.1980b)

valuable photometric observations of RS CVn were started about fifteen years ago.

Up to now, spectroscopic data on RS CVn stars have not been collected in a systematic way, so that they suffer even more of the limitation proper to photometric works. Spectroscopic observations usually refer to flux measurements in chromospheric lines, such as H_α and Ca II H and K. A survey of H_α emission in 28 RS CVn stars has shown that only six stars show H_α continuously - UX Ari and V 711 Tau - or sporadically - HD 86560, HK Lac, AR Lac and SZ Psc (Bopp and Talcott 1978). At present contraddicting evidence on the correlation between line intensity and photometric wave, and on the location and extent of emitting regions, has been presented. Only future observations can settle these controversial points. Instead, line emission enhancements have been reported by several observers in coincidence with the giant 1978 radio flare of V 711 Tau although the photometric light curve shape remained unaffected (see Astron J. 83, 1978).

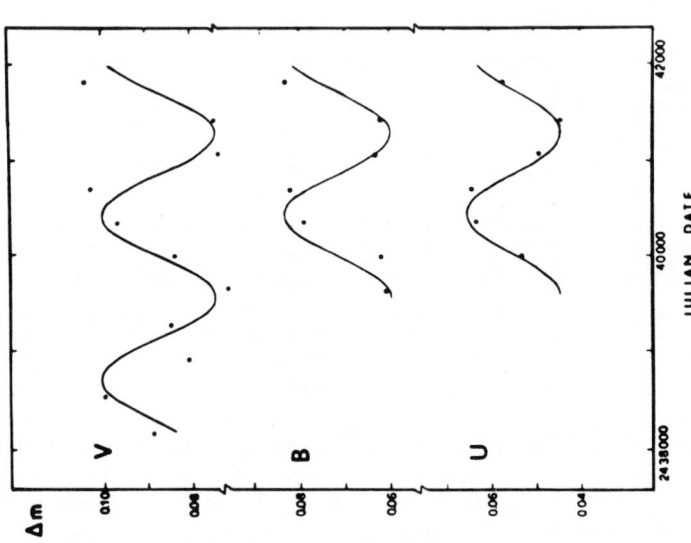

Figure 8. Cyclically variable amplitude of the distortion wave on RS CVn light curve.

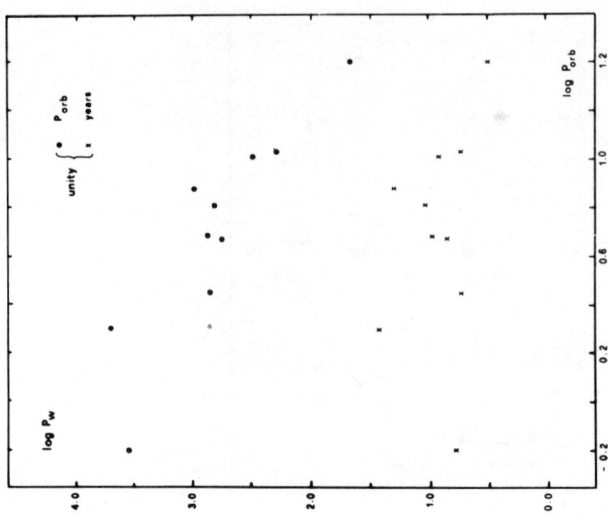

Figure 9. Linear correlation between log P_W; P_W being the travel time over the full light curve in unity of P_{orb}, and log P_{orb} indicating that the wave travels faster as the orbital period P_{orb} increases (dots). The time required for the wave to travel once on the light curve (crosses) appear to be similar, within a factor of two, for all systems.

3.2 Radio, UV and X-ray Observations

Variable radio emission has been recently detected from several RS CVn binaries (Owen and Gibson 1978, Hjellming 1980). The radio fluxes are smoothly varying with time scales of hours to a few days: the general behaviour is that of superposed nonthermal flares rather than of persistent coronal-type thermal emission. The strong radio flare on V 711 Tau detected by Feldman et al.(1978) has prompted an international collaborative effort to observe this exceptional event in any possible wavelength region. Radio observations, which were obtained in several wavelenghths from 1.4 to 86.1 GHz, have shown the emission to be gyrosynchrotron radiation emanating from a region with characteristic dimension of the order of the binary star separation. The degree of circular polarization, which had already been detected in the quiescent phase (Gibson et al. 1978) and was indicative of a large ordered magnetic field of the order of 10 Gauss, increased to 40% with peak value up to 70%. The energy involved in the radio flare was of the order of 10^{17} erg s^{-1} Hz^{-1}, i.e. 10^6 times that in the largest solar events. Near the end of this radio flare, low-dispersion far ultraviolet spectra - covering the wavelength region 1150-1950 A - were obtained by Linsky et al. (1978) with IUE (International Ultraviolet Explorer). The detected high-excitation lines of C II, C IV, Si II, Si IV and N V were suggestive of solar-like flare activity but at levels order of magnitude higher. Additional UV and X-ray observations made with IUE and EINSTEIN satellites during quiescent phases (see Linsky 1980 and Walter et al. 1980), have shown that the atmospheric structure of RS CVn stars - chromosphere, transition region and corona - is only quantitatively different to the solar one. Typically, a factor of ten additional nonradiative heating and huge facular regions covering a large fraction of the stellar surface are inferred. Therefore, the stellar surface being almost entirely covered by active regions, it is not surprising that no modulation due to star rotation results, while highly enhanced emissions are observed: Chromospheric and transition region surface fluxes 25 times and 75-250 times, respectively, larger than the corresponding solar ones have been found from quiescent UV spectra of UX Ari and V 711 Tau (Simon and Linsky 1980). Near the maximum of a large flare of UX Ari, Simon et al. (1980) have detected an additional 5.5 time enhancement of the emission lines with respect to the quiescent level and asymmetric red wings up to 475 Km s^{-1}. These asymmetric profiles can be due to infalling material in the KO IV component along a giant flux tube extending up to several stellar radii. These extreme activity manifestations do not have quantitatively similar solar analogues. In Figure 10 a few representative IUE spectra in the quiescent phase are shown.

Soft X-ray emissions (0.15-3.0 KeV) have been detected from all of the 41 RS CVn systems observed with HEAO-1 and HEAO-2 (EINSTEIN) satellites up to February 1980 (Warter et al. 1980). The X-ray lumi-

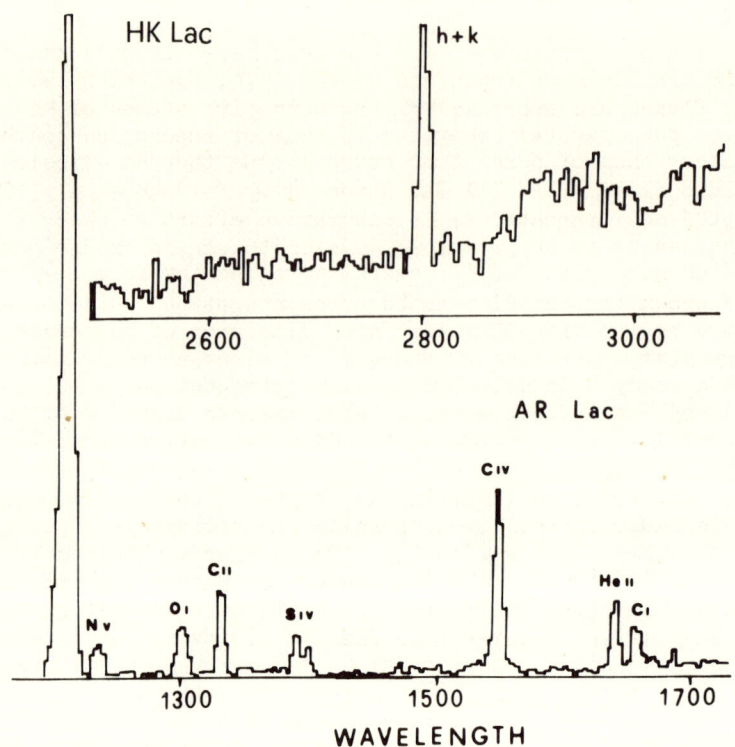

Figure 10. IUE plots of two far-ultraviolet low-dispersion spectra of HK Lac and AR Lac obtained by the Catania group. The spectrum of HK Lac shows the chromospheric emission-line blend of Mg II h and k, in the long wavelength region, and the spectrum of AR Lac, in the short wavelength region, shows high-excitation chromospheric and transition region emission lines of several species.

minosities range from 1×10^{30} erg s^{-1} to 2×10^{31} erg s^{-1}, the latter being comparable with the flaring-Sun level. A peak luminosity of $7 \pm 2 \times 10^{31}$ erg s^{-1} was obtained during a large flare of BD + +61°1211 with total energy output of about 4×10^{37} erg.

In spite of the large ammount of energy involved and postulated extension of coronal loops, the coronae of RS CVn stars seem to be qualitatively similar to the solar one: i.e. a constant large scale structure and a superposed inhomogeneous active one arising from transient active regions. The latter could have produced the X-ray flux modulation of UX Ari, anticorrelated with the photometric wave. No comparable effect is observed in other RS CVn systems. Apart from flare events, the X-ray flux shows also variations of up to one order of magnitude over the few years covered by observations.

One important characteristic of the X-ray luminosity is that the ratio L_X/L_{bol} correlates fairly well with the rotational period (Walter et al. 1980), the most intense sources being the systems with the shortest periods (Figure 11). Ayres and Linsky (1980) have found a similar correlation showing that L_X/L_{bol} increases in the same way as the rotational velocity and this result includes also flare and related BY Dra stars, as well as the Sun (Figure 11).

4. MODELLING AND PROBLEMS

The ultraviolet and X-ray observations, presented in the preceeding section, have shown that, although pure scaling laws from solar phenomena are not always applicable, the bulk of the available evidence indicates that gigantic solar-like structures and activity phenomena occur in RS CVn star atmospheres. These

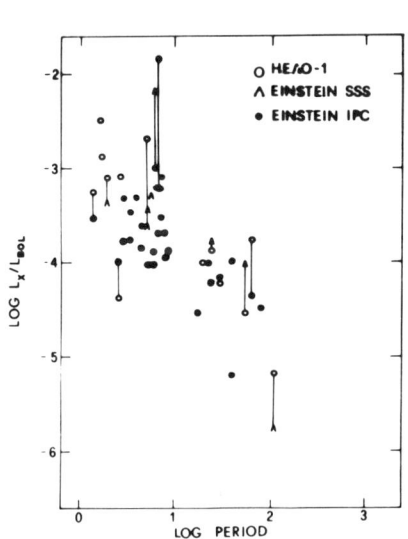 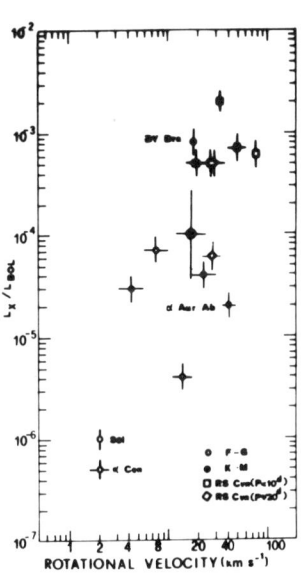

Figure 11. The relative X-ray luminosity L_X/L_{bol} versus the orbital period (left) from Walter et al. (1980) and versus the rotational velocity (right) from Ayres and Linsky (1980) showing that L_X/L_{bol} is higher in more synchronous and highly rotating systems.

observations have given more credit to Hall's (1972) suggestion that the distortion wave in RS CVn light-curves id due to distribution in longitude of photospheric dark spots on the cooler KO IV component, extending approximately 30 degrees in latitude about the equator. As the star rotates almost synchronously with the orbital motion, a variable fraction of the visible stellar surface is covered by spots. For RS CVn, this variable fraction is assumed by Eaton and Hall (1979) to follow the sinusoidal dependence required by the sinusoidal shape of the distortion wave affecting its light curves. The wave migrates towards decreasing orbital phases because the star angular rotation is slightly larger than the orbital one.

Quite generally, since stellar rotation is not exactly locked to the orbital motion, both direct and retrograde migrations are possible and are, indeed, observed (Figure 6). Changes in the migration directions from direct to retrograde (or viceversa), as those presented in Figure 7, can be explained assuming that i) the star rotates differentially, with the corotating latitude somewhat away from the equator, and ii) the photocenter of the spotted region progressively moves from latitudes with slower-than-synchronous rotation to higher-than-synchronous ones (or viceversa) while the active region migrates towards the aquator during the activity cycle, as in the Sun. Using this hypothesis, Blanco et al. (1980b) have determined activity cycles and differential rotation rates in a few RS CVn systems.

The spot model is able to explain qualitatively the numerous photometric peculiarities presented in section 3 and, notably, the correlation presented in Figure 9: the distortion wave moves at a lower rate in the short-period systems because, due to forced synchronization, the difference between orbital and stellar rotation in these systems is smaller than for wider ones with longer orbital periods. However, why the travel times required to cross the full light curve of RS CVn systems (Figure 9), gather around 9 years is not so easily explained.

The spot model is actually able to reproduce quantitatively the observed distortion wave (Figure 12) as also demonstrated by previous spot modeling of the somewhat cooler BY Dra stars (Torres and Ferraz Mello 1973, Bopp and Evans 1973). However, spot modeling invariably leads to some serious indeterminacy of the solution. Bearing in mind: that also bright photospheric facular regions can develop, as found for II Peg (HD 224085) from simultaneous photometry and IUE spectroscopy (Rodonò et al. 1980), and that a dark spotted region in BY Dra has evolved in a bright facular one (Vogt 1975), much caution should be exercised by regarding the spot model only as a working one probably leading in the right direction (Hall 1980).

As a matter of fact, spot modeling (see also Friedmann and

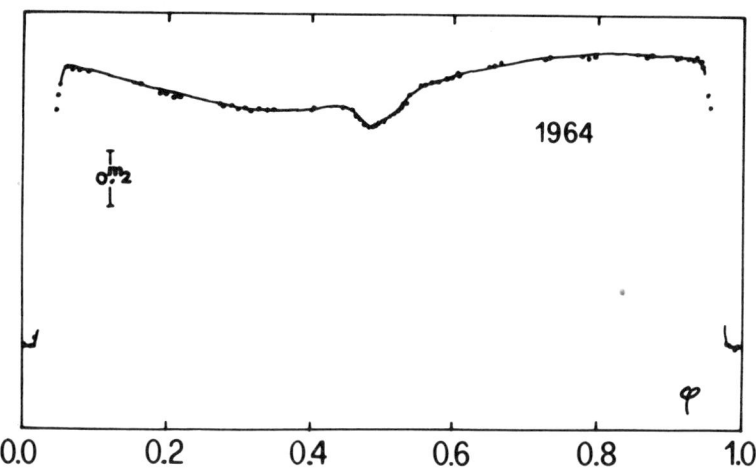

Figure 12. Light curve fit by spot modeling (solid line) from Eaton and Hall (1979) of the 1964 RS CVn light curve (dots) by Chisari and Lacona (1965).

Gurtler 1975) is a curve fitting method with an analytical function which depends on several parameters, namely i) the temperature difference between spots and unperturbed photosphere, which can be either positive (bright spots) or negative (dark spots), ii) the inclination of the star rotation axis to the line of sight, iii) the area and iv) the location of the spotted region. One of the most critical parameters - the inclination of the rotation axis - is generally well known for eclipsing binaries. However, the temperature, size and distribution of spot groups on the stellar surface can be combined in a number of ways to reproduce the observed light curve variations. Therefore simultaneous observations in other wavelength regions or in particular spectral lines are needed to avoid estethically appreciable but meaningless "computer games" in spot modeling using photometric-curve fitting methods exclusively. At present, uniqueness is far from being achieved.

My present criticism of spot modeling does not imply that starspots are not capable and are not the most likely candidates in explaining the numerous photometric peculiarities of RS CVn stars. Actually, the general picture emerging from the observations in different spectral regions, from radio to X-ray wavelengths, indicates that we are most probably dealing with solar-like phenomena in RS CVn as well as in BY Dra stars. Moreover, recent theoretical approaches (see Rodonò 1980) indicate that the energy requirement of the supposedly magnetic stellar activity can be provided by the stressing of emerging toroidal fields generated by the so called αω-dynamos in highly convective, rapidly rotating - i.e. differentially rotating - stars. Both RS CVn and BY Dra stars actually appear to be rapid rotators and have sizable surface convective zones. Unfortunately, despite numerous and qualified observational efforts, no indisputable evidence of magnetic fields on these stars, with the possible exception of II Peg (Vogt 1980), has been presented. This is not very discouraging because of the elusiveness of localized stellar magnetic fields when measured against the relatively bright star background.

5. CONCLUSIONS

I should like to conclude my talk by simply listing observational (-) and theoretical (⁰) topics or questions deserving particular attention in future research on RS CVn binary systems

- Establishing light curve variations for several systems and looking for correlation with relevant parameters, such as spectral type, orbital period and star rotation and depth of convection zones.

- Accurate determinations of orbital period variations.

- Reliable solutions of radial velocity and light curves, with migrating waves completely removed from the latter.

- Measurement of stellar magnetic fields.

- Studies of chromospheric and transition region line variations in the quiet and active phases.

- Studies of X-ray flux variability.

⁰ Are rotation and convection the key parameters of stellar activity, i.e. how do stellar dynamos work?

⁰ Are the long term stability and extension of the suggested spotted active regions in RS CVn stars consistent with solar type activity?

⁰ Can moderate mass loss account for the observed orbital period variations, as indicated by the possibility that convectively driven wind is constrained to corotate out of the Alfven radius (DeCamply and Baliunas 1979)?

⁰ Which is the connection, and the interaction, between differ-

ential rotation and forced synchronization?

⁰ Bearing in mind that the cooler K0 IV component is in, or very near, the Hertzsprung instability zone, dynamical approaches, as pulsation (Popper 1977), oblique rotator plus accretion disc and precessional motion (Catalano and Rodonò 1967) or physical instabilities triggered by variable tidal forces in slightly eccentric orbits, should not be disregarded "a priori" because they do not conform to the main stream of thought.

Acknowledgements.

I should like to thank the Catania Colleagues C. Blanco, S. Catalano and E. Marilli for their kindness in allowing me to present several common observations of RS CVn stars before pubblication and for stimulating discussions.

REFERENCES

Ayres, T.R. and Linsky, T.L.: 1980, preprint.
Blanco, C., Catalano, S., Cerruti-Sola, M., Marilli, E., Rodonò, M., Scaltriti, F. and Strazzulla, G.: 1980a, Workshop on "Stellar Variability", Massalubrense, Italy, in press.
Blanco, C., Catalano, S., Marilli, E. and Rodonò, M.: 1980b, Fifth European Regional Meeting of the I.A.U. on "Variability in Stars and Galaxies", Liegi, July 1980.
Bopp, B.W. and Evans, D.S.: 1973, Monthly Not. Roy. Astron. Soc. 164, 343.
Bopp, B.W. and Talcott, J.C.: 1978, Astron. J. 83, 1517.
Catalano, S., Frisina A. and Rodonò, M.: 1980, I.A.U. Colloquium 88.
Catalano, S. and Rodonò, M.: 1967, Mem. Soc. Astron. Ital. 38, 345.
Catalano, S. and Rodonò, M.: 1968, Pubbl. Osserv. Astrofis. Catania No; 118, 53.
Catalano, S. and Rodonò, M.: 1969, "Non-periodic Phenomena in Variable Stars", I.A.U. Colloquium, L. Detre Ed., 345.
Chisari, D. and Lacona, G.: 1965, Mem. Soc. Astron. Ital.
Cerruti Sola, M., Scaltriti, F., Blanco, C., Catalano, S., Marilli, E., Rodonò, M. and Strazzulla, G.: 1980, in press.
De Campli, W.M. and Baliunas, S.C.: 1979, Astrophys. J. 230, 815.
Eaton, J.A. and Hall, D.S.: 1979, Astrophys. J. 227, 907.
Eggen, O.J.: 1978, Comm. 27 I.A.U. Inf. Bull. Var. Stars No. 1426.
Feldman, P.A., Taylor, A.R., Gregory, P.C., Seaquist, E.R., Balonek, T.J. and Cohen, N.L.: 1978: Astron. J. 83, 1471.
Fracastoro, M.G.: 1965, I.A.U. Colloquium, Veröff, Remeis-Sternwarte Bamberg Bd. IV, Nr. 40, 253.
Friedmann, C. and Gurtler, J.: 1975, Astron. Nachr. 296, 125.

Gibson, D.M., Hicks, P.D. and Owen, F.N.: 1978, Astron. J. 83, 1495.
Hall, D.S.: 1972, Publ. Astron. Soc. Pacific 84, 323.
Hall, D.S.: 1976, I.A.U., Colloquium 29, 287.
Hall, D.S.: 1980, Highlights of Astronomy 5, in press.
Hall, D.S.: 1981, Space Sci. Rev., in press.
Hjellming, R.M.: 1980, Highlights of Astronomy 5, in press.
Larson-Leander, G.: 1980, Highlights of Astronomy 5, in press.
Linsky, J.L.: 1980, Ann. Rev. Astron. Astrophys., in press.
Linsky, J.L. et al.: 1978, Nature 275, 389.
Morgan, J.G. and Eggleton, P.P.: 1979, Monthly Notices Roy. Astron. Soc. 187, 661.
Owen, F.N. and Gibson, D.M.: 1978, Astron. J. 83, 1488.
Popper, D.M.: 1977, Highlights of Astronomy, Part II, 397.
Popper, D.M. and Ulrich, R.K.: 1977, Astrophys J. 212, L131.
Rodonò, M.: 1980, Workshop on "Stellar Variability", Massalubrense, Italy, in press.
Rodonò, M., Romeo, G. and Strazzulla G.: 1980, Second European I.U.E. Conference, 55.
Simon, T. and Linsky, J.L.: 1980, Astrophys. J., in press.
Simon, T., Linsky, J.L. and Schiffer, F.H. III: 1980, Astrophys. J., in press.
Struve, O.: 1946, Ann. d'Astrophys. 9, 1.
Torres, C.A.O. and Ferraz Mello, S.: 1973, Astron. Astrophys. 27, 331.
Vogt, SS.: 1975, Astrophys. J. 199, 418.
Vogt, SS.: 1980, Astrophys. J., in press.
Walter, F., Charles, P. and Boyer, S.: 1980, Proc. of the Workshop on "Cool Stars and Stellar Systems", Cambridge Mass., Ed. A.K. Dupree, in press.
Zeilik, M., Hall, D.S., Feldmann, P.A. and Walters, F.: 1979, Sky and Telescope 57, 132.

OBSERVATIONS OF RS CVn-TYPE BINARY STARS

P. Vivekananda Rao and M. B. K. Sarma

Centre of Advanced Study in Astronomy,
Osmania University, Hyderabad, India.

The star UV Piscium (BD + 6°189) was observed photoelectrically for 54 nights during 1976-77, 77-78 and 1978-79 observing seasons with standard V, B and U filters. The observations were made using an unrefrigerated EMI 6256B photomultiplier attached to the 122 cm reflector of the Japal-Rangapur Observatory. The photocurrent was amplified by means of a 1230 A dc amplifier and recorded on a Honeywell Brown recorder. The star BD + 6°186 was used as a check star. The Δm (check-comparison) was $\sim 0\overset{m}{.}02$ in all the colours which indicates that the comparison star was constant in brightness during the period of observation. The observations of the comparison star were used for determining the nightly extinction coefficients. After applying the extinction corrections, the extra atmospheric magnitude difference Δm (var-comp) were obtained. These Δm's were then transformed to the Johnson and Morgan's standard system using linear transformation relations obtained from the observations of many UBV standard stars. The observations made in UBV during the above observing seasons are plotted in Figures 1a, 1b, 1c, 1d, 1e, 1f, 1g, 1h and 1i. The appearance of these curves indicates that there is a wave-like distortion similar to RS CVn-type stars. The analysis of these light curves is in progress and will be published elsewhere.

The observations of the stars WY Cnc and RZ Eri during the above observing seasons were also made by us using the same equipment as that of UV Psc. The reduction of the data is in progress.

The RS CVn-type eclipsing binary TY Pyxidis was observed photoelectrically for 75 nights with standard B, V filters using the 38 cm refractor of the Nizamiah Observatory during the

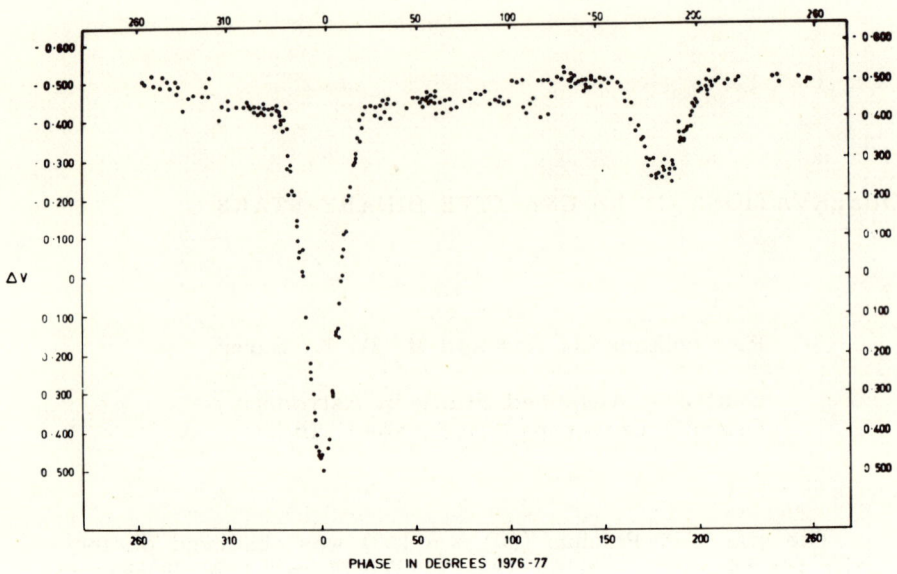

Figure 1a.   Star UV Psc

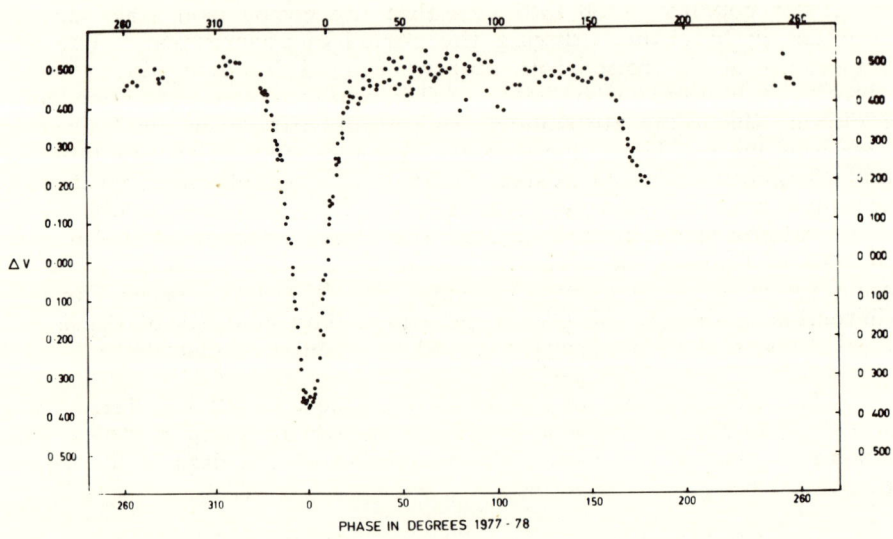

Figure 1b.   Star Uv Psc

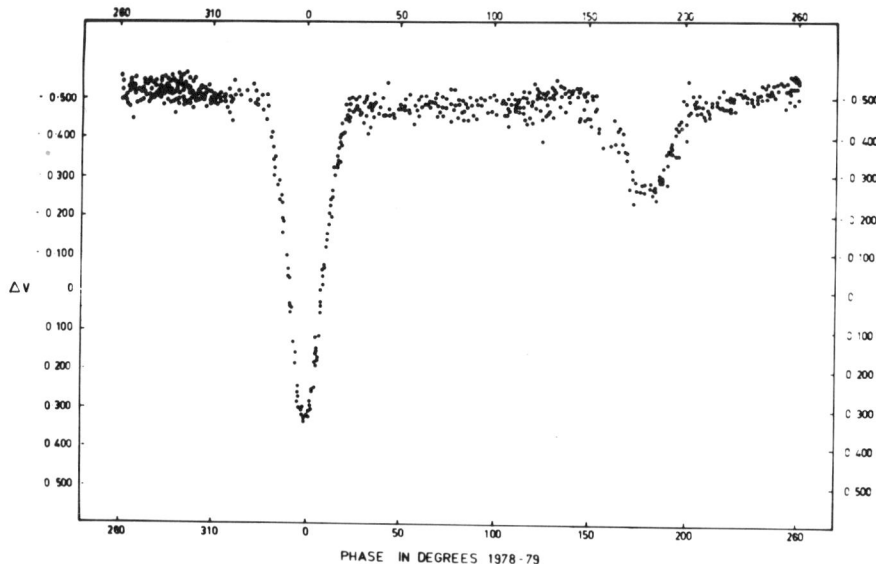

Figure 1c. Star UV Psc

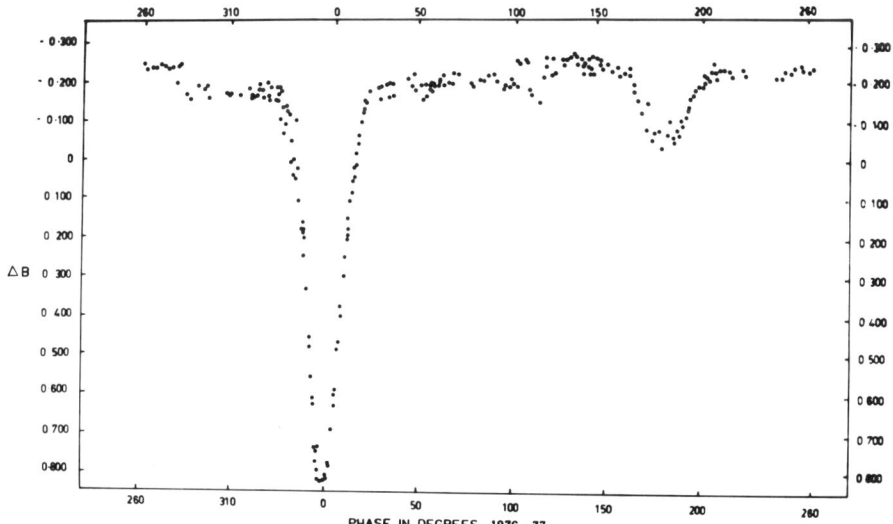

Figure 1d. Star UV Psc

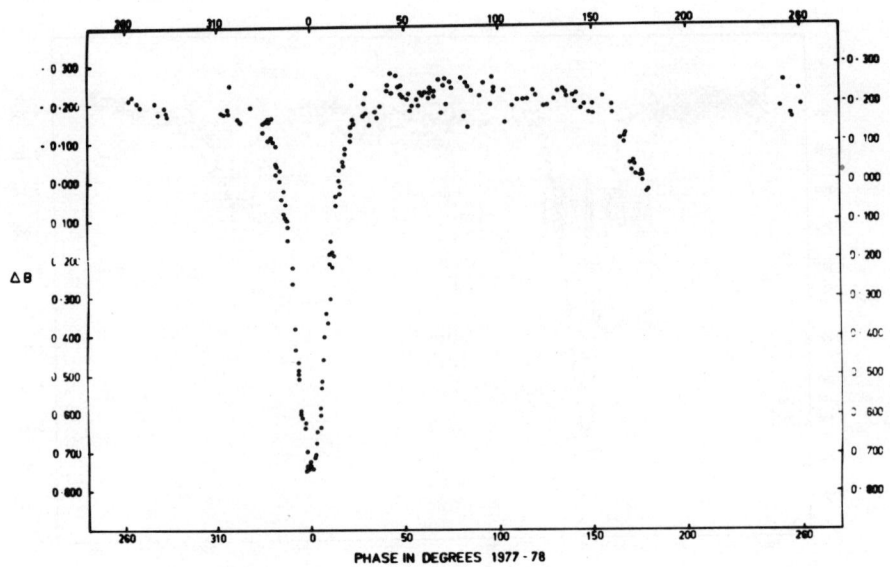

Figure 1e. Star UV Psc

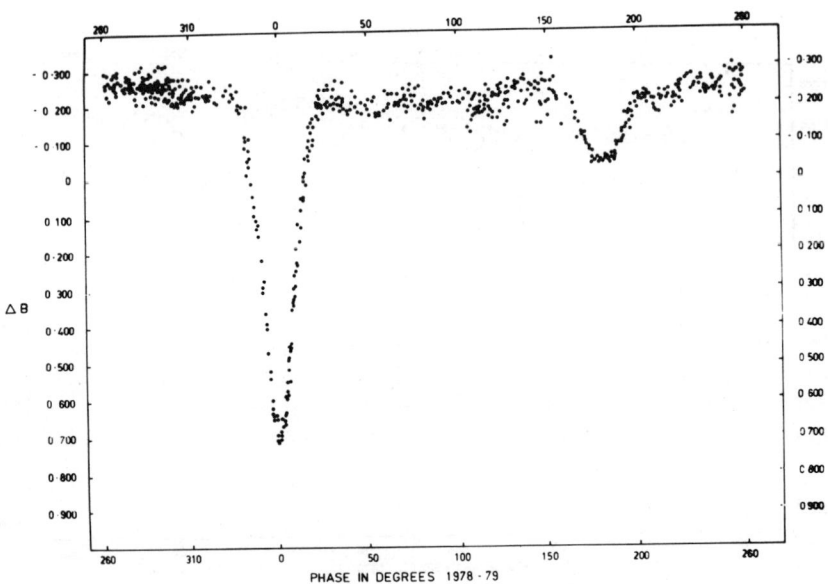

Figure 1f. Stqr UV Psc

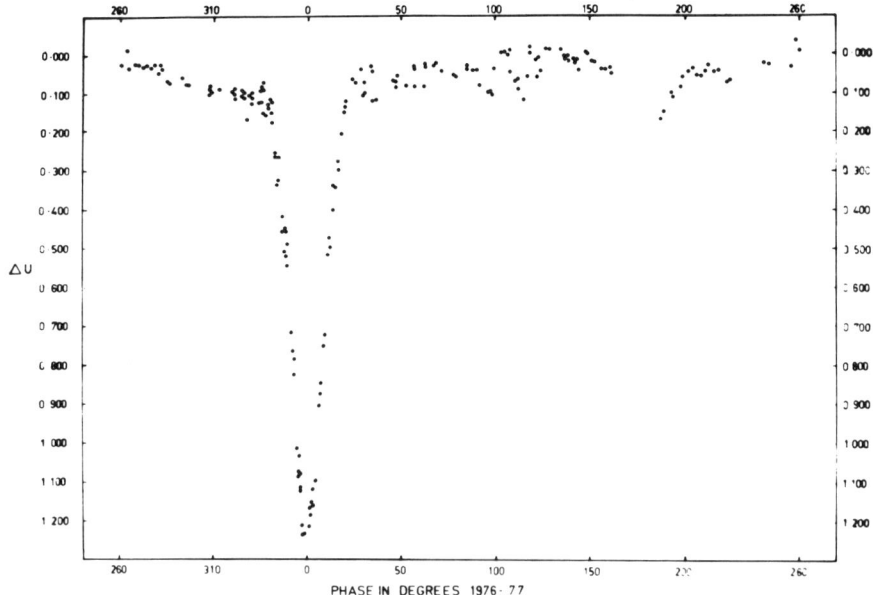

Figure 1g. Star UV Psc

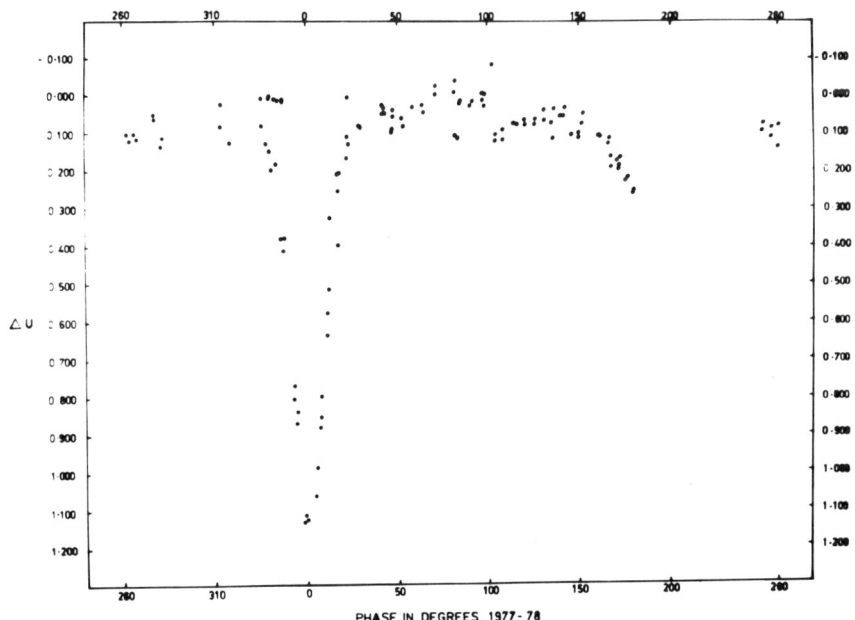

Figure 1h. Star UV Psc

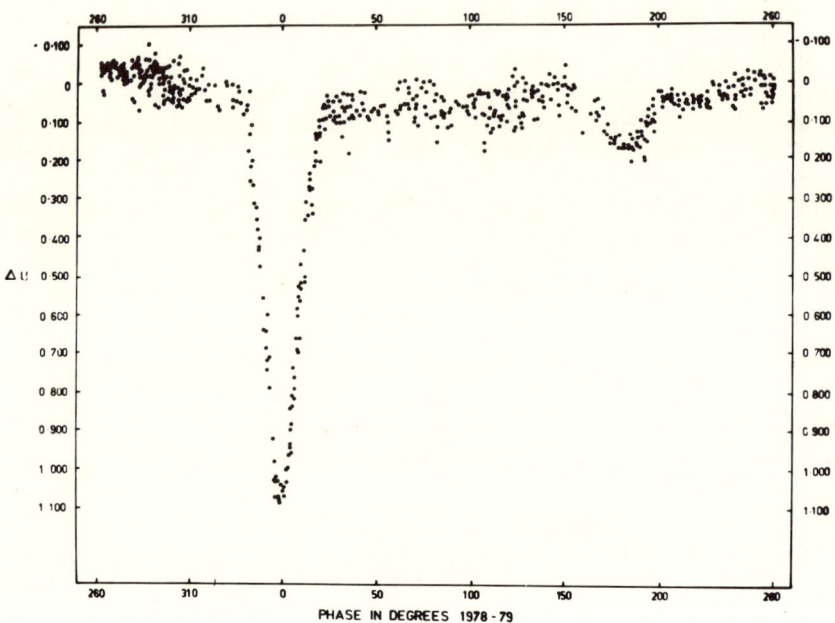

Figure 1i. Star UV Psc

observing seasons 1976-77, 1977-78, 1978-79. The method of observation is on similar lines as that of UV Psc. During the 1976-77 and 1977-78 observing seasons, observations were made using an unrefrigerated IP21 photomultiplier, while in the 1978-79 observing season we made use of the unrefrigerated EMI 9502B photomultiplier. The star HD 77087 was used as a comparison, while the stars HD 77506 and HD 77361 were used as check stars. The r.m.s. error Δm (check-comparison) was $\sim 0\overset{m}{.}02$ in both yellow and blue which indicates that the comparison star is constant in brightness during the period of observation. A total number of 1203 observations in yellow and 1189 observations in blue were obtained. These observations were transformed to the Johnson and Morgan's UBV standard system by observing a number of standard stars and are plotted in Figure 2. The light curve indicates that the epoch given by Strohmeier refers to the secondary minimum and not to the primary minimum. An improved period of $3\overset{d}{.}1985787$ was determined from the present observations. No period variation has been detected from the present study; the spectral type and temperature deduced by us for TY Pyx correspond to G4-6 (5520 ± 100°K) and G5-7 (5400 ± ±00°K) for the primary and secondary respectively. This system does not show any variation in the depths of the minima or a wave-like distortion outside the eclipse, like other RS CVn members. The photometric elements of the system are derived using the Russell and Merrill method. The absolute dimensions of the system are determined by combining the spectroscopic elements determined by Popper and Anderson. The system is classified as detached. Plotting derived temperatures and absolute dimensions of this system in Iben's pre- and post-Main Sequence evolutionary tracks, an age of $1-2 \times 10^7$ years has been derived which is in variance with the ages estimated for other members of the RS CVn group. The components of TY Pyx have properties common to those of T Tauri stars. TY Pyx is a unique member of the RS CVn group with both components having the same mass, radii, temperature, luminosity and probably in pre-Main Sequence contraction evolutionary phase. A paper presenting the solutions and conclusions is being published elsewhere.

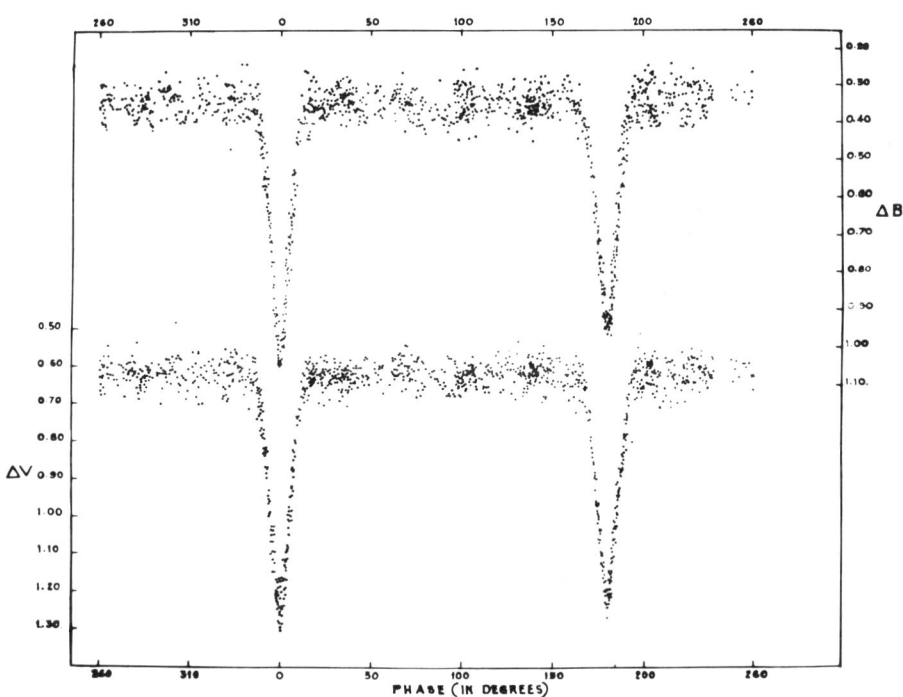

Figure 2. Star TY Pyxidis.

THE SYSTEM RT And AND THE TRANSIT/OCCULTATION QUESTION

S. Mancuso, L. Milano and A. Vittore

Astronomical Observatory of Naples, Italy

E. Budding and D. M. Z. Jassur

Department of Astronomy,
University of Manchester, England.

ABSTRACT

The system RT And is reviewed and recent observational data analysed. With reference to this photometric evidence alone, either a transit or occultation hypothesis for the primary minimum can provide a satisfactory curve-fit with almost the same likelihood, i.e., we cannot clearly resolve between the two possibilities on this basis alone.

The independent spectroscopic identification of the primary as a G0V star, however, allows us to conclude that the transit hypothesis provides a generally self-consistent picture of two close to Main Sequence stars. The presence of known "distortions" in the light curve adds to the uncertainties to some extent, but our treatment shows that the main geometric parameters are little affected on this account.

We consider the problem of the alternative transit or occultation hypothesis in the analysis of eclipsing binary photometric data in a general way, and show that a genuine ambiguity may be possible (in the most common types of close binaries) for a certain restricted range of ratio of the radii ($0.79 \lesssim k \lesssim 1$), if the transit model secondary is a star of low mass. If some independent indication of the primary mass is available, however, Kepler's Law can be used to ascertain whether a given instance is ambiguous. In the case of RT And it is not, and we can safely reject the occultation model.

We are left with a disturbing picture of Main Sequence star(s) exhibiting very erratic behaviour. Some pertinent questions are raised in a final section.

1. PREVIOUS STUDIES OF RT AND

RT And (= BD +52°3383a), an eclipsing binary with period $0.^{d}6289$ is a photometrically peculiar system. The peculiarities can be placed under the following headings: (i) intrinsic night to night fluctuations in brightness of up to $0.^{m}04$; (ii) asymmetric minima (especially the secondary), with shapes and depths which can vary considerably in time, and more so at shorter wavelengths; (iii) pronounced light curve changes ($\gtrsim 0.^{m}1$) on a larger timescale (years); (iv) large variations in period ($\Delta P/P \sim 5 \times 10^{-6}$ over fifty years).

The system was observed visually by Šternberk (1927), and photographically and spectrographically by Payne-Gaposchkin (1946), whose radial velocity curve seems to be the only one published so far. She adopted an occultation (total eclipse) hypothesis for the primary minimum, which apparently gave a better curve fit, and found the ratio of radii (eclipsing to eclipsed) to be ~ 1.9. The masses quoted by Payne-Gaposchkin of 1.5 $M_\odot$ and 0.99 $M_\odot$, should perhaps not be taken to be so accurate, especially as the detection of secondary lines was marginal. The systematic radial velocity is 20 km/s. Nothing was reported about H and K emission by Payne-Gaposchkin though it has been included in a list of stars showing such effects by Eggen (1978), apparently on the basis of evidence in Kron's study (1950). It has thus come to be classified in the "short period" group of stars related to the RS CVn systems (Hall, 1976). H and K emission can be clearly seen on a recent spectrogram of the system obtained at Asiago (see Figure 2).

Gordon (1965) made extensive two colour photometric observations of RT And and pointed out photometric peculiarities. Though, in view of such peculiarities, she was not confident about the matter of accurate element determination, she argued in favour of a transit nature for the primary minimum. This hypothesis has also been supported by Dumitrescu (1973, 1974) who, however, agreed with Payne-Gaposchkin about a slight eccentricity to the system ($e \simeq 0.08$) on the basis of photometric evidence. Keeping in mind the difficulties, both caused by intrinsic variations and the practical matter of obtaining good resolution spectrograms for this short period 9th magnitude system in 1946, the question of a genuine eccentricity should perhaps be treated circumspectly.

However, a concensus seems to have been formed in more recent years that the system is indeed detached and **with** a transit primary (Dean, 1974; Cester et al, 1978; Mancuso et al, 1978, 1979a).

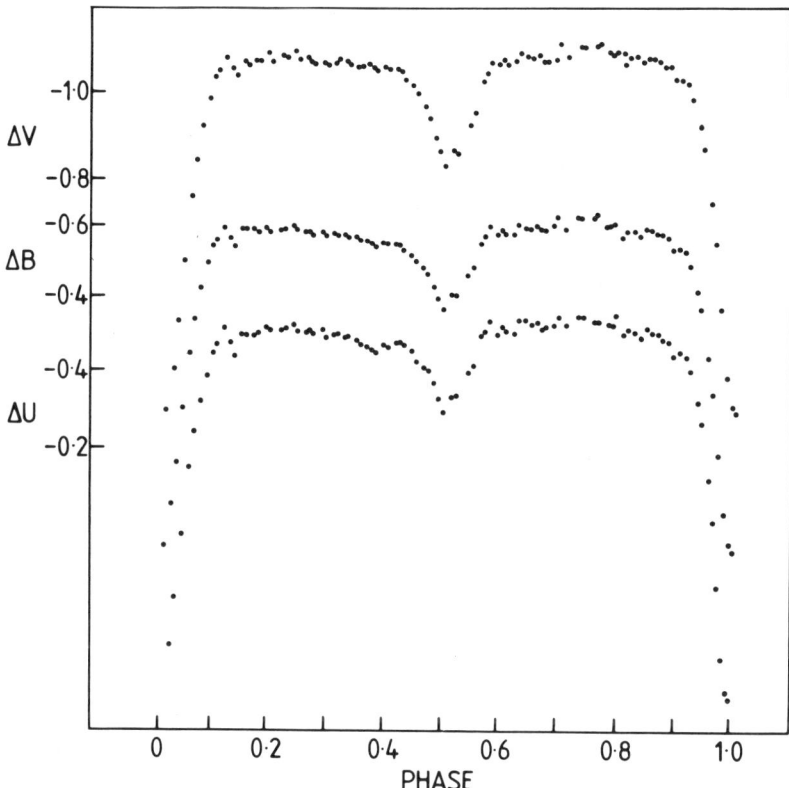

Figure 1   The "normal points" of RT And as observed by Mancuso et al (1978).

It is perhaps still tempting, though, to think in terms of some sort of infall from primary to secondary to maintain a "hot spot" or some other kinetic to radiative energy process to explain the photometric peculiarities (cf., Dumitrescu, 1974; Sadik, 1979), as well as accounting for the period variation (for details of this for RT And, see the thorough study of Williamon, 1974); though no detailed hydrodynamics of such a process has been given for detached systems. The mechanism is much clearer for semi-detached systems which are usually occultation primaries.

Leaving aside more detailed and complete spectrographic evidence, which may, as with WY Cnc (Awadalla and Budding, 1979), be strongly indicative one way or another, we shall in a later section try to examine, in a general way, how far real ambiguities are possible concerning the transit/occultation alternatives on the basis of photometric evidence alone and some fairly simple deductions.

First we re-analyse the relatively recent data on the system of Mancuso et al (1979a), and in a final discussion we shall try to summarize our understanding of RT And, and compare it with other members of the "short period" group of RS CVn stars.

2.1 ANALYSIS OF THE OBSERVATIONS

The V-light curve of RT And observed by Mancuso et al (1978) was first analysed by an optimization method essentially based on that of Budding (1973). The observed points have been grouped into 97 normal points and are shown in Figure 1. Subsequently the B and U observations were similarly reduced and analysed.

a) Photometric Curve-Fitting

First we applied a spherical model (the "eight parameter" program) in order to determine trial values for the more detailed distorted model which was applied subsequently. Starting values for parameters were influenced by the work of other authors, i.e., a transit picture was first assumed and limb darkening coefficients of 0.65 (primary) and 0.75 (secondary) were adopted from Dumitrescu's (1973, 1974) classification of F8V (primary) and K0V (secondary). It was also shown, though, (see next section), that an occultation solution is possible for which starting parameters could be obtained.

The mass ratio ($q = m_2/m_1$), required for the distorted model fitting function, was set at a value of 0.65 and coefficients of gravity darkening and reradiation were calculated from equations given by Kopal (1959), with some slight modifications to the latter, as appropriate to a complete absorption-remission situation.

Since there is clear evidence of asymmetry in the regions out-of-eclipse (the portion of the light curve after secondary minimum is brighter than the preceding region by $0.^m016$), we have attempted to find solutions by assuming the limit of light as corresponding first to the brighter and then the fainter levels. There are slight changes in the elements (relatively deeper eclipses can be effected by increasing the orbital inclination), but the changes are less than the probable errors of the derived elements. The values in

Table 1 Geometrical and Physical Elements of RT And (V Data).

Parameter	Optimal Value (Transit)	Optimal Value (Occultation)
L_1	0.830 ± 0.008	0.720 ± 0.012
L_2	0.170	0.280
$k = r_2/r_1$	0.871 ± 0.017	1.136 ± 0.022
r_1	0.287 ± 0.006	0.250 ± 0.006
i	82°.0 ± 0°.11	79°.9 ± 0°.15
u_1	0.65	0.65
u_2	0.75	0.75
q	0.65	0.65
τ_1	1.1	1.1
τ_2	1.4	1.4
E_1	1.4	1.4
E_2	1.1	1.1
N	97	97
χ^2	109.1	118.4
$\Delta \ell$	0.012	0.012

The four independent parameters are fractional luminosity of the primary L_1, ratio of radii of eclipsing to eclipsed star at the primary minimum k, primary radius in units of the mean separation r_1 and orbital inclination i. The other parameters have been assigned reasonable values. They include linearized limb darkening coefficients $u_{1,2}$, mass ratio q, gravity darkening and reflection factors $\tau_{1,2}$ and $E_{1,2}$. 97 "normal points" (N) have been used in the analysis, which, with an assumed accuracy of 0.012 for each point, results in the χ^2 values shown for each fitting. In the occultation case the mass ratio cannot be indicated *a priori*. We have just retained the figure which applied also (but self consistently) to the transit case. The gravity darkening and reflection factors are not significantly changed by the slightly increased temperature of the occultation secondary. (The formulae for these are given, for example, by Sadik, 1979).

Table 2 Geometrical and Physical Elements of RT And (B and U data).

Parameter	Optimal Value (Transit)	Optimal Value (Occultation)
L_1	0.863 ± 0.019	0.900 ± 0.018
L_2	0.137	0.100
k	**0.884** ± 0.041	0.904 ± 0.043
r_1	0.289 ± 0.006	0.285 ± 0.006
i	82°.2 ± 0°.61	81°.3 ± 0°.66
u_1	0.7	0.85
u_2	0.8	0.9
q	0.65	0.65
τ_1	1.4	1.6
τ_2	1.6	2.0
E_1	2.0	3.0
E_2	1.1	1.1
N	73	73
χ^2	90.4	97.5
$\Delta \ell$	0.015	0.016

The B and U solutions are not as definitive as the V solution; fewer normal points have actually been used, and the scatter, as well as general irregularities, are significantly greater than with V.

Table 3 Adopted Set of Geometric Elements.

r_1 :	0.2870 ± 0.0020,
r_2 :	0.2525 ± 0.0045,
i :	81°.95 ± 0.49.

The elements correspond to an appropriately weighted mean of the different transit solutions. The error assessments (standard deviations) have been determined from inter-comparison of the three sets of values.

Table 1 correspond to appropriate averages.

The results of the alternative occultation solution are indicated in the third column of Table 1. Though the final χ^2 value is slightly greater for this fitting, the difference in "goodness of fit" cannot really be regarded as significant, i.e., the photometry alone is unable to clearly distinguish between these alternatives. We will argue in the next section, however, that the transit alternative is much more likely and, unlike the occultation possibility, leads to a more astrophysically self-consistent picture.

The transit hypothesis solutions for the B and U observations are given in Table 2 and shown diagrammatically in Figure 3c and d. The finally adopted set of geometric elements are given in Table 3.

b) Temperatures.

In Figure 5 we have plotted flux ratios, taken from the convective blanketed series of models of Carbon and Gingerich (1969) for given effective temperature ratios. From the data of Table 1, the flux or surface mean intensity ratio of the two stars in our case is 0.270. If we were to assume an effective temperature of 6300° for the variously reported F8V spectral type of the primary star we could interpolate from the graphs to find an effective temperature of 4790° for the secondary which would correspond closely with the K0V spectral classification obtained by Dumitrescu. Dean's G5 classificication can be associated, in the same way, almost entirely with the lower radius which he found for the secondary star at almost the same luminosity, i.e., pushing up the flux from this star. However, if the two stars are reasonably close to the Main Sequence, we find a much better agreement between our temperature ratio and the corresponding ratio of radii than that of Dean.

The surface mean intensity ratio (J-ratio) involves the square of the ratio of radii, in addition to the overall luminosity ratio, and must therefore be regarded as subject to rather more uncertainty than the more directly determined eclipse parameters (the probability error is about 10% for the present cases). Further support for the foregoing temperatures comes, however, from the B data which yield a J-ratio of 0.199 which is essentially the same as that computed from the Carbon and Gingerich models with the same temperatures. The U observations, on the other hand, are slightly discordant - the predicted models enhance the J-ratio over what one might expect from simple black body considerations by reducing the flux just shortward of the Balmer jump in the higher temperature star. The calculated value is 0.22 - significantly greater than the value coming from the data analysis which is only 0.13. The suggestion is thus an ultraviolet excess for the G0 star - a well-known characteristic of RS CVn systems.

From recent spectrographic study of the star (see next section), a determination of the spectral type (G0V) suggests somewhat lower temperature than the foregoing, i.e., 6050° (primary) and 4650° (secondary). In view of the noticeable brightness changes, however, possible small variability of the determined spectral type need not be surprising.

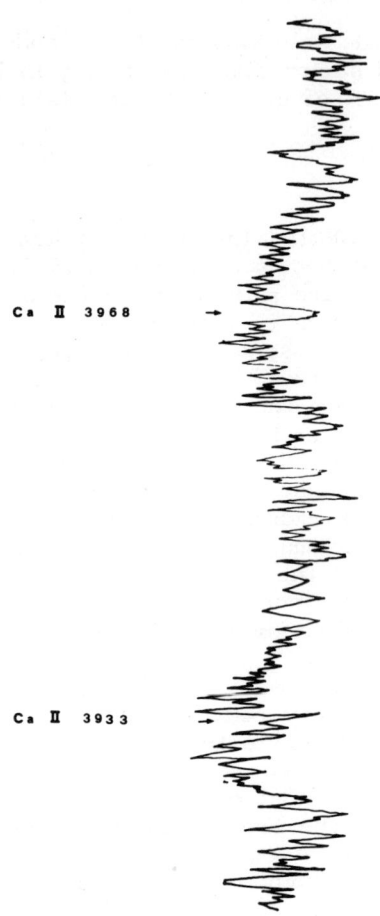

Figure 2   Specimen result of recent spectrophotometry of RT And from observations with the 1.2 m telescope at Asiago. Emission cores in the H and K lines are clearly visible.

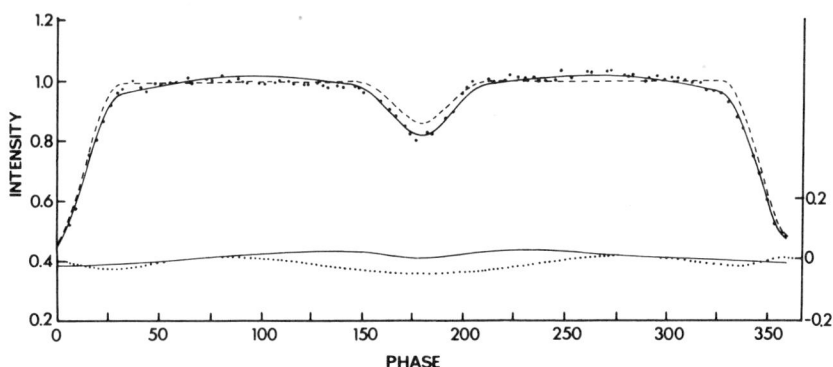

a) V observations (Transit hypothesis)

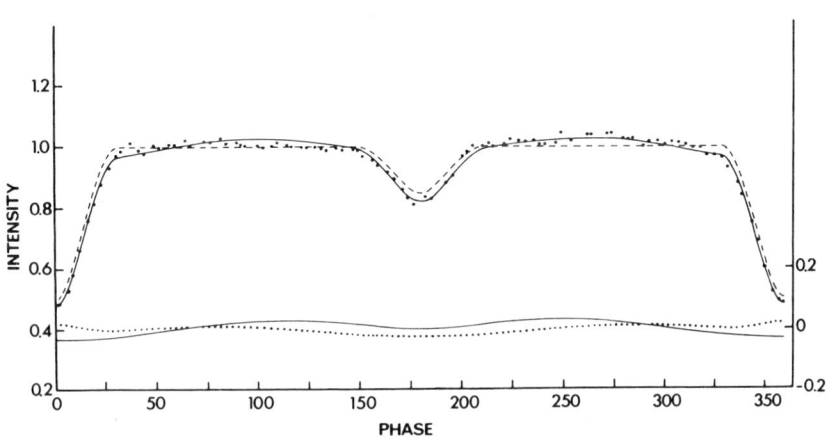

b) V observations (Occultation hypothesis)

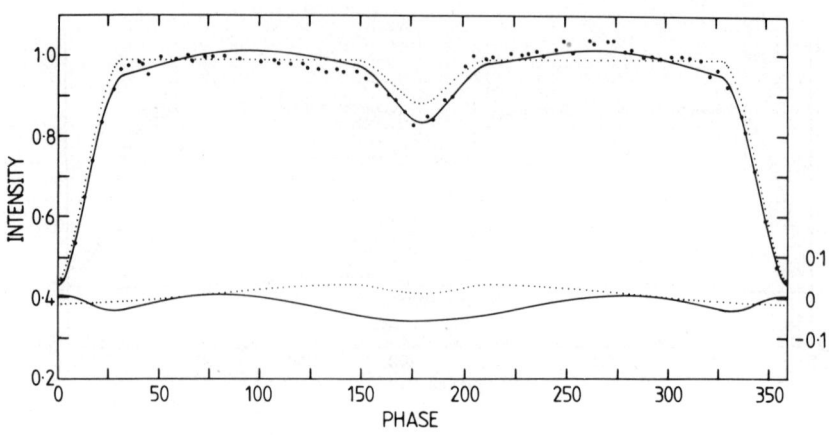

c) B observations (Transit hypothesis)

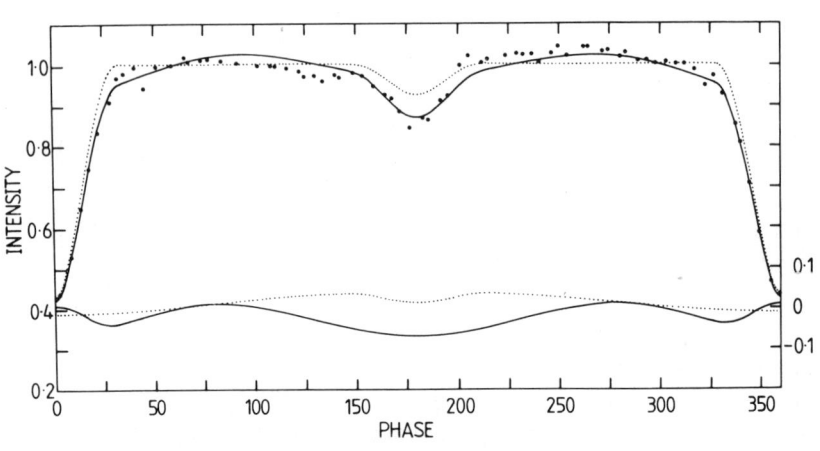

d) U observations (Transit hypothesis)

Figure 3 Optimal curve-fits to the data of Mancuso et al (1978)

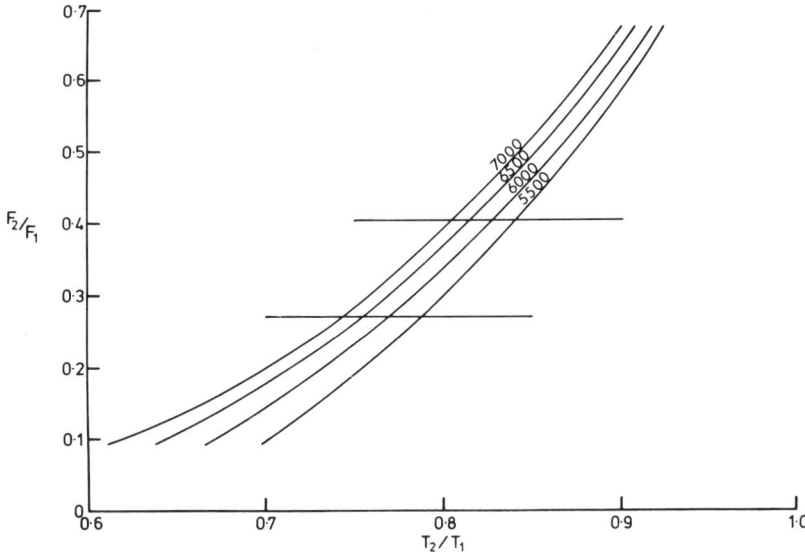

Figure 4 V-band surface mean intensity ratio (secondary to primary) against corresponding temperature ratio for intermediate Main Sequence stars (from Carbon and Gingerich, 1969).

c) Spectral Information

Spectroscopic observations of RT And were obtained at the Cassegrain focus of the 122 cm telescope of the Asiago Astrophysical Observatory during the night of August 31, 1979, with a prism spectrograph (dispersion 60 Å/mm at Hγ) equipped with an RCA image tube. A spectrum (30 minutes of exposure) was secured corresponding to the primary minimum (phase = 0.06) without the contamination of the Moon, and was reduced with a PDS microphotometer of Boller and Chivens model 1010.

A single spectrum is visible indicating a G0V spectral type. This deduced from: 1) the intensity ratio FeI 4143 Å/Hδ, 2) the intensity of CaI 4226 Å, 3) the intensity of H Balmer lines, 4) the structure of G band. Weak and narrow emissions are detectable in the cores of the H and K absorption lines of CaII.

d) Absolute Parameters

Though Payne Gaposchkin obtained somewhat higher values,

the picture we have presented would indicate the sum of the two stellar masses could not be so far from $2M_\odot$. Hence, from Kepler's Law, we can deduce that the separation of the two stars should be about 0.0181 a.u. or 3.87 $R_\odot$, and therefore the primary radius would be 1.11 $R_\odot$ and the secondary 0.97 $R_\odot$, which seems to confirm the Main Sequence picture. The corresponding Main Sequence masses are about $1.2M_\odot$ (primary) and $0.8M_\odot$ (secondary).

During the secondary occultation eclipse the magnitude of the star - i.e., the primary component - is $9^m.21$, while, from the foregoing radius and temperature, its absolute magnitude works out at M_{bol} = 4.28, if we take the bolometric correction for a G0V star to be -0.02. A distance to the system of close to 100 pc is thus indicated.

Using data from Kopal's (1959) book for the radii of the contact surfaces in Roche model stars we deduce, for q = 0.63, the unperturbed radii 0.40 and 0.32 for primary and secondary components, respectively, so that from Table 1 the components are seen to be clearly "detached".

e) Distortion

The classification of RT And as a detached binary, argues against the likelihood of gaseous stream effects as the cause of the photometric distortions. On the other hand, there are indications that the primary could be an intrinsically variable star and produce such effects. The observational arguments for associating the variability with the primary component can be summarized as follows:

1) The contribution of the secondary (K0V) star to the light of the system at secondary minimum is $\lesssim$ 3% of the total light in V and less in B and U, but the depth variation of the secondary minimum is $\gtrsim$ 5% in V, and greater in B and U.

2) As has been mentioned by Mancuso et al (1979b), there is a marked oscillation at phase 0.1 when the primary star comes out of eclipse, and this is more evident in the U-band (see Fig. 1). A large scatter of the observed points at phase 0.9 could mask a behaviour similar to that at phase 0.1.

3) The system becomes bluer/redder when brighter/fainter.

3. TRANSIT AND OCCULTATION ALTERNATIVES

We can have some insight into this question by making a few simple generalisations. With the transit hypothesis the ratio of radii of eclipsing to eclipsed stars (at primary minimum) k < 1. Consider now complete (alternating total and annular) eclipses.

If the depth of the transit eclipse is d_1 in terms of a unit light level outside the minimum, we have for the minimum light at the annular phase

$$L_2 = (1 - k^2)L_1 = (1 - d_1)\phi, \qquad (1)$$

where ϕ is a function that accounts for non-uniformity of brightness of the eclipsed star's surface. Normally $\phi \approx 1$. Then from (1)

$$k^2 = d_1(\frac{\phi}{L_1}) \quad \text{or} \quad k \gtrsim \sqrt{d_1}. \qquad (2)$$

This is a version of the so-called "depth relation". A "shape relation" can also be obtained in terms of the phases of inner and outer contact θ_1 and θ_2 as follows:

$$\frac{\sin \theta_2}{\sin \theta_1} = \sqrt{\frac{r_1^2(1-k)^2 - \cos^2 i}{r_1^2(1+k)^2 - \cos^2 i}} \lesssim \frac{1-k}{1+k}, \qquad (3)$$

where we have introduced the eclipsed star radius r_1 and orbital inclination i. If we combine the inequalities (2) and (3) and simplify a little we find that

$$\theta_2 \lesssim \left(\frac{1 - \sqrt{d_1}}{1 + \sqrt{d_1}}\right) \theta_1, \qquad (4)$$

which must be satisfied for transits.

Let us denote two alternative fractional luminosities as L_1' and L_2' assuming an occultation for the same minimum, i.e., now supposed due to a total eclipse. The depth relation is now simply

$$d_1 = L_1' \qquad (5)$$

so that no such inequality as (2) applies; and, hence, there is no necessary relation connecting the "shape" of the minimum with its depth. In the case of partial eclipses there is no inner contact and we have no inequality such as (4) to discriminate against a false assumption of transit.

We can thus deduce, in a general way, that all transit hypothesis minima can be at least roughly simulated by alternative occultation hypothesis minima (neglecting features like the rounded bottom of an annular eclipse, associated with the relatively small scale effects of limb darkening), though the reverse is not true. We can effect the simulation by simply reversing r_1 and r_2; it will also be required, though, to change the fractional luminosities so that $L_1' \simeq k^2 L_1$.

Let us now confine our attention to those systems where, from spectroscopic or other evidence, we would have good reason to believe that the primary is close to the Main Sequence. The kinds of system one might consider in this way are relatively unevolved close pairs showing transit primaries, or the (semi-detached) Algols, in which the erstwhile secondary is still like a Main Sequence star. Is a genuine ambiguity between transit or occultation hypothesis still possible?

Let us first write a Main Sequence mass radius relation in the form

$$\log R = \alpha \log M \quad \text{(solar units)} . \tag{6}$$

We can also use Kepler's Law in the form

$$\log R_1 = 1/3 \log M_1 + 1/3 \log(1 + q) + 2/3 \log P + \log r_1 + 0.624, \tag{7}$$

where R_1, the primary radius, is in solar radii and the period P is in days. We have introduced the mass ratio $q = M_2/M_1$ which should lie in the range $0 < q \leqslant (R_2/R_1)^\alpha$. If the stars were unevolved we would expect the transit hypothesis to be valid and to be able to insert k^α for q in (7), which could then be combined with (6) to yield absolute parameters which confirm the hypothesis or otherwise. In any case, let us suppose that for some value of q in the range $0 < q \leqslant k^\alpha$ a combination of Equations (6) and (7) will allow a primary which is close to the Main Sequence on the transit hypothesis. The primary (identified) star remains the same with the occultation hypothesis, but if we write M_2' in place of M_2 for the secondary mass on the occultation assumption, Kepler's Law requires that

$$(M_1 + M_2) = k^3 (M_1 + M_2') . \tag{8}$$

Writing q' for M_2'/M_1 we find that

$$q' = \frac{(1 + q - k^3)}{k^3} < k^{-\alpha} . \tag{9}$$

The greatest range for k occurs if $q = 0$, so that ambiguity could occur only for a value of $k < 1$ which satisfies

$$1 - k^3 < k^{3-\alpha} \quad . \tag{10}$$

For Zero-Age Main-Sequence stars α is about 0.6 which would then necessitate $0.77 < k < 1$. This result is little affected by variations of α which could be associated with composition differences or evolution close to the Main Sequence, since the lower limit of k which satisfies (10) is given by

$$k \approx (\tfrac{1}{2})^{\frac{1}{3}} (1 - \frac{\alpha \log_e 2}{3(6 - \alpha)}) \quad , \tag{11}$$

which is insensitive to the feasible range of variation of α. In fact, the inequality in (10) is unnecessarily lax, since the most usual sort of situation in which occultations occur is in Algol systems, where frequency $q' < \tfrac{1}{2}$, and even if mass loss or exchange has not occurred, $q' = fk^{-\alpha}$, where $1 > f = (k/k_1)^\alpha$, k_1 being the ratio of radii when the enlarged component was previously on the Main Sequence. If we are observing an occulting system in which substantial mass loss has still not occurred, $k_1 \approx 1$ is implied in order that sizeable eclipses be detected at all (cf. Morgan and Eggleton, 1979), so that we could with safety raise the lower limit on k to $(\tfrac{1}{2})^{\frac{1}{3}}$ (= 0.79). Obviously, the effect of reducing q' is to push the range of k to a small domain close to unity.

All this assumes that the secondary of the transit primary system is a somewhat peculiar "undersized subgiant" of very low mass. In the more easily understood situation of an unevolved pair which produces a transit primary, q in (9) can be replaced by k^α and then no ambiguity is possible, except in the trivial case of $k = 1$.

Hence, the main result of this is that if a combination of Equations (6) and (7) allows a Main-Sequence primary on the transit hypothesis then it is generally unlikely that an occultation alternative with a Main-Sequence primary is also possible, except perhaps in situations where k is close to unity (where, in any case, the distincton between transit and occultation becomes somewhat blurred), or the eclipsing star of the transit primary is an "undersized subgiant".

The present case of $k = 0.87$ falls into the feasible range of ambiguity at first sight, though if transit and occultation hypothesis were equally likely the mass ratio in the transit case q would have to be at most 0.26 in order to satisfy the inequality (9).

A substitution of the parameters drawn from Section (2) into Equation (7) yields, however, that $q = 0.64$, which is in very good agreement with the Main-Sequence picture of Section (2) and quite out of the range of possible ambiguity. Thus, if we substitute the occultation hypothesis value of r_1 into Equation (7) we obtain an unacceptably high secondary mass ($\sim 1.75 M_\odot$).

4. CONCLUSIONS AND PROBLEMS

The system RT And is undoutedly peculiar, yet reasonable arguments show that its components must both be close to the Main Sequence and be "detached" from their Roche Lobes by substantial distances. In this respect it is quite similar to the recently discussed cases of UV Psc (Sadik, 1979) and WY Cnc (Awadalla and Budding, 1979) and probably also CG Cyg (Jassur, 1980) but because of the closeness of k to unity for the transit solution in the latter case a clear resolution between the two alternative eclipse hypotheses is difficult). The indications are also that it is the primary star in all these systems which is responsible for the irregularities.

The system RT And therefore confronts us with a number of not necessarily independent problems, which we will simply ask about in the present papers as follows:

i) Could the observed phenomena be regarded as some sort of erratic episode in the life of an otherwise quite normal close-to-the-Main-Sequence star? In considering this question the selection effect which weights observers' attention towards inherently interesting stars like close binaries should be kept in mind. Even so, the sort of variation of overall luminosity under consideration ($\gtrsim$ 10% in a year) represents something several orders of magnitude greater than that observed for normal stars away from known instability regions of the H-R diagram.

ii) What is the relationship of this system to the RS CVn stars? Though, as has already been mentioned, Hall (1976) did point out the small subgroup of short period systems composed of intermediate/late type stars, including RT And, which shared some properties, for example CaII emission, with the classical RS CVn stars; the systems differ in the respect of not containing giant or subgiant stars. In this sense they should not show up as the RS CVn-like group in the way discussed by Morgan and Eggleton (1979). Also, though photometric irregularities are definitely observed, the presence of a migrating wave-like distortion has not been clearly demonstrated for RT And.

iii) Are the observed effects definitely linked in some way with the circumstance of (close) binarity? Various authors (e.g.,Young and Koniges, 1977; Shore and Hall, 1980) have pointed to a possible connection between tidal interaction and chromospheric activity which could help explain certain of the RS CVn phenomena. The observational evidence referred to (which is not perfectly direct) does not extend to systems with periods less than one day. Also, there are close binary systems (apart from the W UMa's) falling into the same range of masses, periods, spectral types (e.g., UV Leo) which are not reported to exhibit any of the peculiarities of RT And.

Such questions raised by this intriguing system should make it and similar stars merit further and more detailed attention.

REFERENCES

Awadalla, N. S. and Budding, E.: 1979, Astrophys. and Space Sci., 63, p.479.
Budding, E.: 1973, Astrophys. and Space Sci., 22, p.87.
Carbon, D. F. and Gingerich, O.: 1969, *Theory and Observation of Normal Stellar Atmospheres*, ed. O. Gingerich, p.377.
Cester, B., Fedel, B., Giuricin, G., Mardirossian, F. and Mezzetti, M.: 1978, Astron. Astrophys. Suppl., 32, p.351.
Dean, C. A.: 1974, Publ. Astron. Soc. Pacific, 86, p.912.
Dumitrescu, A.: 1973, Studi Cerc. Astron., 18, 1, p.47.
Dumitrescu, A.: 1974, Studi Cerc. Astron., 19, 1, p.89.
Eggen, O. J.: 1978, Inf. Bull. Var. Stars, No. 1426.
Gordon, K. C.: 1955, Astron. J., 60, p.422.
Hall, D. S.: 1976, IAU Colloq. No. 29, ed. W. S. Fitch, D. Reidel Publ. Co., Dordrecht, p.287.
Jassur, D. M. Z.: 1980, Astrophys. and Space Sci. 67, p.19.
Kopal, Z.: 1959, *Close Binary Systems*, Chapman and Hall, London.
Kron, E. G.: 1950, Publ. Astron. Soc. Pacific, 62, p.141.
Mancuso, S., Milano, L., Russo, G. and Sollazzo, C.: Inf. Bull. Var. Stars, No. 1409.
Mancuso, S., Milano, L. and Russo, G.: 1979a, Astron. Astrophys. Suppl., 38, p.187.
Mancuso, S., Milano, L. Russo, G. and Sollazzo, C.: 1979b (Private Communication).
Morgan, J. G. and Eggleton, P. P.: 1979, Mon. Not. Roy. Astron. Soc., 187, p.661.
Payne-Gaposchkin, C.: 1946, Astrophys. J., 103, p.191.
Sadik, A. R.: 1979, Astrophys. and Space Sci., 63, p.351.
Shore, S. N. and Hall, D. S.: 1980, Proceedings of IAU Symp. No. 88 (ed. M. Plavec et al), *Close Binary Stars: Observations and Interpretation*, p.389, D. Reidel Publ. Co., Dordrecht.
Šternberk, B.: 1927, Prague Publ., 2, Part VII, p.10.
Williamon, R. M.: 1974, Publ. Astron. Soc. Pacific, 86.p.924.
Young, A. and Koniges, A.: 1977, Astrophys. J., 211, p.836.

THE SHORT-PERIOD GROUP OF ECLIPSING BINARIES WITH PROPERTIES SIMILAR TO THE CLASSICAL RS CVn GROUP

L. Milano

Capodimonte Astronomical Observatory of Naples

ABSTRACT

The available data on the short period group of eclipsing binaries with properties similar to the classical RS CVn group are analyzed. The possibility of a new classification of these eclipsing binaries that is not founded on the period criterium established by Hall (1976) is deduced. The new classification is founded on the different types of primary eclipse (occultation/transit) either for the short period RS CVn or classical RS CVn or long period RS CVn. It is also established that more data are needed to confirm the hypothesis quoted above.

In the past a group of eclipsing binaries was defined "having periods of less than one day, in which the hotter component is of spectral class F-G, V-IV and H e K emission is displayed in one or both components". It is also required that "these binaries must not be a contact system, to separate this class from the WUMa binaries" (Hall, 1976).

The object of this lecture is a review of the available data on the eclipsing binaries of the short period group with properties similar to the classical RS CVn group (forthwith I call that class S.P.G., whilst the group of classical RS CVn will be referred to as C.G.). It is to be stressed that I use the definition of S.P.G., quoted above, as a matter of convenience, but I do not attach too much significance to this grouping of C.G. and S.P.G., which is practically based only on different orbital periods (S.P.G. periods less than one day and greater than 0.5 day, C.G. periods between one day and about twenty days).

From the review on the C.G. and S.P.G. by Hall (1976) there is evidence that some of the common properties of both groups are:

a) Light curve variations;

b) Variation of primary and secondary minimum depths;

c) Orbital period variations and displacement of the secondary minimum;

d) Emission lines H and K of Ca II.

I think it will be useful to review star by star the observed peculiarities of the S.P.G. to ascertain either common properties or possible differences of the S.P.G. from the C.G.

The S.P.G. originally included only six stars, and another has been included in the list from the time of Hall's review. The list of stars is shown in Table 1.1.

In the following, particular care has been displayed in the description of the observed properties of one of these stars (RT And), to explain the common difficulties encountered when the presence of a distortion wave on the light curve is doubtful owing to its smallness. Let me turn now to speak about the stars of the S.P.G., beginning with the first in the list, that is:

RT And

There are ten observed light curves of this binary. The first two light curves were obtained one visually (Gadomski, 1928) and the other photographically (Payne-Gaposchkin, 1946), whilst all the others are photoelectric light curves (Gordon, 1948; Dumitrescu, 1973; Dean, 1974; Mancuso et al, 1979a, b). In Figs. 1 and 2 the normal points of all the observed U and B light curves are respectively shown whilst in Fig. 3 there are all the V light curves.

From Fig. 3 it is easily seen that there is a variable trend of the maxima from one epoch to another, before and after the secondary minimum. I computed the differences in the mean level of maxima, i.e., the mean level of the maxima before the secondary minimum minus the mean level of the maximum after the secondary minimum, for all the observed light curves. The results, for the different colours, are shown in Table 2 under the heading M I. From these results a clear trend of the M I's with the wavelengths cannot be argued: in fact the differences are sometimes greater and sometimes smaller for shorter wavelengths, at least in the available light curves. In Table 2 there are also shown the differences of the shoulders of the primary minimum and the secondary

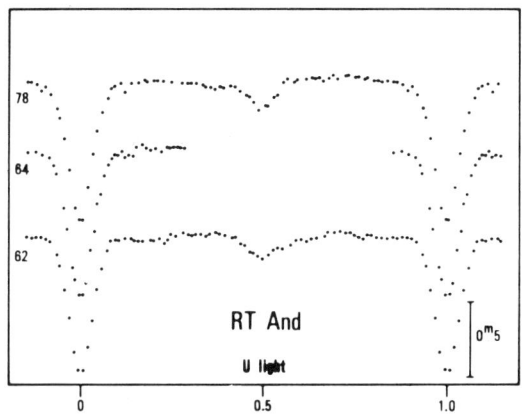

Fig. 1. Normal points of U light curves of RT And

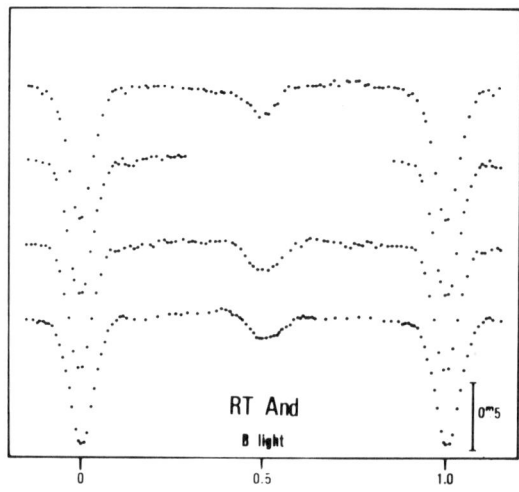

Fig. 2. Normal points of B light curves of RT And

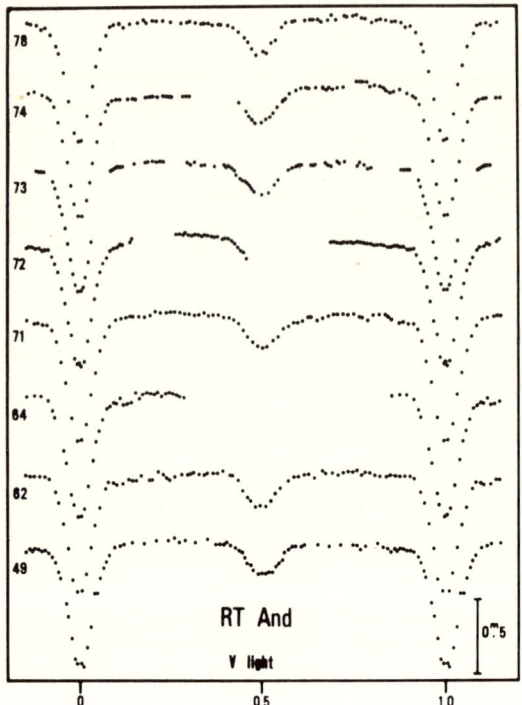

Fig. 3. Normal points of V light curves of RT And

Table 1.1 S.P.G. periods, spectral types, information about H and K emission and types of primary eclipse.

Name	P(Orb) days	Spectral Class hot + cool	H and K	Type of Primary Eclipse
RT And	0.629	F8V + G5 ÷ KOV	Yes	Tr
SV Cam	0.593	G0 ÷ G5V + K4V	?	(Tr)
WY Cnc	0.829	G5V + M2V	Yes	Tr (Oc?)
CG Cyg	0.631	G1V + G9V (G9V+KOV) (Tr)	Yes	Oc (Tr?)
UV Psc	0.861	G2V + KOIV	Yes	Tr
XY UMa	0.479	G2 ÷ G5V + (K5V)	No	(Tr)
ER Vul	0.698	G0V + G5V	Yes	Tr

ones, i.e., the shoulder of the first contact minus the shoulder of the fourth contact, both of primary and secondary minima. I call these differences respectively S I and S II for the primary minimum and the secondary ones. Also in Table 2 the depths of primary and secondary minimum are shown for different epochs of observation under the headings respectively D I and D II for primary and secondary minimum. There is a problem in computing these depths owing to the variable level of maxima and so, following the suggestions of Catalano and Rodono (1967), I referred the depths of the minima to the luminosity between the first and fourth contact of the primary and secondary minimum respectively, to obtain the loss of luminosity due to eclipse effect.

To complete the observed scenario on the light curves of RT And, it is necessary to show the sudden variations of the light curves that take place mainly near the phases 0.1 and 0.9 of the primary minimum and near the phases 0.4 and 0.6 of the secondary one. As one can see from Fig. 4, near these phases there is either a sudden brightening or a kind of oscillation and these morphological features are randomly distributed with the time for the different epochs of observation. There are also fluctuations in the level of maxima up to $0.^m04$ and in the depths of primary and secondary minimum that take place night by night (see Figs. 5 and 6). Having concluded the analysis of the morphology of the light curves, now I shall consider the period variations.

PERIOD

The first considerations on the period of RT And were made in 1955 (Gordon, 1955): she hesitated to attribute the observed variations in the time of primary minimum to a period change. From the analysis by Wood and Forbes (1963) who made a fitting of the (O-C)'s with a third degree polynomial, the period is shown to be diminishing. Subsequently, Kristenson (1967), assuming that the system was semi-detached (Kopal and Shapley, 1956), concluded that "dynamical instabilities are bound to occur that are connected not only with mass transfer between the components, but also with mass motion - presumably in part rhythmical - taking place in the components themselves". Other considerations can be found in Dumitrescu (1973) and Dean (1974); the latter concludes that, due to lack of observed minima between 1911-1923 and 1930-1945 it is not possible to differentiate between the abrupt or continuous nature of period variations. In an extensive study Williamon (1974) collected all published minima of RT And and concluded that the system had abrupt period variations in two epochs. The plot of the (O-C)'s in respect to a linear ephemeris is shown in Fig. 7. Concerning the displacement of the secondary minimum, that some authors found (Kopal and Shapley, 1956; Dumitrescu, 1974), I think that, owing to the variable asymmetry of the

Table 2. Shoulders and depths of primary and secondary minima or RT And

Year	MI U	MI B	MI V	SI U	SI B	SI V	SII U	SII B	SII V	DI U	DI B	DI V	DII U	DII B	DII V
1949	–	–.022	–.034	–	+.025	+.038	–	–.034	–.027	–	+.929	+.943	–	+.122	+.171
1962	+.006	+.003	+.014	–.014	–.023	–.040	+.012	–.031	–.001	+.887	+.809	+.757	+.132	+.181	+.208
1964	–.020	–.013	+.007	–.013	–.018	–.046	–	–	–	+.961	+.879	+.775	–	–	–
1971	–	–	–.020	–	–	+.034	–	–	–.019	–	–	+.823	–	–	+.186
1972	–	–	–.048	–	–	+.027	–	–	–	–	–	+.804	–	–	–
1973	–	–	–.013	–	–	+.026	–	–	–.036	–	–	+.825	–	–	+.183
1974	–	–	+.061	–	–	–.045	–	–	–.025	–	–	+.799	–	–	+.222
1978	+.030	+.028	+.018	–.011	–.005	–.003	–.043	–.024	–.026	+.904	+.871	+.781	+.188	+.186	+.221

secondary minimum, it is not correct to consider the system as having orbital eccentricity, for the obvious reason that one would have a highly variable eccentricity, apart from the fact that it is only the result of a numerical artifice. As a matter of fact, it is very difficult to get reliable times of a secondary minimum either because of its small depth and/or its highly variable degree of asymmetry. To establish the trend, if any, of the displacement of the secondary minimum along time, I computed, by the Kwee and van Woerden method (1956), the phases of secondary minimum of the normal light curves for the different epochs of observations and limiting the points between the phases 0.45 ÷ 0.55. I also computed the phases of secondary minimum for different epochs between the phases 0.4 ÷ 0.6 and the comparison of the results of both these computations are shown in Table 3.

Table 3. RT And Displacement of the secondary minimum (for explanation see the text).

Year	D.04 ÷ 06			D.45 ÷ 55		
	U	B	V	U	B	V
1949	–	+.01208	+.00722	–	-.00904	+.00678
1962	+.00189	+.00363	+.00064	-.00424	+.00108	-.00076
1964	–	–	–	–	–	–
1971	–	–	+.00571	–	–	+.00407
1972	–	–	–	–	–	–
1973	–	–	-.00012	–	–	+.00236
1974	–	–	-.00407	–	–	-.00466
1978	-.00672	-.00677	-.00402	-.00358	-.00446	-.00271

SPECTROSCOPIC OBSERVATIONS

The only spectroscopic data, as far as I know, on RT And are derived from the spectra by Payne-Gaposchkin (1946). This author computed the radial velocity curve and established that only one spectrum is visible, except in some phases outside the minima, where some weak lines may be attributed to the secondary component. Subsequently, Kron (1950) analysed the same spectrograms anew and he detected the presence of very faint H and K emissions of Ca II with a variable intensity at different orbital phases.

The spectral type of RT And from the spectroscopic and photometric data can be stated as follows: the primary component has been classified as F8V (Kron, 1950; Gordon, 1955), whilst the secondary lies between G5 ÷ K0V.

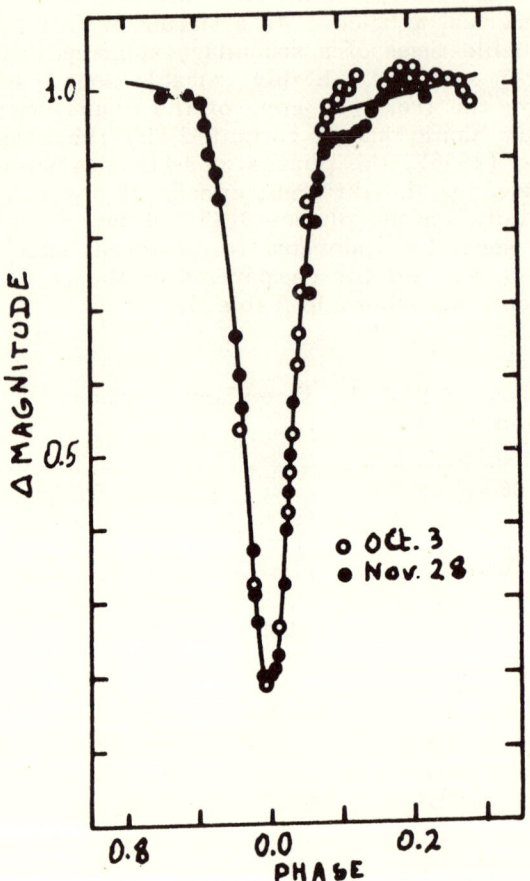

Fig. 4. Yellow light curve of RT And in 1964 between October 3 (empty dots) and November 28 (full dots).

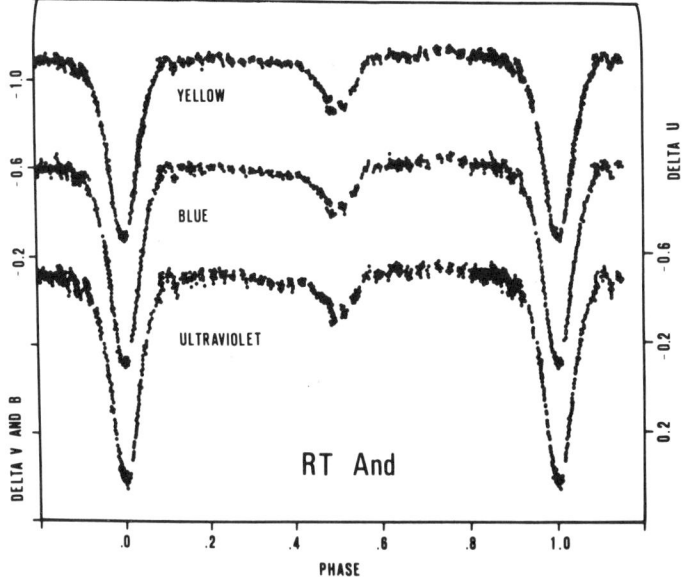

Fig. 5. U-B-V light curves in 1978 with all the observed points of RT And.

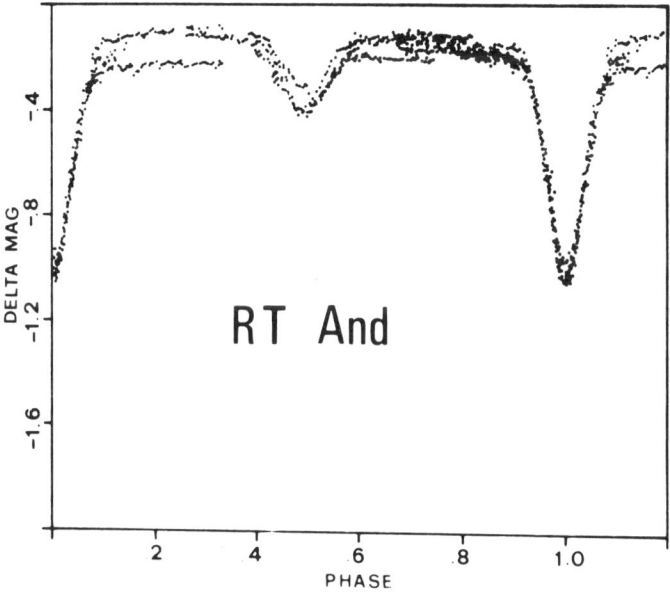

Fig. 6. Yellow light curves of RT And from 1971 to 1974.

SV CAM

There are about 25 observed light curves of SV Cam. The complete references for most of them can be found in a paper by van Woerden (1957) and the last three were published in 1979 (Hilditch et al, 1979). In Fig. 8 the normal points of all the observed V light curves are shown, in Fig. 9 the normal points of all B ones and in Fig. 10 the light curves by Hilditch et al (1979). From these light curves it can easily be seen that there are features like the ones I showed for RT And and they are more pronounced than the ones of RT And. From the analysis of the light curves by van Woerden it is possible to say that there are variations of the depth of secondary minimum between 0.09 and 0.24 magnitudes and between 0.06 and 0.20 respectively, referring the depth to the maxima before the secondary minimum and to the maxima after the secondary minimum. The range of the variations in primary minimum depths is between 0.11 and 0.12 mag.. There is no clear conclusion about the wavelength dependence, apart from a weak dependence, i.e., greater depths with shorter wavelengths, but there are some inconsistencies in the data obtained without filters. The differences between maxima M I are variable by about 0.06 mag., and also in this case there is a weak dependence on the wavelength. The differences between the shoulders S I's are of alternate signs and they might be weakly correlated with M I's. The maxima have a highly variable curvature and the differences between the shoulders and the maxima attain values of up to 0.09 mag. and up to 0.05 mag. respectively for the differences between the maxima before the secondary minimum minus the shoulder near the fourth contact of primary eclipse and for the differences between the maximum after the secondary minimum minus the shoulder near the first contact of the primary minimum.

The phases of the secondary minimum, as for RT And, range from 0.51 to 0.49.

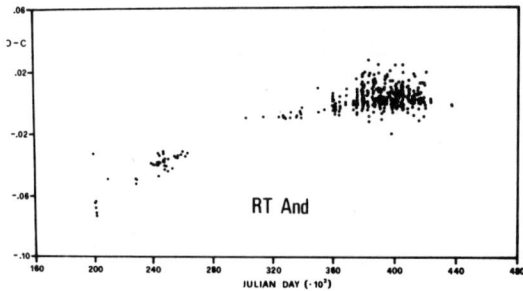

Fig. 7. (O-C) residuals of observed times of minima from a linear ephemeris plotted against Julian date of RT And.

The data I spoke of above are taken from the paper by van Woerden (1957). Van Woerden's conclusion was that the secondary component of SV Cam is probably slowly variable in brightness.

From this starting point an analysis of the intrinsic variability of the secondary component of SV Cam was made by Hilditch et al (1979). The photoelectric observations that these authors used were in 1970, 1977 and 1978.

Hilditch et al stated that:

a) "The shape and depth of primary eclipse (relative to first and fourth contact are the same in each year."

b) "All three light curves coincide to within ±0.01 mag. at mid-secondary eclipse."

From these observations they attributed the light curve variations to intrinsic variability of the secondary component of SV Cam. They analysed the light curves to find out the kind of variability for the secondary component. I shall discuss this point later on in the general analysis of the results for the whole S.P. G.. Concerning the period, also SV Cam has period variations, as one can see from Fig. 11. The nature of these variations was attributed to a third body by Freboes-Condé and Herczeg (1973); Hilditch et al (1979) seem to confirm this hypothesis.

Only one spectroscopic study of this variable exists (Hiltner, 1953); SV Cam displays a single spectrum (even at mid-primary eclipse) and the spectral classification of the components is from G0V to G5V for the primary, whilst for the secondary it appears, from photometric analysis, as a spectrum near K4V. The system, from the u, b, v, y photometry in the Strömgren system, by Hilditch and co-workers (1979), has an ultraviolet excess (c_1 = 0.07 mag. and m_1 = 0.0 mag.) and solar type metallicity.

From a working hypothesis of Hilditch and co-workers, using the mass function by Hiltner (1953), it was estimated that the system has 1 solar mass for the primary and 0.07 $m_\odot$ for the secondary. It is not possible to say anything about the presence of H and K emission lines.

WY CNC

There are five observed photoelectric light curves of this variable (Chambliss, 1965; Oliver, 1974; Sarma, 1976; Awadalla and Budding, 1979), see Figs. 12 and 13.

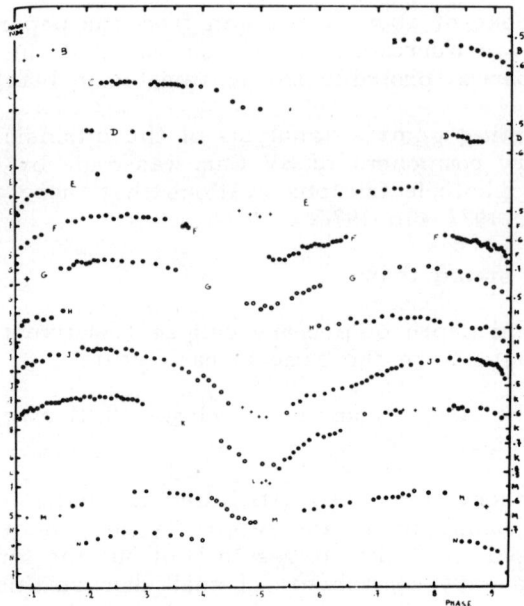

Fig. 8. Light curves of SV Cam; maximum and secondary minimum, 1950-52.

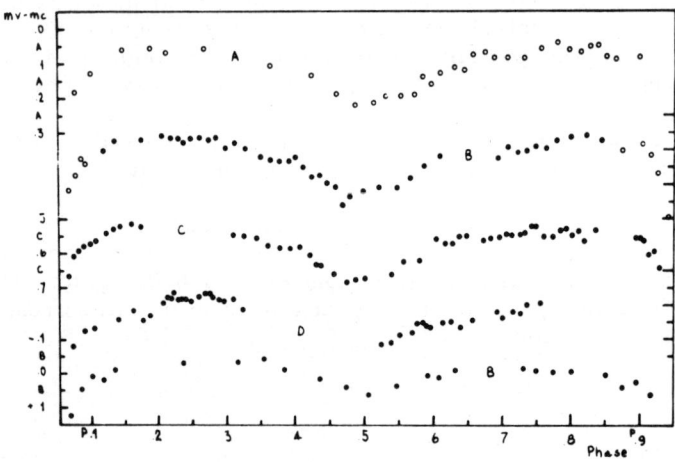

Fig. 9. Light curves of SV Cam at maximum and secondary minimum 1948-49.

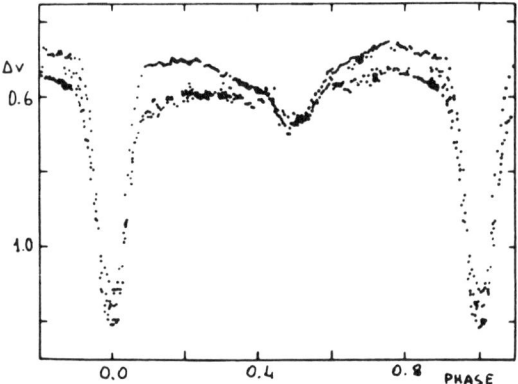

Fig. 10. Light curves of SV Cam 1969/70; 1976/77; 1977/78.

The analysis of the light curves leads to these conclusions:

a) The light curves during the primary minimum are asymmetric.

b) There are significant changes in the depths of the minimum.

c) No shift in the times of primary minimum is detected and by different investigators it has been noticed that the period has remained constant for the past forty years.

Chambliss (1965), Oliver (1974) and Sarma (1976) studied this system and from their light curves it can be deduced that there is a periodic long term variation present on the light curve and its period is about 5.5 years whilst the amplitude appears to be 0.02 mag.

All these investigators attributed the source of the perturbation to one of the components which must be the hottest one. On the other hand the result of an analysis by Popper, as reported by Awadalla and Budding (1978) indicates the source of the emission lines H and K of Ca III to be the primary component and it was a puzzling problem to reconcile this fact with the result of the analysis by Chambliss (1965) that stated the slightly more luminous star remained uneclipsed at the occultation (primary) minimum. Awadalla and Budding (1979) succeeded in getting an agreement on this point, demonstrating the transient nature of the primary minimum. These authors got spectral types for the components G5V + M2V and masses $M_h = 0.93 m_\odot$ and $m_c = 0.53 m_\odot$.

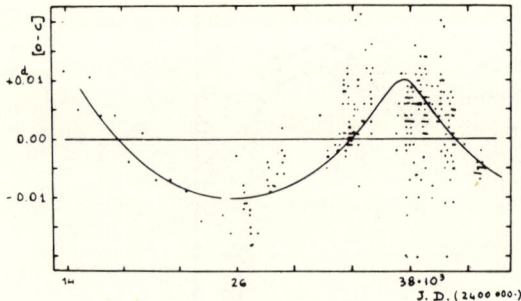

Fig. 11. Plot of (O-C) residuals of observed times of minima from a linear ephemeris of SV Cam. The curve represents a light-time orbit with e = 0.6, ω = 90° and period 23400 day (Hilditch et al, 1979).

CG CYGNI

There are eleven light curves of CG Cygni and from these observations the distortion wave affecting the light curves has been studied by Hall and his co-workers (1979) in a comprehensive study. Other photoelectric observations were made by Jassur (1980). In Figs. 14, 15, 16, 17, 18 there are some examples of light curves taken from the works I quoted above.

The results can be summarized as follows:

a) An increase in the mean light by about 14% in V has been detected since 1965 (see Fig. 19).

b) A sinusoidal distortion wave through the light curve, going at an increasing rate is shown (Fig. 20).

c) The period is variable and has, probably, been continually changing since the late 1960's (see Fig. 21).

d) It is also established that the system has a large apparent infrared excess and a variation in the H and K emission lines of Ca II.

From Jassur's analysis of his photoelectric observations (Fig. 22) it can be seen that he preferred "an occultation solution, on the basis of goodness of fit to the colour curve. It was analyzed also the transit soluton which was discarded because it does not explain the observed colour variations at primary and secondary eclipses very well". Anyway the spectral types of the components are G9V + K0V for the transit hypothesis and G1V + G9V for the occultation one.

The masses are: $M_h = 0.82 m_\odot + M_c = 0.78 m_\odot$ (transit),

and $M_h = 0.82 m_\odot + M_c = 1.04 m_\odot$ (occultation).

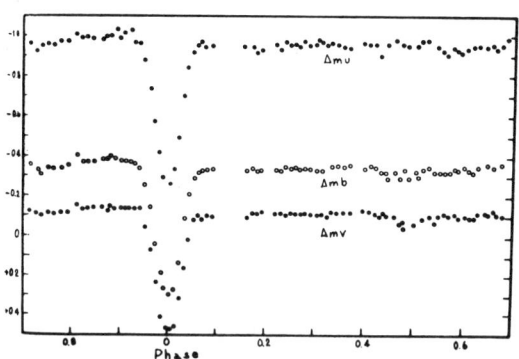

Fig. 12. U-B-V light curves of WY Cnc 1973/74.

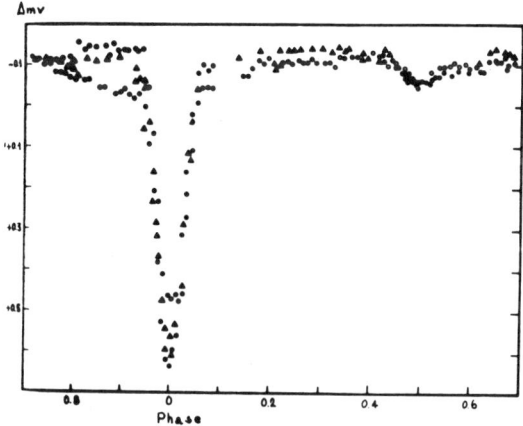

Fig. 13. Yellow light curves of WY Cnc from 1965 to 1974.

UV PSC

There are four observed light curves of UV Pcs (Huth, 1959; Carr, 1960; Oliver, 1979; Sadik, 1979) (see Fig. 23).

The presence of a distortion wave was detected by Oliver (1974) of an amplitude of 0.04 mag.; this feature was confirmed by Sadik (1979). There is a colour dependence of the amplitude of such a wave, that is the amplitude increase at shorter wavelength. In C.G. a colour dependence is observed that is in the opposite sense.

In this star a variable period has also been detected and the presence of H and K emission lines from both the components was ascertained (Popper, 1969; 1976). The masses of the components are $1.0m_\odot$ and $0.9m_\odot$ for the hotter (primary and cooler (secondary) components respectively. The eclipse from the analysis by Sadik comes out a transit. The result is consistent with Carr's solutions, but not consistent with Oliver's ones. The spectral types of the components are G2V + K0IV.

XY UMa

This system can be defined as the "Geiyer's star" owing to the fact that this author made a lot of observations of this eclipsing binary. His results can be summarized as follows:

a) The orbital period of the binary was constant for 20 years.

b) The average brightness of the binary system changed between 1955 and 1975 in a sinusoidal manner by 0.18 mag. in V and 0.20 mag. in B indicating a periodic variation of about 28 to 30 years. The minimum was attained in 1961 and the maximum brightness in 1975.

c) Rather symmetrical light curves of β-Lyrae type, showing a large scatter of the observations in maximum light which exceeds the photometric accuracy by at least five times, ocurred in 1958, 1959 and 1968. Colour index at the primary minimum is redder than at secondary minimum. The secondary minimum appears to be an occultation, so that the hotter component is also the larger one. The maxima differ only slightly, maximum I being brighter in 1958/59 than maximum II. The spectral types of the components result as G2 ÷ G5V + K5V and Geyer stated that the system is a detached one. From a work hypothesis the masses of the components are: $M_h = 0.95m_\odot$ and $M_c = 0.7m_\odot$ (Geyer, 1976, 1977).

During 1975 Lorenzi and Scaltriti (1977) also confirmed the strong short-term variability of the light curves of XY UMa.

ER VUL

As far as I know, only two sets of photoelectric light curves exist for ER Vul (Abrami and Cester, 1973; Northcott and Bakos, 1967). From Fig. 25 it is easily seen that these light curves suffer from "irregular" light variations, of the sort described for the RS CVn binaries. It is necessary to achieve further photoelectric observations, and also the possible period variation has to be ascertained. ER Vul is a double spectrum system, and the components are classified as G0V + G5V (Northcott and Bakos, 1967). The masses of the components are respectively $M_h = 1.07 m_\odot$ and $M_c = 0.98 m_\odot$; the system results as a detached one.

ER Vul was considered to display an orbital eccentricity by both investigators, but this requires, in my opinion, great caution, owing to the variable degree of asymmetry of the secondary minimum.

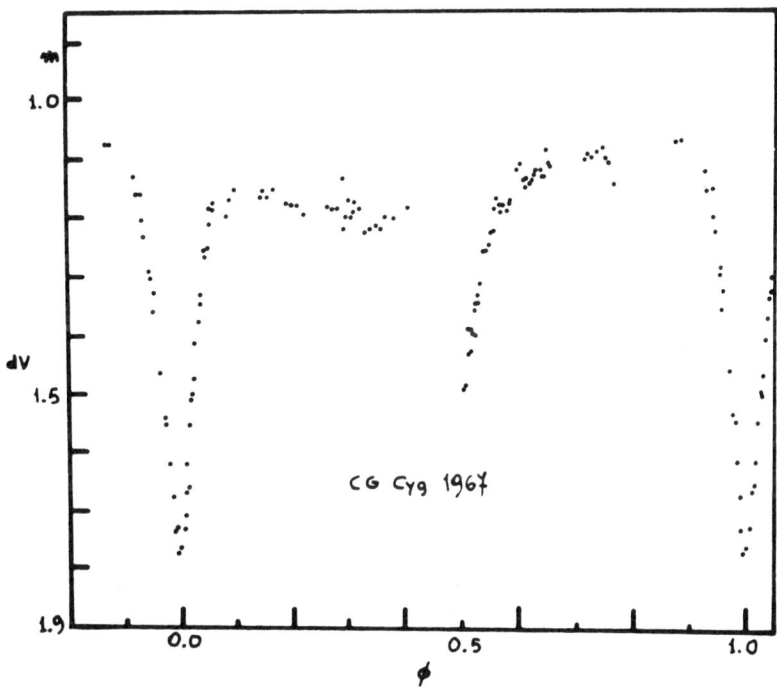

Fig. 14. Yellow light curve of CG Cyg 1967.

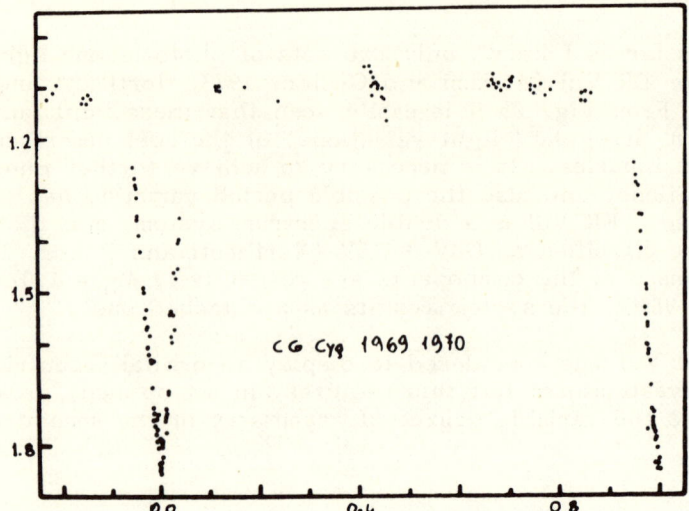

Fig. 15. Yellow light curve of CG Cyg 1969/70.

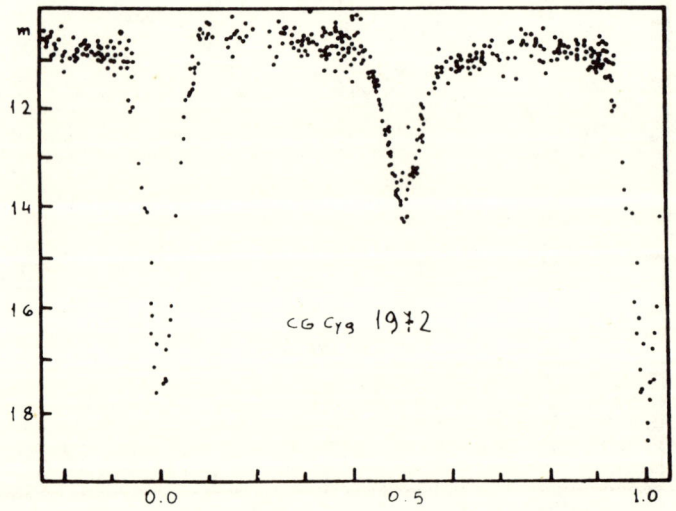

Fig. 16. Yellow light curve of CG Cyg 1972.

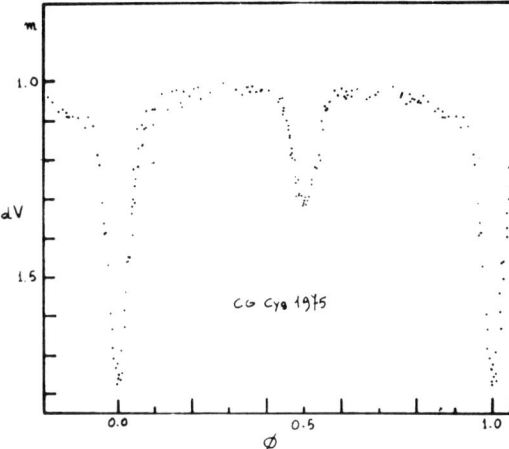

Fig. 17. Yellow light curve of CG Cyg 1975.

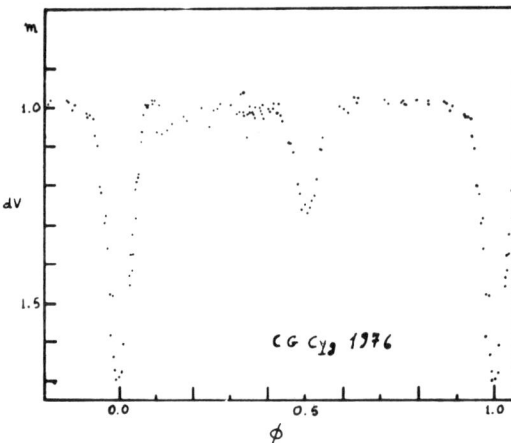

Fig. 18. Yellow light curve of CG Cyg 1976.

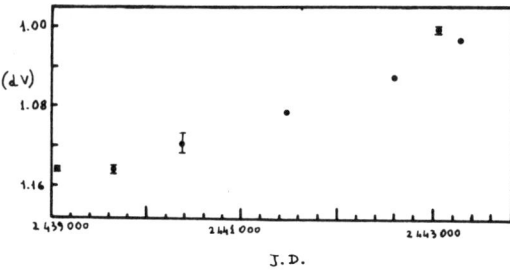

Fig. 19. Mean light level at maximum of CG Cyg plotted against Julian date.

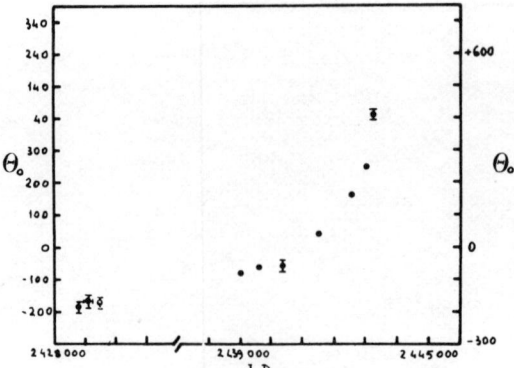

Fig. 20. Phases of maximum contribution to the light of CG Cyg by the distortion wave $\theta_o = \arctan(B_1/A_1)$.

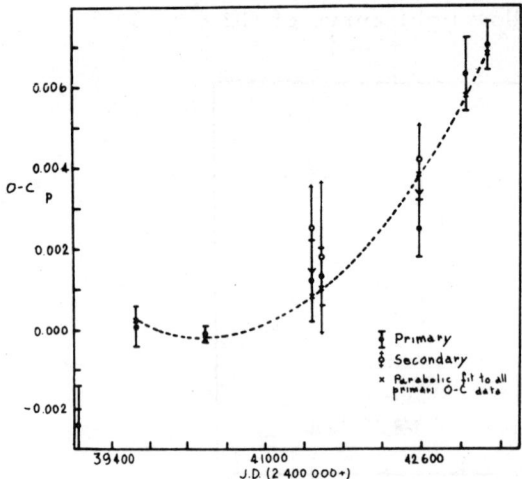

Fig. 21. Observed phases of primary and secondary minima of CY Cyg. The parabola shown is the best fit to all the data post-1965 (Milone et al, 1979).

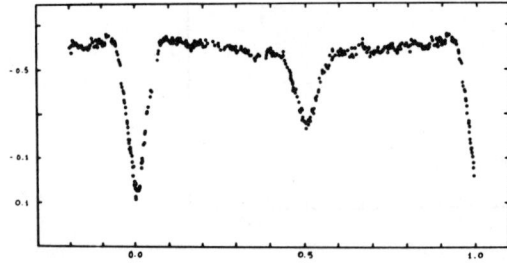

Fig. 22. Yellow light curve of CG Cyg by Jassur (1978).

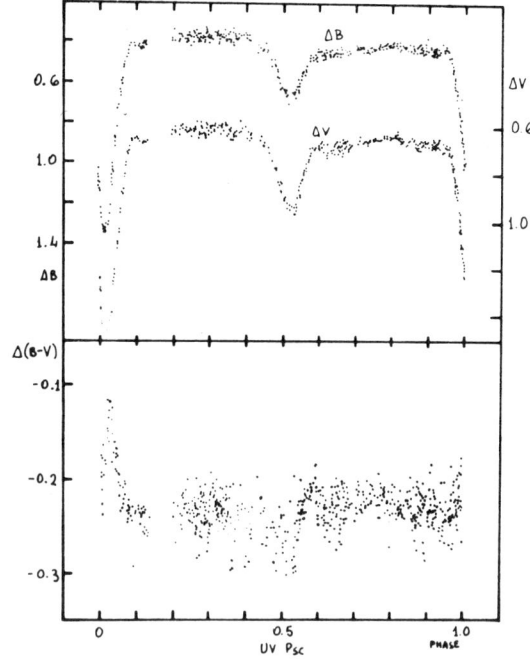

Fig. 23. B and V light curves of UV Psc by Sadik (1979).

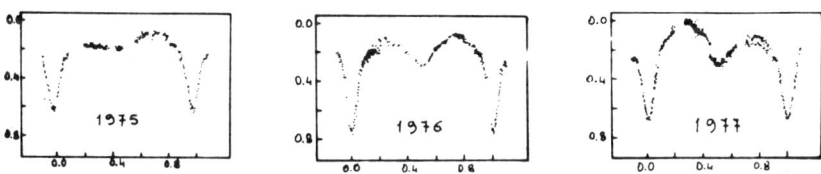

Fig. 24. Yellow light curves of XY UMa.

DISCUSSION AND CONCLUSIONS

From the above data it is evident that there are some difficulties in analysing the light curves of these systems owing to the distortion wave convoluted with the light curve of the eclipsing binary. The method that is currently used to get the deconvolution of the distortion wave from the observed light curve is based on the determination of the coefficients of a truncated five term Fourier series that fits the maxima of the light curve. The assumption of a sinusoidal distortion wave, of the type $l_D = A_D \cos(\theta - \theta_0)$ immediately leads to obtaining $A_D = (\sqrt{A_1^2 + B_1^2})/A_0$ and

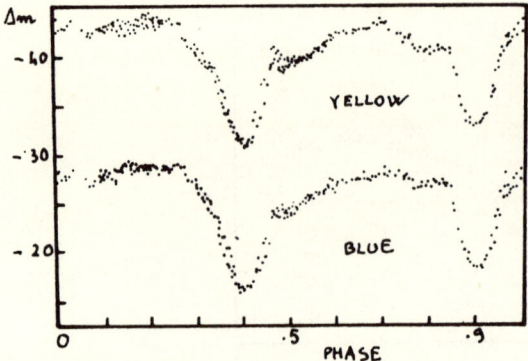

Fig. 25. Yellow and Blue light curve of ER Vul.

$\theta_O = \text{arctg}(B_1/A_1)$.

A_1 and B_1, respectively, are the coefficients of $\cos^\theta$ and $\sin^\theta$ in the Fourier series. It is to be noted that the A_1 coefficient in this series, that, as is well known, comes from the procedure adopted to "rectify" the light curve, is affected both by the distortion wave and the reflection effect. It should be taken into account that for the detached systems the reflection effect is small, but it should be stressed that it is not negligible! In some cases the determineation of both amplitude and phase of the distortion wave can be seriously wrong in this procedure. It should also be noted that the coefficients of the Fourier analysis are correlated in such a way that if significant higher harmonics are neglected, a strong systematic error on the coefficient will be present. Obviously this fact can easily lead to obtaining wrong values of the elements of the distortion wave. Taking these reasons into account, I tried to approach the problem of the determination of the elements of the distortion wave by considering all the known photometric effects acting on the morphology of the light curve of an eclipsing binary and then analysing the residuals between the observed light curve and the computed one. This procedure was applied to a synthetic light curve generated by the computer code of Wilson and Devinney (1971).

A sinusoidal distortion wave of the type $l_D = A_D \cos(\theta - \theta_O)$ was computed. Assuming $A_D = 0.03$ light units and for phases of maximum absorption ranging from 0 to 1 by a step of 0.125 in phase I got the light curves shown in Fig. 26. The next step was the solution of the light curve to get the photometric and geometric parameters from these distorted light curves by the same Wilson and Devinney computer code. Finally, to get the elements of the distortion wave I made a Fourier series expansion of the residuals I got from the Wilson and Devinney solution. The resultant values of A_D and θ_O are practically the same as the starting values of the distortion wave.

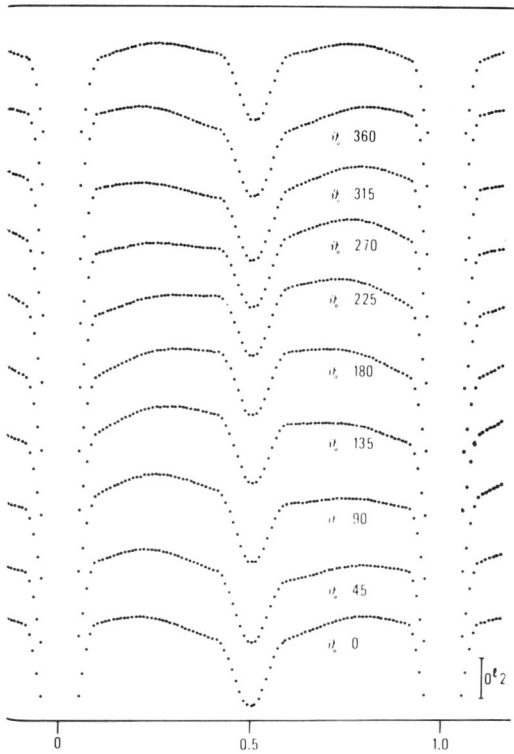

Fig. 26. Synthetic light curves of an eclipsing system with a sinusoidal distortion wave overimposed (for explanation see the text).

To compare the above method with the method of analysis that comes from the procedure of the light curve rectification I also applied this procedure to the light curves affected by the distortion wave and the results are that both the amplitude and phase of maximum absorption are respectively affected by variable errors of about 5% to 15%. At this point it should be stressed that the light curves I used are theoretical ones, so it is obvious that the errors in real cases would be much greater than the ones I have got for the ideal cases I analysed. From the above analysis a possible procedure comes out, but it is premature to outline its details, owing to the great number of parameters that are inherent to the use of a direct method of light curve solution. So caution must be used when the method is applied to a real case.

Having considered in some detail the problem of the determination of the distortion wave elements, let me analyse the data we know on masses, spectral types, periods, etc., for the whole S.P.G..

As one can see from Table 1.1, the S.P.G. has orbital periods that range from about $0.^{d}5$ to $0.^{d}9$, the spectral classes range from F8V to G9V for the primaries and practically from K0V to K5V, apart from a few exceptions, for the secondaries. H and K emission lines of Ca II have been detected for all the components of the S.P.G., except for XY UMa. There is no information on this subject for SV Cam. There is the problem of the determination of the source of H and K emission, in other words which component is responsible for the emission. Practically, we have no information on this subject, apart for WY Cns, UV Psc and for XY UMa.

An interesting fact that comes out from the present analysis concerns the type of primary eclipse for these systems. From Table 1.1 it is easily seen that there is a general tendency for the group to have a transit at the primary eclipse. It has to be ascertained with more accuracy than in the present analysis, but, if confirmed, this fact might be really a clue to distinguish different types of RS CVn stars. It is well known that there are two recognized subcategories of W UMa, called the A type and W type (Binnendijk, 1965; 1970). In the "A" types, the more massive component is covered at primary eclipse whilst "W" types have the lower-mass component covered at primary eclipse. "A" types are well in overcontact whilst "W" type are in thin or marginal contact (Lucy, 1973).

In the same way, I think that, from an observational point of view, it is possible to classify the RS CVn's as compared with the type of primary eclipse and the mass of primary component, all the systems are detached as it is possible to see from Fig. 27, where the crosses and dots respectively show the cooler and hotter components. As one can see, all the components are well below their respective critical radii. In the figure, r_{01} and r_{02} indicate the trend of the critical radii of the Roche lobe in respect to the mass ratio.

From the results I exposed above it is interesting to consider the hypothesis that Hall made (1976) in his review on the S.P.G.. This author stated that the S.P.G. seems to be a subset of the short-period eclipsing binaries with β-Lyrae-type light curves discussed by Lucy (1976). As a matter of fact, Lucy stated that for W UMa of W type it is not possible to achieve a thermal equilibrium so, during their evolution they undergo thermal oscillation about a state of marginal contact. In our case we are faced with systems that are clearly detached and of A-type, using the terminology of

Binnendijk, but only for the type of the primary eclipse. On the other hand, taking into account the spectral range of the components of the S.P.G., the range of total masses (about two solar masses), and their period, we see that the hypothesis that Hall made cannot be completely rejected. From these considerations, the question remains open, however. It is not clear how one can link, from an evolutionary point of view, the two classes of stars. I think that it might be a promising working hypothesis to be pursued to fully verify its validity.

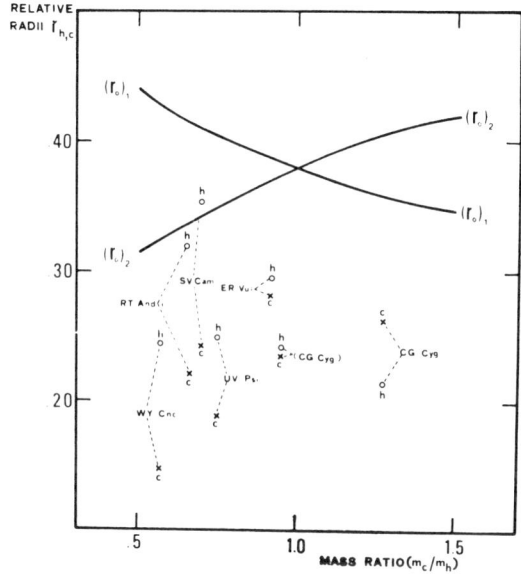

Fig. 27. Relative radii as a function of mass ratio of S.P.G.

What can be said about the models of S.P.G.?

The answer to this question is not easy, because, as C.G. and S.P.G. have been classified as different groups, people tend to give models that in practice are the same as the ones devised for RS CVn's. The current idea is that "large spotted regions are appearing and disappearing without, as far as we can tell, remembering to have a preferential longitude" (Hall, 1976). Maintaining the spot model, an interesting hypothesis has been made on SV Cam by Hilditch, Harland and McLean (1979). These authors showed that SV Cam is composed of a G3V primary slightly evolved above the Main Sequence and a K4V secondary component which also exhibits BY Draconis-type variability. The procedure they adopted to get the deconvolution of the intrinsic light variations of the secondary component relative to the total light of the system is very similar to the one I devised in this lecture. The result is shown in Fig. 28. It is to be noted that the authors found that some light curves can be interpreted in terms of cool spots and some in terms of hot spots. On the other hand the spots are needed to cover about 80° in daimeter of the surface of the secondary to explain the observed luminosity variations. A point that should be stressed now is that much more observational work has to be done on the S.P.G. to get valuable data for modelling as it is possible to see from the Tables 1.2 and 1.3.

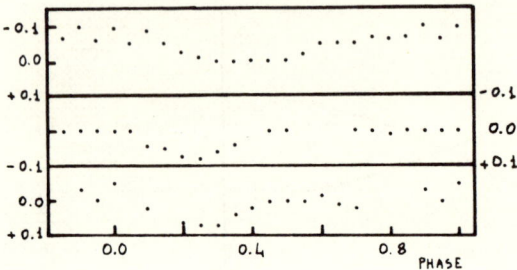

Fig. 28. Intrinsic light variations of the secondary component relative to the total light of SV Cam inferred from the observed light curves and the theoretical model of the eclipsing system by Hilditch et al (1979).

I think that good photometric observations in U, B, V, R, I and spectroscopic monitoring of H and K emission lines along the orbital phase would help us greatly to understand the observed behaviour of the S.P.G.. An accurate work of analysis of the light curve and accurate radial velocity curves have to be made to determine more reliable data on radii and mass ratios of the S.P.G. as well as to detect possible regular variations of these light curves that up to date have been classified as suffering only "irregular" variations. A nice example of this fact is RT And: as one can see in Figs. 29 and 30 in which are shown the wave affecting the light curve and the phase of maximum absorption of a possible 22 years migrating wave. This is only a very preliminary result of an analysis that my colleagues and I are making at the Capodimonte and Teramo Astronomical Observatories on the observed properties of the whole S.P.G.

My concluding remark on the comparison between S.P.G. and C.G. is, that, though these groups are very similar, from the present analysis it comes out that there are some differences, apart from the periods, that might be important. Basically it is important to note the fact that almost all the S.P.G. is constituted

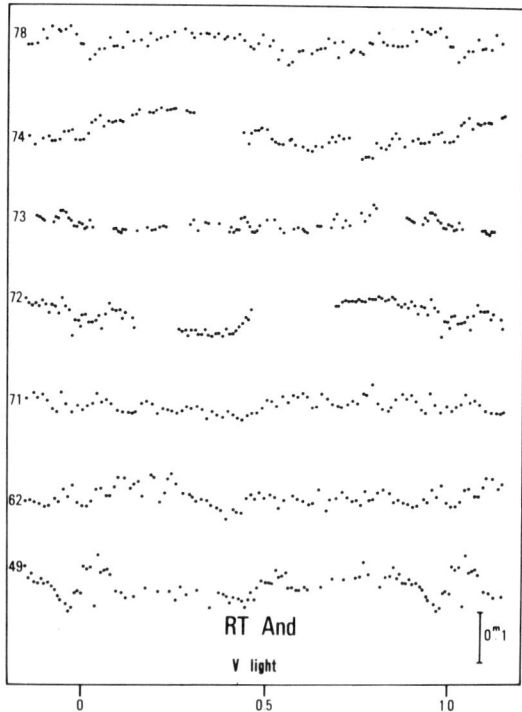

Fig. 29. Amplitude of the distortion wave of RT And plotted against the orbital phases.

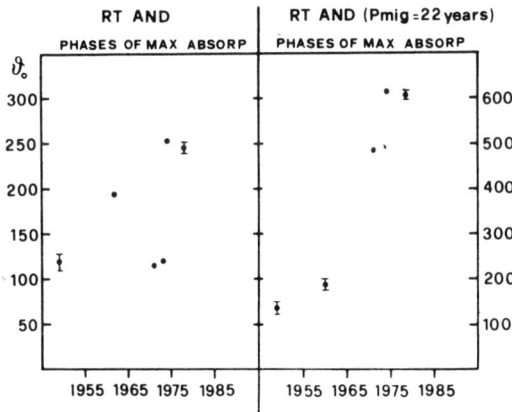

Fig. 30. Phases of the maximum contribution to the light of RT And by the distortion wave $\theta_o = \arctan(B_1/A_1)$. The right side plot was derived adding 36° to the θ_o's after 1965.

by Main Sequence stars and the fact that the type of the primary eclipse seems to be a transit for the S.P.G. and an occultation for C.G.. This last point requires to be tested more accurately owing to the fact that the type of eclipse has been assumed by some authors.

Another point that deserves great care is the determination of the period of the migration wave. Another question that arises from the work by Hilditch and co-workers is: are all the S.P.G. like SV Cam? It is an interesting problem that, in my opinion, deserves great attention.

At this point I conclude my lecture by outlining the work that has to be done in the near future:

1) It is necessary to achieve light curves in more than one colour to ascertain the dependence of the distortion wave from the wavelength and to ascertain possible short-term periodical variation of the light curves.

2) Radial velocity curves are needed for almost the whole of the groups to achieve more reliable absolute dimensions of the components.

3) Monitoring of H and K emission lines along the orbital phase is needed.

4) Photometric analysis of all the light curves is needed to achieve homogeneous results.

5) UV spectra are needed to ascertain the activity, if any, in this spectral range by I.U.E. satellite as it is currently made for C.G..

6) Infrared observations are also needed as it is made for C.G..

Table 1.2. Amplitudes and migration periods of S.P.G..

Name	Wave mag.	P(mig) years	Irregular Lt. C. Var.
RT And	?	?	Yes
SV Cam	?	?	Yes
WY Cnc	0.02	5.5	Yes
CG Cyg	0.07	10.0	Yes
UV Psc	0.04	?	Yes
XY UMa	0.18	28 ÷ 30	Yes
ER Vul	?	?	Yes

Table 1.3 Ultraviolet (δuv) and infrared (δir) excesses of S.P.
G. RS CVn. The answer under the column "Var. P(orb)"
states if the orbital period is variable or not. A question mark indicates that there is no information.

Name	δuv	δir	Var. P(orb.)
RT And	?	?	Yes
SV Cam	Yes	?	Yes
WY Cnc	?	?	No
CG Cyg	?	Yes	Yes
UV Psc	Yes?	?	Yes
XY UMa	?	?	No
ER Vul	?	?	?

Table 1.4 Masses and relative radii of S.P.G.

Name	Masses in solar units hot + cool	Relative Radii
RT And	1.50 + 0.99	0.322 + 0.239
SV Cam	1.0(ass.) + 0.7	0.352 + 0.244
WY Cnc	0.93 + 0.53	0.246 + 0.148
CG Cyg	0.82 + 1.04	0.214 + 0.262
	(0.82 + 0.78)	(0.243 + 0.241)
UV Psc	1.2 + 0.9	0.250 + 0.190
XY UMa	– –	– –
ER Vul	1.07 + 0.98	0.297 + 0.282

ACKNOWLEDGMENTS

The author wishes to thank Mr. S. Marcozzi and Mrs. A. D'Orsi for the help in the computer work, which has been performed with the DEC PDP-11/34 of the Capodimonte Observatory.

This work has been partly supported by Consiglio Nazionale delle Ricerche.

REFERENCES

Abrami, A. and Cester, B.: 1963, Osservatorio Astronomico di Trieste, No. 320.
Awadalla, N. S. and Budding, E.: 1978, I.B.V.S., No.1484.
Awadalla, N. S. and Budding, E.: 1979, Astrophys. Space Sci., 63, p. 479.
Binnendijk, L.: 1965, Klein Veröff. Bamberg, 4, No.40, p.36.
Binnendijk, L.: 1970, in *Vistas in Astronomy*, ed. A. Beer

(Oxford: Pergamon), vol. 12, p.217.
Carr, R. B.: 1969, Ph.D. Thesis, Univ. of Florida (unpublished).
Catalano, S. and Rodono, M.: 1967, Mem. Soc. Astron. Ital., 38, p.395.
Chambliss, C. R.: 1965, Astron. J., 70, p.741.
Dean, C. A.: 1974, Publ. Astron. Soc. Pac., 86, p.912.
Dumitrescu, A.: 1973, Studi Cerc. Astron., 18, (1), p.47.
Dumitrescu, A.: 1974, Studi Cerc. Astron., 19, (1), p.89.
Freboes-Condé, H. and Herczeg, T.: 1973, Astron. Astrophys. Suppl., 12, p.1.
Gadomski, J.: 1974, Acta Astron. Series C, 1, p.21.
Geyer, E. H.: 1976, IAU Colloq. No. 29, p.315, (part 2).
Geyer, E. H.: 1977, Astrophys. Space Sci., 52, p.351.
Gordon, K. C.: 1948, Astron. J., 53, p.198.
Gordon, K. C.: 1955, Astron. J., 60, p.422.
Hall, D. S.: 1976, in *Multiple Periodic Variable Stars*, IAU Colloq. No. 29(Budapest), part 1, p.287, Reidel, Dordrecht.
Hilditch, R. W., Harland, D. M. and McLean, B. J.: 1979, Mon. Not. Roy. Astron. Soc., 187, p.797.
Hiltner, W. A.: 1953, Astrophys. J., 118, p.262.
Huth,   .: 1959, Mitteilungen über veränd. Sterne No. 424.
Jassur, D. M. Z.: 1980, Astrophys. Space Sci., 67, p.19.
Kopal, Z. and Shapley, M. B.: 1956, Jodrell Bank Ann., 1, p.140.
Kristenson, H.: 1967, Bull. Astron. Inst. Czech., 18, p.261.
Kron, G. E.: 1950, Publ. Astron. Soc. Pac., 62, p.141.
Kwee, K. K. and Van Woerden, L.: 1956, Bull. Astron. Inst. Neth., 12, p.327.
Lorenzi, F. and Scaltriti, F.: 1977, Acta Astron., 27, p.273.
Lucy, L. B.: 1973, Astrophys. Space Sci., 22, p.381.
Lucy, L. B.: 1976, Astrophys. J., 205, p.208.
Mancuso, S., Milano, L. and Russo, G.: 1979a, Astron. Astrophys. Suppl., 36, p.415.
Mancuso, S., Milano, L., Russo, and Sollazzo, C.: 1979b, Astron. Astrophys. Suppl., 38, p.187.
Milone, E. F., Castle, K. G., Robb, R. M., Swadron, D., Burke, E. W., Hall, D. S., Michlovic, J. E. and Zissel, R. E.: 1979, Astron. J., 84, p.417.
Northcott, R. J. and Bakos, G. A.: Astron. J., 72, p.89.
Oliver, J. P.: 1974, Ph.D. Thesis, Univ. of California, Los Angeles.
Payne-Gaposchkin, C.: 1976, Astrophys. J., 103, p.191.
Popper, D. M.: 1976, I.B.V.S., No. 1083.
Popper, D. M.: 1979, B.A.A.S., 1, p.257.
Sadik, A. R.: 1979, Astrophys. Space Sci., 63, p.319.
Sarma, M. B. K.: 1976, Bull. Astron. Inst. Czech., 27, p.335.
Van Woerden, H.: 1957, Ann. Sternw. Leiden, 21, p.3.
Williamon, R. M.: 1974, Publ. Astron. Soc. Pac., 86, p.924.
Wilson, R. E. and Devinney, E. J.: 1971, Astrophys. J., 166, p.605.
Wood, F. B. and Forbes, J. E.: 1963, Astron. J., 68, p.257.

PHOTOELECTRIC STUDY OF THE RS CVn-TYPE BINARY
TY PYXIDIS

P. Vivekananda Rao and M. B. K. Sarma

Centre of Advanced Study in Astronomy,
Osmania University, Hyderabad, India

ABSTRACT

The RS CVn type eclipsing binary TY Pyx was observed photoelectrically with standard B, V filters. The light curve indicates that the epoch given by Strohmeier refers to the secondary minimum and not to the primary minimum. An improved period of $3^d.1985787$ was determined from the present observations. No period variation has been detected from the present study. The spectral type and temperature deduced by us for TY Pyx correspond to G5 (5520°K) and G6 (5400°K) for the primary and secondary components respectively. This system does not show any variation in the depths of the minima or a wave like distortion outside the eclipse, like other RS CVn members. The photometric elements of the system are derived using Russell and Merrill's method. The absolute dimensions of the system are determined by combining the photometric elements derived by us with the spectroscopic elements determined by Popper and Anderson. The system is classified as detached. Plotting derived temperatures and absolute dimensions of this system in Iben's pre- and post-Main Sequence evolutionary tracks an age of 10^7 years has been derived which is in variance with the ages estimated for other members of the RS CVn group. The components of TY Pyx have properties common to those of T Tauri stars. TY Pyx is a unique member of the RS CVn group with both components having the same mass, radii, temperature, luminosity and pre-Main Sequence contraction evolutionary phase.

1. INTRODUCTION

The 7th magnitude G type star TY Pyx (BV 811) was found

to be an eclipsing binary with a period of $1\overset{d}{.}6$ by Strohmeier (1967) Later Popper (1969) discovered H and K emission as well as absorption lines of CaII for both the components. He also found that the period determined by him spectroscopically was double that of Strohmeier's.

Hall (1976) has classified TY Pyx as one of the members of the RS CVn binaries on the basis of its group characteristics such as spectral type, mass ratio near unity, H and K emission lines outside the eclipse and period.

Anderson and Popper (1975) observed this system spectroscopically and derived its physical properties which showed that TY Pyx consists of approximately two equal components of masses 1.2 $M_\odot$, radii 1.65 $R_\odot$, and spectral type near G5 (T_e = 5600°K). Further, their results showed that there are two basic differences between TY Pyx and some of the other members of the RS CVn group, viz. (i) the spectral type of its secondary component is earlier than the other members, (ii) the radius of its secondary component is about 1.65 $R_\odot$ while the radii of the secondaries of this group are larger than 2.5 $R_\odot$

Since there exist no published photoelectric light curves for the system and in order to determine more reliable absolute dimensions, and to study its evolutionary status by comparing them with other members of the group, we included this star in our observing programme of the 15-inch Grubb refractor of the Nizamiah Observatory.

2. OBSERVATIONS

This star was observed for 75 nights during 1976-77, 1977-78 and 1978-79 observing seasons in B and V colours of Johnson's system. During the 1976-77 and 1977-78 observing seasons, observations were made using an unrefrigerated 1P21 photomultiplier while in the 1978-79 observing season we made use of an unrefrigerated EMI 9502B photomultiplier attached to the 15-inch Grubb refractor of the Nizamiah Observatory. The photocurrent was amplified by means of a 1230A D.C. amplifier and was recorded on a Honeywell Brown recorder. The star HD 77087 was used as comparison and the stars HD 77506 and HD 77361 were used as check stars during the observing season 1976-77. But in the following successive observing seasons (1977-78, 1978-79) HD 77361 alone was used as a check star. The r.m.s. error for Δm (Check-Comparison) was $\sim 0\overset{m}{.}02$ in both yellow and blue which indicates that the comparison star was constant in brightness during the period of observation. The observations of the comparison star were used for determining the nightly extinction coefficients. After applying the extinction corrections, the extra atmospheric magnitude differences Δm (Var-Comp) were obtained. These Δm's were then transformed to the Johnson and Morgan's standard UBV system using linear transformation relations obtained from the observations of UBV standard stars. The transformation relations gave the following constants shown in Table I for different seasons.

TABLE I

Observing season	Cell used	ε_v	$u_{(b-v)}$
1976-77	1P21	-0.115	1.478
1977-78	1P21	-0.120	1.489
1978-79	EMI 9502B	-0.115	1.537

The V magnitudes and (B-V) colours of the comparison and check stars obtained by us are given in Table II.

TABLE II

Star	HD No.	V	B-V
Comparison	77087	6.25	1.01
Check	77361	6.21	1.14

A total of 1203 observations of TY Pyx in V and 1189 observations in B have been obtained and are being published separately.

3. WAVE DISTORTION AND PERIOD DETERMINATION

In most of the RS CVn-systems we find a variation in the depth of the minima as well as a sinusoidal wave like distortion outside the eclipse. As a consequence of this wave, the minima of the light curves become asymmetric which leads to erroneous determination of the times of minimum. To get the correct times of minimum, one has to remove the effect of the wave (Arnold et al, 1979) and after getting a symmetric curve during minimum, we have to determine the times. In TY Pyx the depths of primary and secondary minimum did not show any variation during the three years (1976-1979) of observation. Similarly the light outside the eclipses has also remained constant during this period. Hence we have concluded that unlike the other RS CVn-type systems, TY Pyx does not have any wave-like distortion in its light, and used the observed light curves for determining the times of minima.

As a result of the observations made during the 1976-77 observing season (Surendranath et al, 1978) it was found that the epoch given by Strohmeier refers to that of the secondary minimum and not to the primary minimum, provided that there is no variation in the depths of the minimum. The observations made during the following successive years confirmed the above finding.

Three times of primary minima and five times of secondary minima derived by us using the method of Kwee and Van Woerden (1956) are listed in Table III:

Primary minimum = Hel.J.D. 2443187.2304 ± 4 +

$3^d.1985787E$ ± 1 .

The (O-C) obtained from the above elements are small, indicating that the period of TY Pyx has remained constant during the period 1976079.

The phases of all the observations made in B and V colours are computed using the above ephemeris and are plotted in Fig. 1.

TABLE III   Times of minima of TY Pyx.

Time of minimum Hel. J. D.	Weight	Type of eclipse	O-C
2443187.230	1	Pr	$-0^d.0004$
2443211.220	1	Sec	0.0003
2443590.252	1	Pr	0.0007
2443598.247	1	Sec	-0.0008
2443889.317	1	Sec	-0.0014
2443905.313	1	Sec	0.0017
2443929.300	1	Pr	-0.0007
2443937.297	1	Sec	-0.0002

4. RECTIFICATION

The light curves in both B and V colours are well observed except for a narrow gap between phase angles 237° to 248°. A visual inspection of the light curves indicates that the light outside the eclipse is fairly constant. The light curves were normal-

PHOTOELECTRIC STUDY OF THE RS CVn-TYPE BINARY TY PYXIDIS

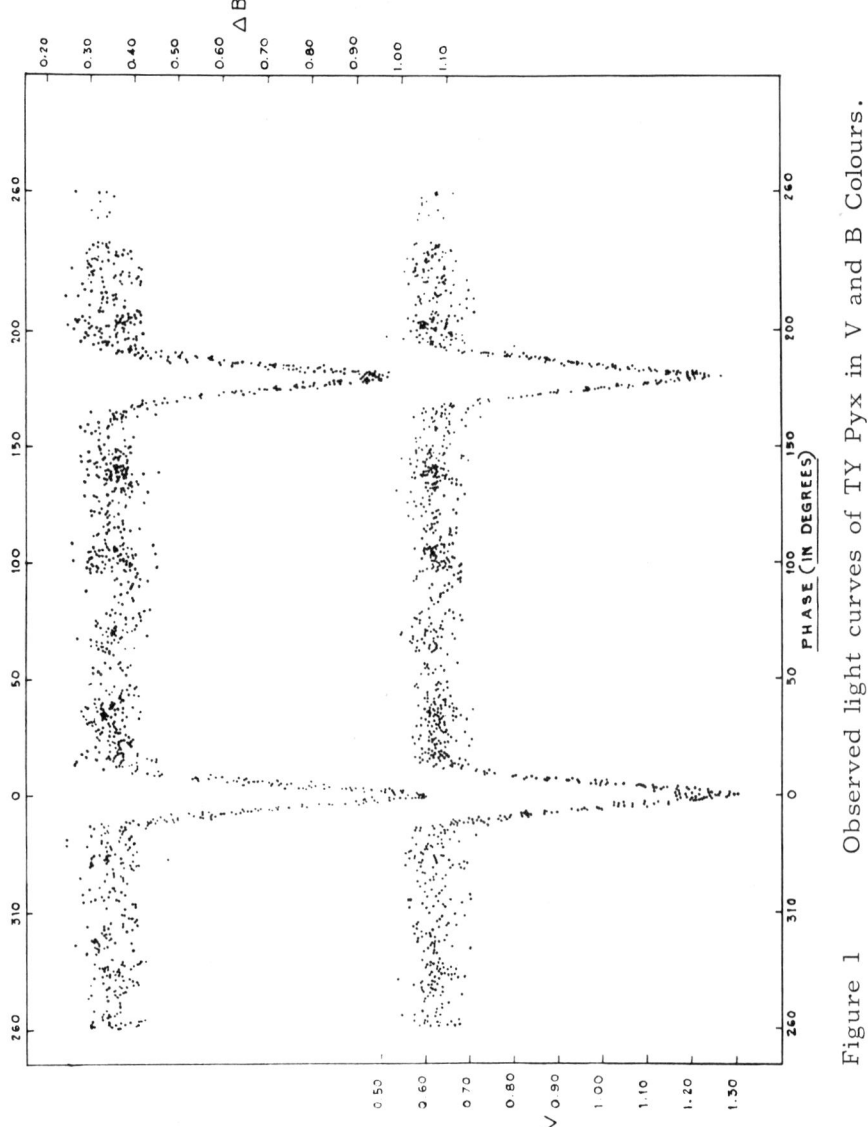

Figure 1 Observed light curves of TY Pyx in V and B Colours.

ized to unit light intensity at maximum by adding $-0^m\!.623$ and $-0^m\!.347$ to yellow and blue Δm's respectively. The observations were grouped to form normal points given in Table IV. The value of θ_e was found to be $15°\!.5$ graphically. The light outside the eclipses was represented by the Fourier expression

$$\ell_{obs} = \sum_{n=0}^{4} A_n \cos n\theta + \sum_{n=1}^{4} B_n \sin n\theta .$$

The values of the coefficients A_n, and B_n and their standard errors were determined by the method of least squares. As the values of A_3, A_4, B_3 and B_4 in both the colours were found to be small compared to their standard errors, another solution was made with terms up to 2θ only. The derived Fourier coefficients are shown in Table V. It can be seen that even with this truncation in the Fourier expression, the derived coefficients A_1, A_2, B_1 and B_2 are small and are comparable to those of their standard errors. Owing to the smallness of the Fourier coefficients, we did not find any necessity to rectify the light curves. Solutions of the light curves were made by using the plot of ℓ_{obs} vs $\sin^2\theta$ for normal points.

TABLE IV

Phase (in deg.)	ΔV (Var-Comp)	No.of Points	Phase (in deg.)	ΔV (Var-Comp)	No.of Points
0.006	$1^m\!.234$	3	7.251	$0^m\!.875$	3
0.116	1.251	2	7.668	0.814	2
0.577	1.253	4	8.533	0.792	3
1.132	1.237	3	9.470	0.743	3
1.327	1.231	3	10.605	0.724	3
1.693	1.191	4	11.563	0.660	2
2.290	1.199	3	12.433	0.653	3
2.599	1.172	3	12.790	0.676	3
2.938	1.174	3	13.422	0.653	4
3.157	1.147	3	14.488	0.617	3
3.298	1.103	3	14.810	0.612	2
3.843	1.065	3	15.379	0.657	3
4.235	1.057	3	15.882	0.611	3
4.579	1.032	3	18.717	0.622	28
4.911	1.005	3	24.987	0.628	24
5.417	0.985	3	30.576	0.624	20
5.768	0.923	3	35.447	0.633	29
5.972	0.998	2	40.226	0.626	18
6.258	0.895	3	45.320	0.613	12
6.883	0.877	3	50.204	0.621	6

(a) Normal Points in Yellow

Phase (in deg.)	ΔV (Var-Comp)	No.of Points	Phase (in deg.)	ΔV (Var-Comp)	No.of Points
55.185	0.619^m	10	177.720	1.155^m	4
62.716	0.613	12	178.486	1.163	4
67.622	0.599	19	179.182	1.202	4
72.174	0.611	16	179.569	1.226	4
77.516	0.630	13	179.989	1.226	3
83.674	0.621	5	180.469	1.197	3
87.649	0.633	4	180.710	1.199	3
90.124	0.640	3	180.931	1.194	3
92.648	0.636	8	181.242	1.187	4
97.811	0.637	21	181.835	1.156	3
102.352	0.627	26	182.032	1.188	4
106.739	0.622	19	182.260	1.117	4
112.994	0.631	13	182.744	1.096	4
117.226	0.618	8	183.204	1.051	3
122.668	0.612	16	183.407	1.059	3
128.387	0.624	9	183.752	1.041	3
132.654	0.618	17	184.148	1.036	4
137.660	0.620	28	184.722	0.999	4
141.997	0.617	22	185.090	0.964	4
147.304	0.618	24	185.552	0.935	3
152.197	0.611	9	185.857	0.939	4
157.762	0.641	11	186.118	0.925	3
162.750	0.638	17	186.566	0.915	3
165.322	0.613	2	187.185	0.886	3
166.372	0.639	4	187.691	0.858	3
167.204	0.656	3	188.112	0.864	3
167.612	0.684	2	188.406	0.844	3
168.382	0.650	2	188.793	0.803	3
169.458	0.735	2	189.214	0.771	3
169.931	0.732	2	189.603	0.780	3
170.735	0.772	3	190.234	0.757	4
171.669	0.831	3	190.533	0.726	4
172.475	0.863	2	191.386	0.694	3
172.762	0.878	2	192.653	0.692	3
173.232	0.903	3	193.530	0.645	2
173.694	0.929	2	193.870	0.626	2
174.207	0.954	4	194.539	0.639	3
174.476	0.978	4	194.826	0.603	2
174.790	0.988	4	195.494	0.645	3
175.433	1.033	4	198.321	0.618	23
175.810	1.039	4	203.461	0.617	28
176.415	1.068	3	207.967	0.643	11
176.689	1.095	3	213.547	0.626	14
177.436	1.113	4	218.952	0.631	13

(a) Normal Points in Yellow (continued)

Phase (in deg.)	ΔV (Var-Comp)	No. of Points	Phase (in deg.)	ΔV (Var-Comp)	No. of Points
223.719	0.615	16	347.135	0.688	3
228.694	0.612	14	347.734	0.699	2
233.706	0.621	15	348.065	0.686	3
237.244	0.606	5	348.594	0.675	3
250.141	0.611	5	348.937	0.712	2
257.904	0.617	9	349.296	0.715	3
261.932	0.625	14	349.758	0.758	4
266.089	0.600	8	350.206	0.708	2
270.078	0.639	3	350.736	0.788	3
273.720	0.624	6	351.223	0.816	3
277.420	0.608	14	351.534	0.819	3
282.560	0.634	17	352.280	0.850	4
287.066	0.624	17	352.764	0.881	4
293.834	0.628	9	353.295	0.894	3
298.338	0.612	14	353.674	0.972	3
303.347	0.599	11	354.100	0.950	3
307.652	0.641	6	354.426	0.992	2
313.702	0.618	13	355.135	1.063	4
318.589	0.623	10	355.625	1.043	3
323.941	0.616	7	356.196	1.117	3
328.777	0.623	13	356.557	1.073	3
333.141	0.608	20	357.347	1.143	3
338.501	0.615	13	357.928	1.190	3
343.314	0.631	11	358.448	1.205	4
345.513	0.605	3	358.804	1.229	4
346.395	0.612	3	359.111	1.225	3
346.752	0.623	2	359.582	1.203	2

(a) Normal Points in Yellow (conclusion)

Phase (in deg.)	ΔB (Var-Comp)	No. of Points	Phase (in deg.)	ΔB (Var-Comp)	No. of Points
0.036	1.027	3	4.175	0.783	3
0.431	1.024	4	4.528	0.749	3
1.209	0.979	3	4.823	0.703	4
1.651	0.958	4	5.488	0.724	2
2.251	0.961	3	5.947	0.704	3
2.671	0.919	3	6.136	0.667	3
2.915	0.840	4	6.331	0.644	3
3.137	0.815	4	7.080	0.611	3
3.420	0.800	2	7.476	0.578	3
3.717	0.827	2	7.745	0.547	3

(b) Normal Points in Blue

Phase (in deg.)	ΔB (Var-Comp)	No. of Points	Phase (in deg.)	ΔB (Var-Comp)	No. of Points
8.706	0.^{m}539	3	166.570	0.^{m}410	2
9.185	0.439	2	167.472	0.397	4
9.646	0.440	2	168.388	0.427	3
10.261	0.422	2	169.416	0.474	3
11.539	0.375	3	170.134	0.427	2
12.679	0.378	4	170.662	0.494	2
13.453	0.360	4	171.566	0.534	3
14.268	0.381	2	172.337	0.567	2
14.834	0.356	4	172.803	0.632	2
15.338	0.400	2	173.343	0.641	3
15.668	0.376	2	173.734	0.627	3
18.277	0.351	26	174.466	0.713	3
24.767	0.353	26	174.639	0.696	3
30.554	0.327	20	174.834	0.739	4
35.345	0.333	30	175.603	0.770	2
40.165	0.343	20	175.788	0.767	3
45.472	0.371	13	176.174	0.805	3
50.432	0.319	6	176.692	0.844	2
55.311	0.351	10	176.873	0.854	3
62.751	0.337	11	177.554	0.805	3
67.668	0.339	19	177.805	0.871	3
72.080	0.354	16	178.155	0.890	2
77.380	0.372	13	178.462	0.933	2
82.756	0.388	5	178.949	0.934	3
87.409	0.353	5	179.252	0.926	3
90.016	0.397	2	179.746	0.927	2
92.378	0.384	8	180.028	0.952	2
97.776	0.345	22	180.324	0.943	2
102.446	0.345	26	180.688	0.907	2
106.785	0.346	20	180.878	0.930	3
112.876	0.346	11	181.077	0.931	3
116.543	0.350	7	181.342	0.923	3
122.281	0.334	15	181.790	0.854	2
128.056	0.342	10	182.180	0.904	4
132.422	0.372	19	182.334	0.850	4
137.770	0.353	29	182.504	0.812	4
142.188	0.338	24	183.197	0.802	4
147.214	0.345	22	183.496	0.756	3
151.896	0.353	8	183.830	0.763	3
157.868	0.315	11	184.398	0.741	4
162.255	0.365	12	184.848	0.726	4
164.307	0.355	2	185.170	0.713	4
164.812	0.326	2	185.497	0.655	3
165.352	0.363	3	185.938	0.652	2

(b)   Normal Points in Blue (continued)

Phase (in deg.)	ΔB (Var-Comp)	No. of Points	Phase (in deg.)	ΔB (Var-Comp)	No. of Points
186.108	$0^m.579$	4	314.037	$0^m.344$	11
186.690	0.557	4	318.551	0.332	9
187.317	0.554	3	323.960	0.340	7
187.789	0.534	2	328.951	0.358	12
188.202	0.477	4	333.171	0.352	19
188.565	0.527	3	338.505	0.341	14
189.173	0.468	4	342.544	0.340	8
189.683	0.440	4	344.646	0.331	2
190.438	0.428	4	345.667	0.354	3
190.731	0.392	4	346.284	0.339	2
191.569	0.363	3	346.709	0.372	3
192.184	0.407	2	347.236	0.332	3
193.433	0.376	3	347.580	0.415	2
194.249	0.316	3	348.240	0.377	4
194.750	0.330	2	348.812	0.397	3
195.198	0.320	3	349.357	0.459	3
198.383	0.346	24	349.763	0.440	3
203.715	0.337	29	350.130	0.464	3
208.143	0.369	11	350.827	0.515	4
213.717	0.326	14	351.279	0.567	3
218.893	0.335	12	351.635	0.550	3
223.689	0.342	17	352.391	0.545	4
228.735	0.322	13	352.676	0.608	3
232.387	0.346	14	353.210	0.582	2
237.087	0.324	7	353.490	0.646	2
250.365	0.322	4	353.758	0.682	2
258.631	0.317	6	354.252	0.671	3
262.124	0.354	19	354.375	0.689	3
266.422	0.333	10	355.092	0.728	2
270.463	0.359	2	355.335	0.757	3
273.902	0.324	6	356.169	0.837	3
277.438	0.357	13	356.410	0.836	3
282.621	0.350	18	357.237	0.914	3
287.370	0.361	17	357.638	0.911	2
294.186	0.319	10	358.220	0.948	4
298.211	0.322	13	358.882	0.987	3
303.305	0.347	12	359.044	1.028	3
307.734	0.383	6	359.208	1.002	3

(b) Normal Points in Blue (conclusion)

TABLE V Fourier Coefficients

	V	B
A_0	1.0010 ± 0.0008	1.0021 ± 0.0008
A_1	+0.0012 ± 0.0011	−0.0013 ± 0.0012
A_2	−0.0014 ± 0.0012	+0.0016 ± 0.0013
B_1	−0.0016 ± 0.0010	−0.0036 ± 0.0011
B_2	−0.0012 ± 0.0010	+0.0026 ± 0.0010

5. SOLUTION

For each filter a plot of the ℓ_{obs} vs $\sin^2\theta$ for normal points was made and a smooth curve is drawn through them. The appearance of the light curves in both V and B colours suggests that the eclipses are partial. Further, in yellow $\chi_{0.8}^{pr} > \chi_{0.8}^{sec}$ which indicated that the primary is an occultation and secondary transit, while the blue curve gave $\chi_{0.8}^{sec} > \chi_{0.8}^{pr}$ suggesting that primary is transit and secondary an occultation. Assuming that the above contradiction is due to free hand curve drawn through the normal points, solutions were made for both possibilities; (i) primary to be transit and secondary to be occultation, and (ii) primary to be occultation and secondary to be transit for both the colours. We used a modification of Wellmann's (1953) method to solve the light curves in both yellow and blue colours. Assuming different values of k, we obtained α_0^{oc} and α_0^{tr} from the depth relation. The χ_n values were calculated for individual normal points using the relation

$$\chi_n = \frac{(1 + kp)^2 - (1 + kp_o)^2}{(1 + kp_{\frac{1}{2}})^2 - (1 + kp_o)^2}.$$

These values of χ_n were used to compute $\sin^2\theta_e$ by the least square solution

$$\sin^2\theta_e = \frac{\sum w_n N(\chi_n/\chi_o) \sin^2\theta_n}{\sum w_n N(\chi_n/\chi_o)^2},$$

where

$$w_n = \left[\frac{1 - \ell_o}{2\alpha_o(1 + kp)\, \partial p/\partial\alpha}\right]^2$$

as given by Kopal (1959), and N is the number of observations in

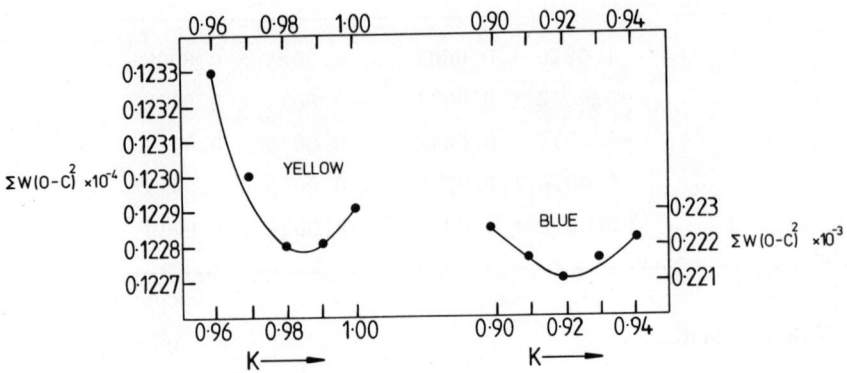

Figure 2a. Plot of $\sum W(O-C)^2$ for primary and secondary combined, for various values of k, with the assumption of primary to be transit and secondary to be occultation for both the colours.

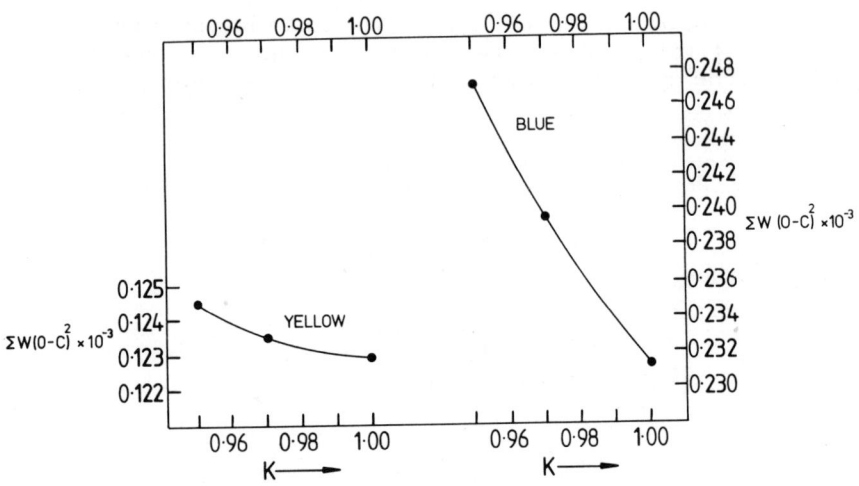

Figure 2b. As Figure 2a, with the assumption of primary to be occultation and secondary to be transit.

a normal point. Since, according to Anderson and Popper (1975), the spectral type of both the components is between G2-G5, the limb darkening coefficients of $x_g = x_s = 0.6$ for V and $x_g = x_s = 0.8$ for B were used to determine the 'p' values from Tsesevich's (1940) tables. Data for both the primary and secondary eclipses were combined together to derive the values of $\sin^2\theta_e$. Then k and $\sin^2\theta_e$ gave the desired elements i, r_g, r_s, L_g and L_s for each assumed value of k. These elements were used for computing a theoretical light curve which gave $(O-C)_i$ for each normal point. By plotting $\sum \omega_i (O-C)_i^2$ against k (where $\omega_i = N_i/\ell_i^2$) we got a minimum at k = 0.984 for V and k = 0.920 for B for the possibility of primary transit, secondary occultation, while the other possibility did not give any minimum even with k = 1. Here N_i is the number of observations in a single normal point.

The plots of $\sum \omega_i (O-C)_i^2$ vs k for the two possible possibilities are shown in Figures 2a and 2b. From the plots it is evident that the assumption primary transit, secondary occultation only, gives a solution. An average value of k = 0.950 is assumed for further discussion.

A correction to the depth $(1 - \ell_o)$ for each minimum was obtained in the following manner. Theoretical values of the light for each observed normal point were calculated using k = 0.950 for both V and B colours. These values gave $(O-C)_i$ in light for each normal point. The correction $\Delta(1 - \ell_o)$ was obtained from

$$\Delta(1 - \ell_o) = - \frac{\sum \omega_i (O-C)_i}{\sum n_i \omega_i},$$

where

$$n_i = \frac{1 - \ell_i}{1 - \ell_o} \quad \text{and} \quad \omega_i = N_i/\ell_i^2.$$

In this way for V a correction of +0.002 for the depth of secondare minimum was obtained. In the case of B the correction to the primary depth amounted to -0.007 and to the secondary minimum it amounted to -0.003. Using these new depths and the new depth relation, $\sin^2\theta_e$, r_g, r_s, L_g and L_s were determined as before for k = 0.950 for both the V and B colours and the averaged values that were adopted are given in Table VI.

The theoretical light curves computed using the elements given in Table VI are shown as continuous lines in ℓ_{obs} vs $\sin^2\theta$ plot of Figures 3a and 3b for V and B, respectively. In these Figures dots represent the observed normal points on the ascending branch and crosses the normal points on the descending branch of the light curves. It can be seen that the fit of the computed light curve to the observed normal points is satisfactory.

TABLE VI Adopted Elements for TY Pyx

Element	V	B
x_g	0.6	0.8
x_s	0.6	0.8
α_o^{tr}	0.885	0.890
α_o^{oc}	0.901	0.914
$1 - \ell_o^{tr}$	0.441	0.457
$1 - \ell_o^{oc}$	0.420	0.418
L_g	0.534	0.542
L_s	0.466	0.458
J_g/J_s	1.03	1.07
k	0.950	
p_o	-0.780	
θ_e	15°330	
r_g	0.137	
r_s	0.130	
i	87°971	

Using the spectroscopic elements given by Popper and Anderson (1975), the absolute dimensions of the system have been determined, and are given in Table VII.

TABLE VII Absolute Dimensions of TY Pyx

Elements	Primary	Secondary
Radius of the orbit	12.26 $R_\odot$	12.26 $R_\odot$
Mass	1.22 $M_\odot$	1.20 $M_\odot$
Radius	1.68 $R_\odot$	1.59 $R_\odot$
Density	0.26 $\rho_\odot$	0.30 $\rho_\odot$

PHOTOELECTRIC STUDY OF THE RS CVn-TYPE BINARY TY PYXIDIS

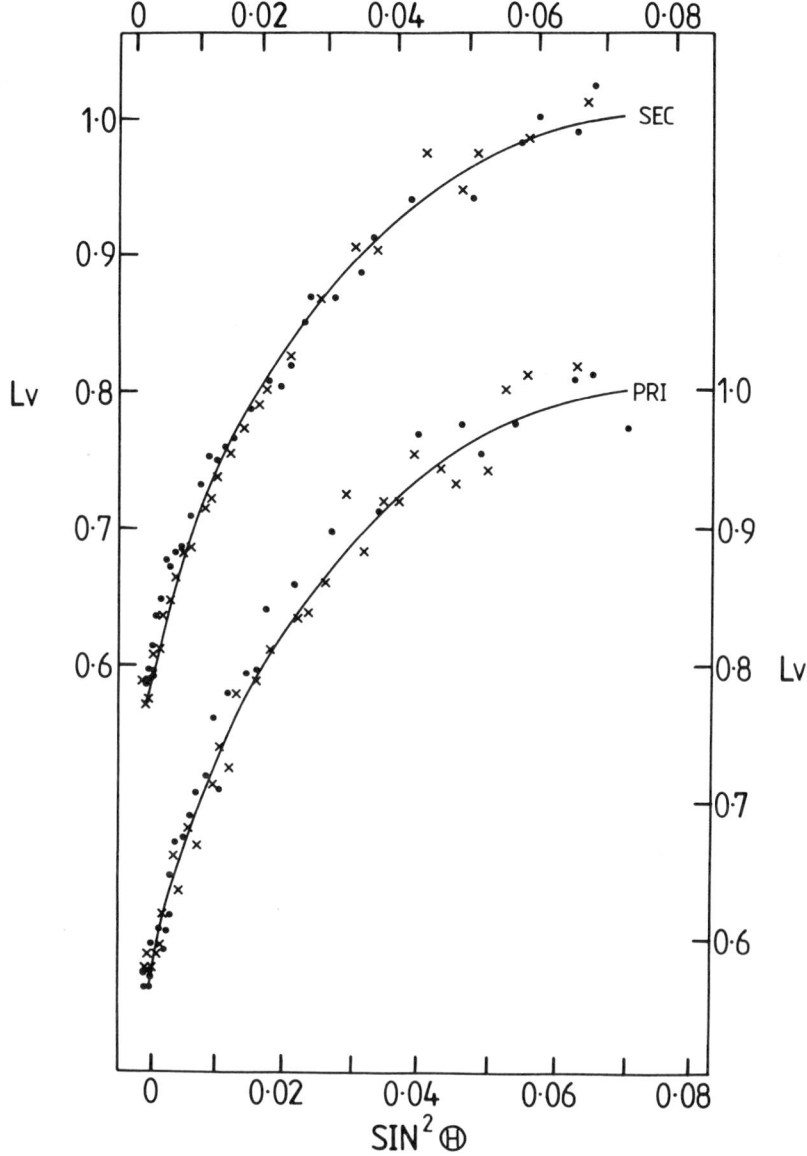

Figure 3a. Computed light curve and the observed normal points for primary and secondary minima in yellow light. The dots represent the observed normal points on the ascending branch and crosses the normal points on the descending branch of the light curve. The theoretical light curve is shown as a continuous line.

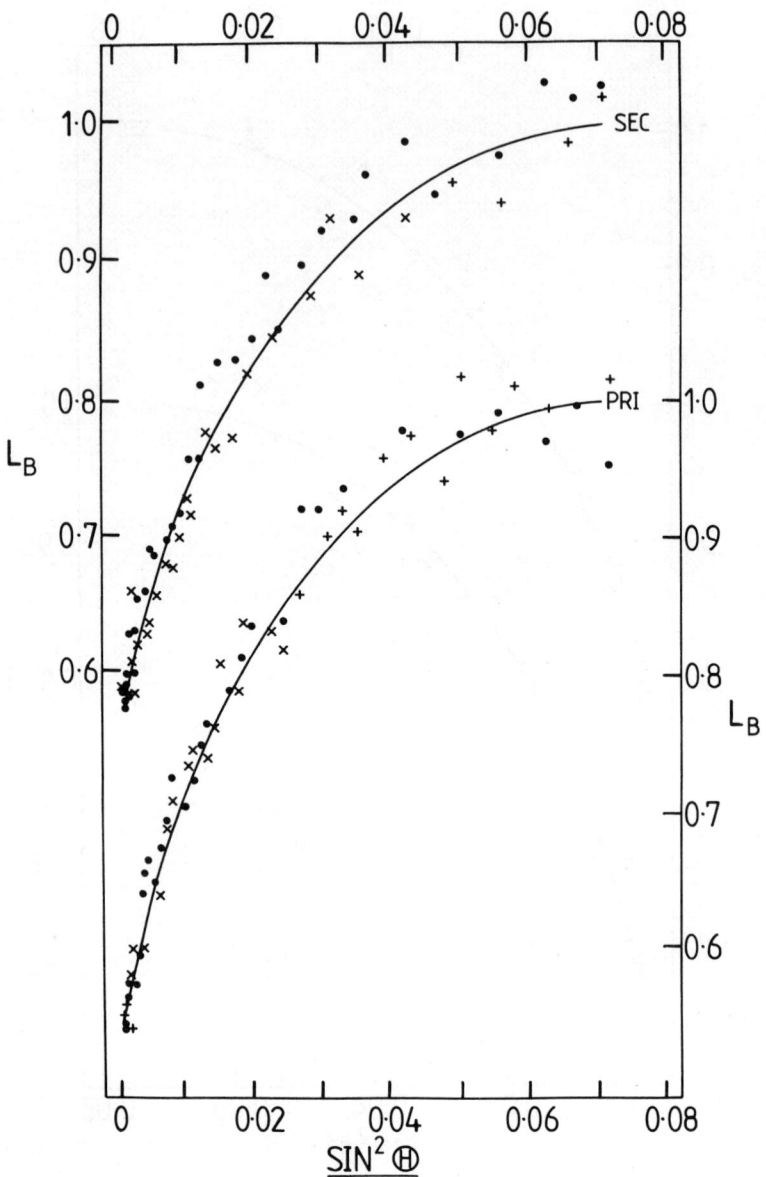

Figure 3b. Computed light curve and the observed normal points for primary and secondary minima in blue light. The dots represent the observed normal points on the ascending branch and crosses the normal points on the descending branch as a continuous line.

6. DISCUSSION

Using the V and (B-V) values of the comparison star, and the differential magnitude corresponding to unit luminosity of maximum light, we obtained V = $6^m.87$ and B = $7^m.61$ for the variable at maximum light. From these magnitudes and the derived L_g and L_s of the primary and secondary components the following magnitudes and colours for the two components were obtained. Here the primary star refers to the hotter and secondary to the cooler components.

	Primary	Secondary
V	$7^m.55$	$7^m.70$
B-V	$0^m.72$	$0^m.76$

Assuming the components of TY Pyx to be close to Main Sequence with no space reddening, from the above colours we get the primary to be G5 (T_e = 5520°K) and secondary to be G6 (T_e = 5400°K) (Allen, 1972). The spectral type of the secondary can also be estimated from the relative depths of the two minima by assuming that the radiation of both the primary and secondary components may be approximated by the Planck's function over the passbands of B and V filters (Wood, 1971). Assuming a temperature T_e = 5520°K (Allen, 1972) for the primary of spectral type G5 we get for the secondary, a temperature T_e = 5450°K in V and T_e = 5440°K in B. The average temperature T_e = 5445°K for the secondary corresponds to a spectral type of G6. As the spectral types of the components derived in the above manner and estimated from the colours agree with one another, a spectral type of G5 (T_e = 5520°K) for the primary and G6 (T_e = 5400°K) for the secondary is justified. This is in agreement with the results of Anderson and Popper (1975) who estimated a temperature T_e = 5600°K for both the components. Using the derived temperatures and radii of the two components, the bolometric luminosities as determined from the Stefan-Boltzmann law

$$L = 4\pi R^2 \sigma T_e^4$$

are

$$\text{Log}(L_{pri}/L_\odot) = +0.37 \quad \text{and} \quad \text{Log}(L_{sec}/L_\odot) = +0.28$$

for the primary and secondary, respectively. From these luminosities, the absolute bolometric magnitudes are determined using M_{bol} = 4.75 - 2.5 log($L/L_\odot$) for the two components which gave M_{bol}(pr) = +3.82 and M_{bol}(sec) = +4.05. If we apply the standard bolometric corrections of $-0^m.07$ and $-0^m.11$ (Allen, 1972) for the Main-Sequence stars of spectral type G5 and G6, the absolute visual magnitudes determined for the two components are M_v(pr) =

+3.89 and M_v(sec) = +4.16. These values are in good agreement with the values of M_v derived from the photometric distance of 0".018 ± 0.008 given by Dworak (1973). From the spectral type and absolute visual magnitudes, we conclude that TY Pyx is slightly above the Main Sequence and below the sub-giant sequence.

From Plavec and Kratochvil's (1964) tables for a mass ratio of 1.0, the radii of Roche lobes are $r_g^* = r_s^* = 0.374$. Comparing these values with $r_g = 0.137$ and $r_s = 0.130$ of TY Pyx derived from the analysis, we conclude that both the components of TY Pyx are well within their Roche lobes. Hence we conclude that TY Pyx is a detached system.

7. EVOLUTION

From the photographic solution of Gaposchkin, Koch (1970) had suggested that the long period RS CVn type binary RZ Cnc may be in the pre-Main Sequence evolutionary phase. Considering the T Tauri-like properties for some of the secondaries of the RS CVn-binaries Hall (1972) has suggested a pre-Main Sequence type evolution for these stars. But Popper and Ulrich (1977) have suggested that the components of RS CVn type binaries might have entered the HR gap, suggesting a post-Main Sequence type evolution and an age of $4-5 \times 10^9$ years for these stars. However, Eggleton (1978) and Guy Morgan and Eggleton (1979) have suggested that these are normal group stars and their peculiarities might be explained as due to observational selection. Since TY Pyx seems to be the only known member in the group with the radii, temperature, luminosity and masses of both the components almost the same, it would be interesting to look into its evolutionary status.

Assuming TY Pyx to be evolving according to the model calculations of Iben (1965, 1967), we have plotted in Figures 4a and 4b the derived values of log $(L/L_\odot)$ vs log T_c for both the components in the pre-Main Sequence and post-Main Sequence evolutionary tracks. These figures show that both the components of TY Pyx fit only the pre-Main Sequence track and not the post-Main Sequence. From these tracks the age of the system seems to be of about $\sim 10^7$ years.

To compare the evolutionary status of TY Pyx with the other RS CVn members, we used the data given in Table I of Popper and Ulrich (1977). Figures 5a, 5b, 5c and 5d show a plot of log $(L/L_\odot)$ vs log T_e for both the secondary and primary components separately on the pre-Main Sequence and post-Main Sequence evolutionary tracks of Iben (1965, 1967). These figures show that the primaries of the other members of this group fall slightly away from the Main Sequence indicating that they may be either in the pre-Main Sequence or post-Main Sequence, while the secondaries definitely show that they are in the post-Main Sequence evolutionary phase. Since primaries may be either in the pre- or post-Main Sequence we determined their ages from both these sequences and compared them with the ages determined for their secondaries. The ages so determined are consistent only if both the primaries and secondaries are in the post-Main Sequence and of ages of about $4-5 \times 10^9$ years. For TY Pyx also Anderson and Popper (1975) have suggested an age of $4-5 \times 10^9$ years from a study of its space motion. But for a star of this age, temperature (5520°K) and mass $1.2 M_\odot$, the radius should be of the order of $2.5 R_\odot$ which the present observations do not permit. Hence we conclude that TY Pyx is definitely in the pre-Main Sequence evolutionary phase and that the high space motion may be due to some other reasons.

The components of TY Pyx have the following common properties:

1. Spectral type of about G5.
2. Exhibit H and K emission outside eclipse.
3. Mass $1.2 M_\odot$.
4. Radii $1.6 M_\odot$.
5. Both are in the pre-Main Sequence contraction phase of evolution.
6. Well within their Roche lobes.

Since the above properties (1 to 5) are similar to those of T Tauri stars (Herbig, 1962), we made a plot of log $R/R_\odot$ vs log T_e for both components in Figure 6. This figure shows that both the components of TY Pyx lie in the HR diagram in the region occupied by T Tauri stars and are very close to T Tauri Star itself.

Since one of the properties of T Tauri is irregular light variation, we tried to detect any variation in the light curve of TY Pyx by taking the deviation between the theoretical and observed light (Normal points) and brought them to the same units of intensity by dividing with the computed ℓ at that phase. As the scatter of the observations from the mean level is found to be comparable to that of the r.m.s. error of (Check-Comparison), we could not detect any significant light variation.

8. CONCLUSIONS

The present photoelectric studies of the system TY Pyx indicate that the epoch given by Strohmeier refers to the secondary minimum and the improved light elements Hel. J.D. $2443187.2304 \pm 3^d.1985787 E$ was determined. No period variation has been detected from the present observations. As none of the components fills its

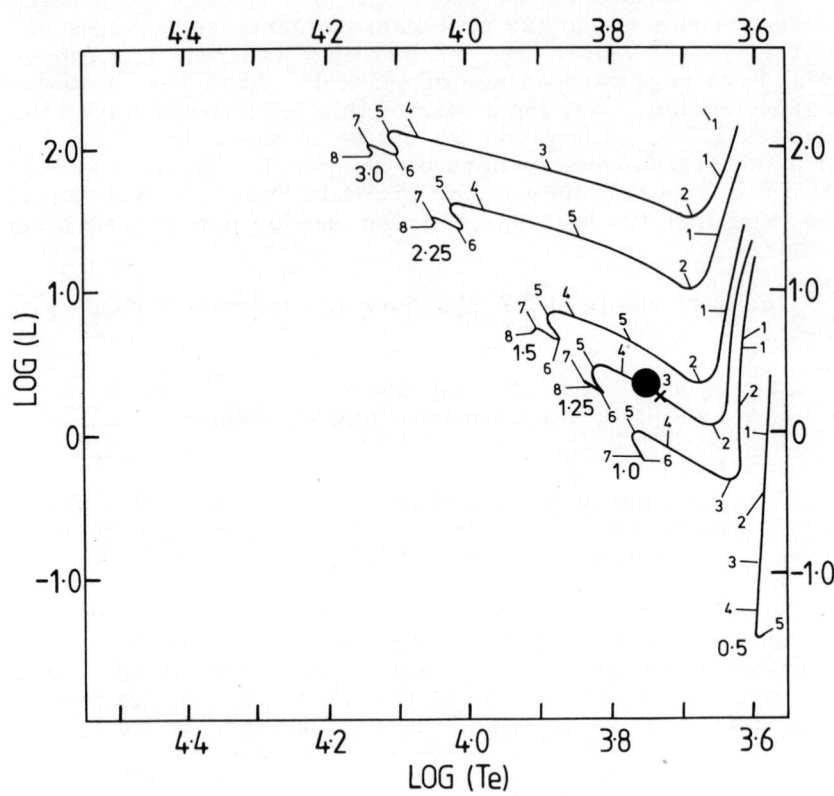

Figure 4a. Position of the components of TY Pyx - primary (mass 1.22 $M_\odot$, represented by ●) and secondary (1.20 $M_\odot$, X) in Iben's (1865) pre-Main Sequence evolutionary tracks.

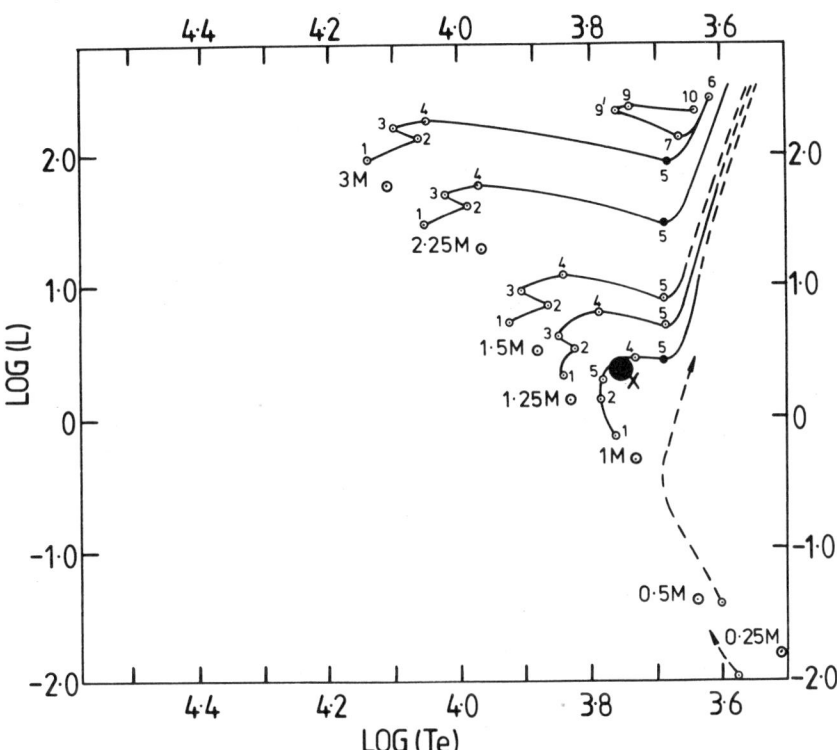

Figure 4b. Position of the components of TY Pyx - primary (mass 1.22 $M_\odot$, represented by O) and secondary (1.20 $M_\odot$, X) in Iben's (1967) post-Main Sequence evolutionary tracks.

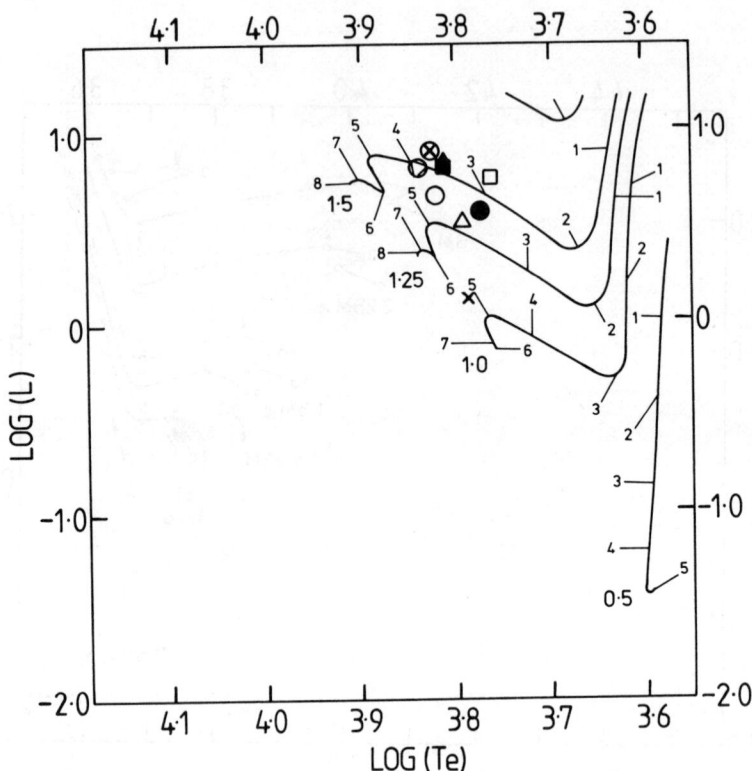

Figure 5a. Position of the primaries of RS CVn binaries in Iben's (1965) pre-Main Sequence evolutionary tracks.

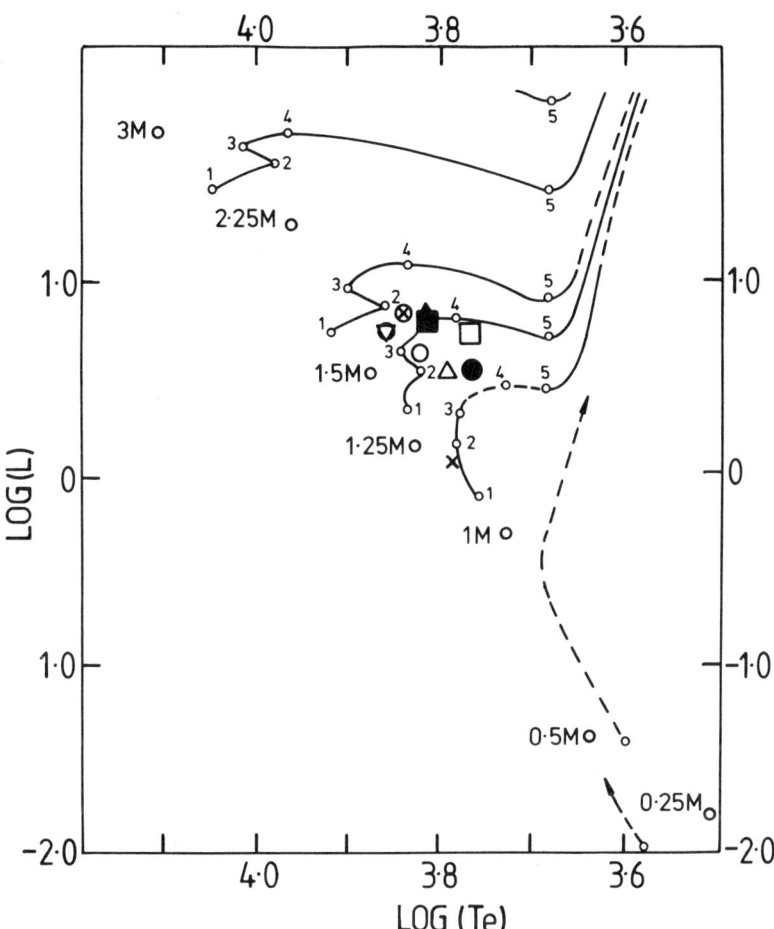

Figure 5b. Position of the primaries of RS CVn binaries in Iben's (1867) post-Main Sequence evolutionary tracks.

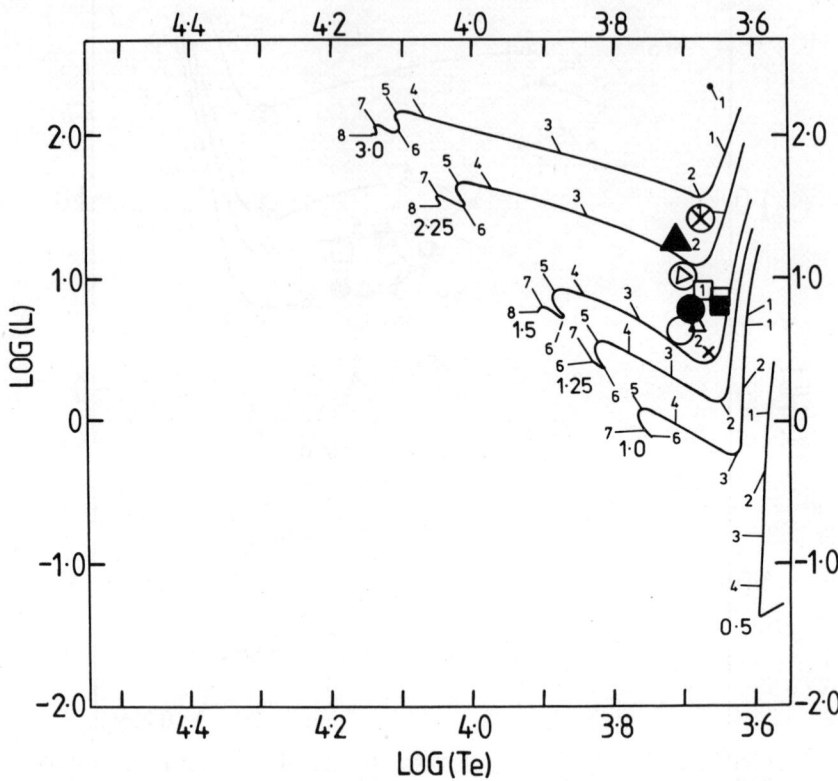

Figure 5c. Position of the secondaries of RS CVn binaries in Iben's (1965) pre-Main Sequence evolutionary tracks.

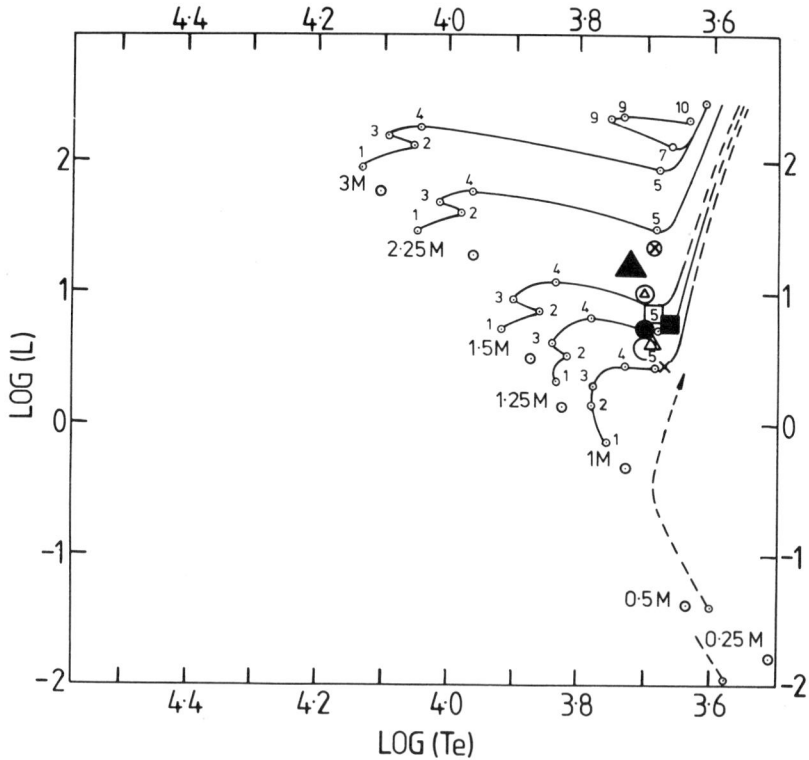

Figure 5d. Position of the secondaries of RS CVn binaries in Iben's (1967) post-Main Sequence evolutionary tracks.

NOTE: The symbols and masses of the primaries and secondaries for different RS CVn binaries used in the above Figures are:

RS CVn	⊘	1.35 $M_\odot$,	1.40 $M_\odot$
UX Com	X	0.95 $M_\odot$,	1.12 $M_\odot$
WW Dra	□	1.40 $M_\odot$,	1.40 $M_\odot$
Z Her	O	1.22 $M_\odot$,	1.10 $M_\odot$
AR Lac	●	1.30 $M_\odot$,	1.30 $M_\odot$
LX Per	△	1.23 $M_\odot$,	1.32 $M_\odot$
SZ Psc	▲	1.33 $M_\odot$,	1.65 $M_\odot$
RW UMa	■	1.50 $M_\odot$,	1.45 $M_\odot$
SS Cam	⊗	1.70 $M_\odot$,	2.20 $M_\odot$

(Arnold et al, 1979)

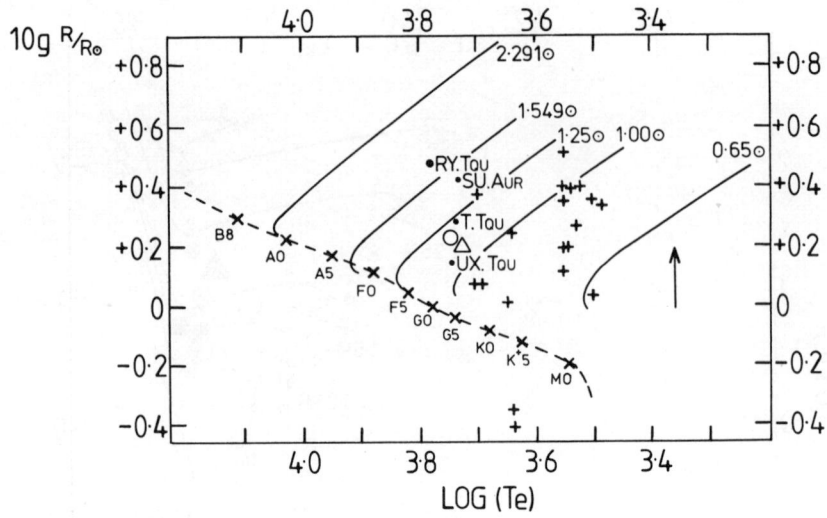

Figure 6. Location of the components of TY Pyx primary (1.22 $M_\odot$, O) and secondary (1.20 $M_\odot$,) in the HR diagram. The positions of some T Tauri stars are also shown.

Roche lobe, TY Pyx is classified as a detached system. Unlike the other RS CVn systems, this star was not found to show any variation in the depth of the minima or a wave-like distortion outside the eclipse. The derived temperatures and absolute dimensions indicate that both the components of TY Pyx are in the pre-Main Sequence evolutionary phase and an age of 10^7 years has been estimated by comparison with Iben's evolutionary tracks, while the other members of the RS CVn group have an age 4-5 x 10^9 years. If TY Pyx is to have an age similar to the other members of the group its radius should be about 2.5 $R_\odot$ which the present observations (R = 1.6 $R_\odot$ and T_e = 5520°K) do not permit. The components of TY Pyx are found to have many properties common to those of T Tauri stars.

TY Pyx seems to be a unique member of the RS CVn group, indicating where both the components have the same mass, temperature, radii, luminosity and are in pre-Main Sequence evolutionary phase. Even though TY Pyx is considerably younger than the other members of the RS CVn group, it has some properties (like spectral type of primary, mass ratio near unity, H and K emission lines, both components within the Roche lobe and period) similar to them. Hence we conclude that the RS CVn group may consist of systems in different evolutionary phases. Discovery of new systems similar to TY Pyx may shed more light on the nature of the RS CVn group stars.

ACKNOWLEDGMENTS

The authors thank Drs. K. D. Abhyankar and N. B. Sanwal for useful discussions. We also thank Sri B. D. Ausekar for his help at the telescope. One of us (PVR) would like to thank the University Grants Commission, New Delhi, for financial assistance during the tenure of this work.

REFERENCES

Allen, C. W.: 1972, *Astrophysical Quantities*, Athlone Press, London, p.206.
Anderson, J. and Popper, D. M.: 1975, Astron. Astrophys., 39, p.131.
Arnold, C. N., Hall, D. S., Montle, R. E. and Stuhlinger, T. W.: 1979, Acta Astronomica, 29, p.243.
Dworak, T. Z.: 1973, Inf. Bull. Var. Stars, No. 846.
Eggleton, P. P.: 1978, Bull. Astron. Soc. India, 6, p.9.
Guy Morgan, J. and Eggleton, P. P.: 1979, Mon. Not. Roy. Astron. Soc., 187, p.661.
Hall, D. S.: 1972, Publ. Astron. Soc. Pacific, 84, p.323.
Hall, D. S.: 1976, Internat. Astron. Union Colloq. No. 29.
Herbig, G. H.: 1962, *Advances in Astronomy and Astrophysics*, vol.1, p.47.
Iben, I.: 1965, Astrophys. J., 141, p.993.
Iben, I.: 1967, Astrophys. J., 147, p.624.
Koch, R. H.: 1970, Proc. Internat. Astron. Union, No.6, p.75.
Kopal, Z.: 1959, *Close Binary Systems*, Chapman and Hall, London.
Kwee, K. K. and Van Woerden, H.: 1956, Bull. Astron. Inst. Neth., 12, p.327.
Plavec, M. and Kratochvil, P.: 1964, Bull. Astron. Inst. Czech., 15, p.165.
Popper, D. M.: 1969, Bull. Amer. Astron. Soc., 1, p.257.
Popper, D. M. and Ulrich, R. K.: 1977, Astrophys. J., 212, L131.
Strohmeier, W.: 1967, Inf. Bull. Var. Stars, No.217.
Surendranath, R., Vivekananda Rao, P. and Sarma, M. B. K.: 1978, Acta Astronomica, 28, p.231.
Tseaevich, V. P.: 1940, Bull. Astron. Inst. U.S.S.R. Acad. Sci. No.45.
Wellmann, P.: 1953, Zeits. f. Astrophys., 32, p.1.
Wood, D. B.: 1971, Publ. Astron. Soc. Pacific, 83, p.286.

BV PHOTOMETRY OF SELECTED RS CVn-TYPE ECLIPSING BINARIES

M. Kurutaç and C. Ibanoğlu

Ege University Observatory,
Izmir, Turkey.

ABSTRACT

In this paper, observations of the RS CVn-type eclipsing binaries RS CVn, Z Her, RT CrB, RT Lac, AR Lac, LX Per and SZ Psc are presented. Their light curves show a wave-like distortion outside eclipses, which is one of the common properties of this type of binaries. The migration periods of some of the binaries are estimated.

1. INTRODUCTION

Recently, a new group of eclipsing binaries has been introduced by Hall (1976), in a paper presented in the IAU Colloquium held at Budapest in 1975. A year later, during the General Assembly of the IAU, an international campaign on the observations of RS CVn-type binaries was initiated and several working groups were formed in order to investigate the problems concerning these binaries.

A group at Ege University Observatory has taken part in the photoelectric photometry working group and the observations of some RS CVn-type binaries commenced in the spring of 1978 and are still being carried on. The eclipsing binaries RS CVn, RT CrB, Z Her, RT Lac, AR Lac, LX Per and SZ Psc are included in the observational program of RS CVn-type binaries. For every program star, the observational data and the results are given and discussed briefly below.

2. OBSERVATIONAL DATA AND RESULTS

Observations of the selected RS CVn-type binaries were made in two colours, B and V, with the 48-cm Cassegrain telescope equipped with unrefrigerated 1P21, but later changed to EMI 9781A, photomultiplier. The differential observations were taken in the sense m = m(comp) - m(var). Individual observations have been corrected for atmospheric extinction and reduced to the Sun's centre.

2.1 RS Canum Venaticorum

This system has been observed from March 25 to August 3, 1979. For comparison star, BD+36°2347 was used as suggested by Hall (1976a). The phases of individual observations have been computed from the following light elements,

$$\text{Min I} = \text{JD Hel. } 2439834.471 + 4\overset{d}{.}79781 \text{ E}.$$

The light curves obtained in B and V by Evren et al (1980) are reproduced and shown in Figures 1 and 2. The wave-like distortion outside eclipses is clearly noticeable. The amplitude of the wave is about $0\overset{m}{.}09$ in B and $0\overset{m}{.}14$ in V. The maximum and minimum of the wave are separated by half a period and fall approximately at the phases $0\overset{p}{.}22$ and $0\overset{p}{.}72$, respectively. The amplitude and the phase of the minimum of the wave agree well with values computed from the equations given by Hall (1972) as

$$\Delta V(\text{max to min}) = -0\overset{m}{.}12 - 0\overset{m}{.}07 \sin[(E + 450(/1800],$$

$$\theta(\text{min}) = 0\overset{p}{.}2 - E/725.$$

2.2 RT Coronae Borealis

This eclipsing system has been observed from April 26 to July 30, 1979. For comparison and check stars, BD+29°2691 and BD+29°2692 were used. The phases of the observations have been computed from the following light elements given in the second supplement to the GCVS (1969),

$$\text{Min I} = \text{JD Hel. } 2428273.28 + 5\overset{d}{.}11712 \text{ E}.$$

The light curves obtained in B and V by Ertan et al (1980) are shown in Figure 3. Although there is a large scattering in the maxima of the light curves, it is still possible to distinguish the wave-like distortion outside eclipses. The amplitudes of the wave are about $0\overset{m}{.}04$ and $0\overset{m}{.}03$ in V and B respectively. The maximum and the minimum of the wave fall at around the phases $0\overset{p}{.}05$ and $0\overset{p}{.}55$ respectively. In order to determine the migration period of the wave, more observations of this system are needed.

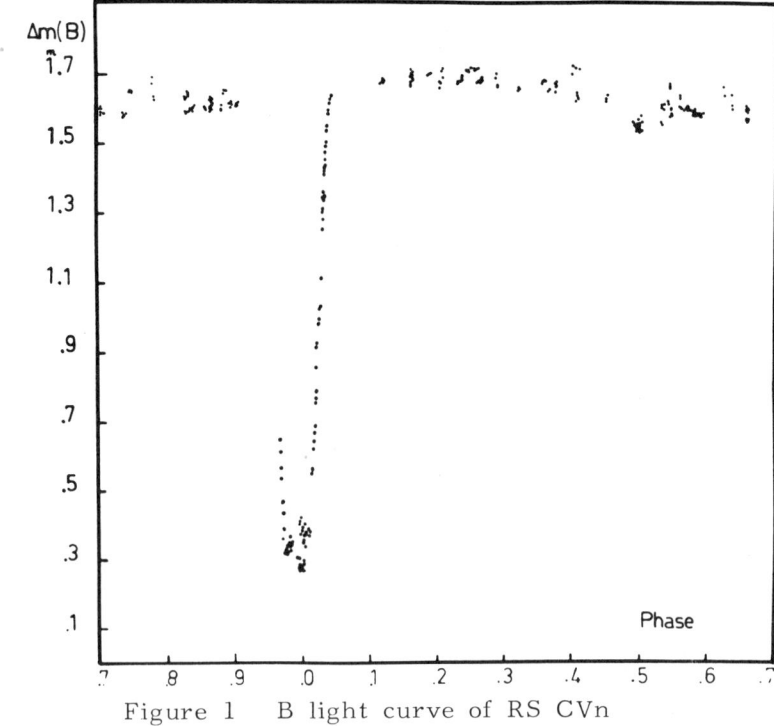
Figure 1 B light curve of RS CVn

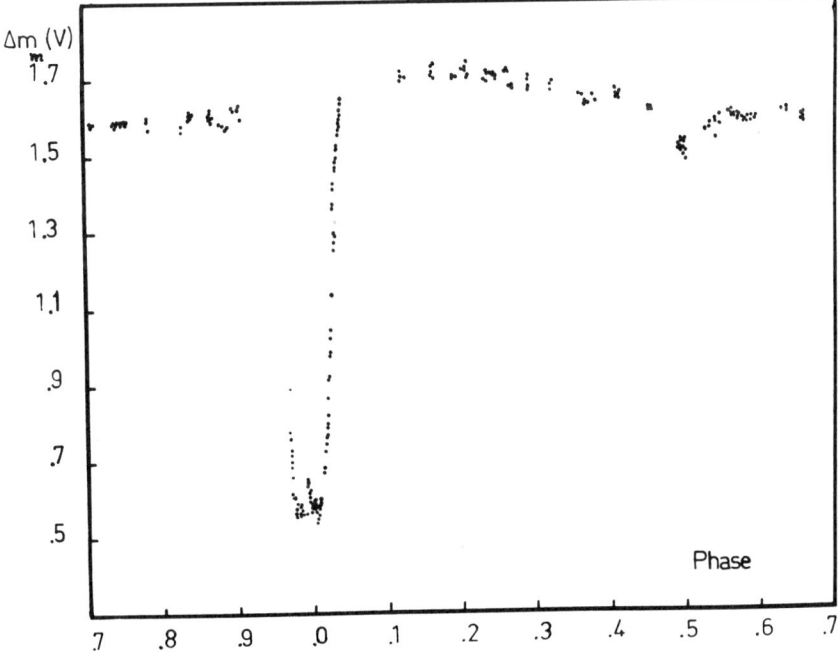
Figure 2 V light curve of RS CVn

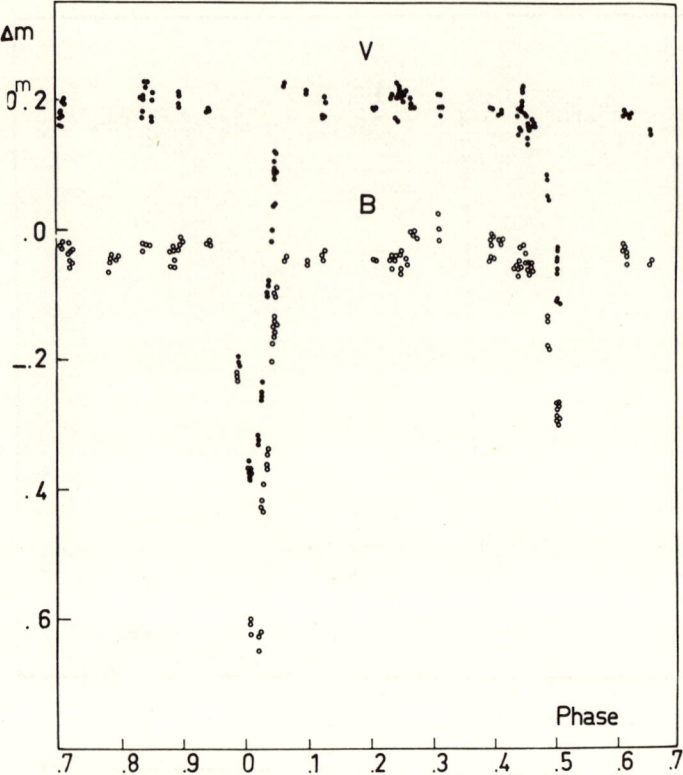

Figure 3   B and V light curves of RT CrB

2.3 Z Herculis

The observations of this system were made in B and V from June 8 to August 18, 1978, and from April 23 to August 20, 1979. For comparison star BD+14°3378 was used as suggested by Hall (1976a). The phases of the individual observations have been computed from the following light elements

Min I = JD Hel. 2441111.8211 + $3\overset{d}{.}9928012$ E .

The light curves of the eclipsing system Z Her obtained in 1978 and 1979 in two colours B and V are shown in Figures 4 and 5. The wave-like distortion outside eclipses is clearly seen and shown in Figures 6 and 7 for 1978 and 1979 respectively. As seen from Figures 6 and 7, the amplitudes of the wave are rather small in both colours, and are about $0\overset{m}{.}02$ in 1978 and about $0\overset{m}{.}03$ in 1979. Moreover, while the minimum of the wave is around the phase $0\overset{p}{.}11$ in 1978, it is around the phase $0\overset{p}{.}95$ in 1979. Hence,

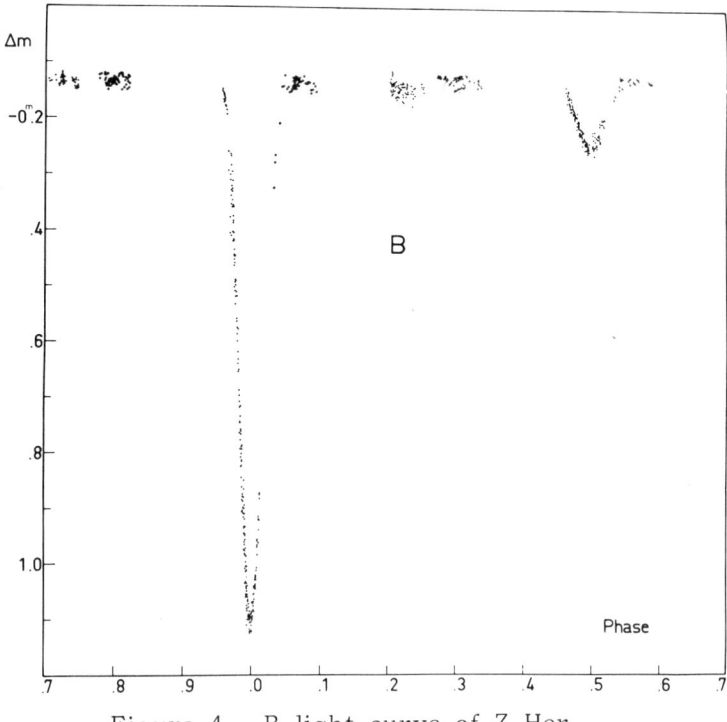

Figure 4 B light curve of Z Her

Figure 5 V light curve of Z Her

a preliminary value for the migration period of the wave has been estimated to be about 6 years.

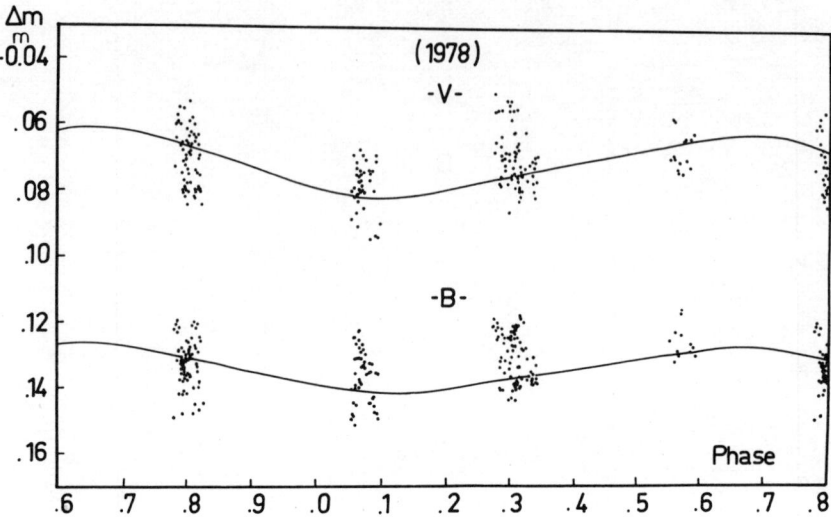

Figure 6   Distortion waves of Z Her in 1978

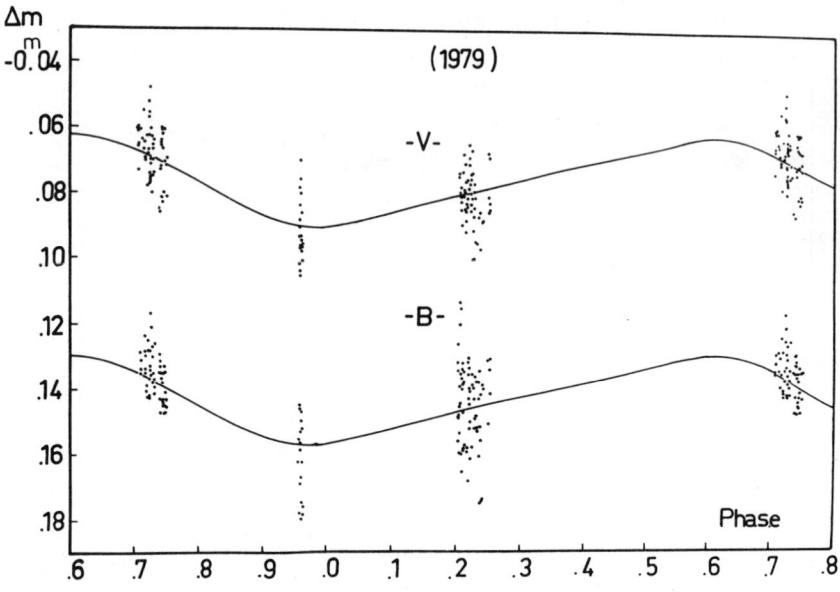

Figure 7   Distortion waves of Z Her in 1979

2.4 RT Lacertae

This system has been observed from July to October in 1978 and from May to November in 1979. For comparison and check stars, BD+43°4108 and BD+43°4109 were used.

The light curves obtained in two colours by Ibanoglu et al (1980) are shown in Figures 8 and 9. The first light curves of the system were obtained by Milone (1968, 1976), first in UBV and later in infrared. The main features noticed in the light curves by Milone were the asymmetry and unequal maxima. These features, although present in the light curves obtained in 1978 and 1979, are inconspicuous. Another feature detected in the present light curves is the noticeable brightening of the system outside eclipses and within primary minimum. The system seemded to brighten about $0^m.12$ in B and $0^m.15$ in V in one year.

The maximum brightnesses at first and third quarters and the minimum brightnesses at mid-primary and mid-secondary with respect to comparison are given in Table 1 for 1965, 1978 and 1979 (reproduced from Ibanoğlu et al, 1980). It should be noted that the brightnesses given in Table 1 for mid-secondary in B for 1965 and in B and V for 1978 are only crude estimates.

TABLE 1
Maximum and Minimum Brightnesses of RT Lac

Year	First Qtr B	V	Third Qtr B	V	Mid-primary B	V	Mid-secondary B	V
1965(a)	$-1^m.32$	$-1^m.56$	$-1^m.20$	$-1^m.45$	$-2^m.20$	$-2^m.50$	$-2^m.05$x	$-2^m.20$
1978(b)	$-1^m.24$	-1.47	-1.18x	-1.43x	-2.13	-2.39	-2.05x	-2.15x
1979(b)	-1.10	-1.33	-1.10	-1.28	-2.04	-2.35	-2.05	-2.20

(a) The values quoted from Milone (1968)
(b) The values quoted from Ibanoğlu et al (1980)
x   Estimated values.

An inspection of Table 1 reveals that while the total brightness of the system remains unchanged at mid-secondary, it varies at mid-primary. According to the preliminary solution given by Milone (1976), K1 companion, the more massive one, is eclipsed at secondary minimum which is an occultation. Therefore, the change in total brightness of the system can be attributed to the cooler (K1 companion).

If we assume the star-spot model proposed by Hall for RS CVn

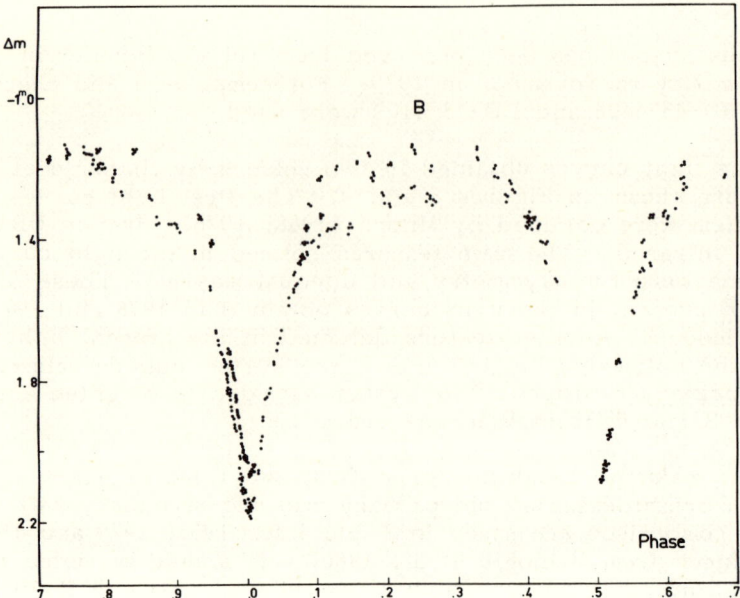

Figure 8 B light curves of RT Lac. Dots denote the observations made in 1978 and the circles denote those made in 1979.

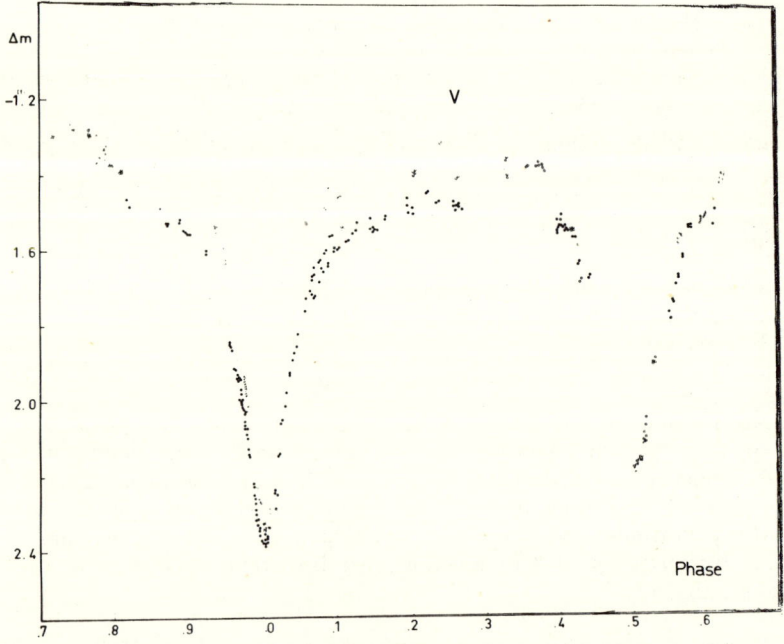

Figure 9 V light curves of RT Lac. Symbols are the same as those used in Figure 8.

type binaries, we can then attribute the brightness increase in RT Lac over a period of one year to the lessening star-spot activity in the primary component. Although the system is brighter in 1979, a slight asymmetry in the maxima of the V light curve suggests that star-spot activity is not completely absent.

From the published data, nothing conclusive could be deduced regarding the period of the migration wave. Several values concerning the migration period, which have been estimated by Hall and Taylor (1971), Hall and Haslag (1976) and Shore and Hall (1978), differ from each other considerably, ranging between 5 and 40 years. Nevertheless, the amplitude of the distortion wave appears to have changed, being smaller in 1978 and 1979, while larger in 1965. However, in order to obtain the most accurate values of migration period and amplitude changes of the distortion wave, accumulated data over a long time are needed.

2.5 AR Lacertae

The eclipsing binary AR Lac was observed in two colours B and V from June to October in 1978 and from May to November in 1979. The light curves obtained by Kurutaç et al (1981) in both colours are shown in Figures 10 and 11.

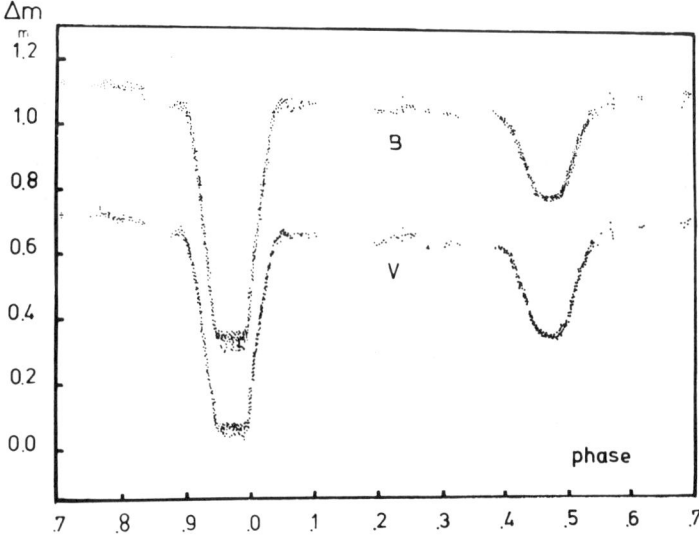

Figure 10 Band V light curves of AR Lac obtained in 1978.

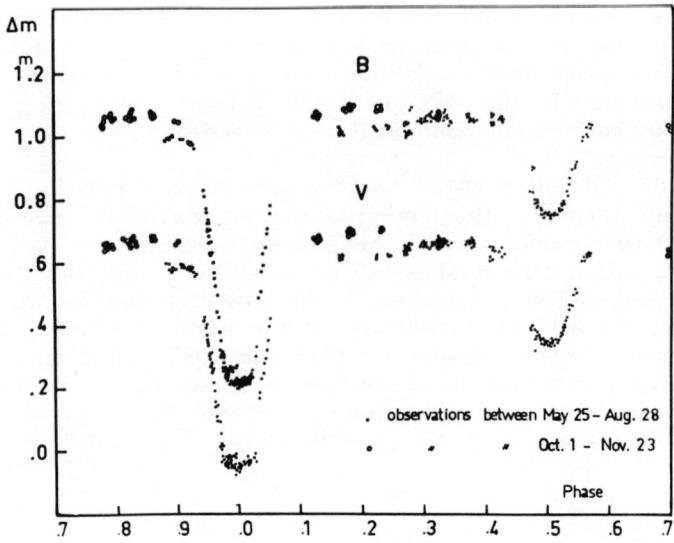

Figure 11   Band V light curves of AR Lac obtained in 1979.

AR Lac shows a significant distortion outside eclipses as well as during totality. A night to night plot of observations during totality is shown in Figures 12 and 13. The depths of primary minima increased in 1978 and continued to increase in the early observing season of 1979, and then started decreasing when it reached the minimum level of brightness during totality in August 1979. This phenomenon can be attributed to the distortion wave moving towards decreasing phases.

The distortion wave outside eclipses, as shown in Figures 14 and 15 is clearly seen in both colours. The amplitudes of the wave in B and V are $0^m.09$ and $0^m.1$ in 1978, and $0^m.12$ and $0^m.13$ in 1979. The minimum of the wave falls at the phase $0^p.40$ in 1978 and around $0^p.0$ in August 1979. Thus, the amount of shift, $0^p.4$, in one year indicates that the migration period of the wave is about 2.5 years on the assumption that no cycles occurred in between two seasons of observations.

2.6 LX Persei

The observations of LX Per were obtained between 2 March and 17 March, 1979 and between 12 July, 1979 and 17 February, 1980. As a comparison star, BD+47°776 was used as suggested by Hall (1976a). The phases of individual observations were computed from the following light elements given in GCVS (1969) as

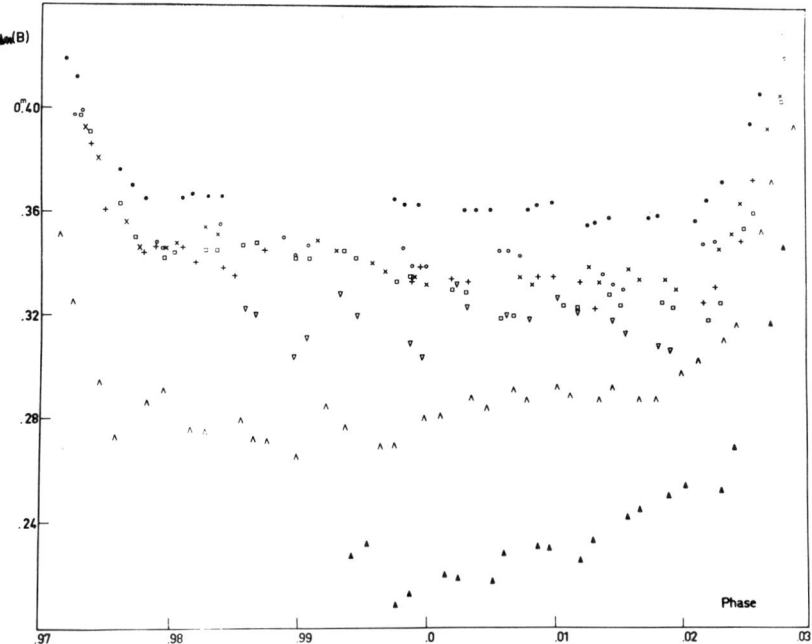

Figure 12 Light variation of AR Lac during totality in B.
Symbols denote the nightly observations:
- • 18 August 1978   + 30 August 1978   ▲ 6 August 1979
- o 24 August 1978   □ 3 Sept 1978   ∧ 28 August 1979
- x 26 August 1978   ▽ 11 Sept 1978

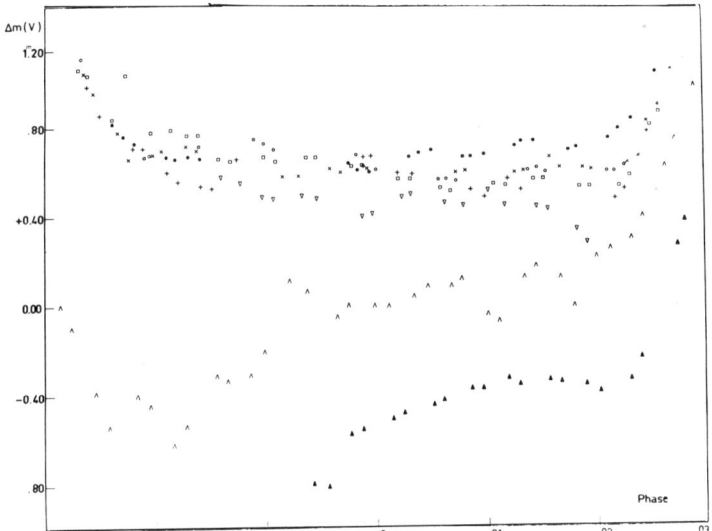

Figure 13 Light variation of AR Lac during totality in V.
Symbols are the same as those used in Figure 12.

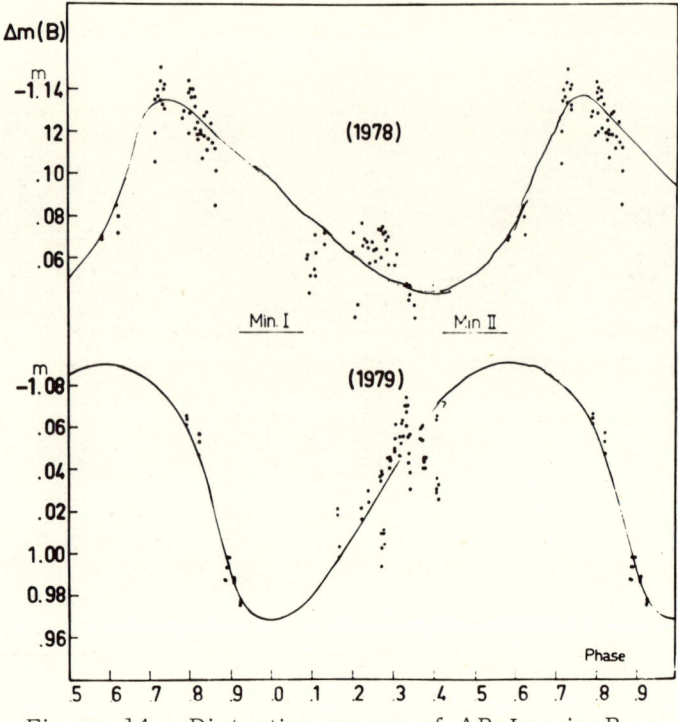

Figure 14 Distortion waves of AR Lac in B

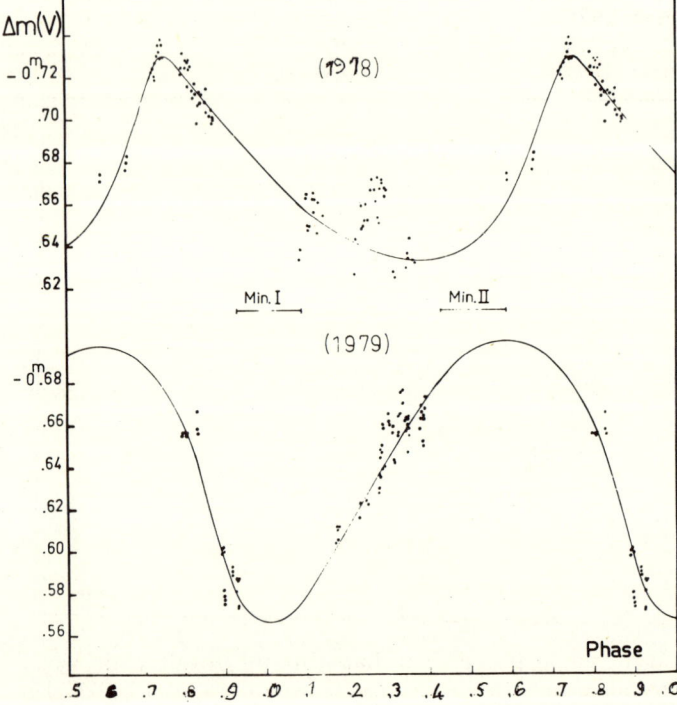

Figure 15 Distortion waves of AR Lac in V

Min I = JD Hel.24 27033.12 + $8^d.038044$ E .

Nightly observations were taken as a mean and plotted against phase. Figure 16 shows the distortion waves in two colours. The numbers adjacent to the individual observations on Figure 16 denote the month and the day of that particular observation made (i.e., March 17 has been shown as 3/17). The dashed curves represent the distortion waves in March 1979 and November 1979. The amplitudes of the wave are about $0^m.04$ and $0^m.06$ in both colours for March 1979 and November 1979 respectively. As seen from Figure 16, the distortion wave moves towards decreasing phases. From the amount of shift $0^p.13$, in seven months the migration period has been estimated to be about seven years. Further observations of this system are planned in future.

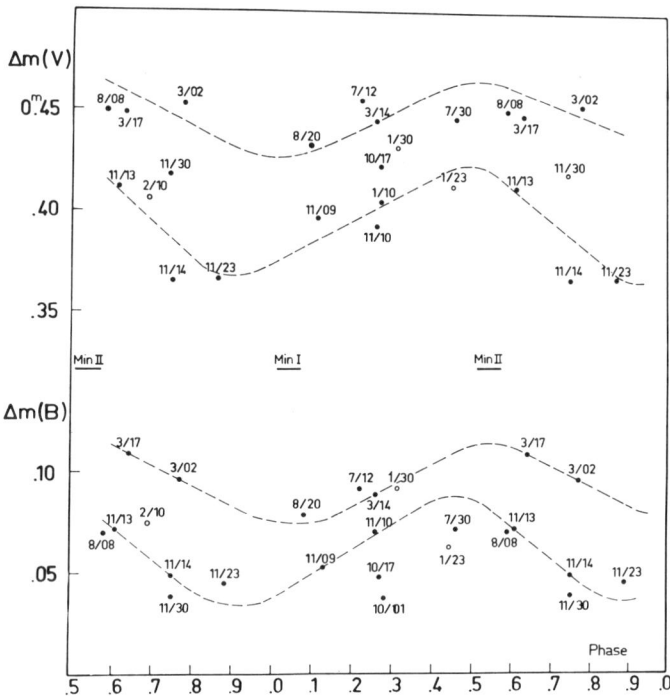

Figure 16   Distortion waves of LX Per in B and V.

2.7 SZ Piscium

The eclipsing binary SZ Psc was observed from July to December, 1979. For comparison star, HD 219150 was used as suggested by Jakate (1979). The light curves obtained in both colours and recently published by Tümer et al (1980) are reproduced and shown in Figure 17.

As seen from Figure 17, the mid-primary is displaced and falls around the phase 0^P05. No observations falling within secondary minimum have been obtained, thus making it difficult to estimate the mid-secondary. However, from the shoulders of secondary minimum it seems that mid-secondary is around the phase 0^P52 and not separated by the half period. This may be the consequence of either an eccentric orbit or a migrating wave as generally proposed by Hall (1976) for RS CVn-type binaries.

The asymmetric and unequal maxima are the most distinguished features noticed in the light curves of SZ Psc. In order to provide an answer to the comparable light variation of SZ Psc in maxima, more observations are greatly needed in future.

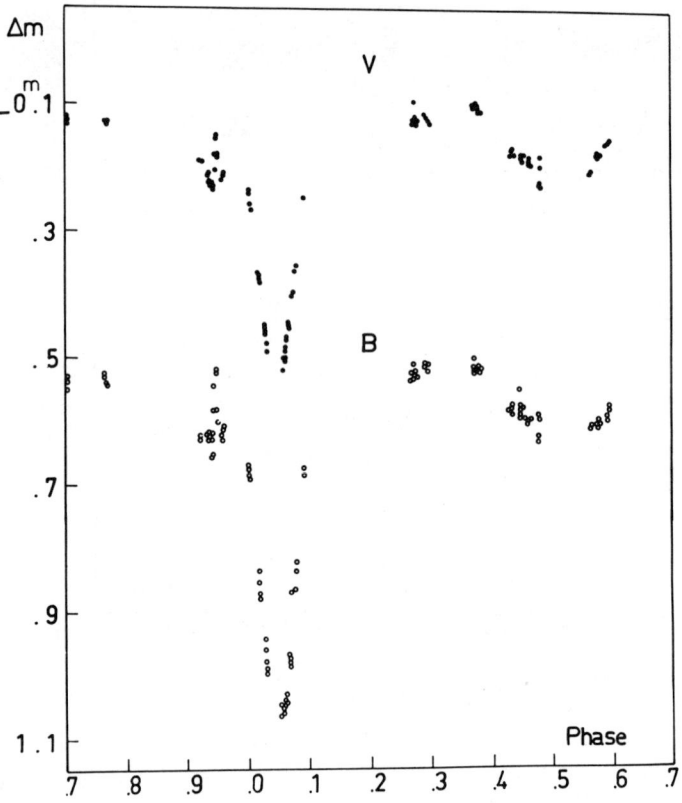

Figure 17 Light curves of SZ Psc in Band V.

Part of this research was supported by the Scientific and Technical Research Council of Turkey.

REFERENCES

Ertan, A. Y., Tunca, Z., Evren, S., Tümer, O., Kurutaç, M. and
    Ibanoğlu, C.: 1980, I.B.V.S. No. 1578.
Evren, S., Tunca, Z., Tümer, O., Ertan, A. Y., Ibanoğlu, C. and
    Kurutaç, M.: 1980, I.B.V.S. No. 1737.
Hall, D. S.: 1972, Publ. Astron. Soc. Pacific, 84, p.323.
Hall, D. S.: 1976, *Multiple Periodic Variable Stars*, I.A.U. Colloquium No. 29, Budapest.
Hall, D. S.: 1976a, I.A.U. Comm. No. 42, Circ. No. 2.
Hall, D. S. and Haslag, K. P.: 1976, *Multiple Periodic Variable Stars*, I.A.U. Colloquium No. 29, Budapest.
Hall, D. S. and Taylor, M. C.: 1971, Bull. A. A. S., 3, p.12.
Ibanoğlu, C., Kurutaç, M., Tümer, O., Evran, S., Tunca, Z. and
    Ertan, A. Y.: 1980, Astrophys. Space Sci., 72, p.61.
Jakate, S. M.: 1977, I.B.V.S. No. 1578.
Milone, E. F.: 1968, Astron. J., 73, p.708.
Milone, E. F.: 1976, Ap. J. Suppl., 31, p.93.
Kurutaç, M., Ibanoğlu, C., Tunca, Z., Ertan, A. Y., Evren, S.
    and Tümer, O.: 1981, Astrophys. Space Sci. (in press).
Shore, S. and Hall, D. S.: 1978, Report at RS CVn Workshop,
    Socorro, N.M.
Tümer, O., Kurutaç, M., Tunca, Z., Evren, S., Ertan, A. Y. and
    Ibanoğlu, C.: 1980, I.B.V.S. No. 1741

OBSERVATIONS AND ANALYSIS OF UU Sge

E. Budding

Department of Astronomy,
University of Manchester, England

ABSTRACT

Raw observational data has been processed, and a corresponding optimal curve fit is presented for the system UU Sge - the hot subdwarf containing the central star of the planetary nebula Abell 63. Parameter values are compared with previous estimates.

The system is required to be rather more massive than expected if the secondary is a Main Sequence star or "contact" component. This requirement can be reduced if the secondary relative radius r_2 is increased over its most probable (photometric) value, or if the secondary is itself somewhat overmassive for its radius.

1. INTRODUCTION

That the central stars of planetary nebulae should be searched for evidence of close binarity was suggested to the author by Z. Kopal a number of years ago. At that time, however, attention had not been drawn to the existence of UU Sge, the short period binary at the centre of planetary nebula Abell 63. Also the photometric capabilities really available to the Manchester group until recently would have been insufficient to carry out accurate or continuous photometry of more than a small number of candidate stars.

With the introduction of an improved detector at the Kottamia Station of Helwan Observatory (Egypt) as a user facility for this group eighteen months ago, significant advances became possible. The 14th magnitude star UU Sge was not difficult to monitor with this system, though, for reasons discussed elsewhere in these

Advanced Study Institute Proceedings by Sedmak, it was only feasible, in the interests of time resolution and accuracy, to select one waveband, for the single channel D.C. mode apparatus employed.

The significance of UU Sge as a short period eclipsing binary containing an early subdwarf component has been pointed out elsewhere (Bond et al, 1978; Budding and Kopal, 1980). The purpose of the present short paper is to add slightly to the account of observations given by Budding and Kopal (1980), such as by inclusion of a finding chart on which various comparison and check stars have been marked (Figure 1), to mention some pre-analysis processing of the raw data and to present a set of parameters which optimize the fit of a function characterising eclipse and proximity effects in close binary systems (based largely on expressions which can be found in Kopal's books (1959, 1979)), along lines described in some detail by Budding and Najim (1980).

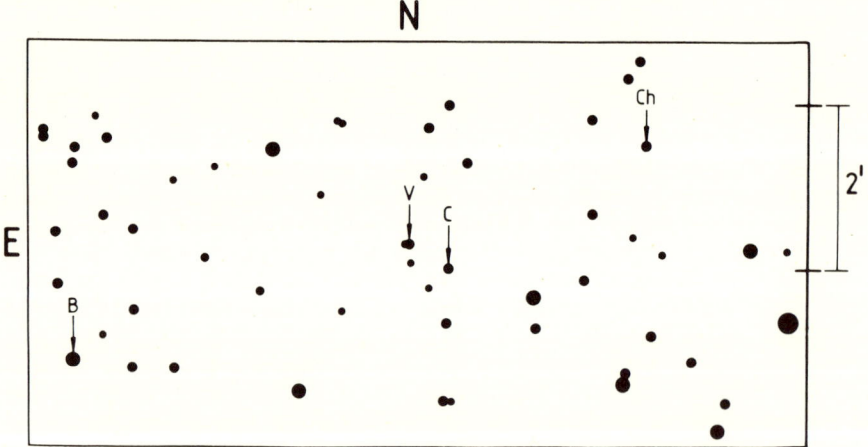

Figure 1 Chart showing the field of the variable UU Sge, marked V on the diagram. Comparison and check star used by the author are marked by C and Ch respectively, while B marks the comparison star used by Bond et al (1978).

2. OBSERVATIONS - PROCESSING

The raw data for the V light curve were presented by Budding and Kopal (1980) (hereafter referred to as BK). The approach to data reduction involved dividing the light curve into three distinct regions: (1) around primary minimum (phase range 0.9 - 1, 0 - 0.1); (2) secondary minimum (0.4 - 0.6); and (3) outside minima (the remaining phase ranges). Within regions (1) and (2) the data were binned in 2° intervals of phase, within region (3) the interval was 4°. The point of this is that theoretical photometric behaviour (on which the fitting function is based) should be relatively showly varying and smooth outside minima. Within minima the variation with phase can be very rapid (even with a discontinuous derivative in certain situations), but the scale of noise in the original data ($\sim 0\overset{m}{.}04$) means that the slight smoothing associated with 2° binning for these phase ranges is unlikely to introduce any significant information loss. The reduction in the number of data points achieved in this way (down to about one-quarter the original) allows a distinct improvement in computer economy and hence flexibility of experimentation with the fitting function.

The actual reduction was effected by taking the arithmetic mean of phase values within each data bin, together with the geometric mean of the corresponding magnitudes.

3. ANALYSIS AND DISCUSSION

The reduced data were analysed with the aid of the "16-parameter" optimization algorithm which has been described by Budding and Najim (1980). In the present case only 8 of the available parameters have actually been optimized, the two eccentricity parameters being regarded as inapplicable, while the two linear limb darkening coefficients were taken from the tables of Al-Naimiy (1978) on the basis of temperature estimates which will be discussed subsequently. The mass ratio and gravity darkening coefficients have very little influence indeed, since the relatively bright and massive subdwarf must be sensibly undistorted and its own reflection coefficient E_1 has no significant role to play in view of the great difference in temperature of the two stars. The semi-empirical parameter E_2, which governs the scale of the very noticeable reflection effect is, however, calculable (particularly since the large scale of the effect makes a correlative simulation of the same effect by a suitable combination of other parameters unlikely). The measure of agreement between observations and fitting function concerning this parameter is encouraging, though one should perhaps be on guard against taking the simple linearized theory, which the use of this parameter implies, too seriously.

The full list of determined parameters is given in Table I and the observational data and corresponding theoretical curves shown in Figure 2.

It will be observed that the solution in BK, calculated on the basis of frequency domain techniques, lies within the (standard deviation) error assessments of the present solution, which has been calculated (as explained in Budding and Najim, 1980) from inversion of the curvature Hessian at optimum. On this basis, BK underestimated the probable scale of such errors.

Table I Optimal Parameter Set

Parameter	Value	Error Assessment
L_1	0.618	0.014
L_2	0.043	0.04
L_3	0.339	0.04
r_1	0.131	0.005
r_2	0.210	0.068
i	88°4	1°9
u_1	0.27	-
u_2	0.71	-
$\Delta \theta_o$	-0°218	0°13
q	0.2	-
T_1	30000	-
T_2	5000	-
τ_1	0.37	-
τ_2	1.32	-
E_1	0.04	-
E_2	12.75	0.40
χ^2	105.0	-
N	116.0	-
$\Delta \ell$	0.28	-

Footnote to Table I:

The parameters are indicared by conventional representation; fractional luminosities L_1, L_2, L_3; relative radii r_1, r_2; inclination i; limb darkening coefficients u_1, u_2; correction to zero phase $\Delta \theta_o$; mass ratio $q(=M_2/M_1)$; temperatures T_1, T_2; gravity-darkening coefficients τ_1, τ_2; reflection coefficients E_1, E_2; goodness of fit criterion χ^2; number of data points N; assumed accuracy of a single data point $\Delta \ell$.

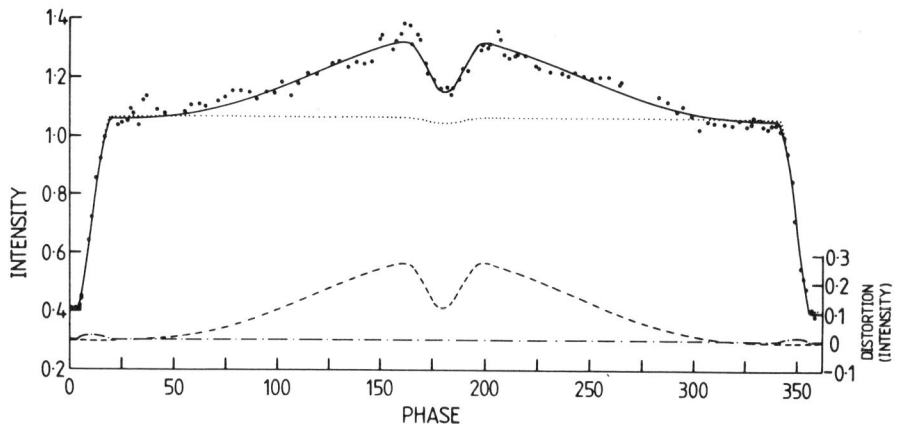

Figure 2 "Normal points" and optimal curve fit for UU Sge. The light distortions shown below the observational data are added to the spherical model (dotted) to give the final curve (continuous). The primary star gives essentially no distortion effect (dot-dashed), the apparent effect in primary minimum is due to the "boundary correction" terms associated with departure of the secondary's outline from circularity.

The solution published by Bond et al is, however, significantly different. The Kitt Peak observations of primary minimum on which their analysis was based are superior to our data for the minimum, though their procedure (which they describe as "preliminary") may have been rather approximate and the primary limb darkening coefficient, which is known to correlate with the geometric elements, was probably overestimated at 0.5. We have not yet analysed Bond et al's data, so cannot say anything definite about whether there could be any real difference in the elements for the two epochs, though the possibility of some intrinsic variability in the system was mentioned in BK.

The slight discrepancy at zero phase $\Delta \theta_o$ suggests that the observed minimum occurred very slightly after the predicted time according to the ephemeris of Bond et al (1978), though the timing accuracy, perhaps influenced by slight asymmetries, is not very high. The epoch of minimum is presented as HJD 244073.3546 ± 0.0054 which is within the probable errors of Bond et al's ephemeris.

The temperature ratios considered in BK are rather less than that estimated by Bond et al. A reasonable compromise occurs with the choice given in Table I. These temperatures produce L_1 and

L_2 values in rather a higher ratio than that following from Table I, but in view of the high uncertainty of L_2 this need not be so worrying. These temperatures reproduce the same value of reflection coefficient E_2 (= 12.8) which if the heating of the secondary could be regarded as a linear perturbation accounts for the scale of the reflection effect (as discussed, for example, in Budding and Ardabili, 1978) according to the formula (black-body approximation)

$$E_2 = \frac{c_2}{4\lambda T_2} \left(\frac{T_1}{T_2}\right)^4 \frac{e^{\frac{c_2}{\lambda T_2}}\left(e^{\frac{c_2}{\lambda T_1}} - 1\right)}{\left(e^{\frac{c_2}{\lambda T_2}} - 1\right)^2}.$$

Since, in fact, the scale of the reflection requires an order of magnitude increase in the mean flux from unilluminated to illuminated hemispheres of the secondary, the entailed surface temperature increase rises to about 100% - hardly what one would normally associate with the scale of a linear perturbation. The quality of the fitting depends on the shape of the "reflection" curve, not only its amplitude, and the consistency of this over practically the whole range of phases suggests that the applied linearization may not be that bad. A more detailed study of this problem should be made in the future.

If the secondary is not far from MS dimensions and we retain 6000° as the temperature of the third star, for the sake of argument, it is required to be about one-fifth the distance to the binary. At 200 pc. the required interstellar reddening seems to be too large (~ 0.4) to go from the intrinsic to observed B-V values ((B-V) $\sim 1^m$ for the optical comparison star). It is very probable, therefore, that this object is one of the numerous cool red dwarfs and essentially unreddened. Any possible variability of this star would inject another source of uncertainty into the analysis.

Arguments with regard to masses and radii are summarized in Figure 3, where curves (a) and (b) show the relationship between $\log(1 + q)$ and $\log R_2$ according to the hypothesis that the secondary is a Main-Sequence star (a), and that it conforms to the contact configuration (b), respectively. The corresponding formulae for these curves, in which r_2 acts as a parameter, are

(a) $\log(1 + q) = 1.5 \log R_2 - \log q + 3 \log r_2 +$ constant,

and

(b) $q \equiv q(r_2) =$ constant.

The lowest (unilluminated) temperature for the secondary that has been considered is consistent with a spectral class about K2, which for a Main-Sequence star corresponds to the vertical dashed line.

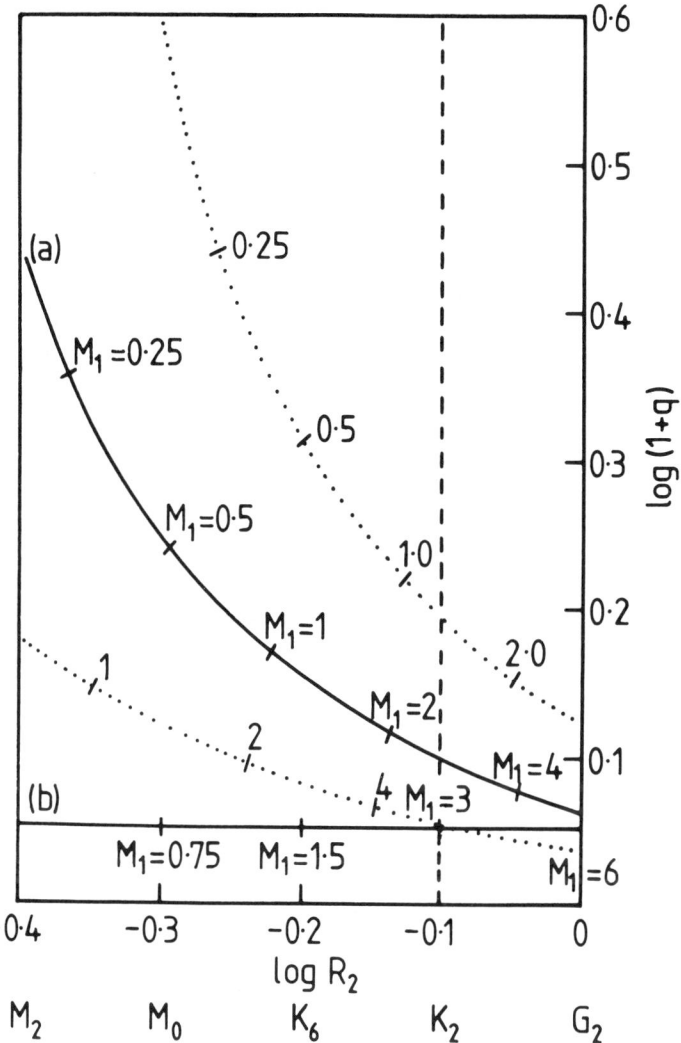

Figure 3 Curves showing the relationship between secondary radius R_2, in solar units, and mass ratio q obtained essentially from combining Kepler's law with (a) the Main-Sequence mass-radius relation (Allen, 1973); and with (b) the r_2:q relation required for the "contact" configuration according to Kopal (1959). The continuous lines adopt the r_2 value given in Table I, while the dotted lines on either side of curve (a) correspond to r_2=0.25 (above) and r_2= 0.17 (below). Appropriate primary mass values, in solar units, are marked at various positions along the curves (symbol M_1 is omitted for the dotted curves). Main-Sequence spectral types are indicated below the log R_2 axis, and the vertical dashed line corresponds to the lowest (unheated) secondary temperature that has been considered reasonably likely.

The photometric solution gives a value of 0.21 to r_2, which, in turn, implies that the primary's mass is somewhere in the rather high-mass region to the right of the dashed line. An undermassive "contact" secondary seems to require still higher total mass values if this star is to have anything like "subgiant" properties.

The requirement for a high mass primary can be eased somewhat if the secondary relative radius r_2 is increased. This is quite possible since r_2 appears to be a rather poorly determined quantity from Table I. Alternatively, or perhaps together with a larger r_2, if the secondary is more massive than a Main Sequence star of the same radius, as envisaged in the evolutionary scheme of Paczyński (1976), the need for a high mass primary can be relaxed.

Bond et al (op.cit.) referred to several objects that may have some evolutionary similarity to UU Sge. Further possibilities are mentioned by Saltzman et al (1980), who have apparently considered the close binary mechanism for planetary nebula production in more detail. In a similar spirit we would call attention to the system LB 3459 (Kilkenny et al, 1978; Paczyński and Dearborn, 1980; Hilditch and Kilkenny, 1980), which appears to show some morphological similarity to UU Sge.

REFERENCES

Allen, C. W.: 1973, *Astrophysical Quantities*, Athlone Press, London.
Al-Naimiy, H. M.: 1978, Astrophys Space Sci., 53, p.181.
Bond, H. E., Liller, W. and Mannery, E. J.: 1978, Astrophys. J., 223, p.252.
Budding, E. and Ardabili, Y. R.: 1978, Astrophys. Space Sci., 59, p.19.
Budding, E. and Kopal, Z.: 1980, Astrophys. Space Sci., 73, p.83.
Budding, E. and Najim, N. N.: 1980, Astrophys. Space Sci., 72, p.369.
Hilditch, R. W. and Kilkenny, D.: 1980, Mon. Not. Roy. Astron. Soc., 192, p.15.
Kilkenny, D., Hilditch, R. W. and Penfold, J. E.: 1978, Mon. Not. Roy. Astron. Soc., 183, p.523.
Kopal, Z.: 1959, *Close Binary Systems*, Chapman and Hall, London.
Kopal, Z.: 1979, *Language of the Stars*, D. Reidel Publ. Co., Dordrecht.
Paczyński, B.: 1976, in IAU Symposium No.73, *Structure and Evolution of Close Binary Systems* (ed. P. Eggleton, S. Mitton and J. Whelan), D. Reidel Publ. Co., Dordrecht,.p.7
Paczyński, B. and Dearborn, D. S.: 1980, Mon. Not. Roy. Astron. Soc., 190, p.395.
Saltzman, J., Livio, M. and Shaviv, G.: 1980, in IAU Symposium No. 88, *Close Binary Stars: Observations and Interpretation* (ed. M.J.Plavec, D.M.Popper and R.K.Ulrich, D. Reidel Publ. Co., Dordrecht, p.571.

THE LIGHT AND PERIOD VARIATIONS OF OO AQUILAE

Osman Demircan and Necdet Güdür

Ege University Observatory,
Izmir, Turkey.

ABSTRACT

Photoelectric B and V. observations of a W Ursae Majoris system OO Aquilae are presented. Its light and period variations are considered. The period was found to have undergone an abrupt change in 1960 of about one part in 10^4. Plausible periodc fluctuations of the times of minima with an (approximate) 6-years period and $0\overset{d}{.}005$ amplitude may be attributed to a nodal regression of the relative orbit. Thus this system is the first and only candidate so far for which the observations of the times of the minima suggest the effects of nodal regression of its orbit.

1. OBSERVATIONS

The 21 nights minimum observations for the eclipsing binary OO Aquilae (BD+8°4224) have been assembled. All these observations which cover the whole light curve were carried out with the Cassegrain telescope at the Ege University Observatory between July, 1968 and September, 1974. Observations of the first 10 nights of the minima which have been used for the study of period changes were carried out without any filter. The subsequent 11 nights observations from September, 1972 onwards were made simultaneously in blue and yellow light. The filtered set of observations refer (approximately) to the standard UBV system. A total of 309 observations have been obtained in each colour. BD+8°4220 was used as comparison star. From its observations, the differential extinction coefficients for each night were derived. The method of Hardie (1962) was used for this purpose, and observations were corrected for the differential extinction. The corrections were unneces sarily small, because the comarison star is in close proximity of OO Aql.

The phases were evaluated with the light elements (cf. Kukarkin et al, 1971) given as

Min I JD Hel. 2440 522.294 + $0.^{d}5067887$ E.

The observations are listed in Table I. The light curve shows that the eclipses are partial. The shape and the depth of the minima are almost free from the wavelength of observations (see B-V curve). The depths of the minima are almost the same which causes confusion in definition of the primary and secondary minima; for this confusion see, for example, Pohl (1969). The maximum loss of light ($1-\lambda$) of about one magnitude in the primary eclipse appears still the largest one observed as yet for W UMa systems. The maximum light at both quadratures appears to be the same. The light curves in both wavelength are variable in time which causes problems for the analysis of the light changes. For this reason we did not attempt to solve for the eclipse elements. The eruptive prominences occurring on one of the components may be responsible for the variability of the light curves and may also be for the period changes.

2. PERIOD CHANGES

Observations of the minima of this binary have been carried out at the Ege University Observatory since 1968, and a total 9 primary, 11 secondary minima were observed (10 of them in both B and V colours while the rest are unfiltered observations). Together with these, 437 individual times of minima were collected (see Table II). The columns 1 and 2 in Table II have been evaluated from the light elements given in the first section above. The meaning of the columns is self-explanatory. The O-C's in this table indicate the variability of the period. The times of the two different types of minima are within the errors of the observations, which eliminates at once the possibility of apsidal motion in the system; and so it does any eccentricity of the orbit (this result is, in fact, obvious if we assume that the system is a truly contact binary).

A plot of the times of the minima against the epoch bears out two different features; one is the changing slope in the ±960's; and the other is the plausible periodic form of the whole curve. The change in the slope of the curve is decidedly caused by an abrupt decrease in the period of the system in the 1960's with a magnitude of about one part in 10^4 of the period. If we assume that a large eruptive prominence was responsible for this abrupt change in the period, then the active component should be the more massive one. Morover, if the ejected material is captured by the less massive component, we may assume that the total mass and the angular momentum of the system are conserved and in this case,

$$\frac{\Delta P}{P} = 3\frac{2\mu-1}{\mu(1-\mu)} \frac{|\Delta m_1|}{m}$$

(cf. Kuiper, 1941; and many other sources), where $m_{1,2}$ are the masses of the mass ejecting and the mass receiving components, respectively; $m = m_1 + m_2$, $\mu = m_2/m$, and ΔP is the change in the period produced by the transfer of the amount of mass Δm_1. The observed value of $\Delta P/P$ for OO Aql is about 10^{-4}. If the total mass $m_1 + m_2 = 2.5\ m_\odot$ according with Lucy's (1968) A model for W UMa systems, the quantity μ is somewhat arbitrarily assumed to be $2/3$ (it must lie between 0.5 and 1) then

$$10^{-4} = (9/5)\Delta m_1$$

or

$$\Delta m_1 = 5.56 \times 10^{-5}\ m_\odot$$

is obtained for the matter ejected from the more to the less massive component of OO Aql in about 1960. It may also be speculated that if Lucy's (1976) revised theory about the contact nature of these systems is valid, then the year 1960 may be an important year in the history of OO Aql; a "contact on" or "contact off" phase may have started at that time.

The plausible periodic fluctuations of the times of minima show a period of about six years; which may be attributed to the presence of a third body in the system, or a nodal regression of the relative orbit. If, for example, we assume that the hypothetical third body is a solar-like star, it must be in close proximity (~ 5 AU) of the eclipsing pair. The possibility of its presence will be left as an open question. We shall next turn our attention to a more exciting second possibility for the cause of the periodic fluctuations in the O-C diagram: namely, a nodal regression in the relative orbit. If this is the case, it would mean (cf. Kopal, 1978; p.260) that we deal with a very young system, in which secular action as a dissipative process did not have time yet to "rectify" the axis of rotation of both components. For this binary we can safely postulate with the aid of the colour-temperature relation (cf., e.g., Allen, 1976, p.206) for the normal stars that the components of OO Aql may still be in the pre-Main Sequence stage which supports the fission theory for the formation of binaries.

Thus, it seems most likely that OO Aql is the first and only candidate so far whose observations of the times of minima transpires the nodal regression in its orbit. The period of regression is found to be about six years with an amplitude (which according to Kopal's theory is always finite even for circular orbits) of about 0.005 days (cf. Kopal, 1978).

3. CONCLUSIONS

The properties of OO Aql which can be deduced from the observed light and period variations have been discussed to some extent in the first and second sections of the present paper. The eruptive prominences occurring on the more massive star were mentioned as the primary cause of both the changes in the light and the period. The active component is more probably the eclipsing one in the primary minimum. The main reason which led us to this conclusion was the more visibility of the light changes in the primary minimum, where mostly the eclipsing star is photometrically effective. If this reason turns out to be only statistical (since we have less observations on the secondary minimum), but not real, then in the above conclusion the active component might be eclipsed in primary minimum. For confirmation of this idea some further observations, especially in the secondary minimum, are required.

Another outcome of the period changes concerns the behaviour of the O-C difference of the minimum observations. It is known that the small random variations in the intervals between successive minima can build up to a large, apparently systematic, variation; since we also know that the variable light curve during eclipse could evidently introduce some random errors into the determination of the time of minimum. Thus, the observed behaviour of the O-C diagram may not be real; and, for this reason, especially the periodic changes of the period should not be considered as definitely established until they have been observed over many more cycles. In this connection, for the new observations of the minimum times the new elements

$$\text{Min I JD Hel. } 24\ 42\ 218.51607 + 0.^{d}5067848\ E$$

have been derived by the least squares from ten minimum observations (given in Table II) of the authors. A final comment concerns the abrupt decrease in the period of the system estimated to have taken place around 1960. An inspection of the periods listed in the literature for the system OO Aql permits us to speculate that the extended continuous existence of the eruptive prominences on the more massive component is more acceptable, rather than one large eruption in 1960. The eruptions may have started in 1955 and lasted about 10 years.

REFERENCES

Allen, C. W.: 1973, *Astrophysical Quantities*, (3rd ed.), The Athlone Press, Univ. of London.
Ashbrook, J.: 1952, Astron. J., 57, p.259.
Ashbrook, J.: 1953, Astron. J., 58, p.171.

Baldwin, M. E.: 1973, I.B.V.S. No. 795.
Binnendijk, L.: 1968, Astron. J., 73, p.32.
Braune, W. and Quester, W.: 1962, Astron. Nachr., 286, p.209.
Braune, W. and Hübscher, J.: 1967, Astron. Nachr., 290, p.105.
Domke, K. and Pohl, E.: 1952, Astron. Nachr., 281, p.113.
Dueball, J. and Lehmann, P. B.: 1965, Astron. Nachr., 288, p.167.
Flin, P.: 1971, I.B.V.S. No. 584.
Flin, P.: 1972, I.B.V.S. No. 740.
Hardie, R. H.: 1962, *Astronomical Techniques in Stars and Stellar Systems*, Vol. II, Univ. of Chicago Press, p.178.
Herczeg, T.: 1972, I.B.V.S. No. 699.
Kizilirmak, A. and Pohl, E.: 1971, I.B.V.S. No. 530.
Kizilirmak, A. and Pohl, E.: 1974, I.B.V.S. No. 937.
Klimek, Z.: 1973, I.B.V.S. No. 779.
Kopal, Z.: 1978, *Dynamics of Close Binary Systems*, D. Reidel Publ. Co., Dordrecht and Boston.
Krobusek, B. A. and Mallama, A. D.: 1975, I.B.V.S. No. 954.
Kuiper, G. P.: 1941, Astrophys. J., 93, p.133.
Kukarkin, B. V., Kholopov, P. N., Efremov, Yu. N., Kukarkina, N. P., Kurochkin, N. E., Medvedeva, G. I., Perova, N. B., Pskovsky, Yu. P., Fedorovich, V. P. and Frolov, M. S.: 1971, *First Supplement to the 3rd Edition of the General Catalogue of Variable Stars*, Mowcos.
Lucy, L. B.: 1968, Astrophys. J., 151, p.1123.
Lucy, L. B.: 1976, Astrophys. J., 205, p.208.
Mallama, A. D., Skillman, D. R., Pinto, P. A. and Krobusek, B. A.: 1977, I.B.V.S. No. 1249.
Pohl, E. and Kizilirmak, A.: 1964, Astron. Nachr., 283, p.69.
Pohl, E.: 1969, I.B.V.S. No. 391.
Pohl, E. and Kizilirmak, A.: 1972, I.B.V.S., No. 647.
Pohl, E. and Kizilirmak, A.: 1975, I.B.V.S., No. 1053.
Pohl, E. and Kizilirmak, A.: 1977, I.B.V.S., No. 1358.
Robinson, L. J.: 1965a, I.B.V.S., No. 111.
Robinson, L. J.: 1965b, I.B.V.S., No. 114.
Robinson, L. J.: 1965c, I.B.V.S., No. 119.
Robinson, L. J.: 1966, I.B.V.S., No. 154.
Robinson, L. J.: 1967, I.B.V.S., No. 180.
Robinson, L. J. and Ashbrook, J.: 1968, I.B.V.S., No. 247.
Rudolph, R.: 1960, Astron. Nachr., 285, p.161.
Stephan, C. P.: 1977, I.B.V.S., No. 1350.
Stephan, C. P.: 1978, I.B.V.S., No. 1502.

Table I Unrectified observational points of OO Aquilae.

JD Hel	Phase	m(B)	m(V)
4 41 571.2749	0.8585	0.360	0.765
.2813	0.8711	0.330	0.730
.2856	0.8796	0.290	0.735
.2910	0.8902	0.290	0.695
.2972	0.9025	0.240	0.660
.3032	0.9143	0.180	0.595
.3083	0.9244	0.120	0.535
.3163	0.9401	0.015	0.445
.3227	0.9528	−0.090	0.350
.3285	0.9642	−0.160	0.250
.3347	0.9765	−0.265	0.150
.3401	0.9871	−0.360	0.065
.3451	0.9970	−0.400	0.015
.3507	0.0080	−0.350	0.075
.3571	0.0207	−0.245	0.165
.3624	0.0311	−0.170	0.255
.3674	0.0410	−0.090	0.350
.3735	0.0530	0.025	0.430
.3789	0.0637	0.065	0.475
.3849	0.0755	0.145	0.560
.3914	0.0883	0.200	0.625
.3980	0.1014	0.250	0.665
.4045	0.1142	0.285	0.695
.4105	0.1260	0.300	0.735
.4161	0.1371	0.365	0.740
.4257	0.1560	0.360	0.770
24 41 626.2324	0.3011	0.410	0.815
.2386	0.3133	0.405	0.810
.2439	0.3238	0.405	0.805
.2503	0.3364	0.360	0.770
.2566	0.3488	0.355	0.740
.2630	0.3615	0.340	0.705
.2699	0.3751	0.295	0.685
.2758	0.3867	0.265	0.670
.2817	0.3984	0.225	0.620
.2892	0.4132	0.155	0.565
.2952	0.4250	0.105	0.500
.3019	0.4382	0.010	0.425
.3076	0.4495	−0.075	0.325
.3149	0.4639	−0.170	0.235
.3204	0.4747	−0.265	0.145
.3265	0.4868	−0.345	0.055
24 41 890.2956	0.3530	0.390	0.780
.2979	0.3575	0.370	0.770
.3021	0.3658	0.355	0.745

JD Hel	Phase	m(B)	m(V)
24 41 890.3088	0.3791	0.330	0.740
.3112	0.3838	0.320	0.710
.3156	0.3925	0.290	0.680
.3179	0.3970	0.250	0.650
.3221	0.4053	0.225	0.640
.3252	0.4114	0.205	0.620
.3337	0.4282	0.075	0.525
.3381	0.4369	0.030	0.455
.3425	0.4455	−0.020	0.390
.3446	0.4497	−0.050	0.365
.3503	0.4609	−0.140	0.280
.3593	0.4787	−0.300	0.115
.3672	0.4943	−0.450	0.000
.3757	0.5111	−0.450	0.000
.3784	0.5164	−0.400	0.030
.3836	0.5266	−0.330	0.115
.3889	0.5371	−0.220	0.220
.3912	0.5416	−0.140	0.270
.3964	0.5519	−0.090	0.330
.3998	0.5586	−0.060	0.390
.4028	0.5645	−0.015	0.415
.4073	0.5734	0.050	0.470
.4099	0.5785	0.080	0.495
.4156	0.5898	0.145	0.560
.4189	0.5963	0.185	0.600
24 41 922.2812	0.4672	−0.218	0.230
.2870	0.4787	−0.318	0.118
.2909	0.4864	−0.383	0.065
.2960	0.4964	−0.442	0.011
.3001	0.5045	−0.440	0.008
.3065	0.5171	−0.370	0.053
.3104	0.5248	−0.275	0.130
.3142	0.5323	−0.199	0.198
.3195	0.5428	−0.120	0.284
24 41 940.2667	0.9564	−0.105	0.315
.2707	0.9643	−0.175	0.263
.2745	0.9718	−0.250	0.190
.2786	0.9799	−0.315	0.120
.2825	0.9876	−0.380	0.040
.2868	0.9960	−0.435	0.000
.2909	0.0041	−0.450	−0.005
.2947	0.0116	−0.397	0.035
.2990	0.0201	−0.345	0.095
.3040	0.0300	−0.245	0.168
.3081	0.0381	−0.190	0.240
.3120	0.0458	−0.145	0.290
.3159	0.0535	−0.065	0.335

JD Hel	Phase	m(B)	m(V)
24 42 218.3698	0.7119	0.450	0.830
.3737	0.7196	0.460	0.865
.3772	0.7265	0.490	0.860
.3809	0.7338	0.485	0.855
.3827	0.7374	0.480	0.870
.3869	0.7456	0.475	0.870
.3888	0.7494	0.470	0.860
.3948	0.7612	0.485	0.875
.3968	0.7652	0.490	0.880
.4039	0.7792	0.455	0.855
.4056	0.7825	0.440	0.855
.4113	0.7938	0.480	0.850
.4148	0.8007	0.480	0.860
.4203	0.8116	0.450	0.850
.4229	0.8167	0.425	0.840
.4275	0.8258	0.425	0.815
.4300	0.8307	0.430	0.805
.4354	0.8413	0.390	0.780
.4371	0.8447	0.375	0.780
.4391	0.8486	0.370	0.780
.4423	0.8550	0.370	0.765
.4438	0.8579	0.360	0.770
.4456	0.8615	0.355	0.760
.4496	0.8694	0.325	0.740
.4527	0.8755	0.305	0.730
.4553	0.8806	0.290	0.705
.4602	0.8903	0.265	0.680
.4620	0.8938	0.260	0.660
.4640	0.8978	0.235	0.650
.4678	0.9053	0.210	0.620
.4702	0.9100	0.190	0.595
.4741	0.9177	0.150	0.580
.4761	0.9217	0.125	0.540
.4812	0.9317	0.070	0.500
.4857	0.9406	0.000	0.425
.4874	0.9440	−0.010	0.400
.4918	0.9526	−0.075	0.330
.4938	0.9566	−0.110	0.300
.4993	0.9674	−0.205	0.205
.5011	0.9710	−0.245	0.195
.5054	0.9795	−0.330	0.100
.5074	0.9834	−0.350	0.085
.5106	0.9897	−0.405	0.030
.5122	0.9929	−0.415	0.015
.5166	0.0016	−0.450	−0.010
.5194	0.0071	−0.435	0.000
.5235	0.0152	−0.365	0.060
.5253	0.0187	−0.335	0.105

JD Hel	Phase	m(B)	m(V)
24 42 218.5267	0.0215	−0.300	0.125
.5313	0.0306	−0.250	0.185
.5328	0.0335	−0.230	0.210
24 42 247.3200	0.8367	0.395	0.810
.3302	0.8568	0.365	0.790
.3349	0.8661	0.350	0.750
.3396	0.8754	0.330	0.725
.3474	0.8908	0.285	0.682
.3521	0.9000	0.240	0.645
.3568	0.9093	0.215	0.622
.3762	0.9476	−0.045	0.375
.3807	0.9565	−0.125	0.290
.3858	0.9665	−0.220	0.210
.3938	0.9823	−0.365	0.088
.3983	0.9912	−0.420	0.015
.4030	0.0005	−0.473	−0.020
.4096	0.0135	−0.390	0.050
.4157	0.0255	−0.290	0.170
.4206	0.0352	−0.200	0.230
.4252	0.0443	−0.130	0.300
.4325	0.0587	−0.020	0.392
.4372	0.0680	0.040	0.472
.4412	0.0759	0.095	0.516
.4466	0.0866	0.140	0.563
.4514	0.0960	0.185	0.597
.4556	0.1043	0.213	0.640
.4605	0.1140	0.245	0.677
.4659	0.1246	0.287	0.702
.4706	0.1339	0.310	0.718
.4767	0.1459	0.342	0.733
.4819	0.1562	0.370	0.778
24 42 252.2961	0.6556	0.425	0.820
.3025	0.6683	0.440	0.810
.3088	0.6807	0.465	0.830
.3142	0.6913	0.485	0.835
.3203	0.7034	0.500	0.840
.3265	0.7156	0.525	0.850
.3328	0.7280	0.520	0.845
.3390	0.7403	0.505	0.860
.3449	0.7519	0.510	0.860
.3600	0.7817	0.490	0.855
.3662	0.7940	0.500	0.835
.3799	0.8210	0.450	0.815
.3861	0.8332	0.430	0.780
.3925	0.8458	0.400	0.745
.3985	0.8577	0.380	0.740

JD Hel	Phase	m(B)	m(V)
24 42 252.4050	0.8705	0.350	0.710
.4111	0.8826	0.305	0.675
.4140	0.8883	0.280	0.655
.4202	0.9005	0.240	0.610
.4280	0.9159	0.170	0.565
.4341	0.9279	0.090	0.495
.4404	0.9404	0.000	0.415
.4474	0.9542	−0.140	0.300
.4581	0.9753	−0.315	0.120
.4664	0.9917	−0.470	0.010
.4734	0.0055	−0.520	−0.050
.4794	0.0173	−0.400	0.050
.4854	0.0292	−0.285	0.155
.4918	0.0418	−0.175	0.270
.4997	0.0574	−0.030	0.365
.5098	0.0773	0.105	0.480
.5156	0.0888	0.150	0.535
.5213	0.1000	0.205	0.590
.5271	0.1114	0.250	0.610
24 42 301.2703	0.2920	0.490	0.885
.2761	0.3034	0.465	0.860
.2826	0.3162	0.455	0.850
.2909	0.3326	0.425	0.790
.2979	0.3464	0.405	0.780
.3049	0.3602	0.375	0.755
.3126	0.3754	0.335	0.715
.3229	0.3958	0.270	0.660
.3318	0.4133	0.160	0.585
.3477	0.4447	−0.055	0.385
.3576	0.4642	−0.210	0.205
.3638	0.4765	−0.325	0.100
.3881	0.5244	−0.340	0.110
.3955	0.5390	−0.195	0.250
.4077	0.5631	−0.025	0.405
.4164	0.5802	0.105	0.510
.4214	0.5901	0.150	0.560
.4286	0.6043	0.215	0.630
.4341	0.6152	0.250	0.675
.4402	0.6272	0.315	0.715
.4450	0.6367	0.330	0.735
.4514	0.6493	0.375	0.775
.4571	0.6606	0.400	0.805
24 42 307.2562	0.1034	0.245	0.680
.2614	0.1137	0.280	0.680
.2654	0.1215	0.295	0.710
.2691	0.1288	0.315	0.725

JD Hel	Phase	m(B)	m(V)
24 42 307.2726	0.1358	0.355	0.745
.2764	0.1433	0.380	0.790
.2812	0.1527	0.395	0.810
.2852	0.1606	0.420	0.830
.2897	0.1695	0.435	0.840
.2935	0.1770	0.430	0.860
.3034	0.1965	0.465	0.870
.3077	0.2050	0.450	0.845
.3116	0.2127	0.490	0.850
.3159	0.2212	0.485	0.845
.3202	0.2297	0.475	0.835
.3245	0.2382	0.485	0.845
.3285	0.2461	0.480	0.850
.3325	0.2539	0.485	0.865
.3368	0.2624	0.495	0.865
.3408	0.2703	0.505	0.885
.3446	0.2778	0.490	0.895
.3486	0.2857	0.465	0.880
.3530	0.2944	0.465	0.865
.3566	0.3015	0.440	0.840
.3598	0.3078	0.450	0.825
.3633	0.3147	0.400	0.795
.3663	0.3206	0.440	0.820
.3694	0.3268	0.425	0.800
.3727	0.3333	0.420	0.820
.3764	0.3406	0.400	0.845
.3796	0.3469	0.370	0.760
.3830	0.3536	0.380	0.760
.3866	0.3607	0.360	0.755
.3899	0.3672	0.335	0.730
.3942	0.3757	0.315	0.720
.3982	0.3836	0.305	0.685
.4038	0.3946	0.270	0.665
24 42 311.2327	0.9499	−0.065	0.375
.2368	0.9579	−0.130	0.295
.2409	0.9660	−0.210	0.215
.2448	0.9737	−0.285	0.155
.2477	0.9795	−0.350	0.095
.2502	0.9844	−0.380	0.065
.2538	0.9915	−0.435	0.005
.2568	0.9974	−0.455	−0.030
.2619	0.0075	−0.420	−0.015
.2656	0.0148	−0.385	0.040
.2697	0.0229	−0.310	0.125
.2742	0.0317	−0.210	0.225
.2791	0.0414	−0.135	0.295
.2832	0.0495	−0.080	0.350

JD Hel	Phase	m(B)	m(V)
24 42 311.2892	0.0613	0.020	0.425
.2941	0.0710	0.105	0.500
.2989	0.0805	0.110	0.545
.3034	0.0894	0.170	0.580
.3080	0.0984	0.205	0.630
.3124	0.1071	0.245	0.665
.3167	0.1156	0.285	0.705
.3207	0.1235	0.280	0.705
.3253	0.1326	0.355	0.750
.3292	0.1403	0.370	0.765
.3334	0.1486	0.370	0.780
.3380	0.1576	0.395	0.800
.3431	0.1677	0.420	0.825
.3466	0.1746	0.425	0.830
.3507	0.1827	0.445	0.840
.3556	0.1924	0.460	0.845
.3594	0.1999	0.480	0.860
.3633	0.2076	0.480	0.875
.3681	0.2170	0.475	0.890
.3716	0.2239	0.480	0.895
.3753	0.2312	0.500	0.900
.3794	0.2393	0.515	0.900
.3832	0.2468	0.520	0.905
.3883	0.2569	0.520	0.890
.3942	0.2685	0.510	0.895
.4002	0.2804	0.505	0.900
.4043	0.2885	0.485	0.915
.4080	0.2958	0.480	0.870
.4115	0.3027	0.470	0.860
.4146	0.3088	0.455	0.840

Table II. The times of minima of OO Aql.

JD Hel.	Min	O − C	Meth.	Observer	References
24 26 892.060	I	−0.152	v	N. Florga	Binnendijk, 1968
26 953.382	I	−0.151	v	D. Ya. Martynoff	"
27 213.366	I	−0.150	v	"	"
27 365.409	I	−0.144	v	F. Lause	"
27 697.356	I	−0.143	v	"	"
33 891.426	I	−0.045	v	A. Jahn	Domke et. al, 1952
33 895.468	I	−0.057	v	"	"
33 895.472	I	−0.053	v	"	Binnendijk, 1968
33 900.536	I	−0.057	v	"	Domke et. al, 1952
33 925.37418	I	−0.05131	pe	K. K. Kwee	Binnendijk, 1968
34 194.48196	I	−0.04833	pe	"	"
34 205.632	I	−0.048	v	J. Ashbrook	Ashbrook, 1952
34 206.646	I	−0.047	v	"	"
34 207.662	I	−0.045	v	"	"
34 209.684	I	−0.050	v	"	"
34 211.708	I	−0.053	v	"	"
34 213.736	I	−0.052	v	"	"
34 216.784	I	−0.045	v	"	"
34 242.633	I	−0.042	v	"	"
34 244.662	I	−0.040	v	"	"
34 247.696	I	−0.047	v	"	"
34 248.708	I	−0.049	v	"	"
34 249.726	I	−0.044	v	"	"
34 251.752	I	−0.045	v	"	"
34 273.539	I	−0.050	v	"	Ashbrook, 1953
34 600.42550	I	−0.04254	pe	K. K. Kwee	Binnendijk, 1968
35 309.423	I	−0.042	v	F. Dörr	Rudolph, 1960
35 309.426	I	−0.039	v	R. Rudolph	"
35 310.434	I	−0.045	v	J. Müller	"

JD Hel.	Min	O − C	Meth.	Observer	References
35 310.441	I	−0.038	v	R. Rudolph	Rudolph, 1960
35 311.457	I	−0.036	v	J. Müller	"
35 311.460	I	−0.033	v	R. Rudolph	"
35 313.485	I	−0.035	v	"	"
35 313.485	I	−0.035	v	P. B. Lehmann	"
35 379.365	I	−0.037	v	F. Dörr	"
35 379.369	I	−0.033	v	R. Rudolph	"
35 380.383	I	−0.033	v	"	"
35 380.384	I	−0.032	v	W. Quester	"
35 380.389	I	−0.027	v	F. Dörr	"
36 074.443	II	−0.020	v	R. Rudolph	Binnendijk, 1968
36 712.494	II	−0.016	v	"	Braune et. al, 1962
37 111.598	I	−0.008	v	J. Ashbrook	Robinson, 1965c
37 112.604	I	−0.016	v	"	"
37 113.617	I	−0.016	v	"	"
37 118.690	I	−0.011	v	"	"
37 128.576	II	−0.007	v	"	"
37 132.620	II	−0.018	v	"	"
37 147.576	I	−0.012	v	"	"
37 148.588	II	−0.014	v	"	"
37 172.660	I	−0.014	v	"	"
37 173.4210	II	−0.0133	pg	P. Ahnert	Binnendijk, 1968
37 191.4100	II	−0.0154	pg	"	"
37 201.548	II	−0.013	v	J. Ashbrook	Robinson, 1965c
37 217.517	I	−0.008	v	"	"
37 475.480	I	−0.000	v	P. B. Lehmann	Dueball et. al, 1965
37 508.650	II	−0.024	v	J. Ashbrook	Robinson, 1965c
37 518.546	I	−0.011	v	"	"
37 539.573	II	−0.016	v	"	"

JD Hel.	Min	O - C	Meth.	Observer	References
37 544.395	I	-0.009	v	P. B. Lehmann	Dueball et. al, 1965
37 579.367	I	-0.005	v	W. Braune	"
37 579.368	I	-0.004	v	P. Krüger	"
37 879.6377	II	-0.0066	pe	B. B. Bookmyer	Binnendijk, 1968
37 915.6215	II	-0.0048	pe	L. Binnendijk	"
37 929.5597	I	-0.0033	pe	"	"
37 932.6007	I	-0.0030	pe	"	"
37 932.348	II	-0.002	v	J. Masuch	Dueball et. al, 1965
37 932.352	II	+0.002	v	M. Fernandes	"
37 932.354	II	+0.004	v	W. Braune	"
37 933.363	II	-0.001	v	J. Masuch	"
37 934.373	II	-0.004	v	F. Lehmpfuhl	"
37 934.373	II	-0.004	v	J. Masuch	"
37 934.377	II	-0.000	v	W. Braune	Dueball et. al, 1965
37 946.283	I	-0.004	v	"	"
37 948.312	I	-0.002	v	J. Dueball	"
37 948.313	I	-0.001	v	W. Braune	"
37 967.314	II	-0.006	v	"	"
38 183.462	I	-0.002	v	W. Quester	Braune et. al, 1967
38 204.494	II	-0.004	v	V. Orlovius	Pohl et. al, 1964
38 238.453	II	+0.002	v	H. Bode	"
38 238.452	II	+0.001	v	V. Orlovius	"
38 239.466	II	+0.002	pe	E. Pohl	"
38 336.266	II	+0.005	v	V. Orlovius	"
38 611.449	II	+0.002	v	H. Peter	Braune et. al, 1967
38 612.461	II	0.000	v	"	"
38 640.337	II	+0.003	v	H. Marx	"
38 640.337	II	+0.003	v	H. Peter	"
38 641.6028	I	+0.0017	pe	L. Binnendijk	Binnendijk, 1968
38 645.6576	I	+0.0022	pe	"	"

JD Hel.	Min	O − C	Meth.	Observer	References
38 661.367	I	+0.001	v	H. Marx	Braune et. al, 1967
38 662.387	I	+0.008	v	H. Peter	"
38 927.687	II	+0.004	v	C. Ricker	Robinson, 1965a
38 928.704	II	+0.007	v	D. Williams	"
38 931.745	II	+0.007	v	"	"
38 936.813	II	+0.008	v	R. Monske	"
38 937.824	II	−0.006	v	"	"
38 960.631	II	+0.007	v	"	Robinson, 1965b
38 961.637	II	−0.001	v	"	"
38 962.656	II	+0.004	v	"	"
38 963.670	II	+0.005	v	"	"
38 964.674	II	−0.005	v	D. Williams	Robinson, 1965a
38 966.715	II	+0.009	v	"	"
38 967.720	II	+0.001	v	R. Monske	Robinson, 1965b
38 970.757	II	−0.003	v	"	"
38 972.791	II	+0.004	v	"	"
38 983.432	II	+0.002	v	W. Braune	Braune et. al, 1967
38 987.489	II	+0.005	v	G. Haas	"
38 987.490	II	+0.006	v	H. Marx	"
38 996.614	II	+0.007	v	R. Monske	Robinson, 1965b
38 997.623	II	+0.003	v	"	"
38 999.647	II	0.000	v	"	"
39 000.660	II	−0.001	v	D. Loring	"
39 001.673	II	−0.001	v	W. Grady	"
39 001.682	II	+0.008	v	M. Baldwin	"
39 002.693	II	+0.005	v	R. Monske	Robinson 1965b
39 003.704	II	+0.002	v	W. Grady	"
39 033.353	I	+0.004	v	W. Braune	Braune et. al, 1967
39 036.651	II	+0.008	v	M. Baldwin	Robinson, 1965c

JD Hel.	Min	O – C	Meth.	Observer	References
39 046.276	II	+0.004	v	K. Löcher	Braune et. al, 1967
39 051.346	II	+0.006	v	M. Seidl	"
39 062.245	I	+0.009	v	H. Peter	"
39 269.767	II	+0.001	v	R. Monske	Robinson, 1966
39 271.793	II	0.000	v	"	"
39 286.487	II	-0.003	v	A. Howell	"
39 287.505	II	+0.002	v	R. Monske	"
39 287.755	I	-0.002	v	"	"
39 288.765	I	-0.005	v	"	"
39 289.786	I	+0.002	v	"	"
39 291.805	I	-0.006	v	D. Williams	"
39 293.841	I	+0.003	v	M. Baldwin	Robinson, 1967
39 297.636	II	-0.003	v	R. Monske	Robinson, 1966
39 298.652	II	-0.001	v	"	"
39 299.665	II	-0.001	v	"	"
39 300.679	II	-0.001	v	"	"
39 316.649	I	+0.005	v	M. Baldwin	Robinson, 1967
39 317.660	I	+0.003	v	"	"
39 317.661	I	+0.004	v	R. Monske	Robinson, 1966
39 318.672	I	+0.001	v	"	"
39 319.689	I	+0.005	v	M. Baldwin	Robinson, 1967
39 320.701	I	+0.003	v	"	"
39 322.725	I	0.000	v	R. Monske	Robinson, 1966
39 322.7288	I	+0.0037	pe	L. Binnendijk	Binnendijk, 1968
39 327.7967	I	+0.0037	pe	"	"
39 328.809	I	+0.002	v	R. Monske	Robinson, 1966
39 337.677	II	+0.002	v	"	Robinson, 1967
39 341.7338	II	+0.0041	pe	L. Binnendijk	Binnendijk, 1968

JD Hel.	Min	O – C	Meth.	Observer	References
39 342.747	II	+0.004	v	R. Monske	Robinson, 1967
39 350.602	I	+0.003	v	M. Baldwin	"
39 352.626	I	0.000	v	R. Monske	"
39 355.664	I	−0.002	v	M. Baldwin	"
39 355.665	I	−0.001	v	Curtis Anderson	Baldwin, 1973
39 659.743	I	+0.003	v		"
39 661.775	I	+0.008	v		"
39 671.650	II	+0.001	v	M. Baldwin	Robinson et. al, 1968
39 672.667	II	+0.004	v		"
39 674.690	II	0.000	v		"
39 675.707	II	+0.004	v		"
39 677.730	II	−0.001	v	M. Baldwin	Robinson et. al, 1968
39 678.739	II	−0.005	v	Curtis Anderson	Baldwin, 1973
39 679.760	II	+0.002	v	M. Baldwin	Robinson et. al, 1968
39 685.846	II	+0.007	v	Curtis Anderson	Baldwin, 1973
39 694.709	I	+0.001	v	M. Baldwin	"
39 695.724	I	+0.002	v	S. Cook	Robinson et. al, 1968
39 696.736	I	+0.001	v		"
39 696.738	I	+0.003	v	Curtis Anderson	Baldwin, 1973
39 699.779	I	+0.003	v	T. Cragg	"
39 701.805	I	+0.002	v	M. Baldwin	"
39 716.751	II	−0.002	v	R. Monske	Robinson et. al, 1968
39 729.676	I	0.000	v	Curtis Anderson	Baldwin, 1973
39 730.685	I	−0.005	v		"
39 734.746	I	+0.002	v		"
39 735.756	I	−0.002	v	R. Monske	Robinson et. al, 1968
39 737.783	I	−0.002	v	M. Baldwin	Baldwin, 1973
39 744.631	II	−0.002	v	W. Lowder	Robinson et. al, 1968
39 757.808	II	+0.005	v	M. Baldwin	Baldwin, 1973
39 765.658	I	0.000	v	Curtis Anderson	"

THE LIGHT AND PERIOD VARIATIONS OF OO AQUILAE

JD Hel.	Min	O – C	Meth.	Observer	References
39 800.623	I	−0.004	v	M. Baldwin	Baldwin, 1973
39 801.637	I	−0.003	v	"	"
40 046.677	II	+0.004	v	Curtis Anderson	"
40 047.688	II	+0.002	v	"	"
40 048.701	II	+0.001	v	"	"
40 052.751	II	−0.003	v	"	"
40 053.771	II	+0.003	v	T. Cragg	"
40 054.782	II	+0.001	v	"	"
40 056.811	II	+0.003	v	M. Baldwin	"
40 058.834	II	−0.002	v	"	"
40 066.686	I	−0.005	v	R. Monske	"
40 067.702	I	−0.002	v	"	"
40 068.4638	II	−0.0009	pe	Ibanoglu, Kurutac	Pohl, 1969
40 069.728	I	−0.004	v	M. Baldwin	Baldwin, 1973
40 070.740	I	−0.005	v	"	"
40 081.648	II	+0.007	v	J. Bortle	"
40 096.588	I	−0.003	v	"	"
40 113.575	II	+0.006	v	K. Simmons	"
40 115.596	II	0.000	v	J. Bortle	"
40 126.748	II	+0.003	v	T. Cragg	"
40 128.775	II	+0.003	v	M. Baldwin	"
40 134.605	I	+0.004	v	J. Bortle	"
40 151.580	II	+0.002	v	M. Baldwin	"
40 156.644	II	−0.002	v	"	"
40 366.4544	II	−0.0021	pe	Ibanoglu, Kurutac	Pohl, 1969
40 417.648	I	+0.006	v	J. Bortle	Baldwin, 1973
40 418.660	II	+0.004	v	W. Hampton	"
40 442.722	I	−0.006	v	M. Baldwin	"
40 443.736	I	−0.006	v	"	"

JD Hel.	Min	O − C	Meth.	Observer	References
40 453.625	II	+0.001	v	Carl Anderson	Baldwin, 1973
40 454.639	II	+0.001	v	J. Bortle	"
40 455.655	II	+0.004	v	R. Monske	"
40 811.4152	II	−0.0018	pe	O. Demircan	Kizilirmak et. al, 1971
40 817.494	II	−0.004	pe	C. Ibanoglu	"
40 825.3515	I	−0.0021	pe	N. Gudur	"
40 858.291	I	−0.004	pe	E. Pohl	"
41 152.487	II	+0.001	v	L. Frasinski	Flin, 1971
41 152.491	II	+0.005	v	L. Krzanik	"
41 153.500	II	+0.001	v	A. Letkowski	"
41 153.503	II	+0.004	v	A. Soska	"
41 155.524	II	−0.002	v	"	"
41 161.3518	I	−0.0028	pe	Gudur, Akinu	Pohl et. al, 1972
41 179.3457	II	+0.0002	pe	Sengonca, Karacan	"
41 182.3845	II	−0.0018	pe	Akinu, Sengonca	"
41 187.4531	II	−0.0011	pe	H. Sengonca	"
41 472.527	I	+0.004	v	K. Löcher	BBSAG Bul., 1972, No. 3
41 483.414	II	−0.005	v	R. Germann	"
41 487.473	II	0.000	v	H. Peter	"
41 490.519	II	+0.005	v	K. Löcher	"
41 500.396	I	0.000	v	"	"
41 501.413	I	+0.003	v	H. Peter	No. 4
41 503.440	I	+0.003	v	"	"
41 503.495	I	+0.004	v	K. Löcher	"
41 512.560	I	+0.001	v	"	"
41 536.381	I	+0.003	v	"	"
41 539.408	I	−0.009	v	L. Frasinski	Flin, 1972
41 539.422	I	+0.003	v	K. Szlachcic	"
41 540.433	I	+0.001	v	H. Peter	BBSAG Bul., 1972, No. 5

THE LIGHT AND PERIOD VARIATIONS OF OO AQUILAE

JD Hel.	Min	O – C	Meth.	Observer	References
41 542.453	I	−0.007	v	L. Frasinski	Flin, 1972
41 544.480	II	−0.007	v	Z. Klimek	Klimek, 1973
41 550.574	I	+0.006	v	K. Löcher	BBSAG Bul. 1972, No. 5
41 555.382	II	0.000	v	R. Diethelm	"
41 555.387	II	+0.004	v	H. Peter	"
41 555.388	II	+0.005	v	R. Germann	"
41 565.520	II	+0.002	v	K. Löcher	"
41 571.3468	I	+0.0002	pe	O. Demircan	Kizilirmak et. al, 1974
41 574.392	I	+0.005	v	H. Peter	BBSAG Bul.,1972, No. 5
41 586.295	II	−0.002	v	R. Germann	"
41 587.313	II	+0.003	v	K. Löcher	"
41 595.416	II	−0.003	v	R. Diethelm	"
41 604.292	II	−0.002	v	R. Germann	No. 6
41 605.294	I	−0.007	v	K. Löcher	"
41 621.265	I	0.000	v	R. Germann	"
41 622.272	II	−0.007	v	"	"
41 623.290	II	−0.002	v	K. Löcher	"
41 623.291	II	−0.001	v	H. Peter	"
41 625.314	II	−0.006	v	R. Germann	"
41 626.331	II	−0.002	pe	O. Demircan	Kizilirmak et. al, 1974
41 637.232	I	+0.003	v	K. Löcher	BBSAG Bul., 1972, No. 6
41 639.255	I	−0.001	v	R. Germann	"
41 657.239	II	−0.008	v	"	"
41 673.210	I	−0.001	v	K. Löcher	BBSAG Bul., 1973, No. 7
41 675.240	I	+0.002	v	R. Germann	"
41 806.488	I	−0.009	v	"	"
41 814.602	I	−0.003	v	"	"
41 845.511	I	−0.008	v	K. Löcher	No. 10
41 850.586	I	−0.001	v	"	"

JD Hel.	Min	O − C	Meth.	Observer	References
41 859.459	II	+0.003	v	K. Löcher	BBSAG Bul., 1973, No. 10
41 866.546	II	−0.005	v	"	"
41 877.447	I	0.000	v	"	"
41 890.3716	II	+0.0005	pe	O. Demircan	Kizilirmak et. al, 1974
41 892.401	II	+0.004	v	K. Löcher	BBSAG Bul., 1973, No. 10
41 894.426	II	+0.002	v	R. Germann	" No. 11
41 904.569	II	+0.009	v	R. Diethelm	" No. 11
41 916.473	I	+0.003	v	K. Löcher	"
41 922.2987	II	+0.0009	pe	Ib, Tn, Er	Kizilirmak et. al, 1974
41 924.325	II	0.000	v	R. Germann	BBSAG Bul., 1973, No. 11
41 931.417	II	−0.003	v		
41 940.2886	II	−0.0002	pe	Ib, Gl, Tn	Kizilirmak et. al, 1974
41 941.302	I	0.000	v	R. Germann	BBSAG Bul., 1973, No. 11
41 942.313	I	−0.003	v	K. Löcher	"
41 943.328	I	−0.002	v	"	"
41 953.458	I	−0.007	v	"	"
41 959.294	II	+0.001	v		No. 12
41 961.326	II	+0.005	v	R. Germann	"
41 975.263	I	+0.006	v	"	"
42 010.227	I	+0.001	v	"	"
42 134.639	II	−0.003	v	K. Löcher	No. 14
42 200.524	II	−0.001	v	"	"
42 201.541	II	+0.003	v	"	"
42 214.466	I	+0.005	v	R. Germann	BBSAG Bul., 1973, No. 16
42 232.709	I	+0.003	v	Mallama	Krobusek et. al, 1975
42 233.458	II	−0.008	v	H. Peter	BBSAG Bul., 1973, No. 16
42 234.724	I	−0.009	v	Mallama	Krobusek et. al, 1975
42 235.744	I	−0.003	v	"	"
42 235.750	I	+0.003	v	Krobusek	
42 247.4023	I	−0.0004	pe	N. Güdür	Pohl et. al, 1975

JD Hel.	Min	O - C	Meth.	Observer	References
42 248.667	II	-0.003	v	Krobusek	Krobusek et. al, 1975
42 250.699	II	+0.002	v	"	"
42 251.450	I	-0.007	v	R. Diethelm	BBSAG Bul., 1975, No. 16
42 251.453	I	-0.004	v	R. Germann	"
42 252.470	I	-0.0005	pe	N. Güdür	Pohl et. al, 1975
42 253.490	I	+0.006	v	K. Löcher	BBSAG Bul., 1973, No. 16
42 256.786	II	+0.008	v	Krobusek	Krobusek et. al, 1975
42 260.580	I	+0.001	v	K. Löcher	BBSAG Bul., 1973, No. 16
42 261.588	I	-0.004	v	"	BBSAG Bul., 1974, No. 17
42 263.367	II	0.000	v	R. Germann	"
42 264.635	I	+0.001	v	Krobusek	Krobusek et. al, 1975
42 265.647	I	0.000	v	"	"
42 266.414	II	+0.007	v	H. Peter	BBSAG Bul., 1974, No. 17
42 296.306	II	-0.002	v	R. Germann	"
42 296.311	II	+0.003	v	K. Löcher	"
42 301.3758	II	+0.0001	pe	Güdür, Demircan	Pohl et. al, 1976
42 302.381	II	-0.008	v	R. Germann	BBSAG Bul., 1974, No. 17
42 303.400	II	-0.003	v	R. Diethelm	"
42 304.412	II	-0.004	v	H. Peter	"
42 311.2580	I	-0.0001	pe	Güdür, Demircan	Pohl et. al, 1976
42 324.687	I	-0.001	v	Krobusek	Krobusek et. al, 1975
42 337.609	I	-0.002	v	"	"
42 339.376	II	-0.009	v	R. Germann	BBSAG Bul., 1974, No. 18
42 354.330	I	-0.005	v	H. Peter	"
42 367.254	II	-0.004	v	R. Germann	"
42 369.283	II	-0.002	v	"	"
42 384.237	I	+0.001	v	K. Löcher	No. 19
42 385.256	I	+0.007	v	R. Germann	No. 18
42 402.220	II	-0.007	v	R. Diethelm	No. 19

JD Hel.	Min	O – C	Meth.	Observer	References
42 403.244	II	+0.004	v	R. Diethelm	BBSAG Bul., 1974, No. 19
42 455.694	I	+0.001	v	K. Löcher	"
42 473.684	II	0.000	v	"	BBSAG Bul., 1975, No. 22
42 510.670	II	-0.009	v	"	"
42 525.626	I	-0.004	v	"	"
42 526.643	I	0.000	v	"	"
42 535.508	II	-0.004	v	H. Peter	"
42 549.452	I	+0.003	v	R. Germann	"
42 551.467	I	-0.009	v	"	"
42 558.574	I	+0.003	v	K. Löcher	No. 23
42 569.469	II	+0.002	v	H. Peter	"
42 572.517	II	+0.009	v	K. Löcher	"
42 577.573	II	-0.002	v	"	"
42 589.488	I	+0.003	v	"	"
42 601.654	I	+0.003	v	Krobusek	Mallama et. al, 1977
42 602.659	I	-0.003	v	"	"
42 603.670	I	-0.005	v	"	"
42 604.443	II	+0.008	v	K. Löcher	BBSAG Bul., 1975, No. 23
42 606.460	II	-0.002	v	R. Germann	"
42 606.465	II	+0.003	v	R. Diethelm	"
42 607.485	II	+0.009	v	H. Peter	"
42 607.4755	II	-0.0006	pg	P. Anhert	Anhert, 1975
42 620.401	I	+0.002	v	R. Diethelm	BBSAG Bul., 1975, No. 23
42 621.399	I	-0.004	v	"	"
42 625.465	I	-0.002	v	R. Germann	"
42 652.332	I	+0.005	v	"	"
42 669.301	II	-0.003	v	H. Peter	No. 24
42 692.372	I	+0.009	v	L. Löcher	"
42 708.323	II	-0.004	v	"	"

THE LIGHT AND PERIOD VARIATIONS OF OO AQUILAE 437

JD Hel.	Min	O - C	Meth.	Observer	References
42 709.600	I	+0.006	v	Krobusek	Mallama et. al, 1977
42 710.351	II	-0.003	v	R. Germann	BBSAG Bul., 1975, No. 24
42 725.308	I	+0.004	v	K. Löcher	"
42 738.224	III	-0.002	v	M. Kissling	"
42 738.226	III	-0.002	v	T. Wüthrich	"
42 738.231	III	+0.003	v	M. Baumann	"
42 738.232	III	+0.004	v	H. Peter	"
42 758.243	I	-0.003	v	R. Germann	BBSAG Bul., 1976, No. 25
42 774.209	III	-0.001	v	K. Löcher	"
42 878.612	III	+0.004	v	"	" No. 27
42 879.617	III	-0.005	v	"	"
42 897.608	I	-0.005	v	"	"
42 906.737	I	+0.002	v	Stephan	Stephan, 1977
42 926.503	I	+0.003	v	H. Peter	BBSAG Bul., 1976, No. 28
42 934.604	I	-0.004	v	K. Löcher	"
42 944.492	II	+0.002	v	"	"
42 956.402	I	+0.002	v	R. Germann	"
42 957.416	I	+0.002	v	H. Peter	"
42 958.436	I	+0.009	v	R. Germann	"
42 959.442	I	+0.001	v	K. Löcher	"
42 960.4541	II	-0.0003	pe	Gr, Si	Pohl et. al, 1977
42 961.474	I	+0.004	v	H. Peter	BBSAG Bul., 1976, No. 29
42 974.388	II	-0.003	v	K. Löcher	"
42 975.403	II	-0.002	v	R. Germann	"
42 990.355	I	0.000	v	"	"
42 993.393	I	-0.003	v	"	"
42 997.453	I	+0.003	v	H. Peter	"
43 008.598	I	-0.001	v	Krobusek	Mallama et. al, 1977
43 010.629	I	+0.002	v	"	"

JD Hel.	Min	O – C	Meth.	Observer	References
43 015.441	II	0.000	v	H. Peter	BBSAG Bul., 1976, No. 29
43 029.376	I	-0.002	v	"	No. 30
43 030.391	I	0.000	v	K. Löcher	"
43 043.305	II	-0.009	v	R. Germann	"
43 057.254	I	+0.003	v	K. Löcher	"
43 061.311	I	+0.006	v	A. Royer	"
43 061.314	I	+0.009	v	R. Germann	"
43 062.319	I	0.000	v	A. Royer	"
43 069.414	I	0.000	v	"	BBSAG Bul., 1977, No. 32
43 219.676	II	-0.001	v	K. Löcher	" No. 33
43 231.588	I	+0.002	v	"	"
43 251.604	II	-0.001	v	R. Germann	"
43 311.411	II	+0.005	v	"	"
43 327.376	I	+0.007	v	K. Löcher	No. 34
43 328.384	I	+0.001	v	"	"
43 340.561	I	+0.005	v	"	"
43 347.389	II	+0.001	v	R. Germann	"
43 348.399	II	-0.002	v	"	"
43 348.654	I	-0.001	v	Stephan	Stephan, 1978
43 358.538	II	+0.001	v	K. Löcher	BBSAG Bul., 1977, No. 34
43 364.373	I	+0.008	v	A. Royer	"
43 365.377	I	-0.002	v	"	"
43 366.396	I	+0.004	v	"	"
43 367.406	I	0.000	v	"	"
43 657.549	II	+0.007	v	K. Löcher	BBSAG Bul., 1978, No. 37
43 658.553	II	-0.001	v	"	"
43 671.482	I	+0.003	v	"	"
43 671.483	I	+0.004	v	H. Peter	"
43 673.510	I	+0.004	v	K. Löcher	"

THE LIGHT AND PERIOD VARIATIONS OF OO AQUILAE

JD Hel.	Min	O − C	Meth.	Observer	References
43 678.578	I	+0.004	v	K. Löcher	BBSAG Bul., 1978, No. 37
43 689.475	II	−0.005	v	R. Germann	"
43 703.405	I	−0.002	v	R. Germann	No. 38
43 703.412	I	+0.005	v	K. Löcher	"
43 706.452	I	+0.005	v	H. Peter	"
43 723.426	II	+0.001	v	H. Peter	BBSAG Bul., 1978, No. 38
43 735.339	I	+0.005	v	R. Germann	"
43 738.378	I	+0.003	v	R. Diethelm	"
43 739.391	I	+0.002	v	R. Germann	"
43 754.338	II	−0.001	v	R. Diethelm	No. 39
43 755.352	II	0.000	v	"	"
43 756.364	II	−0.002	v	"	"
43 756.374	II	+0.008	v	R. Germann	"
43 776.384	I	0.000	v	"	"
43 787.282	II	+0.002	v	"	"
43 788.289	II	−0.005	v	"	"
43 789.308	II	+0.001	v	R. Diethelm	"
43 790.321	II	0.000	v	"	"
43 791.339	II	+0.005	v	R. Germann	"
43 791.341	II	+0.007	v	H. Peter	"
43 805.278	I	+0.007	v	R. Germann	"
43 806.289	II	+0.004	v	"	"
43 823.269	II	+0.007	v	"	"
43 978.599	I	+0.006	v	"	BBSAG Bul., 1979, No. 40
44 059.430	II	+0.004	v	R. Diethelm	No. 43
44 060.447	II	+0.008	v	K. Löcher	No. 44
44 079.442	I	−0.002	v	H. Peter	"

PHOTOELECTRIC OBSERVATIONS OF ECLIPSING BINARY SYSTEMS AT KRYONERION OBSERVATORY, GREECE.

P. G. Niarchos

Department of Astronomy, University of Athens,
Athens, Greece.

ABSTRACT

During a programme of photoelectric observations of eclipsing binaries, the W UMa-type systems of AH Virginis and VW Cephei have been observed in two wavelengths (V and B) in the years 1977 - 1979. The light curves of the systems are presented and some of their characteristics and peculiarities are discussed.

1. AH VIRGINIS

a. Introduction

AH Virginis is an eclipsing binary of W UMa-type with a period of 0.4075 days. The history of AH Virginis has been summarized by Binnendijk (1960). A lot of visual, photographic and photoelectric observations of this system have been made since its discovery by Guthnick and Prager in 1929. Systematic observations of AH Vir have been made by Binnendijk in the years 1955 and 1957. It has been pointed out by a number of investigators, e.g., Binnendijk (1960), Herczeg (1962), Purgathofer and Procházka (1967) and Bakoš (1971, 1977) that the light curve and the period of AH Vir are variable. The primary minimum is an occultation. Spectroscopic observations of the system were made by Chang (1948).

b. The Observations

The system was observed for four consecutive nights (19-22) in March of 1977 and for three nights (29-31) in March of 1979, using the 48-inch Cassegrain reflector at the new Kryonerion Station of the National Observatory of Athens (Contopoulos and Banos, 1976). The telescope was used together with a two-beam multi-mode photometer (Goudis and Meaburn, 1973). The three intermediate pass-band filters used were selected to be in close accordance with the standard colour system U, B, V. As comparison star we used BD + 12 2443, as check star BD + 12 2434.

The ephemeris used for the reduction of the observations is (Bakos, 1977).

$$\text{Min I} = \text{HJD } 2442549.6969 + 0.4075292E \tag{1}$$

A total of 264 B and 264 V observations were obtained (each observation consists of two individual measurements). The B and V light curves of AH Vir are shown in Figure 1.

c. Discussion

The light curves of AH Vir are subject to considerable changes. A great deal of variations exist in both the heights of maxima and the depths of minima. The magnitude difference between the extremes of the maxima and the minima is significant especially for the maxima. This difference shows, for instance, that the outer and inner envelope of max. I has a width of 0.045 mag. in the yellow and 0.040 mag. in the blue. The observations also show a very small displacement of the maximum corresponding to phase 0.75. As a rule, the variations in the blue light are greater than in the yellow. However, the observed light curves show maxima of the same height and symmetrical minima.

2. VW CEPHEI

a. Introduction

VW Cephei is an eclipsing binary of W UMa-type which shows large variations in the light curve (Kwee, 1966a). Systematic photoelectric observations of the system were made during the IAU campaign of 1959, published by Kwee (1966a). The system was also observed photoelectrically by Cristescu (1978) and Hopp et al. (1979). Spectroscopic observations were made at Mount Palomar Observatory and at David Dunlap Observatory during the IAU campaign of 1959. No satisfying solution of the light curve of the system has been obtained so far.

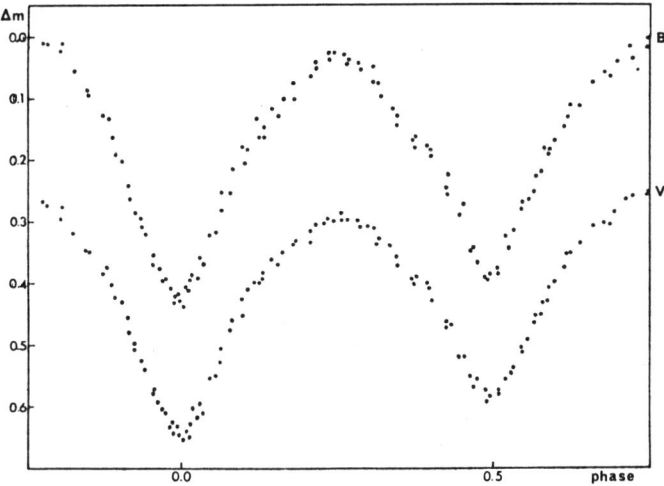

Fig. 1 The B and V light curves of AH Virginis.

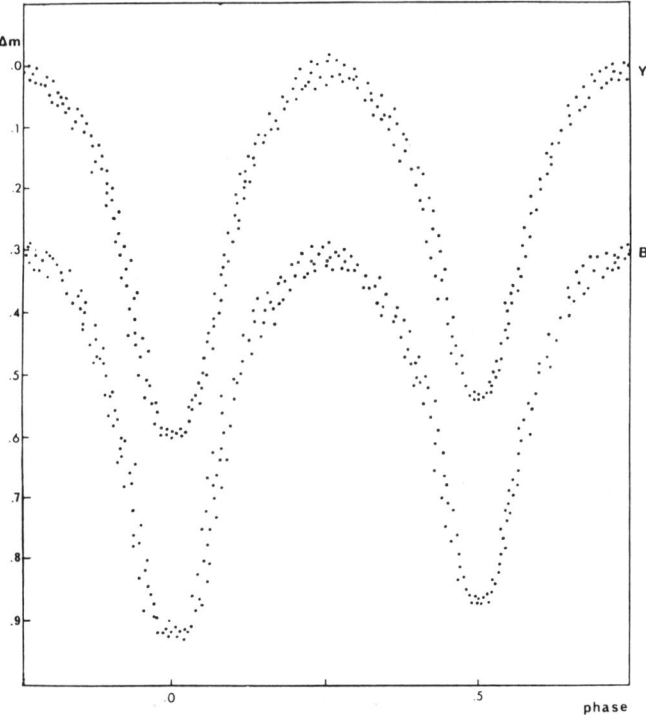

Fig. 2 The B and V light curves of VW Cephei

b. The Observations

The observations of VW Cep were made on July 20-23, 1979, with the same telescope and the same photometer we used to observe AH Virginis. As comparison star we used BD + $74°$ 889, as check star BD + $75°$ 726. The observations of July 20-21 and those of July 22-23 cover the whole light curve, while those of July 21-22 cover the phase interval from 0.0 to 0.65 and from 0.87 to 1. The following ephemeris (Cristescu, 1978) was used for the reductions of the observations

$$\text{Min I} = \text{HJD } 2443448.2663 + 0.2783176 E \qquad (2)$$

A total of 116 B and 116 V observations were obtained (each observation consists of two individual measurements). The B and V light curves of VW Cep are shown in Figure 2.

c. Discussion

The eclipsing system VW Cep has both a variable period and a variable light curve (Kwee, 1966a). Many authors have looked for a long term and a short term periodic variation of the period. The long term variation was generally interpreted as due either to the presence of a third component or to dynamical interactions between the two components. The short term variations of the period of VW Cep have been interpreted by Kwee (1966a, b) as due to an inhomogeneous cloud of circumstellar absorbing material revolving at a very small distance around the system with a period about 3.5% longer than ther period of the component stars.

According to Schmidt and Schrick (1955) the following three values were determined from our observations to describe the variability of the light curve around the extrema

	B	V
$\Delta M_1 = M_{minI} - M_{minII}$	+0.042	+0.060
$\Delta M_2 = M_{maxI} - M_{minII}$	−0.385	−0.332
$\Delta M_3 = M_{maxII} - M_{minII}$	−0.358	−0.297

The heights of the two maxima are not equal. Their difference is about 0.035 mag. and 0.030 mag. for the V and B bands respectively. A shoulder appears at about $0^P.65$ (Hopp et al, 1979) in the V-band, while another one is present at $0^P.11$ in both bands. An unusual scatter in B and V is observed around the primary minimum

REFERENCES

Bakoš, G. A.: 1971, IAU Coll. 15, p. 293.
Bakoš, G. A.: 1977, Bull. Astron. Inst. Czech., 28, p. 157.
Binnendijk, L.: 1960, Astron. J., 65, p. 358.
Chang, Y. C.: 1948, Astrophys. J. 107, p. 96.
Contopoulos, G. and Banos, C.: 1976, Sky and Telescope, 51, p. 154.
Cristescu, C.: 1978, I.B.V.S. No. 1383.
Goudis, C. and Meaburn, J.: 1973, Astrophys. Space Sci., 20, p.149.
Guthnick, P. and Prager, R.: 1929, Beob. Zirk. No. 13, p. 32.
Herczeg, T.: 1962, Veröff. Univ. Sternw. Bonn, No. 63.
Hopp, U., Witzigmann, S., and Kietel, M.: 1979, I.B.V.S., No. 1599.
Kwee, K. K.: 1966a, Bull. Astron. Inst. Neth. Suppl. 1, p. 245.
Kwee, K. K.: 1966b, Bull. Astron. Inst. Neth. 18, p. 448.
Purgathofer, A. and Procházka, F.: 1967, Mitteil. Univ. Sternw. Wien 13, p. 151.
Schmidt, H. and Schrick, K. W.: 1955, Zeitschrift f. Astrophys. 37, p. 73.

STEPANIAN'S STAR

Peter Rovithis

National Observatory of Athens,
Athens, Greece.

ABSTRACT

The light curve of the new variable star that Stepanian discovered in 1979 is given based on photoelectric observations made at Kryonerion Astronomical Station of the National Observatory of Athens during April 1980. The variable is an eclipsing binary of short period and the differences between primary minimum and a mean value at maximum were found to be $2^m\!.425$ in B and $2^m\!.925$ in V.

1. INTRODUCTION

After Stepanian (1979) had reported the discovery of a new variable star in Serpens ($\alpha = 15^h 35^m 44^s$, $\delta = +19°01'30"$), many investigators were interested in it. The star was found to be around 14^m on plates obtained at Byurakan with a 40" Schmidt-camera with a low-dispersion objective prism in the spring of 1979, while it appeared to be very faint - around 18^m - on the POSS prints of April 1950. According to Wenzel (1979) the inspection of almost one thousand Sonneberg sky patrol plates showed that the maxima of the star's light, which reach $13^m\!.2$ pg are not infrequent and thus its brightening was not due to one eruption of an unknown date.

Stepanian (1979) reports that on slit spectra taken with the Byurakan 2.6m telescope on June 6, 1978 there were very strong and broad emission lines (HeII 4686, HeI 6678, 5875, 4471, 3888, Hα - Hε). Moreover, three grating spectrograms of 140 Å/mm taken with the image tube device at the 2m telescope of Tautenburg

Observatory, at Wenzel's request, confirm Stepanian's results showing that the star's spectrum is characterized by broad ($\approx 50\text{Å}$) emission lines of Hα, Hβ, Hγ HeI 5875 and HeII 4686 typical of U Geminorum stars.

A search of the plate collection at the Harvard College Observatory gave the B magnitudes from 1897 to 1979 and according to M. Liller (1980) the corresponding light curve is reminiscent of that for R Coronae Borealis stars.

Horne et al (1980a) report that they had discovered Stepanian's star to be a new eclipsing cataclysmic variable. Three eclipses observed by them in 34 nights require a period of $3^h 48^m 11^s \pm 1^s$ and epoch HJD = 2444293.0243. Later, Horne et al (1980b) refine the period to $3^h 48^m 08^s \pm 0^s\!.5$ from 6 further eclipses observed in March 1980. Finally, Margon et al (1980) report that spectrophotometric and photometric observations of this star made at Lick and Kitt Peak Observatories reveal several extraordinary variations synchronous with the 228-min period.

2. THE OBSERVATIONS

Photoelectric observations of Stepanian's star were carried out in B and V during April (19/20 and 20/21) 1980). They were made using a two-beam, multi-mode, nebular-stellar photometer (Goudis and Meaburn, 1973) attached to the 48-inch Cassegrain reflector at Kryonerion Station (Contopoulos and Banos, 1976) of the National Observatory of Athens.

The filters used are in close accordance to the standard ones and dry ice was used to freeze the photomultiplier. As comparison star we used the one marked with an A in Figure 1, which gives the variable's field.

Reduction of the observations have not been made, firstly because of the absence of photometric elements for both stars (colour correction) and secondly because of the small difference in air mass between the two stars (atmospheric extinction correction).

Using the ephemeris formula given by Horne et al (1980a), we produced the light curve of Figure 2, which represents a lot of irregularities.

3. DISCUSSION

In the light curve of Figure 2 the duration of the primary minimum is 40 min and it is 20 min at the middle depth of the

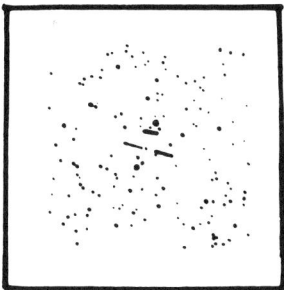

RA = $15^h 35^m 44^s$
Decl. = $+19° 01' 30"$

Figure 1. The field of Stepanian's star (after Stepanian, 1979). The underlined star is that which served as comparison.

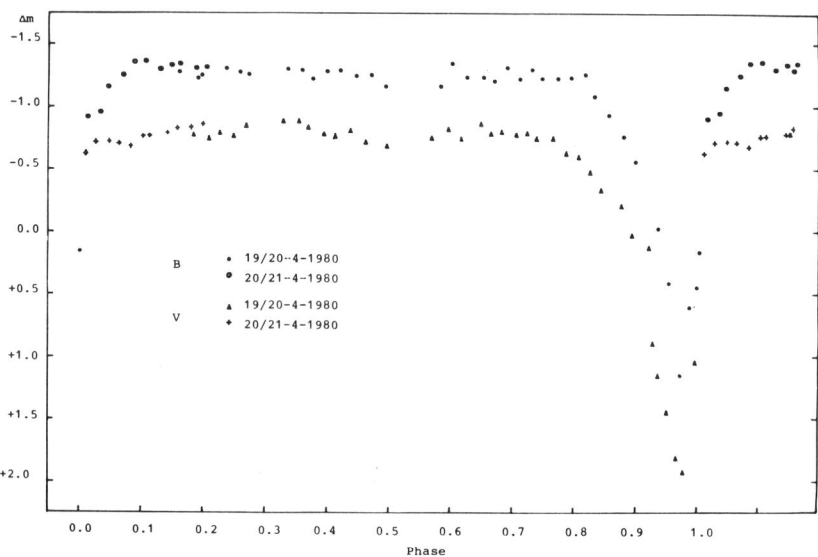

Figure 2. The light curve of Stepanian's star in B and V.

eclipse in both B and V colours. This is in good agreement with Horne et al's (1980a) results.

The descending and ascending branches of primary minimum are of different shape and there are a lot of anomalies outside eclipses. Unfortunately, we do not have observations around secondary minimum.

Using Horne et al (1980a) and Horne et al (1980b) ephemeris formulae, the following values for the residuals $(O-C)_a$ and $(O-C)_b$ were found respectively for the observed primary minimum:

Hel. J. D.	$(O-C)_a$	$(O-C)_b$
2444349.5908	-0.0028	-0.0070

REFERENCES

Contopoulos, G. and Banos, C.: 1976, Sky and Telescope, 51, p.154.
Goudis, C. and Meaburn, J.: 1973, Astrophys. Space Sci., 20, p.149.
Horne, K., Cohen, J., Oke, J. B. and Mochnacki, S. W.: 1980a, I.A.U. Circ. No.3462.
Horne, K., Margon, B. and Africano, J.: 1980b, I.A.U. Circ. No.3466.
Liller, M. H.: 1980, I.B.V.S. No.1743.
Margon, B., Dowers, R. and Szkody, P.: 1980, I.A.U. Circ. No.3465.
Stepanian, I. A.: 1979, I.B.V.S. No.1630.
Wenzel, W. W.: 1979, I.B.V.S. No.1720.

NOTE ON THE ECLIPSING BINARY AW URSAE MAIORIS

Katarzyna Otmianowska

Astronomical Observatory, Jagiellonian University, Kraków, Poland.

The eclipsing binary AW UMa (BD+30°2163) was discovered in 1963. Then it was observed by Kalish in 1965. In 1967, the variable was classified by Eggen as W UMa-type A contact binary. Considerable attention was paid to the star and its light curve was analyzed since 1970 by Mochnacki and Doughty, Lucy, Al-Naimiy and Woodward. They have shown that, of all known contact systems, this binary exhibits the shallowest total eclipses and if it is a contact binary, its mass-ratio is extremely low (q = 0.08 ± 0.01).

In 1971-1974, Dworak and Kurpinska undertook two series of observations of AW UMa, and found that the period of the eclipsing binary is equal to 0.43873231 ± 5 days. The light curves are slightly asymmetric. On a comparison of observations of secondary minimum by Dworak and Kurpinska or Kalish and Paczynski, one can prove that the secondary minimum changes its depth. A decrease of period was indicated by Kurpinska (1980), and it is equal to $5^d 4 \times 10^{-6}$ E. Observations carried out since 1978 have shown that the period of AW UMa changed suddenly its vaue.

A fuller discussion and determination of the new values of actual elements are given in the paper of Kurpinska (1980).

REFERENCES

Al-Naimiy, H.: 1978, Astrophys. Space Sci., 56, p.219.
Dworak, T. Z. and Kurpinska, M.: 1975, Acta Astron., 25, p.417.
Kalish, M. S.: 1965, Publ. Astron. Soc. Pacific, 77, p.36.
Kurpinska, M.: 1980, IBVS (in press).
Paczynski, B.: 1964, Astron. J., 69, p.124.
Woodward, E. J.: 1980, Astron. J., 85, p.50.

ULTRAVIOLET SPECTRA OF INTERACTING BINARIES

Margherita Hack
Osservatorio Astronomico, Trieste

ABSTRACT

The results of the ultraviolet observations of several spectroscopic interacting binaries are reported. A common characteristic is the presence of an expanding envelope where lines of higher ionization (C IV, N V, Si IV) than one could expect from the photospheric stellar temperature are found

INTRODUCTION

To understand the importance of observing ultraviolet spectra of interacting binaries, it is useful to review briefly what additional information we can expect to obtain from this range of the electromagnetic spectrum, and what are the main characteristics of the astronomical ultraviolet satellites which have been, or are still operating.

The importance of studying the ultraviolet range of the spectrum consist in the possibility of deriving the physical characteristics of extended outer atmospheres in single stars and of extended envelopes in close binaries, at a greater distance from the photosphere than can be obtained from the visual and photographic region. This is possible because almost all the resonance lines of the abundant ions fall in the ultraviolet, and the outer rarefied layers are optically thick in these strong lines, but optically thin in the continuum.

The only strong ground level lines observable in the visual

spectrum are the Na I lines and the Ca I and Ca II resonance lines while in the ultraviolet we observe strong ground level and metastable lines of the following ions (to cite only the most abundant): C I, C II, C III, C IV, N I, N II, N IV, N V, O I, O VI, Mg II, Si I, Si II, Si III, Si IV, S I, S II, S III, S IV, P I, P II, P III, P IV, P V, Cr II, Mn II, Fe II, Fe III. In early-type stars these are observed as absorption lines formed in the outer layers. The main results are that almost all the most luminous stars have expanding outer layers at velocities greater than escape velocity (stellar winds) and present ionization states much higher than those one could expect at the stellar photospheric temperature (superionization). In cool stars no appreciable continuum is left at $\lambda < 2000$ A; it becomes, therefore, possible to detect the chromospheric and coronal lines in emission.

The satellites employed for studying the ultraviolet spectrum (i.e. the range included between the Lyman discontinuity at 913 A and the limit at which our atmosphere cuts out almost entirely the stellar radiation, at 3100 A) have been the OAO-2 (NASA), giving low resolution spectra (RP $\sim$ 20 A) from Lyα to 3400 A; TD -1 (ESRO), with the two experiments S 59 (RP $\sim$ 2 A, spectral ranges 2070-2160; 2500-2600; 2780-2880) and S2/68 (RP $\sim$ 40 A, spectral range 1360-2540, and a broad band photometer, passband 310 A, centered at 2740 A); the OAO-3, better known as Copernicus (NASA) which has a high resolution scanning spectrometer, operating in the low resolution mode (0.2 A, 913-1500 A and 0.4 A, 2000-3000 A) and in the high resolution mode (0.05 A, 913-1500 A and 0.1 A, 2000-3000 A) and the still operating International Ultraviolet Explorer (IUE) a joint venture of NASA-ESA-UK, with an echelle spectrograph; the high resolution mode has RP $\sim$ 0.15 A and the low resolution mode $\sim$ 7 A); the spectral ranges are 1170-2000 A and 1900-3200 A.

ULTRAVIOLET OBSERVATIONS OF EARLY-TYPE INTERACTING BINARIES

Beta Lyrae : its main peculiarity is to have a massive companion ($m_2 \geq 8.5$ ⊙) with no observable spectrum. It is in a fast mass-exchange phase.

The characteristics of the system are the following:

M_V system = -3.9, d = 300 pc (as can be derived from the membership of the eclipsing binary in a multiple system containing normal-type stars). Hence the absolute visual magnitude of the primary (as we will call the star whose spectrum we observe, which

is very probably the less massive of the two) is included between
−3.9 and −3.2. The primary spectrum is B8.5 II. The mass function
$f(m) = 8.5\ \odot$. Hence for $\sin i = 1$, $m_1 = m_2 = 34$. Values of $m_1/m_2 > 1$
give unacceptably high values of the masses. The period is
$P = 12.9$ days and $dP/dt = 19$ sec/y; this constant lengthening of
the period has been observed since 1785. There is spectroscopic
evidence of mass-loss from the system (gas flowing from L_2) and
of mass transfer through L_1. Different estimates of mass-loss are
obtained: $\dot{m} = 10^{-6}\ \odot/y$ from radio flux, $10^{-4}\ \odot/y$ from optical
estimates, $5\ 10^{-5}$ from the infrared excess, 5×10^{-6} from the lengthening of the period, assuming that no mass is lost from the system,
an hypothesis which has not been verified.

Wilson (1974), from fitting observed to computed light curves,
derives $R_1 = 13\ R\odot$, R_2 equat. $= 25\ R\odot$, R_2 pole $= 6 \div 9\ R\odot$,
$a_1 \sin i = 45\ R\odot$. The flattening of the secondary (flattened star
or disk, whichever it is explains the continuous visibility of
the primary.

The ratios of the depths of the two eclipses (expressed in
fluxes) is $d_2/d_1 \sim 0.6$ between 6500 and 3500 A. The spectrum of
the primary gives $T_1 = 11500$ K; it follows that $T_2 = 9000$ from
the ratios of the fluxes, since the same areas are eclipsed during
primary and secondary eclipses. From these values of the temperatures and the values of the radii derived by Wilson, it follows
that L (B8.5)/ $L_2 = 2.25$. It is hard to understand why no trace
of the secondary spectrum is visible. A disk rotating around a
secondary with m_2 of 10 to 20 solar masses will have a velocity
of 300 to 450 Km/sec, and the spectral lines may be washed out
by rotational broadening. From the relation $d_2/d_1 = A_2/A_1 =$
$= B(\lambda_i\ T_1)/B(\lambda_i\ T_2)$ (where d, A and B are the depth, the area of
the light curve enclosed between the line of maximum luminosity
and the curve from beginning to the end of eclipse, and the Planck
function, respectively) it follows that the color of the secondary body corresponds to spectral type early-F. This was the story
prior to ultraviolet observations by satellite.

The first ultraviolet observations were made with OAO-2 and
with TD-1, S2/68; the most striking results were the following:
the small depths of the two minima at λ 1920 A and the fact that
the secondary minimum becomes deeper than the primary one at
λ 1380 A (Kondo, McCluskey and Houck, 1971), and the strange shape
of the energy distribution curve, which presents a strong bump at
λ 2000 (Hack, 1974). Both these facts can be explained by the
presence of strong emitting matter surrounding the system or one
of the two components, as was partly confirmed by high resolution

observations obtained with Copernicus (Hack et al. 1974, 1975, 1976, 1977). Observations obtained at the two minima and the two quadratures in 1973, in 1974, and during 13 consecutive days in June 1975, brought out the following facts: the far UV is dominated by P Cygni contours (with strong emission wings) of the resonance lines of the abundant ions; the expansional velocity of the absorption core is about -150 Km/s and depends on the ionization potential. Especially noticeable is the presence of the resonance lines of N V, Si IV and C IV (the latter observed with IUE, since the sensibility of Copernicus at $\lambda 1500$ is very low) and the absence of the resonance lines of O VI. IUE observations have shown that $\lambda 1640$ He II and also the semi-forbidden line $\lambda 1908$ C III are definitely absent. Only ground-based or metastable lines are present at $\lambda < 1500$, and these do not vary appreciably with the phase. Absorption lines formed in the photosphere of the B8.5 star (especially Fe II) are observable mainly in the near UV range and present orbital radial velocity shifts in agreement with those observed in the visual range. No feature which could be clearly attributed to the secondary was observed, with the exception of λ 1175 C III, which presents a radial velocity curve in antiphase with that of the primary, a gamma velocity of about -250 Km/s and a semiamplitude K2 = 70 Km/s which gives $m_2/m_1 = 2.7$ The presence in the visual spectrum of metastable lines of He I and the absence of forbidden lines and of the semiforbidden line 1908 C III makes it possible to deduce that the density of the expanding envelope surrounding the system is larger than 10^{10} cm^{-3} and smaller than 10^{12} cm^{-3}. The absence of excited lines like 1640 He II and the presence of resonance lines of highly ionized elements suggests a collisional ionizing mechanism and a kinetic temperature which ranges from 10^4 (H I, He I) to $2\ 10^5$ (N V). The ultraviolet light curves derived from Copernicus data confirm and extend the results obtained with OAO-2. The relevant facts are the following: the depth of the primary eclipse at 2800 and 2600 A is less than expected, and this can be explained by the presence of non-eclipsed material in the vicinity of the primary, as suggested by the emission of Mg II at 2800 A and of Fe II at 2600 A. The very low depths of both primary and secondary eclipse at 2000-1900 A indicate the presence of non-eclipsed emitting material in the extended envelope; this is revealed by the strong emissions of Fe III which are concentrated at those wavelengths. At shorter wavelengths ($\lambda \lesssim 1400$) and in the infrared ($\lambda > 2\mu$, Jameson and Longmore, 1976) the primary minimum becomes less deep than the secondary. Now, inversion in the relative brightness per unit surface area cannot

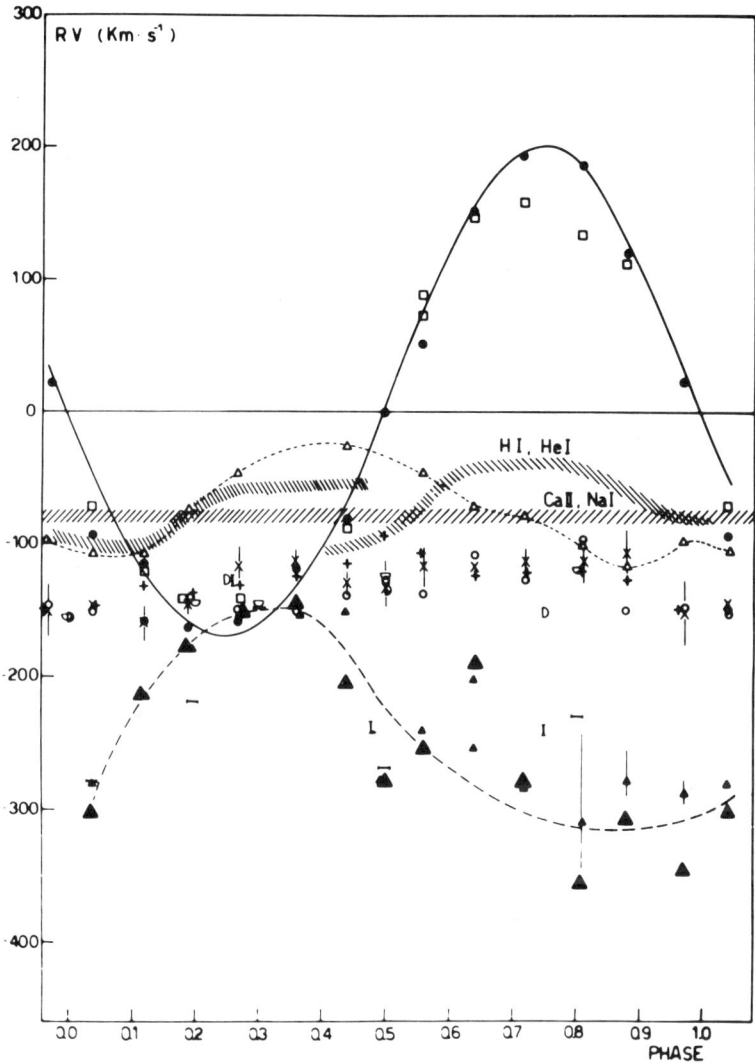

Figure 1. - Radial velocity, absorption lines. The full line is the orbital velocity curve derived from FeII, SiII, MgII lines in the visible range. ●: stellar lines; ○ :resonance lines SiIV; ◻, subordinate lines of MgII; + resonance lines MgII; ✗ excitation lines FeII, △ , SiIII $\lambda\lambda$ 1301, 1303; ▲, CIII λ 1175. The shaded areas give the radial velocities of the envelope components of Hα and $\lambda\lambda$ 6678, 5875 HeI lines, and of the circumstellar components of CaII and NaI resonance lines.

occur if the two stellar energy curves can be roughly approximated by planckian curves. Hence: a) either the secondary energy curve cannot be approximated by a planckian curve, or b) the brightness of the secondary body is not uniform and the non-uniformity becomes relevant in the infrared and ultraviolet, or c) substantial circumstellar clouds are present, or d) all these causes together are operative. Now we know that a contribution from circumbinary plasma is certainly present at some wavelengths (2800 Mg II, 2600 Fe II, 2000-1900 Fe III), but at $\lambda \sim 1380$ the regions where the continuous flux has been measured are not contaminated by emissions; hence hypothesis c) does not apply. Hypothesis b) has been explored. We expect a bright spot to be formed in the disk surrounding the companion, at the point where the stream of matter from the primary impinges on the disk. Calling the area of the spot (in units of the eclipsed area) $\underline{a}$, we can write the following set of equations:

$$\left(\frac{d_2}{d_1}\right)_{\lambda_i} = \frac{B(\lambda_i, T_{Disk}) + \underline{a}\, B(\lambda_i, T_{spot})}{B(\lambda_i, T_1)}$$

By using the light curves at all wavelengths not disturbed by the presence of strong emission lines, and putting $T_1 = 11500$ K, it is found that $T_{disk} = 8500$ K, $T_{spot} = 15900$ K and $\underline{a} = 0.1$ in the ultraviolet and visual range and there are increasing values of $\underline{a}$ in the infrared, up to $\underline{a} = 0.55$ at $8.6\,\mu$. The behavior in the infrared can be understood, because the free-free opacity of the ionized gas increases with λ^3; hence the area of the spot increases out of the orbital plane, where the density is lower.

In conclusion, the model we obtain for Beta Lyrae on the basis of ultraviolet, visual and infrared observations can be summarized as follows:

An extended expanding envelope, where the C IV, N V and Si IV resonance lines are formed, surrounds the whole system (no evidence of partecipation in the orbital motion has been observed); the kinetic temperature increases outward from 10^4 (shell lines of H I and He I) to $2 \cdot 10^5$ (N V). This suggests the existence of two regions that we will call "the chromosphere" with $T \sim 10^4$, $N_e \sim 10^{12}$ cm^{-3}, and "the corona" with $T \sim 10^5$, $N_e \sim 10^{10}$ cm^{-3}. If we plot the expansional velocities of the absorption cores versus the width of the whole line (emission wings and absorption core) measured in the continuum (in Km/s), we observe that the "chromospheric lines" (H I, He I, C II, Si II, Al II, Al III, Mg II,

Fe II, Fe III, Ni III) have expansional velocity increasing outward from the value of -120 Km/s given by the fainter lines to a maximum of about -190 given by the strongest lines of Fe III and decreasing again to the value of -100 given by Hα. The latter, because of the high optical depth in this line, must be observable farther away than the other lines, a fact confirmed by its larger width, produced by the largest velocity gradient of the whole of the layers where it is formed. The "coronal lines" have expansional velocities of about -160 -180 Km/s and are broader than the "chromosperic" lines, a fact that may indicate that the linear rotational velocity of the envelope is increasing outward. However, the total width of the C IV lines is smaller than that of the N V lines. It seems improbable that the region of formation of the C IV lines is less extended than that of formation of N V lines, because of the larger cosmic abundance of carbon, the lower degree of ionization of this ion and the comparable gf values. It is more reasonable to assume that both the expansional velocity and the rotational velocity in the "corona" decrease outward passing from -180 and 700 Km/s for Si IV to -160 and 500 for N V and to -150 and 350 for C IV.

The radial velocities of the "chromospheric" and "coronal" lines at the epoch of the second quadrature are generally less negative by 20 to 30 Km/s than at the epoch of the first quadrature, probably because of the presence of streams and condensations in the vicinity of the primary. Instead, the sharp resonance lines of Ca II and Na I give strictly constant expanding velocities of -80 Km/s relatively to the primary, suggesting that they are formed in a cooler outward stellar wind. The presence of an extended relatively cool envelope is suggested also by thermal radioemission.

A disk surrounding the companion, where the line 1175 C III is formed, is expanding at -250 Km/s. Three energy sources feed the disk: electromagnetic radiation from the primary; gaseous flow, corpuscolar radiation (producing shocks, hot spots) from the primary (mass transfer); electromagnetic radiation and other forms of energy from the secondary. The disk absorbs these forms of energy and becomes luminous itself. The rotational velocity of the disk required to balance the gravitational force ranges from 300 to 450 Km/s, for m_2 ranging from 10 to 20 m$\odot$. Hence it should be difficult to detect absorption lines formed in the thick disk, except for the strongest ones, such as 1175 C III, which is probably formed in the outer parts of the disk and excited by a collisional mechanism like friction between the inner and outer layers of the disk rotating at different velocities. This can also be the source of energy for the observed expansion. The spot on the disk at the point where the stream from the primary impinges on the secondary can explain the infrared and ultraviolet excess of the secondary.

The present observations do not make it possible to decide which is the nature of the companion. As discussed by Hack et al. (1977), the companion, which has a mass larger than 9 m☉ and probably equal to 15 m☉ and a radius smaller than 9 R☉, may equally well be a black hole accreting the thick disk, which is the only observable body, or a B-type main sequence or giant star, completely embedded in the thick disk. In the first case the system should be in the second phase of mass exchange and might represent the evolutionary stage following that probably represented by Cyg-X1. The latter should be at the beginning of the phase of mass-exchange from the hot star to the black hole, and therefore no X-ray absorbing disk has yet been formed; Beta Lyr should be at the end of this phase, and a thick disk capable of absorbing all X-rays and reradiating them in the ultraviolet and visible has formed. In the second case, we will be faced with the more common phase of first mass exchange from the originally more massive star to the companion.

<u>Upsilon Sagittarii</u>: Upsilon Sgr is a single-lined spectroscopic and eclipsing binary, characterized by a very peculiar visible spectrum showing features similar to that of Beta Ori (B8 Ia), but with much stronger lines of He I, N II, Ne I, Si II, P II, S II and much fainter Balmer lines, and metallic lines similar to that of Epsilon Aur (F0 Ia). All these lines have the same orbital radial velocity, and are therefore formed in the atmosphere of the same star. A quantitative analysis by Hack and Pasinetti (1963) gives H/He = 1/40 and C/N = 1/20 suggesting that the star has lost its hydrogen-rich envelope and shows evidence of the products of the CNO cycle. The underabundance of hydrogen has produced a transparent atmosphere simulating an F0 Ia star in its outer layers and a B8 Ia star lower down. Photometric observations by Gaposchkin (1945) indicate that the star whose spectrum is observed in the visual range is eclipsed at secondary eclipse, hence the companion should be hotter and smaller than the primary. The characteristics of the system are the following: P = 137.96 days, f(m) = 1.677, sin i = 47°, and assuming m_1/m_2 = = 0.5, m_1 = 4.8, m_2 = 9.64, while for m_1/m_2 = 1 m_1 = m_2 = 17. Values of $m_1/m_2 > 1$ give improbably high values of the masses.

Ultraviolet spectra of Upsilon Sgr were obtained with

TD-1, S2/68 experiment, and with Copernicus and IUE at the epoch of the minima and of the quadratures. The continuum clearly shows the presence of a hot companion and a reddening of about + 0.30; a very good fit of the observations is obtained with a composite spectrum formed by adding the flux of Alpha Cyg (A2 Ia) reddenned for E(B-V) = + 0.31 to the flux of Zeta Oph (O9.5 V, E(B-V) = 0.31) reduced by a factor of 35. Hence, if the primary is an A-type supergiant, M_V = - 7 or - 8, as indicated by the widths of the central emissions in the Mg II resonance lines, the companion should be an O9 dwarf which at λ 1500 has about the same luminosity as the primary, but in the visual is about 100 times less luminous than the primary. The value of F2 can be determined univocally since at each wavelength we can write one equation

$$F_{\lambda \text{total observed}} = F_{\lambda 1 \text{ correct.}} \times 10^{-0.434 K_\lambda E(B-V)} + F_{\lambda 2 \text{ corr.}} \times 10^{-0.434 K_\lambda E(B-V)}$$

where $F_{\text{total obs.}}$, $E(B-V)$ are known, F_1 can be derived by the observations of the near UV and visual where the contribution of F_2 is negligible, and, therefore, F_2 can be derived.

The line spectrum of Upsilon Sgr in the near UV is what we can expect from an early A supergiant; the far UV spectrum on the contrary, presents strong absorption resonance lines of C IV, N V, Si IV, with profiles characteristic of an expanding shell with a terminal velocity of about -700 Km/s, more negative by about 100 Km/s at the epochs of the minima than at the quadratures, suggesting a flow of matter through the lagrangian points L2 and L3. These lines can be formed in the outer atmosphere of the secondary or in an extended envelope surrounding the whole system, as the absence of any detectable orbital radial velocity shift suggests. However, they are deeper at the epoch of the secondary eclipse when the companion is in front, than at the other phases, confirming that the contribution of the spectrum of the companion is more important at this phase. One striking characteristic of the profiles of the resonance lines of the highly ionized elements is that the intensity left at $v = v_{\text{term}}$ is almost zero, this indicating that the total optical depth of the envelope is high (Castor and Lamers 1979). But the higher the total optical depth is, the higher the expected intensity of the emission wing; this, on the contrary, is completely or almost completely missing in the spectra of Upsilon Sgr. The only explanation is that a strong

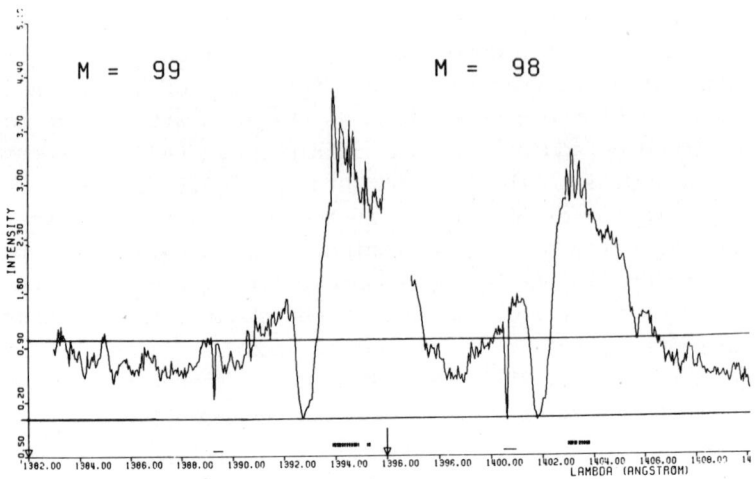

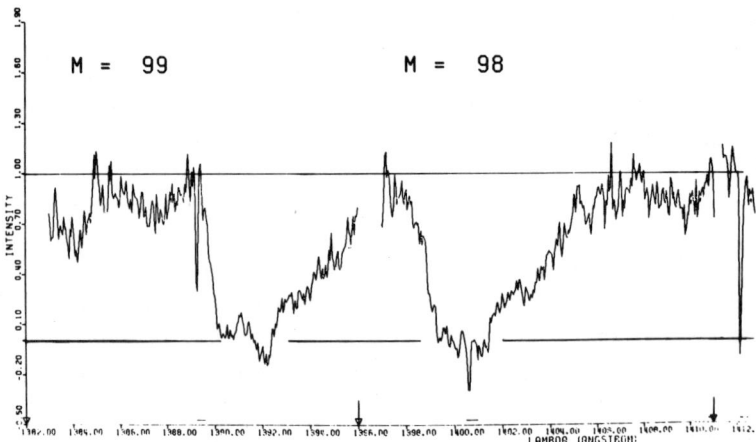

Figure 2. - The profiles of the SiIV resonance lines. Top: Beta Lyr (phase 0.51); bottom: Upsilon Sgr (phase 0.02).

absorption stellar line is present, and the theoretical computations for such a case fit our observations. Now the presence of photospheric absorption lines of C IV and N V has been observed only in early O-type stars, suggesting a spectral type earlier than O7 for the companion of Upsilon Sgr.

In conclusion, the high hydrogen deficiency of the primary suggests that it has lost a large amount of mass, and that this has been partly transferred to the companion as indicated by the gaseous stream that one sometimes observes at the epoch of secondary minimum, and has also produced the extended envelope associated with the hot companion. If the companion is equally massive or more massive than the primary, the orbital shift is less than 0.4 A in the far UV and can be detected with difficulty on such broad lines with such a complex structure. Because of the large quantity of mass lost from the primary, it seems reasonable to assume that $q = m_1/m_2 < 1$; for a value of $q \sim 0.5$ the values $m_1 = 4.8$ and $m_2 = 9.6$ are compatible with the observed spectral characteristics.

Other interacting binaries which have been observed in the ultraviolet and which present emission lines of N V, C IV, Si IV, etc., were discovered by Plavec and Koch (1979).
The peculiar eclipsing binaries W Ser, W Cru, V 367 Cyg, SX Cas, RX Cas are probably in a stage of rapid mass transfer from a cool giant to an object which in the visual region appears as a giant or supergiant A, F or G. A hot spectrum and lines from highly-ionized atoms are present in all the UV spectra of these stars. According to the model proposed by Plavec, the "yellow" giant is actually an extended, optically thick disk surrounding a hot star.

Another very hot single-lined spectroscopic binary HD 206267,06 has been observed with IUE by Burger, Hack and De Loore (1980). Stellar lines of He II and three- and four-times ionized metals are present; especially all the stronger lines of Fe V have been identified as broad, rather strong features. The profiles of these lines do not vary but a radial velocity variation is observable at the various phases. The continuum, after dereddening for $E_{b-V} = 0.37$, does not fit any theoretical models corresponding to reasonable values of T_e. It appears that $T_e \lesssim$ 20000 K, which is not acceptable for an O6 star, showing absorption lines of Fe V. Strong, broad absorption lines of N V and C IV with red emission wings are present; the profiles are typical of an expanding envelope; $v_{terminal}$ is about 3000 Km/s. The Si IV lines are formed in the photosphere . Numerous interstellar or circumstellar lines are present: among them, the resonance lines of C IV and Si IV (of circumstellar origin?) and those of C I, N I, and the molecular lines of $C^{12}O$ and $C^{13}O$.

In conclusion, all interacting binaries which have been observed in the far ultraviolet show evidence of extended expanding envelopes having a degree of ionization much higher than could be expected from the photospheric temperatures of the stars composing the system.

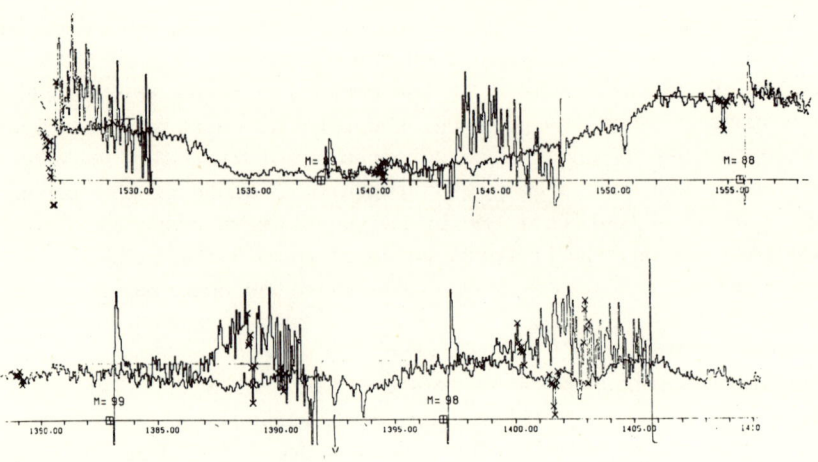

Figure 3. - The profiles of CIV (top) and SiIV (bottom) lines in HD 206267.

REFERENCES

Burger, M., De Loore, C. and Hack, M., 1980 IAU Colloquium No.55, in press.
Castor, J.I. and Lamers, H.J.G.L.M., 1979 Astroph. J. Suppl. 39, 481.
Hack, M., 1974 Astron. Astroph. 36, 321.
Hack, M., Hutchings, J.B., Kondo, Y., McCluskey, G.E., Plavec, M. and Polidan, R.S., 1974, Nature 249, 534.
Hack, M., Hutchings, J.B., Kondo, Y., McCluskey, G.E., Plavec, M. and Polidan, R.S., 1975 Astroph. J. 198, 453.
Hack, M., Hutchings, J.B., Kondo, Y., McCluskey, G.E. and Tulloch, M.K., 1976 Astroph. J. 206, 777.
Hack, M., Hutchings, J.B., Kondo, Y. and Mc Cluskey, G.E., 1977 Astroph. J. Suppl. 34, 565.
Hack, M. and Pasinetti, L., 1963 Contr. Obs. Milano-Merate No. 215.
Jameson, R.F. and Longmore, A.J., 1976, M.N.R.A.S. 174, 217.
Kondo, Y., McCluskey, G.E. and Houck, T.E., 1971 in Scientific Results from OAO-2, ed. A.D. Code (Washington: NASA SP-310), p. 485.
Plavec, M. and Koch, R.H., 1979 The first Year of IUE, p. 80.
Wilson, R.E., 1974 Astroph. J. 189, 319.

U CEPHEI VIEWED FROM MARATEA

Alan H. Batten

Dominion Astrophysical Observatory, Victoria, B.C.
Herzberg Institute of Astrophysics   Canada

The variability of U Cephei was discovered by Ceraski (1880) on the night of June 23rd, 1880. We are, therefore, within three weeks of the centenary of the discovery, and it is a good time to take stock of 100 years of progress in our attempts to understand this star. In many ways U Cephei might be regarded as a typical Algol system, but in important respects it is unusual, if not unique. It is typical in the deep primary and (at normal wavelengths) barely perceptible secondary eclipses. These, in turn, indicate the equally typical relationship of a hotter brighter primary component that is nevertheless smaller than its companion. Other typical characteristics are the distorted velocity-curve and the resulting spurious eccentricity. Unusual characteristics are the degree of that distortion, the persistence of a marked distortion of the light-curve, and above all the continually increasing period. I stress this last feature, because few systems show it (β Lyrae is the most noted example) and it is in some ways the most certain thing we know about the system.

In the 100 years that we have known it, U Cephei has certainly been important for our understanding of Algol systems. It was the first system for which the orbital eccentricities derived from the light-curve and the velocity-curve showed a significant discrepancy. Struve based his earliest studies of gas streaming in binary stars on this system and a few others. Under his direction, Hardie (1952), made the first attempt at separating the spectra of stream and star, and thus at deducing some of the characteristics of the stream. Finally, it is the first Algol-type system in which eruptive events - presumably connected with transfer of mass within, or loss of mass from, the system - have been observed with modern instrumentation by several independent workers.

Let us fix our ideas of the basic properties of the system. Unfortunately, the primary spectrum in Algol-type systems usually contains broad distorted lines that are hard to measure. Even when they have been measured, there is no certainty that the resulting radial velocity represents that of the centre of mass of the star. In some systems, as I have recently found for U CrB, there are weak lines in the primary spectrum that give reliable velocities. No lines of this kind can be seen in the photographic region of the spectrum of U Cephei, although Kondo et al (1979) suspect that there may be some in the far UV. I have had to do the best I could by studying those few hydrogen-line profiles that appear undistorted. Secondary spectra of Algol systems are normally not visible in the photographic region, except during the total phase of the primary eclipse. They often can be recorded in the red or infrared. Popper, working photographically, and Lambert and Tomkin, with a reticon, have made important contributions to our knowledge of several Algol systems. Popper felt that no reliable measurements could be made photographically of the secondary D lines in the spectrum of U Cephei, while Tomkin is accumulating reticon observations. At present, my own attempts to measure the D lines and Hα in the secondary spectrum give the best available values for the masses, although they may soon be superseded. They are (Batten, 1974) $m_B \sin^3 i = 4.2 \pm 0.6 m_\odot$ and $m_G \sin^3 i = 2.8 \pm 0.5 m_\odot$.

The solution of the light-curve is also difficult and uncertain, because of the well-known distortions in it. Progress has been made since my 1974 paper. Hall and Walter (1974) and Markworth (1979) have made more thorough studies of the light-curve than I did. They agree that the orbital inclination and the radius of the primary component are both somewhat smaller than Dugan (1920) and I found. The radius of the secondary and the masses are scarcely affected ($\sin^3 i$ is still nearly unity) so the secondary fills its Roche lobe. The new results remove, or substantially reduce, the discrepancy I found between the observed and calculated eclipse depths, and I think we should accept them. They give radii of around $2.3 R_\odot$ and $4.7 R_\odot$ for the B and G stars respectively. (I am assuming the spectral tupes of B7V and GIII-IV that I derived in 1974.)

As I have said, the period change is one of the best determined facts about this system. Minima observed up to 1972 gave a mean rate of increase of $\Delta P/P = 5.5 \times 10^{-9}$ (about 1 second/year). Departures from this mean rate are obvious, but some care is needed in their interpretation. We now know that the period decreased suddenly in 1972 (Crawford and Olson, 1979), and apparently increased again, by a lesser amount, in 1974. Crawford (1979) has also shown that sometimes the distortion of the light-curve can cause the time of mid-eclipse to be wrongly estimated.

We cannot go back over the decades of visual observations and decide how often this has happened, but we do know now that some fluctuations in the period changes are only apparent, and we need not devise a dynamical explanation for every detail of the (O-C) diagram. My own interpretation of the diagram was to postulate a steady process within the system (mass transfer) responsible for the mean rate of increase in the period, and to suppose that some erratic and intermittent process was making the period increase now faster, now slower than the mean rate. Such alternating random period variations are found in virtually all semi-detached systems - thus I suggested that U Cephei differed from these only in that random changes in its period were superposed on a steady increase.

Biermann and Hall (1973), however, proposed an attractive theory of period changes that would require only one process operating in all systems. They suppose that mass transfer occurs in discrete bursts which usually lead to a sharp *decrease* in period. This occurs because the mass-receiving star cannot immediately accept the angular momentum of the incoming matter, which, therefore, is stored in a disk rotating around the star, or in the outer layers of the star itself. The matter is gradually absorbed by the star proper, and its angular momentum is returned to the orbit, thus increasing the period. Period *increases*, therefore, should be gradual: only the *decreases* are abrupt. The only difference between U Cephei and other systems is that in the former enough mass and angular momentum are transferred for the net long-term effect to be a progressive increase in period. The theory is attractive in many ways. The very idea of a sudden period change implies that mass transfer is not a steady process. Many systems are known to have disks rotating around the primary - we know now that U Cephei sometimes has such a disk. Mass-receiving components are often rotating rapidly - as is especially the one in U Cephei - and this rapid rotation may well be confined to the outer layers. The theory also has the advantage that period changes in all Algol systems are explained by one mechanism. Shortly after Biermann and Hall published their theory, U Cephei itself (to which they had applied) obliged with a spectacular outburst, and the period was observed to decrease. Thus an abrupt period decrease was observed, and it appeared at first to be associated with clear evidence of mass transfer. We still have to determine, however, whether or not period increases are gradual. A collection of (O-C) diagrams, such as that compiled by Frieboes-Condé and Herczeg (1973) for example, gives the impression that some period *increases* are abrupt, although the scatter is often too large for one to be sure, and, as those authors point out, the appearance of an (O-C) diagram is affected by the base ephemeris chosen for its presentation. More important, Crawford and Olson (1979) find evidence for a sudden decrease) ($\Delta P/P = 3 \times 10^{-5}$)

in the period of U Cephei in 1972 (before the main activity began) followed by a sudden but appreciably smaller increase ($\Delta P/P = 6 \times 10^{-6}$) late in 1974.

Quantitative interpretation of period changes is more controversial. If the mean rate is interpreted as being entirely the result of *conservative* mass transfer, the rate required by the masses adopted is about $2 \times 10^{-6} m_\odot$/year. If it were caused entirely by *isotropic* mass loss from the system, about twice as much matter would be involved. The mechanism proposed by Biermann and Hall, if it acts conservatively, requires about 10 times the rate of mass transfer ($2 \times 10^{-5} m_\odot$/yr) as does the simpler method. We now have some evidence (from UV spectrograms) for mass leaving the system (Kondo et al, 1979) so that it is very difficult, without a fuller knowledge of details, to relate the changes in period to the amount of mass either leaving the system or being transferred within it. It is puzzling that although part of the period change is clearly intermittent, as are events like those of 1974-5, circumstellar matter is always present (although not always in the form of a disk) and both the light-curve and the velocity-curve are *always* distorted and, except during outbursts, always in much the same way.

Even when the system is quiescent, the light during totality is not always constant. I believed in 1974 that this was primarily because of variations in the light of the secondary star. Hall and Garrison (1972) had earlier suggested that the source of variation was the circumstellar matter. The events of 1974 and 1975 leave no doubt that variations in the circumstellar matter play a considerable role, but during some eclipses, the variations observed during totality would be difficult to explain by this agency. If the subgiant really is throwing matter across to its companion (or out of the system) it *must* be varying, and Olson (1980) has produced evidence that the G-star does sometimes vary: he believes that the outer hemisphere brightens before an outburst. For this reason, I am sceptical of the attempt by Hall and Walter to relate the varying slope during totality with the theory of period changes put forward by Hall and Biermann. Hall and Walter say that when the stream is at its strongest, matter should accumulate on the leading side of the primary star, and, as this accumulation emerges from eclipse (ahead of the primary star itself) the light received from the system will increase, even during totality. They believe the stream will be at its strongest about the time of a sudden decrease in period, and that the occurrence of upward slopes will be correlated with the times of such decreases. I do not find the correlation they present very convincing. Crawford and Olson (1979) suggest that period changes and the *visible* evidence of mass transfer are not correlated. Moreover, the observations by Huffer and Code published in my 1974 paper show that the slope during totality can change very quickly. I doubt if there is a *simple* relation between the shape of the light-curve, quantity of circumstellar matter and behaviour.

These variations during totality are probably closely linked in origin to the ultraviolet excess. The 1974-5 outburst makes it clear that the greater part of this excess also comes from the circumstellar matter, and I have to retract my earlier suggestion that a substantial part of it comes from the G8 star. Hardly any origin other than circumstellar matter can account for the rapid variations observed in September and October 1974. As late as 1972, it could still be seriously argued that this ultraviolet excess might be due to metal underabundance. Rapid variations in the excess rule this out, as did the work of Baldwin (1973) and Parasarathy et al (1979). A more general and cogent argument for ascribing the excess to circumstellar matter is provided by the recent work of Plavec (1980). It is not yet fully published, but a preliminary account of it is given in the Proceedings of last year's symposium in Toronto. He has obtained IUE spectra of several close binaries, and finds that many of them (W Ser, V367 Cygni, W Cru, β Lyr and SX Cas) have ultraviolet continua much stronger than would be expected from the spectral classification made in the visual-photographic region of the spectrum. We know that all these systems contain shells or streams, and Plavec ascribes the UV continuum to a plasma between the stars. He believes the systems to be in the rapid phase of Case B mass transfer. He did not observe such a continuum in the spectra of U Cephei and other Algol systems - but they were not active when he observed them.

The outburst of 1974-5 has stimulated studies of U Cephei, and it should now be described. It is unclear how many separate incidents are involved. The first signs of activity were when Bakoš and Tremko (1973) observed an eclipse in 1969 that was 20 minutes late. Dynamically this is impossible, since neighbouring eclipses occurred at close to the predicted times, but it may be at least partly explained by distortion of the light-curve: the total phase was short. A few weeks later, I detected very weak Balmer emission during an eclipse. Next, in the spring of 1972, Naftilan (1975) saw much stronger Balmer emission during eclipse, although he did not publish this information until after the 1974 outburst. All was normal in the fall of 1973: we observed no emissions, and Coyne (1974) failed to detect any polarization of clearly stellar origin in the light of U Cephei. Early the following summer (1974), Rhombs and Fix (1976) observed variations in the UV flux received from U Cephei, and then Plavec and Polidan observed strong Hα emission, while Baldwin, quite independently, found very strong Balmer emission at Victoria. We also found a very distorted light-curve at Victoria, but much better and more extensive photometric observations (which confirmed ours) were obtained by Olson. Piirola (1975) also observed polarization changes. The system appeared to have quietened down by the end of the year, when the most favourable observing season (for North America) had passed, but the activity had renewed by the fall of 1975, although it was in some respects different. Matters eventually settled down again, although Olson has recently reported new signs of photometric activity. I have not yet had a chance to look for emission lines. The analysis of period changes made by Crawford and Olson (1979) is of interest in this connection. It suggests that the major decrease occurred in 1972, before the apparently most active time and that the obvious observational manifestations of 1974 were accompanied by only a modest increase.

In a series of four papers (1976, 1978, 1979, 1980), not all of which are yet published, Olson has tried harder than anyone else to interpret these observations. He has established a "normal" or undisturbed light-curve of the system, and studied the departures of particular light-curves from this. These departures he ascribes mainly to circumstellar matter (in the form of both disk and stream) but also partly, as we have seen, to intrinsic variations of the subgiant. He analyzes variations in the out-of-eclipse light-curve too, and they have proved very important. Prominent dips are sometimes observed at $0\overset{p}{.}2$ and $0\overset{p}{.}6$, especially in UV light. Olson has found that the simple assumption that the stream affects the light-curve only by electron scattering is insufficient. He needs to take account of both absorption and emission (as I did in interpreting the spectrum) by the optically thick stream and disk, radiation from a hot spot ($\sim 50,000°K$) where the stream hits the disk, and even diminished radiation from cool spots ($\sim 10,000°K$) on the surface of the B7 star. Stream and disk behave like extensions of the stellar atmosphere (and their thickness perpendicular to the orbital planes, at times, an appreciable fraction of the radius of the B star). It is not easy to treat them by model-atmosphere methods, however, because they are not hydrostatically stable. Observations show that the disk can collapse in about 10 days, if the stream is cut off, and form again in a comparable time. Olson finds asymmetries in the disk; the trailing edge is about 500°K hotter than the leading, although this difference may not be significant. Olson estimates a minimum disk mass of the order of 2×10^{-10} $m_\odot$, and a mass-transfer rate of from 0.3×10^{-6} to 2×10^{-6} $m_\odot/yr$.

REFERENCES

Bakoš, G.A. and Tremko, J.: 1973, Bull. Astr. Inst. Csl., 24, p. 298.
Baldwin, B.W.: 1973, Publ. Astr. Soc. Pacific, 85, p.714.
Batten, A.H.: 1974, Publ. Dominion Astrophys. Obs., 14, p.191.
Biermann, P. and Hall, D.S.: 1973, Astron. Astrophys., 27, p.249.
Ceraski, W.: 1880, Astr. Nachr., 97, p.319.
Coyne, G.V.: 1974, Ric. Astr. Specola Astr. Vatic., 8, p.475.

Crawford, R.C.: 1979, Publ. Astr. Soc. Pacific, 91, p.111.
Crawford, R.C. and Olson, E.C.: 1979, Publ. Astr. Soc. Pacific, 91, p.413.
Dugan, R.S.: 1920, Contr. Princeton Univ. Obs., No. 5.
Frieboes-Condé, H. and Herczeg, T.: 1973, Astron. Astrophys. Suppl., 12, p.1.
Hall, D.S. and Garrison, L.M.: 1972, Publ. Astr. Soc. Pacific, 84, p.552.
Hall, D.S. and Walter, K.: 1974, Astron. Astrophys., 37, p.263.
Hardie, R.H.: 1952, Astrophys. J., 112, p.542.
Kondo, Y., McCluskey, G.E. and Stencel, R.E.: 1979, Astrophys. J., 233, p.906.
Markworth, N.L.: 1979, Mon. Not. Roy. Astr. Soc., 187, p.699.
Naftilan, S.A.: 1975, Publ. Astr. Soc. Pacific, 87, p.321.
Olson, E.C.: 1976, Astrophys. J., 204, p.141.
Olson, E.C.: 1978, Astrophys. J., 220, p.251.
Olson, E.C.: 1979, Preprint.
Olson, E.C.: 1980, Preprint.
Parthasarathy, M., Lambert, D.L. and Tomkin, J.: 1979, Mon. Not. Roy. Astr. Soc., 186, p.391.
Piirola, V.: 1975, Inf. Bull. IAU Comm. 27, No.1061.
Plavec, M.: 1979, UCLA Preprint, No. 86.
Rhombs, C.G. and Fix, J.D.: 1976, Astrophys. J., 209, p.821.

CERTAIN ASPECTS OF THE PROBLEM OF MASS TRANSFER IN SEMI-DETACHED BINARIES

E. Budding

Department of Astronomy,
University of Manchester, England.

ABSTRACT

The subject of the paper is the "Roche lobe overflow" mechanism in semi-detached close binary systems, with especial reference to classical Algols. The approach has been influenced to a large extent by the work of Lubow and Shu (1975), but certain topics not followed up by those authors are re-examined. These fall under the following headings:

a) Initial conditions on the stream from the contact component,
b) The flow problem in the orbit region,
c) Some possible non-conservative effects in the stream,
d) The effect of non-zero inclination of the rotation axis of the detached star, and
e) The role of radiation pressure and/or a stellar wind from the detached component.
A final section briefly refers to observational work.

With reference to headings (a) and (b), advantageous possibilities of the use of the method of characteristics devoted specifically to stream calculations are pointed out and specimen results presented. Under heading (c) Lubow and Shu's equations are generalized to include a phenomenological approach to turbulence effects by means of frictional terms. Results are compared with numerical calculations of various authors. The slight generalization associated with (d) suggests complications to extensive disk formation while the estimates of section (e) imply a cut-off effect at intermediate B type primaries for classical Algol Roche lobe overflow through L_1 binaries. Potential "Algols" with early type primaries are likely to evolve quite differently. The consideration of observational work is largely restricted to narrowband photometry and polarization studies. There are few findings which, as yet, give unequivocal details, but there is evidence that the predictions of Lubow and Shu are not always exactly fulfilled, which can therefore be used to support alternative considerations, such as those explored in the present work.

INTRODUCTION

Problems connected with mass transfer effects in close "semi-detached" binary systems have received considerable attention since the early discussions of the "Algol paradox" (see, e.g., Paczynski's (1971) review; or Proceedings of IAU Symposia No. 73 (Cambridge, 1976) and No. 88 (Toronto, 1979)). There have developed a variety of approaches and applications for this work, though one linking idea is that of "Roche lobe overflow". The operation of the mechanism is sometimes taken for granted in the more general context of binary evolution with mass loss/transfer, though details should be of great significance in relation to the understanding of currently semi-detached systems. Moreover, there is some area of broad agreement between the results of numerical programs for the hydrodynamic problem like those of Prendergast and Taam (1974); Lin and Pringle (1976); Sørensen et al (1975); Flannery (1975), and semi-analytical treatments, a notable example of which was that of Lubow and Shu (1975, hereafter referred to as LS), though certain differences remain.

The approach of the latter study permits a more analytical enquiry into certain areas of the discussion, and it is this which provides subject matter for the present paper. LS presented a complete discussion of the Roche lobe overflow mechanism, which was divided into four main areas, starting with the subsonic flow in the contact component and arriving at a picture of an accretion disk around the detached component - apparently to give a self consistent whole which could be related with certain observational evidence.

All aspects of this problem cannot be said to have been fully explored, however, and we wish to consider certain topics, either referred to briefly by LS or not followed up, to see what there may be in the way of alternative possibilities or ideas. We shall follow the broad plan of subdivisions into the various flow regions, and in particular discuss the following points: (a) initial conditions on the stream from the contact component, (b) the flow problem in the orbit region, (c) some possible non-conservative effects in the stream, (d) the effect of non-zero inclination of the rotation axis of the detached star, and (e) the role of radiation pressure and/ or a stellar wind from the detached component. A final section will consider briefly observational possibilities relating to the topics discussed and a "classical Algol" picture will be underlying most of the discussion when a real context is sought.

1. INITIAL CONDITIONS ON THE STREAM FROM THE CONTACT COMPONENT

There is a general suspicion that the transfer of matter between components in a close binary system entails some sort of asymmetry, i.e., the transferred fluid tends to fall behind the detached component as this object moves in its orbit about the system centre, and matter veers towards the following hemisphere. The idea can be intuitively associated with the effect of inertia in a rotating coordinate system or the role of the Coriolis force.

In order to start their flow off on the right path (so to speak) LS appeal to the action of the Coriolis force in the flow system in the outer envelope of the contact component, making an analogy with the well known geostrophic approximation of meteorology where the velocity field is obtained from the requirement that the Coriolis force should just balance the pressure gradient. LS argue that this gradient should be directed along a meridian towards the poles, and from the equation

$$\rho u_\perp = \frac{\hat{\xi}}{2e_{z\xi}} \times \nabla_\perp(\varepsilon^2 \rho) \qquad (1.1)$$

($\hat{\xi}$ denotes unit vector in the direction of increasing ξ, $e_{z\xi}$ is the cosine of the angle between $\hat{z}$ and $\hat{\xi}$, other details of the notation are given in LS, or, in any case, will be explained in the next section), deduce a net flow in the sense counter to the orbital revolution.

Actually, (as mentioned by LS) this approximation must break down near the equator, a region which includes the important inner Lagrangian point (to which we also refer as L_1) - the starting point for the stream into the "Roche lobe" of the detached star, since the Coriolis force effectively disappears in this zone. Whether or not there could be a significant surface current in the required sense is not clear - the detailed nature of possible mechanisms controlling the equatorial current distribution in rotating fluid masses is likely to be a complex subject (see, for example, Greenspan, 1968; and references cited therein). What may happen in a classical Algol subgiant component cannot be regarded as settled at the present time, but from the reference to meridional circulation it is evident that more than just the mass transfer process is involved in shaping the flow regime in the atmosphere of this star.

Here, however, it is only pointed out that, though, as will be seen later, forces acting upon the stream in the orbit region tend to have a restorative character about a particular direction, thus giving some significance to a certain deflection angle value, the actual direction taken by a stream in the orbit region is sensitive to its initial conditions at the entrance to this region. This is basically because travel times in the orbit region are short compared to the characteristic time associated with the net restorative action. The LS deductions about stream conditions at the exit from the L_1 region are based on the properties of their particular solution, which is chosen to be self consistent with various asymptotic representations. In actuality, though, the density cannot be infinite at the L_1 point and particulars of the character of the flow system in the subsonic wedge must influence what happens further upstream. In this paper dynamics of the stream will be regarded as an initial value problem in the supersonic wedge. Our response to the general problem referred to by LS will be to consider varying boundary conditions on the initial segment.

2. THE FLOW PROBLEM IN THE ORBIT REGION

A common starting point to consideration of the flow of matter past the vicinity of L_1 has been to follow the "straight line" solution of the two-dimensional particle motion approximation. For ballistic trajectories the equations of motion assume the form (see, e.g., Kopal, 1959)

$$\ddot{X} - 2\dot{Y} = -\frac{\partial \phi}{\partial X} ,$$
$$\ddot{Y} + 2\dot{X} = -\frac{\partial \phi}{\partial Y} , \qquad (2.1)$$

where ϕ, the potential, is of the form

$$\phi = \phi_0 - \frac{(1 - \mu)}{((X - \mu)^2 + Y^2)^{\frac{1}{2}}} - \frac{\mu}{((X - \mu + 1)^2 + Y^2)^{\frac{1}{2}}} - \tfrac{1}{2}(X^2 + Y^2); \qquad (2.2)$$

X and Y being rotating rectangular coordinates in the plane of the two stellar centres, of fractional mass μ for the detached component (in the direction of negative X), $1 - \mu$ for the "contact" star; and centred on their centre of gravity. ϕ_0 is an arbitrary constant. The time is taken to be the reciprocal of the angular velocity Ω and the distance unit the separation of the two stars (d in absolute units), so that scaling constants are removed from this basic form of what is equivalent to a version of the classical restricted three-body problem. (All works with which the author is familiar, that have addressed the Roche lobe overflow problem in detail, have separated variables to give a two-dimensional treatment of the dynamic problem while vertical structure is, at best, regarded as a perturbed hydrostatic problem. This may be too crude a formulation for certain contexts, particularly where high energies or thick accretion disks are involved (see, e.g., remarks made by Rees, 1976), though the restricted problem which we shall give attention to can itself support much further detailed study.)

The "straight line" solution for a small disturbance from L_1 is of the form

$$x = ae^{\lambda t}, \quad y = mae^{\lambda t}, \quad u_x = \lambda ae^{\lambda t}, \quad u_y = m\lambda ae^{\lambda t} ; \qquad (2.3)$$

where a, m and λ are constants, and, for convenience, we have transferred to a coordinate system, x, y centred on L_1. u_x, u_y are components of velocity vector $\underline{u}$; m and λ are not arbitrary but set by the form of the potential which, with the new coordinate scheme can be written for small displacements about L_1 as

$$\phi = \phi_o - (2A + 1)\frac{x^2}{2} + (A - 1)\frac{y^2}{2} \qquad , \qquad (2.4)$$

where A is a positive constant given by

$$A = \frac{1 - \mu}{|(X_{L_1} - \mu)|^3} + \frac{\mu}{|(X_{L_1} - \mu + 1)|^3} , \qquad (2.5)$$

X_{L_1} being the X coordinate of the inner Lagrangian point L_1. For the "Roche model" this can be shown to depend only on the mass ratio (see, e.g., Kopal, 1959, p.134). The constant a is the surviving one of four arbitrary constants which should be required in general; the other three are removed by the condition that $u_x \to 0$, $u_y \to 0$ as x, y $\to$ 0; the particle tending exponentially to a state of rest at L_1 along a straight line trajectory as $t \to -\infty$. The direction of the emergent motion is given by m . It can be specified as

$$m(= \tan \theta_s) = \frac{\lambda^2 - (2A + 1)}{2\lambda} \quad (<1) \quad , \qquad (2.6)$$

where $\lambda^2 = \frac{1}{2}\{(A - 2) + (9A^2 - 8A)^{\frac{1}{2}}\}$. Note here that the deflection angle θ_s depends only on the mass ratio. Appreciation of the role of the Coriolis force requires a special thought, since the latter, though generally associated with a sideways deflection of the kind apparently given in this solution, is proportional to the velocity. In fact, the particular solution chosen is just that one which equilibriates the Coriolis force to the sideways component of gravity (modified to include the centrifugal force). This can be easily seen to be possible as long as both gravity and forward velocity can be adequately represented as linearly proportional to the distance travelled. The solution (2.3) evidently assures the latter, but this was obtained on the basis of a Taylor expansion for the potential (2.4) which stops short at terms of the second order in x. The clear satisfying of this equilibrium relation therefore breaks down as the trajectories curve away from the straight line solution under the influence of higher order terms in the expansion. In fact, gravity rises more rapidly than the Coriolis force, and can later give rise to some centripetal action on the stream. It is, however, this sideways equilibrium condition which identifies the direction θ_s as discernible direction of "flow" in the event of many particles being released from L_1 in a "stream", since in other directions a non-zero sidways velocity gradient would either produce a more rapid lateral separation of particles, or, as actually happens close to this direction, effect a restorative motion towards it: in the case of an initial ejection below θ_s due to a net gravitational attraction towards the central axis and in the case of an ejection above θ_s due to a Coriolis repulsion from it. In the direction θ_s the stream spreading is stationary and particles are able to coherently travel along.

LS generalized the same argument to the fluid flow case, and allowing the pressure term to be of second order in the approximation considered, showed, perhaps not surprisingly, that when substituted as a particular solution with what is effectively the "straight line" particle trajectory for the stream direction that the fluid equations could be satisfied to first order.

The motions of fluid elements in a continuous medium are, however, subject to constraints not present for individually moving test particles; for example a subtle influence on the motion of a fluid can be exercised through its characteristic property of vorticity. This can be seen if, instead of, for example, substituting a particular form of solution into the absolute vorticity free (frictionless) momentum equations, we consider the vorticity equation as a separate expression. The momentum equations just referred to can be usefully processed by multiplying them scalarly by $\underline{u}$, and combining the resulting form with the continuity equation, prior use having been made of an equation of state to write the pressure variation in terms of a density variation. (The isothermal argument of LS can be adhered to for convenience in this discussion - alternative generalizations need not affect the present purposes). Bernoulli's integral appears as a consequence of the first operation and the second leaves us with a convenient quasi-linear form in which the velocity derivatives are expressed in terms of velocities and potential gradients only.

In a similar way a vector multiplication of the momentum equations by $\underline{V}$, producing the so-called vorticity equation, and subsequent combination with the continuity equation establishes Kelvin's circulation theorem for the absolute vorticity, which may then provide another equation to close the system for the two velocity components.

The problem thus formulated, which is one of supersonic flow in the region past L_1, is practically in a classical form for treatment by the method of characteristics (cf., Forsythe and Wasow, 1960), which therefore, since economic and accurate algorithms are available (e.g., Smith and McCall, 1970), offers interesting possibilities for computer experimentation.

The momentum equations can be given in the form (two dimensional discussion)

$$(\underline{\nabla} \times \underline{u}) \times \underline{u} = -\underline{\nabla}(B) - 2(\hat{z} \times \underline{u}) \quad , \tag{2.7}$$

where B is the Bernoulli integral – the fluid motion counterpart of the Jacobi integral for the particle approximation – and is given by

$$B = \tfrac{1}{2}|\underline{u}|^2 + \phi + \varepsilon^2 \log \sigma \tag{2.8}$$

(where ε is the velocity of sound in the adopted units, in which the orbital velocity is unity), which would be constant along a streamline in the absence of shocks.

The continuity equation can be simply written as

$$\underline{\nabla}(\sigma \underline{u}) = 0 \quad , \tag{2.9}$$

where σ is a two-dimensional version of the more usual density function, ρ.

Let us now denote by p, q, r and s the velocity derivatives $\dfrac{\partial u_x}{\partial x}$, $\dfrac{\partial u_x}{\partial y}$, $\dfrac{\partial u_y}{\partial x}$ and $\dfrac{\partial u_y}{\partial y}$; and consider the integration of the equations of motion from a given boundary. The continuity equation affords us one linear relation of the form

$$p + s = f_1 = -\underline{u} \cdot \underline{\nabla}(\log \sigma) \quad , \tag{2.10}$$

where we take f_1 to be a known function along the initial boundary.

The relative vorticity $(\underline{\nabla} \times \underline{u})$ can be written as

$$r - q = f_2 \quad , \tag{2.11}$$

where f_2 is related to the absolute vorticity F_2 in the adopted units simply by $F_2 = f_2 + 2$. Kelvin's circulation theorem can be obtained, by the previously mentioned operations, as

$$(\underline{u} \cdot \underline{\nabla})\frac{F_2}{\sigma} = 0 \quad , \tag{2.12}$$

which leads us to expect absolute vorticity to tend to a low value upstream if that is what is happening to the stream density.

Carrying out the scalar multiplication by $\underline{u}$ and effecting a combination with the continuity equation, referred to before, we obtain

$$(u_x^2 - \varepsilon^2)p + u_x u_y (r + q) + (u_y^2 - \varepsilon^2)s = f_3 \quad , \tag{2.13}$$

where

$$f_3 = -(u_x \frac{\partial \phi}{\partial x} + u_y \frac{\partial \phi}{\partial y}) \quad .$$

With this and Equation (2.11) we have two equations only, though involving four unknown partial derivatives. The method of characteristics consists in finding direction specifiable with respect to the coordinate axes, in which these four reduce to a pair of total derivatives, so that Equations (2.11) and (2.13) become a pair of ordinary differential equations.

The basic pair of equations we are dealing with are of the form

$$ap + bq + br + cs = f_3 \quad ,$$
$$r - q = f_2 \quad , \tag{2.14}$$

which can be easily shown to reduce to the canonical form (see, e.g., Forsythe and Wasow, 1960)

$$ap_{\sigma_1} + (b - \sqrt{b^2 - ac})r_{\sigma_1} = f_3 - f_2\sqrt{b^2 - ac} \quad ,$$
$$ap_{\sigma_2} + (b + \sqrt{b^2 - ac})r_{\sigma_2} = f_3 + f_2\sqrt{b^2 - ac} \quad , \tag{2.15}$$

where p_{σ_i} and r_{σ_i} (i = 1,2) denote total derivatives $\frac{du_x}{d\tau_i}$, $\frac{du_y}{d\tau_i}$, with respect to a variable τ_i measuring distance along the characteristic curve. The gradients of these characteristics are given by

$$\sigma_i = \frac{b \pm \sqrt{b^2 - ac}}{a} \tag{2.16}$$

Substituting for a, b and c in Equation (2.16) we observe

$$\sigma_i = \frac{u_x u_y \pm \varepsilon\sqrt{u^2 - \varepsilon^2}}{u_x^2 - \varepsilon^2} , \qquad (2.17)$$

with $u^2 = u_x^2 + u_y^2$.

If we consider the case where ε is very small (the limit considered by LS) these characteristic directions tend to the same value $\sigma_{1,2} = u_y/u_x$. The local segments of characteristic curves, which bestraddle a local segment of a stream line and meet on that stream line at the apex of the local Mach Cone, in this case are being drawn out and almost parallel to the stream lines - in effect tending towards "the ballistic trajectories of free particles" (LS). In the transonic limit, which may be more applicable in the region of L_1, $\varepsilon \lesssim |\underline{u}|$, and we find, with $u_x^2 + u_y^2 \sim \varepsilon^2$,

$$\sigma_i \simeq -\frac{u_x}{u_y} . \qquad (2.18)$$

The characteristics fan out to be almost perpendicular to the flow lines, with the corresponding implication that progress by numerical integration into the flow field should be slow.

Substituting the appropriate values for a, b and c in Equations (2.15) we have,

$$(u_x^2 - \varepsilon^2)p_{\sigma_1} + (u_x u_y \mp \varepsilon\sqrt{u^2 - \varepsilon^2})r_{\sigma_i} = f_3 \mp f_2 \varepsilon\sqrt{u^2 - \varepsilon^2} . \qquad (2.19)$$

Let us compare these equations with the ballistic trajectory limit formed from Equations (2.7), (2.8), (2.9), (2.10) and (2.11) by dropping the pressure or density variation term, i.e.,

$$u_x p + u_y r - (2 + f_2)u_y = -\frac{\partial \phi}{\partial x} ,$$

$$u_x q + u_y s + (2 + f_2)u_x = -\frac{\partial \phi}{\partial y} ;$$

$$p + s = 0 ,$$

$$r - q = f_2 . \qquad (2.20)$$

The four equations can be solved for any of the derivatives; for example

$$r = -\frac{(u_y\phi_x + u_x\phi_y) + 2(u_y^2 - u_x^2) + f_2 u_y^2}{u^2} . \quad (2.21)$$

Now consider an initial outflow from a short segment perpendicular to the x axis in the vicinity of L_1, for which $u_y = 0$ and $q = 0$. From the earlier discussion we could expect the Coriolis force to introduce a lateral deflection, so that after some distance $u_y \neq 0$. But consider now

$$\delta u_y = r\delta x + s\delta y = r\delta x + (f_1 - p)\delta y . \quad (2.22)$$

Since δy is initially zero at the centre of the stream, and can only become non-zero if $\delta u_y \neq 0$, the question of any first order deflection of the stream depends on the behaviour of the coefficient of δx along the x axis. By the method of characteristics approach we have initially $\sigma_i = \frac{\pm \epsilon}{\sqrt{u^2 - \epsilon^2}}$ and so from symmetry in Equation (2.19) we can see that r is simply determined by the relative vorticity f_2. However, from Equation (2.21) $r = -2$ initially, whatever value of relative vorticity is substituted into Equation (3.21), (though $r = -2$ is, in fact, equivalent to zero *absolute vorticity* in the method oc characteristics approach). This latter result shows up in the asymptotic solution of LS for the orbit region, which thus has asymptotic resemblance to the system (2.20). Why does this apparent dichotomy occur and what is its significance?

In order to appreciate this we should realize that the system (2.20) is, in fact, overdetermined, since there are only two real unknowns, i.e., u_x and u_y, the coordinate system being at our disposal. Hence the first two of Equations (2.20) alone are sufficient in order to calculate the orbits of particles in the restricted three-body problem (subject to suitable initial conditions) which is what the physical assumptions leading to this system mean the equations must be referring to. In the way we have approached it, however, it would seem that we are referring to classes of simultaneous trajectories, rather than the calculation of an individual particle orbit. The two additional equations then act as constraints on such trajectories - that they should not cross each other and be characterized by a certain given tendency to curve. From this point of view, the last of Equations (2.20) does not have the meaning of a vorticity equation, in the sense that the rotation of an individual particle is not required, through continuity, to influence the state of motion of any other particle, as in a fluid.

If we take a more fluid based point of view we must expect, as the method of characteristics indicates, that the relative vorticity implied by synchronous rotation will tend to make the outflowing gas retain its circulation (with respect to stationary axes) about the source, thereby allowing it to begin to move in such a way that the net direction of flow apparently avoids the effect of Coriolis deflection, at least initially. Such vorticity conditions can be reflected in the data provided along the initial line chosen. The subsequent fall of density allows the relative vorticity to decline towards -2, as argued by LS, so that the two approaches will coincide at low density (i.e., upstream from L_1).

In Figures 1 and 2 we present some results of applying the method of characteristics to the flow equations (2.11) and (2.12).

The method of characteristics has been applied iteratively to construct successive sections of the stream. In principle the number of solution points calculated after one iteration is less than the number of data points on the original initial line - by two in the case of the direct version of the method in two dimensions (Massau's method) and by four in the case of Smith and McCall's algorithm, which features extrapolation to the discretization limit. This modification allows the solution points to have an accuracy $O(h^3)$ while Massau's method alone has corresponding accuracy $O(h)$, where h is the subdivision increment on the initial line. The way the solution is continued is to run a smooth fitting polynomial through the solution points and extrapolate for the four extra points required (two on either side). This builds in an assumption that the stream has a smooth character across each section. In reality, disturbances from the boundary region of the stream orifice would propagate into it along the surrounding Mach cone. But this does not necessarily affect the method, since the initial line chosen can be arbitrarily narrow within such an orifice. If no shocks cross the stream path over the integration region the assumption of a smooth velocity profile seems quite reasonable, physically.

If irregularities do propagate into the stream from the boundary the gross character of the stream should not be altered until after $\sim n/2$ iterations, where n is the number of data points on the original initial line. Actually, our results contain typically this number of iterations in the "straight line" portion of the stream where no gross irregularities are present. Numerical instabilities do, however, tend to occur after the stream curves inward around the detached component, or where a noticeable shear becomes evident; and this can be expected since the method is known to become imprecise if the initial line is stretched so that the angle between it and the flow direction becomes small.

The results broadly confirm the findings of LS on stream properties - there is a "straight line" region starting from the vicinity of L_1, followed by a curving in of the stream and some indication of a tendency to circulate around the detached component. For low injection angles the strong sideways deflecting effect

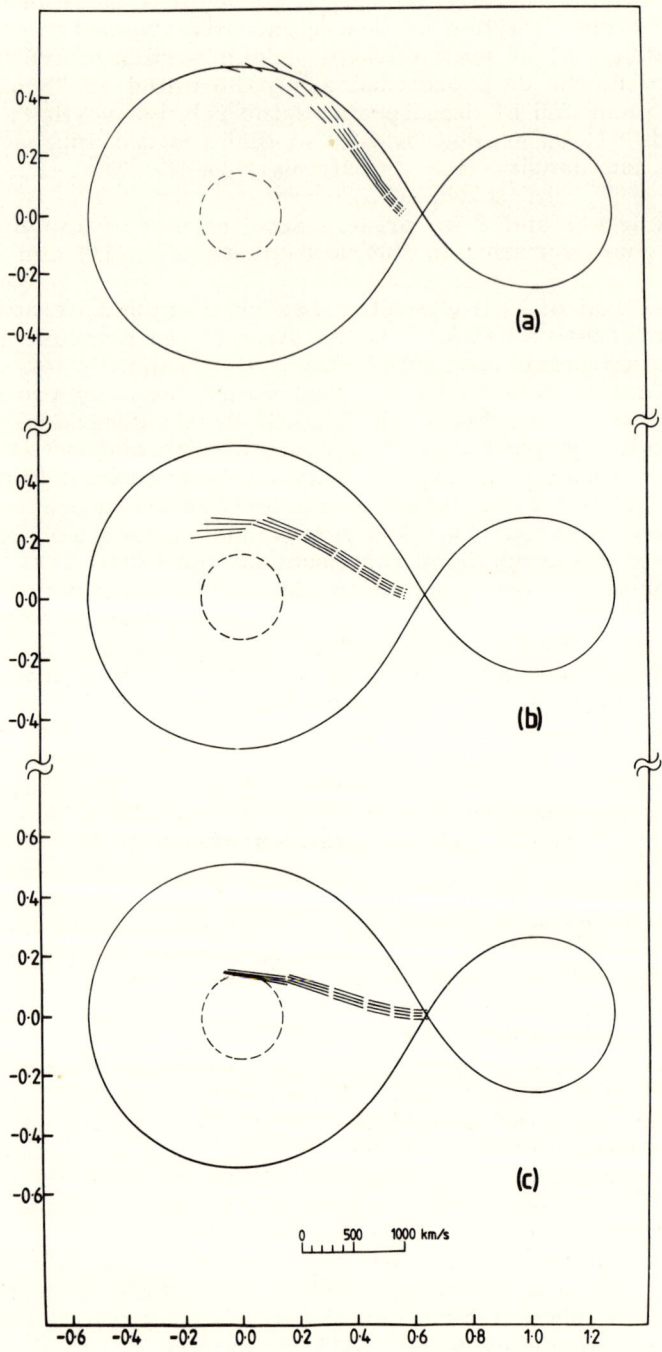

of the Coriolis force is very evident, but for wider injection angles this does not occur and there is no great forward acceleration of the stream. The constricting effect of the net restorative force is evident for the two lower (in absolute magnitude) initial outflow angles, though this is not observed at $\theta_s = -40°$. Introduction of a drag effect in the stream can be seen to reduce the constricting effect of the Coriolis force as discussed in the next section.

One point which was not emphasized by LS, but which must be of some significance in reality, is the high sensitivity of the density reduction to the given value of the parameter ε. This can be understood by consideration of the Bernoulli constant. The density term in Equation (2.8) appears as the decrement of lost potential behind acquired kinetic energy. This decrement itself is not very sensitive to ε since velocity values, proceeding into the orbit region, must tend to the positionally dependent only (for given mass ratio) limit of the particle approximation, but it is divided by ε^2 before appearing as the negative argument of the exponent giving the density reduction. Plainly this introduces a strong dependence on ε which can vary between $\sim 10^{-2}$ and $\sim 10^{-1}$ over the considered range of actual situations; i.e., corresponding to a range of two or more orders of magnitude in the e-folding distance of the density reduction.

◀──── Figure 1

Streams calculated by the method of characteristics as explained in the text. The conditions are intended to simulate a typical classical Algol consisting of two stars of masses $6M_\odot$ (primary) and $1.5M_\odot$ (subgiant). The separation is $\frac{1}{8}$ a.u., so the period is just under 5 days. The velocity of sound is taken to be 11.78 km s^{-1} (corresponding to a temperature of about $10^{4}°$), so that the parameter $\varepsilon = 0.048$ which is slightly less than the Rossby number at the start of the flow. The LS deflection angle = $-20°51$. In (a) and (b) the stream is started some little way into the primary Roche lobe (the beginning of the "orbit" region) with ejection angles of $-40°$ (a) and $-20°$ (b) respectively.

The stream is indicated by lines drawn from suitably selected characteristic intersection points in the flow field. The lines shown are proportional in length and direction to the velocity vectors at these points, i.e., the start (right hand) points of the lines actually lie in the stream (though the other end points do not necessarily). Numerical instabilities tend to occur a few iterations after the last set of lines thus indicated. It may be observed that the stream flow in case (c) tends to be deflected to $\sim -20°$ as a result of the Coriolis action, though the actual path taken differs appreciably from case (b). The dotted circle marks the outline of a Main Sequence $6M_\odot$ star on the same scale.

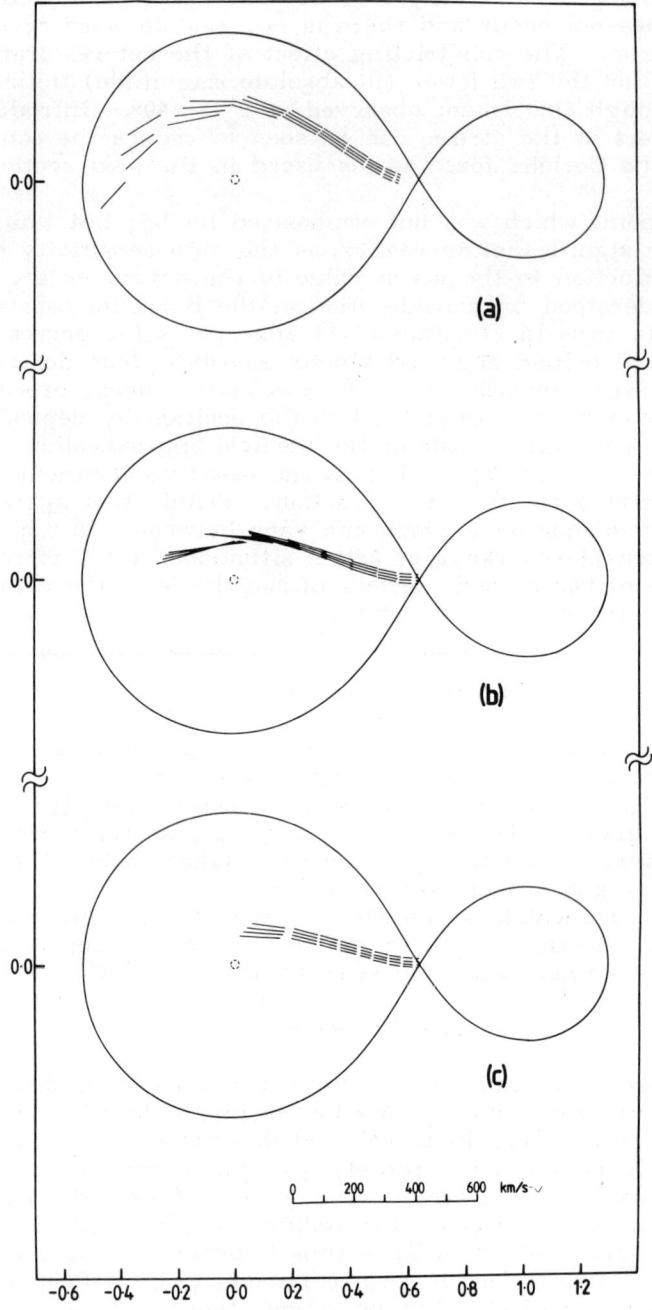

In considering the general application of these methods of characteristics procedures to the real situation of the outflow the following qualifications should be kept in mind.

(1) The method of characteristics can only give us information about a finite domain within the fluid material. In fact, it can be expected that numerical difficulties with the algorithm used to integrate the stream will occur if the stream is deflected through $\sim 90°$. If this happens it must be possible to construct an initial line somewhere in the fluid for which one Mach line at one end point is parallel to the other Mach line at the other end, hence an instability, or at least ambiguity in the direction in which to integrate. The implication is that if there existed returning flow lines somewhere in the outspill the method of characteristics could not show them if the flow was outward on the initial line chosen.

(2) Kelvin's circulation theorem, which requires the number of absolute vortex lines per unit area to remain constant, can be established only for inviscid, homentropic flow. The flow field under consideration has passed through a transonic region of narrow cross sectional area where turbulence effects are likely to become singificant, though in view of the converging-diverging geometry of the "nozzle", (cf., Nariai, 1967) and steep pressure drop in the supersonic region a normal shock would not be expected in this transition to supersonic flow. To have a more convincing description of the behaviour of the vorticity in this region the

---

◄──────── Figure 2

The calculations are similar to those shown in Figure 1 except that the separation is increased to 1 a.u. (or equivalently parameter ε is increased to 0.144, which could also be achieved by retaining the same separation as in Figure 1 but increasing the stream temperature by an order of magnitude). The stellar masses remain the same. Signs of incipient instability are evident in the last group of plotted velocity vectors in case (a), which may be otherwise compared with Figure 1 (b), since the initial angle is again -20°. In a similar way case (b) may be compared with Figure 1 (c). The stream of 2 (b) evidently appears to experience a greater constriction than 1 (c). This can be accounted for by the slower rate of density decline at higher ε (see text). The forward acceleration draws in the stream cross section in response to the mass conservation integral (cross section x velocity x density = constant), and the density function is more sensitive to ε than the velocity increase.

In Figure 2 (c) a drag effect in the stream has been simulated by reducing the calculated velocity increment at each successive stream section by a constant fraction (0.9). It is evident that deflection and constriction effects are appreciably curtailed by this modification (cf. section 3 of this paper).

conditions affecting its diffusion need to be examined more closely.

(3) The adopted method neglects any interaction between stream material and material already accreting around the detached component. In this we are consistent with the LS separation of these regions for the purposes of analytical discussion, and so long as there are distinct regimes where a particular flow pattern predominates this must represent an acceptable approximation. We shall argue later that the transition between "stream" and "disk" regimes is liable to be sudden, but at the same time recognize the difficulties which would apply to a single initial boundary method of characteristics approach to the general problem (cf., Flannery's (1975) criticism of Biermann's (1971) study).

(4) The hydrodynamic approach to the problem, as given so far, though in a sense more physically advanced than the particle trajectory calculations, need not be more realistic unless it can be shown that the effect of the neglected viscosity term is insignificant on the considered stream. The LS stream has a characteristic width of order ϵd (typically $10^{-2} < \epsilon < 10^{-1}$) and initial velocity $\epsilon \Omega d$, while the density is of order $\epsilon^{-3} \dot{M}/\Omega d^3$, where $\dot{M}$ is the mass transfer rate in gm s^{-1}. A normal (particle interaction) Reynold's number for the transonic region is given as

$$R_e = \frac{\dot{M}}{\epsilon d \mu} .$$

For the most commonly met "slow time scale" (Paczyński, 1971) type mass transfer situations in classical Algols we could expect at least $\dot{M} \sim 10^{-9} M_\odot$/year. Corresponding rough estimates for d and plasma viscosity μ can be put at 10^{12}cm and 10^{-5} poise (the particle viscosity for a low density hydrogen plasma at $\sim 10^{4}$°K may be calculated, for example, from formulae quoted by Kopal, 1968), from which we obtain a numerical value of R_e in excess of 10^{11}, i.e., particle viscosity is negligible. If we put $\epsilon \simeq 10^6/V$, where we take a nominal velocity of sound as 10^6cm/s and V is the orbital velocity, we can write $\rho_{L_1} \simeq 2N\, M_\odot^2\, 10^{-(x+33)}/d^3$, where N is the number of solar masses that the system mass amounts to, and x is the (negative) index of solar masses transferred through L_1 per year. For typical separations in the range $\sqrt[3]{100} . 10^{11}$ to $\sqrt[3]{100}.10^{12}$cm we find ρ_{L_1} ranging from $5.10^{-(x+1)}$ (short periods) to $5.10^{-(x+4)}$ (long periods). If this is to be comparable to reversing layer densities in typical subgiants, x should fall in the range 7 to 4 which seems to imply < Kelvin rather than "slow" time scales for the Algol phenomenon. Of course, it is binary evolution rather than the requirement to be observed as a contact system which will control the mass transfer, and quite possibly the actual "critical Roche surface" lies somewhere rather above the subgiant photosphere. On the other hand mass transfer rates of order $10^{-7} M_\odot$/year seem to have some support from the observed scale of period variations, which can be attributed to the sort of mechanism under consideration (cf., Batten, 1973, Ch.9; Söderhjelm, 1980), though the scale of energy release and certain properties of the transferred material does not square easily with the close to Main Sequence spectrum and luminosity of classical Algol primaries (Packet, 1980).

In any case it is evident that a strong turbulence must be present in the outflow region, which may then share some of the character of a turbulent jet (Abramovich, 1963), with the added complications of compressibility and hypersonic flow. We shall consider this in a little more detail in the next section.

3. SOME POSSIBLE NON-CONSERVATIVE EFFECTS IN THE STREAM

Since the early work of Landau (1944) and others there has been a well developed literature on the subject of turbulent jets, although when we add the complications of compressibility and supersonic flow it is evident that any detailed treatment must be a matter of considerable complexity.

Keeping in mind the types of solution found for flow in fluid jets at high Reynolds number (cf., Bachelor, 1967), it is tempting, as a preliminary exercise, to insert frictional terms in the momentum equations of LS, on the general expectation that in some respects turbulence effects can be simulated by the idea of an "eddy viscosity". Utilizing the two dimensional s, n coordinate scheme introduced in LS to be used in the orbit region we can then write for the full equations of motion

$$\left(\frac{R}{R+n}\right) u_s \frac{\partial u_s}{\partial s} + u_n \frac{\partial u_s}{\partial n} + \frac{u_n u_s}{R+n} = -\left(\frac{R}{R+n}\right)\left(\frac{\partial \phi}{\partial s} + \frac{\varepsilon^2}{\sigma}\frac{\partial \sigma}{\partial s}\right)$$

$$+ 2u_n + \nu\left[\frac{R^2}{(R+n)^2}\frac{\partial^2 u_s}{\partial s^2} + \left(\frac{R_n}{(R+n)^3}\frac{\partial u_s}{\partial s} - \frac{Ru_n}{(R+n)^3}\right)\frac{\partial R}{\partial s}\right.$$

$$\left. + \frac{2R}{(R+n)^2}\frac{\partial u_n}{\partial s} + \frac{\partial^2 u_s}{\partial n^2} + \frac{1}{(R+n)}\frac{\partial u_s}{\partial n} - \frac{u_s}{(R+n)^2}\right]$$

$$\left(\frac{R}{R+n}\right) u_s \frac{\partial u_n}{\partial s} + u_n \frac{\partial u_n}{\partial n} - \frac{u_s^2}{R+n} = -\left(\frac{\partial \phi}{\partial n} + \frac{\varepsilon^2}{\sigma}\frac{\partial \sigma}{\partial n}\right) - 2u_s$$

$$+ \nu\left[\frac{\partial^2 u_n}{\partial n^2} + \left(\frac{R_n}{(R+n)^3}\frac{\partial u_n}{\partial s} + \frac{Ru_s}{(R+n)^3}\right)\frac{\partial R}{\partial s}\right.$$

$$-\frac{2R}{(R+n)^2}\frac{\partial u_s}{\partial s} + \frac{R^2}{(R+n)^2}\frac{\partial^2 u_n}{\partial s^2} + \frac{1}{(R+n)}\frac{\partial u_n}{\partial n} - \frac{u_n}{(R+n)^2}\Bigg] ,$$

(3.1)

where R is the local radius of curvature (negative, when projected in the direction of positive n), and ν is the effective viscosity coefficient per unit density.

We can consider that as the fluid exits the secondary wedge through the "nozzle" in the vicinity of L_1, slower moving fluid, which banks up around the central outspill with insufficient forward momentum to carry it into the adjoining Roche lobe, acts as the source of boundary layer turbulence which can propagate through the boundary regions of the stream. In the boundary layer approximation we have

$$\frac{\partial u_s}{\partial s} \ll \frac{\partial u_s}{\partial n} \; ; \; \frac{\partial^2 u_s}{\partial s^2} \ll \frac{\partial^2 u_s}{\partial n^2} .$$

Also, for convenience, let us confine our attention to an initial region and take curvature effects in the stream to be small. The frictional terms then become, corresponding, respectively, to the s and n directions

$$\nu \left(\frac{2R}{(R+n)^2}\frac{\partial u_n}{\partial s} + \frac{\partial^2 u_s}{\partial n^2} + \frac{1}{(R+n)}\frac{\partial u_s}{\partial n} - \frac{u_s}{(R+n)^2} \right)$$

and

$$\nu \left(\frac{1}{(R+n)}\frac{\partial u_n}{\partial n} + \frac{\partial^2 u_n}{\partial n^2} + \left(\frac{R}{R+n}\right)^2 \frac{\partial^2 u_n}{\partial s^2} - \frac{u_n}{(R+n)^2} \right) .$$

Proceeding in the same way as LS we can consider a variation in the n direction $n_1 = \varepsilon^{-1} n$, implying a relatively narrow lateral spread of the stream. Then expanding about the stream centre

$$\phi(s,n) = \phi_o(s) + \varepsilon n_1 \left(\frac{\partial \phi}{\partial n}\right)_o + \tfrac{1}{2}\varepsilon^2 n_1^2 \left(\frac{\partial^2 \phi}{\partial n^2}\right)_o ; \quad (3.2)$$

with similar expansions for the components of $-\underline{\nabla}\phi$ (see LS for details). We write also

$$u_s = u_{s0}(s) + \varepsilon u_{s1}(s,n_1); \; u_n = \varepsilon u_{n_1}(s,n_1) \; ; \sigma = \varepsilon^{-1}\sigma_{-1}(s,n_1).$$

If we substitute now in the s equation, the lowest order term is an expression in ε^{-1} which suggests that

$$\frac{\nu}{\varepsilon}\frac{\partial^2 u_{s_1}}{\partial n_1^2} = 0 \quad .$$

On the face of it, this would seem to contradict the whole concept of the boundary layer retardation (cf. the foregoing inequalities), and the implication which we can draw is that in order that a boundary layer be limited to an ε domain about the stream ν must be $\lesssim O(\varepsilon)$. The coefficient ν can be argued to be of the order of a random (turbulent) velocity multiplied by an interaction length. This latter length has to be less than the stream width, and therefore we can write

$$\nu = \frac{u_{s_o}\varepsilon}{R_e} , \quad (3.3)$$

where R_e is a dimensionless number which expresses the ratio of inertial to frictional momentum-length products. The notion of the boundary layer implies that R_e is of order unity close to the boundary while it should become large near the stream centre.

The zero order in ε equations of motion now becomes

$$u_{s_o}\frac{\partial u_{s_o}}{\partial s} = -\left(\frac{\partial \phi}{\partial s}\right)_o + \frac{u_{s_o}}{R_e}\frac{\partial^2 u_{s_1}}{\partial n_1^2}$$

and

$$-\frac{u_{s_o}^2}{R} = -\left(\frac{\partial \phi}{\partial n}\right)_o - 2u_{s_o} + \frac{u_{s_o}}{R_e}\frac{\partial^2 u_{n_1}}{\partial n_1^2} \quad . \quad (3.4)$$

We now consider that the LS solution will be approximately maintained at the stream centre, which will be in keeping with a trial solution for u_{s_1} of the form

$$u_{s_1} = a_1(s)n_1 + a_2(s)n_1^2 + a_3(s)n_1^3 + \ldots \quad (3.5)$$

The stream centre agreement implies $a_1(s) = \alpha(s)$ in the notation of LS. The higher order coefficients should require tailoring into the higher order perturbation expansions, and what can be legitimately given as boundary conditions. Let us for the present construct a simple model boundary layer with $a_i = 0$ ($i > 2$). As n_1 approaches unity u_{s_1} should be of order $-u_{s_o}/\varepsilon$. A reasonable possibility is therefore

$$u_{s_1} = \alpha(s)n_1 + (\alpha(s) - \frac{2u_{s_o}}{\varepsilon})n_1^2 \quad . \quad (3.6)$$

The form becomes equivalent to the LS solution in the initial "straight line" portion of the orbit region where $\alpha \to 2$ and $u_{s_o} \geqslant \epsilon$. The first of Equations (3.4) becomes

$$u_{s_o} \frac{\partial u_{s_o}}{\partial s} = \lambda^2 s + \frac{2u_{s_o}}{R_e}(\alpha - \frac{2u_{s_o}}{\epsilon}) \qquad . \qquad (3.7)$$

Various features of this equation can be pointed out. Most obvious is the frictional drag which reduces the effective accelerating force on the stream ($\lambda^2 s$) - a result which could be intuitively expected. Away from the stream centre, where R_e decreases, this drag increases, though the equation cannot be satisfied for a small u_s ($\geqslant \epsilon$) (outer boundary) unless $\hat{s}$ tends to point along an equipotential surface. The implied stream spread would then have to be too great and a role for the higher order terms in (3.5) seems to be required.

Let us consider now the n-direction equation in (3.4) which, in the straight line approximation, in the absence of friction, shows the evident balancing of the Coriolis force against gravity. We can take this condition to continue to identify the stream centre line, in the immediate vicinity of which the lateral flow can have the same linear character as the LS solution (i.e., the term in $\partial^2 u_{n_1}/\partial n_1^2$ is small). We can now make another general deduction from the two zero order momentum equations. If the stream centre velocity is reduced, as implied by (3.7), the Coriolis force will not push as hard on the stream. The balance direction considered previously can therefore move inward towards the detached star along a line which will be less inclined to the line of centres. Formally we have

$$-\left(\frac{\partial \phi}{\partial n}\right)_o = -\frac{3}{2} A \sin 2\theta_s s \quad (> 0 \text{ since } -\frac{\pi}{2} < \theta_s < 0) \quad .$$
$$(3.8)$$

Plainly, since

$$\left(\frac{\partial^2 \phi}{\partial \theta_s \partial n}\right)_o < 0 \quad ,$$

the balancing force is reduced by increasing (i.e., making less negative) θ_s. This effect will be greater the greater is the reduction in stream velocity associated with the drag.

Hence we deduce that the boundary layer on the side of negative n (more negative θ_s) will be pulled into the centre of the stream, i.e., it will not grow as rapidly. On the other hand, the stream would tend to spray out on the side of positive n. To some extent, this was already implied by the asymmetric velocity distribution of (3.6). The drag effects will have to be greater on this side of the stream to bring the forward velocity, which at first tends to increase at positive n, down to the lower value associated with the boundary region.

The first order in ε equations are

$$u_{s_1}\frac{\partial u_{s_0}}{\partial s} + u_{s_0}\frac{\partial u_{s_1}}{\partial s} + u_{n_1}\frac{\partial u_{s_1}}{\partial n_1} + \frac{u_{n_1}u_{s_0}}{R}$$

$$= -n_1\frac{\partial^2\phi_o}{\partial n \partial s} + 2u_{n_1} + \frac{u_{s_0}}{R_e}\left(\frac{1}{R}\frac{\partial u_{s_1}}{\partial n_1} - \frac{u_{s_0}}{R^2}\right)$$

(3.9)

$$u_{s_0}\frac{\partial u_{n_1}}{\partial s} + u_{n_1}\frac{\partial u_{n_1}}{\partial n_1} + \frac{n_1 u_{s_0}^2}{R^2} - \frac{2u_{s_0}u_{s_1}}{R}$$

$$= -n_1\frac{\partial^2\phi_o}{\partial n^2} - \frac{1}{\sigma_{-1}}\frac{\partial \sigma_{-1}}{\partial n_1} - 2u_{s_1} + \frac{u_{s_0}}{R_e}\left(\frac{1}{R}\frac{\partial u_{n_1}}{\partial n_1}\right) ;$$

while the corresponding form of the continuity equation remains the same as in LS. The viscous terms apparently do not disturb the character of the LS solution near the stream centre at low curvature ($R \to \infty$), where we may therefore expect the same parabolic vertex in the density distribution. Away from the stream centre this will be disturbed by the asymmetric effects already referred to. A more detailed picture could be made by matching terms in the perturbation expressions and known approximations for turbulent jet flows. We have not done this, but in view of the exploratory nature of the approach to turbulent viscosity effects here the realism which would attend such an exercise may be subject to some doubt. Moreover, another complication is present which is not contained in the equations considered so far.

Mention has already been made of the disk which is built up by the accumulation of inflowing material. The role of viscosity in such disks has been considered by various authors, including Pringle and Rees, 1972; Lightman, 1974a, b; and Lynden-Bell and Pringle (1974). The latter authors emphasized the general effect of an inflow of most of the matter accompanied by a net outflow of angular momentum carried away by a relatively small amount of matter. In the context of classical Algols we do not expect extensive disks to be formed, nor are these observed. The main reason for this is the relatively large size of primary components compared to estimated initial radii of disk forming regions. In many cases the minimum distance of a particle following the "straight line" (initially) trajectory to the detached star's centre of mass is less

than this star's radius. Viscous action of the boundary layer
between the stellar envelope and the circulating accreted material
is likely to be particularly effective in depleting the disk region
of any high density accumulation. Nevertheless, when this occurs
there is likely to be some outward diffusion of angular momentum.

The LS treatment requires, in its formulation, that the disk
region be separate from the orbit region. However, no compelling
reasons have been given why this configuration should be maintained, and the considerable dissipative turbulence which must be
generated in the impact region seems to naturally entail the outward flow of angular momentum along the lines envisaged by Lynden-Bell and Pringle.

The action of a disk on an incoming stream could be accounted
for, appealing to particle concepts, by adding terms of the type

$$\frac{R}{R+n} \frac{\eta \sigma_d u_d^2 \sin \chi}{\sigma_s \lambda_s} \quad \text{and} \quad -\frac{\eta \sigma_d u_d^2 \cos \chi}{\sigma_s \lambda_s}$$

(χ is the same angle of incidence referred to in LS), respectively,
to the right hand sides of the s and n equations. Here u_d
represents the velocity of material in the disk of density σ_d, σ_s
being the stream density which, for the sake of simplification,
is supposed to be "thin" so that the interaction length λ_s over
which the momentum of disk particles would be "absorbed" is
greater than the width of the stream. η represents some factor
indicating how efficiently momentum could be transferred from disk
to stream. The variation of density σ_d with disk radius r follows an $\exp(-(r/r_o)^2)$ form in the feasible solutions of Lynden-
Bell and Pringle, where r_o in close binary systems $\langle r_d$ (the LS
outer disk radius) according to the stability arguments of LS; u_d
is essentially Keplerian, i.e., $\propto r^{-\frac{1}{2}}$. Without going into further
details we could anticipate the effect of such terms would be to
reduce any tendency to curve inwards or spread on the inward
side of the stream, particularly further upstream as the interaction between disk and stream becomes stronger and more radial.

Though the closeness of particle to fluid formulations for the
orbit region of the stream has already been shown, the more fluid
based point of view again presents us with some physical differences. Interaction between disk and stream in the supersonic
flow regime implies the presence of shock fronts, and from the
time such interactions become appreciable the whole character of
the fluid description changes, i.e., the flow is no longer isentropic
and stream properties are discontinuous. The separation of stream
and disk regimes could then be a matter of shock front location,
but the exponential growth of disk density suggests that the
stream, as considered from the outflow region, becomes effectively
wiped out over a short distance and a separation of the two

basic entities, stream and disk, may not be a bad first approximation.

It may be possible to observe some of the foregoing effects in numerical calculations which have been presented in the literature. At a first glance calculations like those of Lin and Pringle (1976); Prendergast and Taam (1974); Flannery (1975); and Sørensen (1976), based on a Roche lobe overflow model which is more or less consistent with the LS formulation, are in general agreement with the LS calculations. The two models of Lin and Pringle, corresponding to mass ratios of 0.4 and 2.5, which should therefore give the same stream deflection angle on the LS model, are, however, slightly but noticeably different. According to our measurements on the published diagrams, which should be accurate to about 1°, θ_s = -17° for the former and -21° for the latter. These models have fairly dense disks, and a larger (or more dense at a given relative distance from the mass centre) disk for the second model. The LS value of the deflection angle is -20°. A possible explanation is therefore a relatively stronger viscous action in the stream of the first model, while the second stream shows restoration by the pressure of the disk material. The curtailing effect of this pressure on any tendency to spread inwards can be more clearly seen in the results of Prendergast and Taam, though since their disks are relatively weak it is only in the outlying regions of the stream that the effect is obvious. Prendergast and Taam's diagrams appear to be in good overall agreement with the LS data, though there is a noticeable effect on the deflection angle when the condition for secondary synchronism is relaxed ($3/4$ synchronous in their case B(i)), so that, for example, if there could be a significant equatorial current affecting the effective angular velocity of the mass losing star (cf. section 1 of this paper) an appreciable change of outflow angle could be anticipated. Flannery's published diagrams indicate initial stream deflection angles which are measurably almost the same as the LS values, though this is less true of the calculations of Sørensen et al, where the initial and boundary conditions for these calculations may make them depart somewhat from the classical Roche lobe overflow problem. With all of these numerical calculations we appear to have to rely on the authors' discretion about the attainment of a steady state. Nevertheless, it seems clear that any departure from the LS stream characteristics, at least in the early part of the orbit region, is only small. In particular, deflection angle values are within a few degrees of their predicted values and viscosity effects within the stream itself appear to be relatively slight, in terms of potentially observable effects.

4. NON ZERO INCLINATION OF THE ROTATION AXIS OF THE DETACHED STAR

The approach to disk formation considered by LS utilized the known existence of quasi-circular periodic orbit solutions to the restricted three body problem (those of class i - the direct orbits around the primary being of special significance to the present considerations) and thus the notion that the disk is in the orbital plane is implicit in their disucssion. The complicating factor of possible non zero inclination of the rotation axis of the detached star to that of the orbital revolution should not go unnoticed, even if this is too difficult to treat in detail here.

A possible end-product which might be visualized in this circumstance is that of an equatorial disk (inclined to the orbital plane), such as has been conjectured in models of HZ Her, for example, (Katz, 1973; Roberts, 1974; Petterson, 1975). The problem of vertical stability of orbits in the restricted problem has been considered by Hénon (1973, 1974) and others, and it has been shown, for the mass ratios studied, that horizontal stability breaks down well before vertical stability for most of the types of orbit considered, including those direct orbits around the primary which approximate to two body circular orbits as the primary to particle separation decreases. This seems quite reasonable, intuitively, since the mean value of the disturbing force in the plane is always greater than in the z-direction for these orbits, and the perturbation equations for horizontal and vertical stability are uncoupled. An equatorial disk about an inclined primary could therefore remain with the same sort of size as calculated on the stability argument of LS, once it has been formed and if it was free from further disturbance associated with continued infall in the plane. Here, however, the complications raised by this seemingly slight generalization of the underlying model become apparent. In the first place, continued infall in the plane is an essential feature of the steady state description of the model. An inclined disk would be subject to a disturbing torque associated with the secondary star, and would therefore precess. A strictly steady state is clearly not consistent with this picture.

But a deeper and more important matter is whether an LS type equatorial disk can form at all in the given conditions. A separation between rotation and revolution axes in a close binary system suggests a relatively large amount of rotational energy in the detached star, if it can avoid being forced into the synchronism with orbital period which characterizes the formal description of the classical prototype model. Such rapidly rotating primaries in the systems under consideration are not unknown - U Cep is an often quoted example. Since a total predomination of orbital over rotational angular momentum vectors is not supposed for such a situation, neither need there be any forcing of these vectors into parallelism. Under these circumstances the predominating distortion of figure of the primary could be an equatorial bulge, especially if the secondary is the low mass object typical of classical Algols.

If we return to the mechanism considered by LS, it is clear that even if the disk could get started in the orbital plane it would be subject to a torque from the equatorial bulge and thus tend to precess about the rotation axis. The mean precession rate $\dot\phi$ for a rapidly rotating disk can be shown to be of order (cf., Goldstein, 1959)

$$\dot\phi \sim \frac{\text{Potential associated with disturbing torque}}{\text{Rotational angular momentum}}$$

$$\sim \eta\mu \frac{\Omega_1^2}{\Omega_d}\left(\frac{r_1}{r_d}\right)^3 \cos i = \eta\mu^{\frac{1}{2}}\Omega_1^2 \frac{r_1^3}{r_d^{3/2}} \cos i ,$$

where η is a numerical factor, which is $3/4$ for a uniform circular disk of radius r_d, rigidly rotating with angular velocity Ω_d inclined at angle i to the equator of the star of radius r_1 which rotates with angular velocity Ω_1; all being expressed in the natural units of the binary. It is a condition of the approximation that the above expression is small, though plainly it cannot be all that small in this instance. For many of the well known classical Algols of periods of a few days only r_1/r_d is a quantity of order 10^{-1} to 10^0, when r_d is given by the stability consideration used by LS. A fairly irregular net motion is therefore implied. The disk would have to be quite well separated for the precession rate to sink to $\sim 10^{-2}$, though LS considered that "hundreds to thousands" of orbital periods might be required before even the plane disk stabilized. If the stabilization period is greater than the precession period it is difficult to visualize the easy formation of a steady disk in the plane. If some equatorial disk can eventually settle down non-conservative effects must be called into play to dampen out the angular momentum components not parallel to the rotation axis. The resulting drop in angular momentum implies a net movement inward of most of the orbiting material - the result being that equatorial disks in inclined systems, if they can be formed at all, should be rather less in size than the values given by LS ($\Delta r_d \sim O(r_d \sin i)$) if other aspects of their disk formation mechanism remain valid. If outward diffusion of angular momentum occurs through the action of viscosity it is the density rather than the size of the disk which will be curtailed.

5. THE POSSIBLE ROLE OF RADIATION PRESSURE AND/OR A STELLAR WIND FROM THE DETACHED COMPONENT

Gas or particles moving towards the detached component in

the manner considered so far are likely to experience forces not only due to the rotating frame of reference or the gravitational attraction. These are generally of a repulsive type, and may be of a radiative (radiation pressure) or collisional (stellar wind) nature.

In the classical Algol picture, the detached star is generally appreciably more massive and has a higher surface temperature than the contact component. It is usually a Main Sequence star in the type range (roughly) B0 - F0. In this situation it is no great restriction to confine attention, where repulsive mechanisms are concerned, to the detached star. We can expect, on general grounds, that collisional repulsion will be much more effective for the ionized component of the entrant gas stream, while radiation pressure may be appreciable on neutral material.

In the vicinity of L_1 the sort of density considered previously, i.e., $\rho \sim 10^{-8}$ gm/cm^3 compares with reversing layer densities of subgiants and since the degree of (hydrogen) ionization in such layers does not vary so greatly with luminosity or type over the expected range of mean parameter values for these stars (see the well known "grid" of cool star models by Carbon and Gingerich (1969)) a representative initial value in the range 10^{-4} to 10^{-5} could be reasonably assumed. Of course, conditions near L_1 are rather different than in the comparison sources; apart from dynamical differences, which are difficult to treat, the gravity may drop considerably near the outflow region though the effect of reduced pressures may be offset, or at least initially, with the lower effective temperatures associated with gravity darkening.

The central stream density in the orbit region does not decrease by much more than an order of magnitude in the LS models and neither will the dilution factor change by much more than this, so that if we start with the low ionization levels expected in the subsonic wedge these two factors alone could not increase the ionization by much more than one order of magnitude along the stream path. In this respect, there is a noticeable difference with the calculations of the present paper, which could allow such a density fall off that full ionization could be attained along the stream path if anything like a Saha ionization was valid. What, in any case, may be highly effective in creating the conditions where plasma interactions could occur is turbulence, which would substantially increase the ionization, as well as allow the stream to spread out to present a greater cross section to an outflowing stellar wind.

For the sort of primaries in the given picture we would expect any such wind, if significant at all, to be radiatively driven, and therefore before considering possibilities of collisional push out we should pay attention to the direct radiative effects. It will quickly transpire that radiation pressure alone can be very influential in defining the range of models in which the Roche lobe overflow mechanism considered so far could be operative.

In order to study the radiation pressure problem thoroughly the radiation transfer equation should be solved for the stream. It is possible and useful, however, to form some general quantitative ideas. A key quantity to have in mind is the ratio μ' of the radiation pressure which acts on a typical particle in the stream to the gravitational attraction acting upon it at the same time. We can express this ratio as

$$\mu' = \frac{L_\nu \sigma'}{4\pi c G M_1 m} , \qquad (5.1)$$

where σ' is some mean effective cross section for a particle of mass m to the energizing radiation L_ν coming from the primary of mass M_1 with velocity c. We may note that the inverse square law affects both forces and so no radial dependence is directly involved in (5.1). Taking the view that Equation (5.1) will, in the present case, apply essentially to neutral ground state hydrogen near the stream origin, so that the radiation L_ν refers predominantly to scattered Lyman α, we can write an expression for μ', making a simple linearization for the flux scattering coefficient product integral, as

$$\mu' = 6.1 \alpha\beta \, \exp(-\frac{11.9}{T_4})(R_1/R_\odot)^2 (M_1/M_\odot)^{-1} \times 10^5, (5.2)$$

where α is a factor which accounts for Lyman α absorption prior to its incidence on the stream, i.e., in the source radiation, and β is an aspect or transparency factor, allowing for the fact that all the atoms in the stream are not equally exposed to the primary energizing radiation. As we have previously compared stream densities with those of the intermediate/late type subgiant reversing layers the optical depth through the stream is at least of order unity (at least initially) and we may expect β to be of order 10^{-1} or less, depending on geometrical factors. T_4 is the temperature in units of $10^4 °K$ of the primary stellar surface, while other symbols have their conventional meanings.

In Table 5.1 we give some values of $\mu'/\alpha\beta$ for some Main-Sequence primaries.

It is difficult to give an accurate value of α without a much more detailed treatment, since apart from the basic stellar atmosphere L_α profile there are complications associated with the possible presence of a disk, and Doppler displacement of the resonance wavelength due to the stream motion. For the earlier types α may rise to be of order unity, but may sink to 10^{-1} or less for

Table 5.1

M.S. Type	$\mu'/\alpha\beta$
B0	$2.9.10^4$
B5	$6.3.10^1$
A0	7.2
A5	$7.4.10^{-1}$
F0	$6.7.10^{-2}$

stars where hydrogen recombination is a major contributor to the photospheric continuum. But, in a general way, we may deduce that radiation pressure is likely to be of considerable importance for binaries where the detached component is earlier than intermediate B type. In any such system a stream from the contact secondary could experience very great difficulty in making its way towards the primary and be quickly repelled out of the system. For stars between B5 and A0 radiation pressure is still important and liable to affect the picture given by LS in ways which we shall consider presently. For classical Algols with primaries later than A0 the effects of radiation pressure can probably be neglected at least in a first order discussion. In parallel with the swift drop in radiation pressure with advancing type we expect also the scale of a stellar wind to similarly decline.

A number of well known classical Algols have primaries falling in the intermediate range B5-A0 (for example, Algol itself, U Cep and U Sge) and radiation pressure effects require to be taken into account. Some understanding of this can be made on the basis of the parameter μ', which was utilized in the discussion of radiation pressure modified Roche surfaces by Schuerman (1972), keeping in mind the close connection between stream characteristics and Roche surface geometry. (The more generalized work of Vanbeveren, 1977, 1978; does not affect the essentials of the present argument.)

The classical Algols are typified by Schuerman's exemplary calculations with $\mu = 0.8$ (rather than the other set, which, for $\mu < 0.5$, exhibit some topological differences). For very small values of μ' we can regard the effects on the inner Roche surfaces as similar to those produced by a decrease in the relative mass of the primary. The two gravitational terms on the right hand side of Equation (2.2) will dominate the expression for small X, Y and these will change their ratio to each other by the factor μ'. Of course, the external Roche envelopes become formally dominated by the centrifugal term when $X^2 + Y^2 \gtrsim 1$ and the shape of these will be little affected by radiation pressure.

Initially, therefore, as μ' increases from zero, the primary critical Roche lobe will begin to decrease while the secondary increases. Matter which is shielded from radiation pressure effects by lying in the shadow cone of the secondary would presumably continue to respond to the same layout of equipotentials as if μ' were zero. Even without the topological peculiarity pointed out by Schuerman, of the disappearance of the inner "contact" point L_1 for some value of $\mu' < 0.5$ ($\mu > 0.5$), it becomes clear that for $\mu' \gtrsim 10^{-1}$ it is easier for matter expanding in response to evolutionary processes within the secondary to spill out behind this star at L_2 than to climb up towards the primary through an L_1 point moved out to higher potential energy, which in any case becomes ruled out for relatively low values of μ' for the mass ratios of typical Algols. Even for small values of μ' ($\lesssim \varepsilon$) the initial conditions on the stream must be appreciably different from the radiation pressure free model. In the deeper layers of the secondary envelope from which the outflowing material originates, radiation pressure from the primary cannot be supposed to have any significant effect and the secondary photosphere must remain in more or less the same place for any of the considered primary surface temperatures. We can interpret this as a diminution of the transparency factor β of Equation (5.2) to be insignificant at significant optical depths, which should be located relatively close by on the secondary side of the original L_1 point. Obviously the effective accelerating force beyond L_1 rises more slowly and in a more complicated way than in the purely gravitational problem, but to fix ideas we may consider that as the stream establishes itself in a small region beyond L_1 β increases to some representative value for the orbit region and the radiation pressure becomes effective. If the stream has picked up sufficient momentum during its early formation some of its matter could reach the modified primary Roche lobe, but as long as the stream is within an ε-domain of some equivalent L_1 point its forward velocity cannot greatly exceed sonic. If, following the motion of a fluid element, there is a rapid transition to circumstances where the effective L_1 point, corresponding to the typical β value considered, has moved to a distance $>2\varepsilon$ nearer to the primary while the fluid element is still within an ε domain of the original L_1 point, it will tend to fall back to the secondary with insufficient energy for escape.

We can summarize with the following general deductions:

1. The character of Equation (5.2) suggests a swift rise in the effectiveness of repulsive mechanisms on gaseous streams in classical Algols with advancing primary spectral type. A cut-off effect on the Roche lobe overflow mechanism is therefore expected at some spectral type of primary, feasibly around B5.

2. If a disk can be formed along the discussed lines in the presence of radiation pressure, its outer radius could not be made significantly larger by the inclusion of a radiation pressure term alone. This is partly because of the restriction on feasible μ' values, or if it is somehow possible to overcome this, the entailed reduction of the modified primary lobe at larger μ' would make for a smaller disk on the LS arguments.

3. If repulsive mechanisms prevent outflow at L_1 an outlet for matter in the expanding secondary envelope can be found at slightly higher potential energy at L_2. What happens in the region beyond L_2 is a subject that calls for a more detailed treatment than can be gone into here (cf., Nariai, 1975, 1977). However, arguing in a general way, the forces tending to ensure uniformity of the rotation, which must rely on viscous interaction at some stage, can no longer dominate over the ever increasing centrifugal force in the synchronous frame. The innermost of the external Roche surfaces is always of approximately circular section in the X, Y plane centred on the centre of gravity of the system, with radius about equal to the separation of the two stars. Hence we could expect that matter collecting here could remain roughly stationary in the rotating frame, since such an angular velocity would correspond to the Keplerian value if the mass of both stars were replaced by a point mass at the system centre of gravity. Beyond this boundary, however, Keplerian orbits would be slower and one possibility could be the build up of an external ring in slow retrograde rotation with respect to the binary system. Various possibilities may occur in relation to the development and properties of such an external structure, about which there has been much discussion, generally in relation to interacting close binary systems other than classical Algols. We note here that radiation pressure provides a natural means whereby such an external structure could be formed from a type of Algol system in which there is a rather early type primary.

4. In view of the possible great importance of radiation pressure driven effects on gas dynamics in the Roche lobe overflow problem where there is a sufficiently early type primary, a full theoretical treatment, taking into account details of the radiative transfer, ionization, excitation and geometry of the transferred material, is highly desirable.

6. OBSERVATIONAL EFFECTS AND CONCLUDING REMARKS

General observational material which can be brought to bear on the Roche lobe overflow problem (cf., e.g., Piotrowski et al, 1974; Walter, 1973; and numerous other studies initiated long ago by Otto Struve and his collaborators, and referred to in Batten's (1973), and Sahade and Wood's (1978) books) supports the general picture of a gaseous stream starting off in the region of L_1 and being deflected towards the following hemisphere of the detached primary. Also observational indications of rings and disks were noted long before deeper theoretical treatments became available, perhaps partly because of computational requirements if hydrodynamic calculations are to be undertaken.

Impressive though the earlier observational results may have been, it could be commented that many of the earlier techniques could not be stretched to give more precise information of a type which could confirm or deny more detailed aspects of theoretical work, some of which have been referred to in preceding sections of this paper. This may continue to be the case for some time to come, though it is the purpose of the present section to point out certain more recent developments which could allow a more incisive kind of evidence to be furnished.

There are two observational methods which should have a particular application to some of the effects considered so far, namely, narrow band photometry and polarization observations. Even simultaneous combination of these two techniques to further enhance interpretative capabilities has been suggested (McLean, 1980). The possibilities of narrow band photometry for studying circumstellar matter in close binary systems were indicated in a notable paper on Algol by Guinan et al (1976). Some formulae relevant to shells and disks were presented and discussed by Budding and Marngus (1980). The effect of the stream itself was not included in the latter paper, which also neglected any finite optical depth effects. Again a detailed treatment could become quite complicated, probably requiring numerical integration to account for a realistic model description. Some efforts along these lines were made in the paper of Prendergast and Taam (1974). Here we shall only note some general lines on which analysis could be directed to relate observations with the presence of a stream of the kind already discussed. At phases such that

$$|\sin(\psi + \theta_s)| \lesssim \frac{r_e}{r_e + R_{L_1}},$$

where ψ is the orbital phase and r_e is the fractional radius of the eclipsing body, R_{L_1} being the distance from the detached star's centre to the L_1 point, also measured in units of the mean stellar separation; the stream will be fairly completely eclipsed. Let us note the implied asymmetry of the effect which centres around $\psi = |\theta_s|$. The ingress leaves some emitting region of the stream uneclipsed until after the passage of the shadow cylinder of the secondary across the primary, so that the line index or some other measure of equivalent (absorption) width should have a deep minimum prior to the eclipse of this region. Just before egress a similar minimum may be observed as the "disk" re-emerges, but the depth should now be somewhat less in relation to the relative strength of stream to disk emissions and as a direct consequence of the underlying stream asymmetry associated with the Coriolis action already discussed in previous sections. Such effects are apparent in the Hβ observations of Olson (1976) of U Cep, and perhaps also with Kaitchuk et al's (1980) Hα data on RW Tau (1980) as well as Bolton and Zubrod's (1980) Hα data on β Per.

The conical character of the Roche lobe in the vicinity of L_1 implies that the stream is essentially uneclipsed by the secondary for phases in the range $|\psi| \gtrsim 60°$. Outside of its eclipse the aspect factor of the stream gradually increases until about phase $90° - \theta_s$, suggesting a steady increase in any radiation contribution due to the stream in this phase range if there is any significant optical depth to the material. In the phase range $90 - \theta_s$ to $180 - \theta_s$, some part of the stream should be eclipsed by the primary component, whereafter the stream and impact region become visible. Maximum emission effects due to the stream based material outside of eclipses could be expected for phases around $270° - \theta_s$.

Polarization methods offer a complementary and potentially very informative method of analysis of the distribution of light scattering material in the vicinity of close binary systems. Brown et al (1978) produced general expressions for the total intensity and linear polarization Stokes parameters of the spatially integrated light from a static, optically thin, Thomson scattering envelope of arbitrary distribution, illuminated by point sources. Though these expressions form the basis for diagnosing polarization effects in close binary systems where some mass transfer process of the type considered in this paper is occurring, they do not take explicit account of the effects of eclipses, though this is so often a feature of known classical Algol systems (perhaps it could be considered a definitive feature in a certain sense).

In the context of classical Algols there would seem to be a distinct difference in the effects of the stream and disk regions, at least as far as we retain an LS type of picture of these regions. The disk region produces a constant polarization $Q = \tau_o(1 - 3\gamma_o) \sin^2 i$ (using the notation of Brown et al) except during eclipses. The stream region, however, can produce a generally elliptical form in the Q, U plane, traced out twice per orbit, with a relatively large variation of Q for $i \approx 90°$. The phases at which Q has its largest values enable stream asymmetry with respect to the line of centres to be examined.

Polarization variations which could be associated with this type of effect is shown in the accompanying diagrams for the

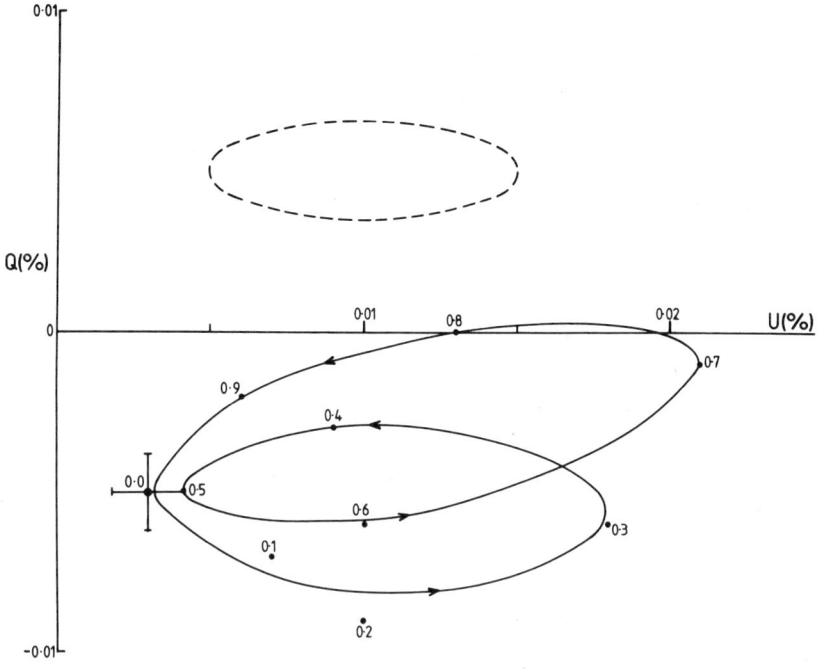

Figure 3

Data points taken from Rudy and Kemp (1980)'s polarization observations of β Per, suitable averaged, have been plotted in the Q:U plane. The Q U parameters given require a further rotation from the instrumental orientation to correspond to the Q, U parameters in the natural frame of the binary, as used in the discussion of Brown et al (1978). However, such a rotation preserves the eccentricity of the mean ellipse, which is a reasonably expectable (simple model) locus for such data points and which can be used to determine an inclination value. The dotted ellipse shown for comparison corresponds to a generally accepted inclination value of 82°. It is evident that the double quasi-elliptical track of the Q, U values with phase approximately confirms this inclination value. The scale of this locus is definitely greater in the second half of the orbital cycle when, if stream distributions have any resemblance to those shown in Figures 1 and 2, a greater amount of scattering material ought to be visible.

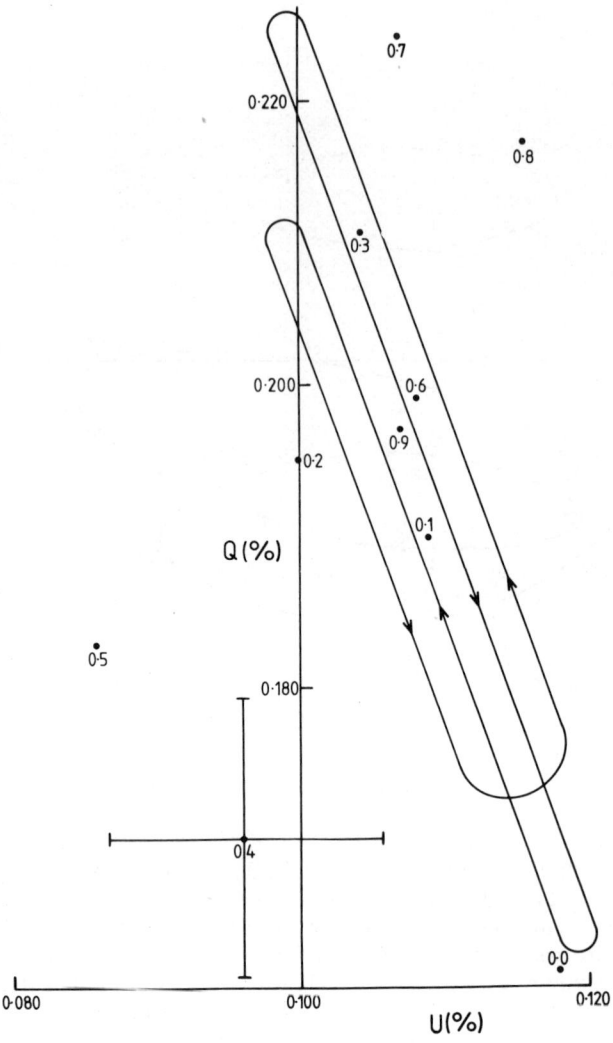

Figure 4

Similar data to Figure 3, also taken from the work of Rudy and Kemp (1978), here showing the variation of Q and U for U Sge. According to the simple model the locus should, in this case, degenerate to a straight line (i = 90°). Here, however, more complications seem to be clearly present, despite the decreased accuracy of observation. A general trend is indicated by the sketched locus (where again a greater excursion is suggested in the second half of the orbital cycle), though the points at phases 0.4 and 0.5, when appreciable regions of the stream may be eclipsed, are clearly at significant variance with any such possible trend.

classical Algol systems β Per and U Sge. It is clear that there is much information in the polarization data which could be perhaps modelled in an exploratory way by reference to a standard scheme like that of the LS description, though, taken in isolation, such data is liable to proportionally large interpretative uncertainty as has been pointed out by Simmons et al (1980). At least the data are consistent with the presence of gaseous streams of the supposed type, and from the generally greater scale of variation in the second ellipse traced out during the phase range 0.5 to 1.0 we can interpret another confirmation of the Coriolis deflection.

To summarize and conclude, the aim of this paper has been to fix attention on the so-called Roche lobe overflow mechanism and, in particular, guided by the path clearing work of Lubow and Shu and others, to look further at details and possible complications or difficulties with the mechanism. The classical Algols form a group of astronomically familiar objects about which a considerable amount of observational information, of more conventional as well as newer kinds, has been amassing. Such objects should provide the natural testing ground for ideas on mass loss and transfer in close binary systems, which, in a general way, have somehow become implanted into most current schemes for close binary evolution, and which therefore, because of their importance to such schemes, deserve to be examined and tested as far as possible.

Observations, such as those referred to in this section, as well as other evidence well discussed in some of the references quoted, can be taken as giving general support to some of the broad predictions of theory. The polarization evidence corroborates very well the picture of a gaseous stream and even shows the presence of a Coriolis action on such a stream, while narrowband or spectrographic results have many times been interpreted in terms of matter circulating around the detached component. What is now required is further and closer scrutiny to see if more details of the stream and disk picture can be confirmed or quantified. What is the exact shape of the stream and how wide is it? What can be observed about its temperature or ionization structure? How far does the disk reach? Are densities quite consistent with more general mass transfer/evolutionary theories? The present state of observational work seems to be capable of giving more definitive answers, though the recent presentation of observational results at Toronto are far from being in clear agreement either with each other or with theoretical predictions. Streams need not be narrow nor necessarily travel along the LS deflection angle. Disks can be thick or thin and may reach out to the outer Roche envelope or be confined further in. Hopefully the general situation will become clearer with continued efforts of this exciting area of research into close binary systems.

REFERENCES

Abramovich, G. N.: 1963, *The Theory of Turbulent Jets* (USSR), MIT Press.
Bachelor, G. K.: 1967, *An Introduction to Fluid Dynamics*, Cambridge Univ. Press.
Batten, A. H.: 1973, *Binary and Multiple Systems of Stars*, Pergamon Press, Oxford and New York.
Biermann, P.: 1971, Astron. Astrophys., 10, p.205.

Bolton, C. T. and Zubrod, D. J.: 1980 in *Close Binary Stars: Observation and Interpretation*, (ed. M. J. Plavec, D. M. Popper and R. K. Ulrich), D. Reidel, Dordrecht, p.225.
Brown, J. C., McLean, I. S. and Emslie, A. G.: 1978, Astron. Astrophys., 68, p.415.
Budding, E. and Marngus, N.: 1980, Astrophys. Space Sci., 67, p.477.
Carbon, D. F. and Gingerich, O.: 1969, in *Theory and Observation of Normal Stellar Atmospheres*, (ed. O. Gingerich), MIT Press, p.377.
Flannery, B. P.: 1975, Astrophys. J., 201, p.661.
Forsythe, G. E. and Wasow, W. R.: 1960, *Finite-Difference Methods for Partial Differential Equations*, John Wiley and Sons, Inc., New York.
Goldstein, H.: 1959, *Classical Mechanics*, Addison-Wesley Publ. Co., Reading, Mass., (Ch. 5).
Greenspan, H.P.: 1968, *The Theory of Rotating Fluids*, Cambr. Univ. Press.
Guinan, E., McCook, G., Bachmann, P. and Bistline, W.: 1976, Astron. J., 81, p.57.
Hénon, M.; 1973, Astron. and Astrophys., 28, p.415.
Hénon, M.: 1974, Astron. and Astrophys., 30, p.317.
Kaitchuk, R. H., Honeycutt, R. K. and Mufson, S. L.: 1980, in *Close Binary Stars: Observation and Interpretation*, (ed. M. J. Plavec, D. M. Popper and R. K. Ulrich), D. Reidel, Dordrecht, p.233.
Katz, J. I.: 1973, Nature, Phys. Sci., 246, p.87.
Kopal, Z.: 1959, *Close Binary Systems*, Chapman and Hall Ltd., London.
Kopal, Z.: 1968, Astrophys. Space Sci., 1, p.411.
Landau, L.: 1944, Doklady Acad. Sci. (USSR), 43, p.286.
Lightman, A. P.: 1974a, Astrophys. J., 194, p.419.
Lightman, A. P.: 1974b, Astrophys. J., 194, p.429.
Lin, D. N. C. and Pringle, J. E.: 1976, in *Structure and Evolution of Close Binary Systems*, (ed. P. Eggleton, S. Milton and J. Whelan), p.237.
Lubow, S. H. and Shu, F. H.: 1975, Astrophys. J., 198, p.383.
Lynden-Bell, D. and Pringle, J. E.: 1974, Mon. Not. Roy. Astron. Soc., 168, p.603.

Nariai, K.: 1967, Publ. Astron. Soc. Japan, 19, p.564.
Nariai, K.: 1975, Astron. Astrophys., 43, p.309.
Nariai, K.: 1977, Publ. Astron. Soc. Japan, 29, p.263.
Olson, E. C.: 1976, Astrophys. J., 204, p.141.
Paczyński, B.: 1971, Ann. Rev. Astron. and Astrophys., 9, p.183.
Packet, W.: 1980, in *Close Binary Stars: Observation and Interpretation* (ed. M. J. Plavec, D. M. Popper and R. K. Ulrich), D. Reidel, Dordrecht, p.211.
Petterson, J. A.: 1975, Astrophys. J.(Lett.), 201, p.L16.
Piotrowski, S. L., Ruciński, S. M. and Semeniuk, I.: 1974, Acta Astron., 24, p.389.
Prendergast, K. and Taam, R,: 1974, Astrophys. J., 189, p.125.
Pringle, J. E. and Rees, M. J.: 1972, Astron. Astrophys., 21, p.1.
Rees, M.: 1976, in *Structure and Evolution of Close Binary Systems*, (ed. P. Eggleton, S. Mitton and J. Whelan), p.225.
Roberts, W. J.: 1974, Astrophys. J., 187, p.575.
Rudy, R. J. and Kemp, J. C.: 1978, Astrophys. J., 221, p.200.
Sahade, J. and Wood, F. B.: 1978, *Interacting Binary Stars*, Pergamon Press.
Schuerman, D. W.: 1972, Astrophys. Space Sci., 19, p.351.
Simmons, J. F. L., Aspin, C. and Brown, J. C.: 1980, preprint submitted to Astron. Astrophys. See also article by the same authors, 1980, in *Close Binary Stars: Observation and Interpretation*, (ed. M. J. Plavec, D. M. Popper and R. K. Ulrich), D. Reidel, Dordrecht, p.343.
Smith, R. R. and McCall, D.: 1970, Comm. A. C. M. (Algorithm 392 (D3)), 13, p.567.
Söderhjelm, S.: 1980, in *Close Binary Stars: Observation and Interpretation*, (ed. M. J. Plavec, D. M. Popper and R. K. Ulrich), D. Reidel, Dordrecht, p.217.
Sørensen, S. A.: 1976, Prog. Theoret. Phys., 56, p.1484.
Sørensen, S. A., Matsuda, T. and Sakurai, T.: 1975, Astrophys. Space Sci., 33, p.465.
Vanbeveren, D.: 1977, Astron. Astrophys., 54, p.877.
Vanbeveren, D.: 1978, Astrophys. Space Sci., 57, p.41.
Walter, K.: 1973, Astrophys. Space Sci., 21, p.289.

THE STUDY OF APSIDAL MOTION IN ECLIPSING BINARIES

A. Giménez

Instituto de Astrofísica de Andalucía,
Granada, Spain.

ABSTRACT

The possibility of empirical determination of information on stellar structure by means of a study of the effect of apsidal motion in eccentric eclipsing binary systems is critically examined. Four conditions are imposed on observational material in order that useful results be determinable. 55 known eclipsing binaries have been found to meet the requirements, and subsequent processing steps are reported.

Light and radial velocity curves are not the only source of information provided by close binaries. The prediction of apsidal motions has been known since long ago as a necessary, although not sufficient, condition for the correctness of our knowledge of the structure of the stellar interiors. Many papers and efforts have been already dedicated to this kind of studes and some of the most important contributions to the subject were made by investigators who are present in this conference. Very recently, Monet (1980) published a detailed analysis of the apsidal motion evidence in spectroscopic binaries enhancing once more the importance of eclipsing binaries to disentangle the rather poor situation concerning the comparison with the theoretical models. Actually, there exists a significant systematic discrepancy between theory and observations that cannot be neglected.

The aim of the present communication is the determination of apsidal motion parameters from eclipsing binaries keeping random and systematic errors small enough for a safe comparison to be made. We summarize the conditions and procedures to be fulfilled in order to achieve this accuracy.

To justify further work on the observational analysis of apsidal motion in eclipsing binaries we have to be sure that more accurate data are really needed and that the obtained data really have the required accuracy. The first point has been repeatedly claimed in previous works as one of the main sources for the discrepancies, while the second is the main result of this investigation.

The physical causes for the movement of the line of the apses can be reduced to the existence of a perturbing potential due to the well-known facts that the stars are distorted by rotation and tides as well as the general relativistic correction to the two-body problem equations, of importance for large masses. As a result of the perturbations we have to take into account the variation of the orbital parameters with the time. Among these, the line of the apses is the easiest to observe and is secularly changing. Assuming coplanar rotation and no tidal lag we can express the variation of the periastron position in terms of seven parameters: the eccentricity of the orbit, e , and the masses, m , the relative radii, r , and the rotational ratios with respect to synchronization, g , for both components. Moreover, it depends obviously on the internal density dsitribution constants k_j and, since we can observationally determine the seven mentioned parameters from a proper analysis of the light and radial velocity curves and high dispersion spectra, we can invert the equation and have an empirical estimation of the values for k_j. It will be of course necessary to obtain the period of apsidal motion and the orbital anomalistic period from the analysis of eclipse timings and we present a procedure for this purpose in a subsequent section.

From the point of view of the models we can compute for any theoretical configuration the k_j coefficients by numerical integration of the Radau equation (Kopal, 1978). Therefore, we also have a function defined by the models in the form

$$k_j = k_j(m,X,t) ,$$

where m is the mass of the star, X the vector representing the chemical composition and t the age as reckoned from the Main Sequence (ZAMS).

It is evident, from the considerations expressed above, that the possibility exists for a good test of the theory with the observations, but several problems remain in both sides that could be claimed as the origin of the discrepancies already detected.

1. Observational problems:

a) Consideration of apparent displacements in the position of the minima as due to apsidal motion.
b) Accuracy with which the seven observational parameters are really obtained.
c) Accurate determination of the apsidal and anomalistic periods.

2. Theoretical problems:
   a) Time dependence of k_2.
   b) Use of different computational methods.
   c) Influence of the opacities.

Let us see now how we tried to deal with each of the above-mentioned subjects.

With respect to the observational problems, the set of systems to be analyzed with possibilities of being safely used for the comparison with the theory, was selected from a list of more than a hundred binaries found in the literature as probable candidates. The following conditions were imposed to them in order to avoid complications or wrong determinations:

1) Both eclipses to be deep enough for accurate measurement and not only primary minima.

2) Eccentricity to be large enough to give an appreciable amplitude of the variations.

3) Light curve not to be distorted by flares or other effects introducing complications in the analysis.

4) Both components to be detached from their Roche lobes.

Do such systems, fulfilling all the four conditions, actually exist? After a careful search within the preliminary list we found 55 candidates already presented (Giménez and Delgado, 1980). For many of these systems observations are still needed. Some of them cannot be analyzed because of the lack of light curves, others because of radial velocities, others because of both, while still others do not have enough times of minima to determine the apsidal motion parameters.

The determination of accurate absolute parameters is nowadays available according to the new treatment of the observations and the use of appropriate methods for the measurements as already pointed out by Andersen et al (1979) or Popper (1980). A careful analysis of the light and radial velocity curves even provides an independent check of the theoretical models or the determination of age and chemical composition of the system by a fitting method given by the group of Copenhagen.

The analysis of the observations in order to determine the apsidal motion parameters needs further insight and a new sequential method has been developed that will be explained in detail in a subsequent paper. We summarize here the main improvements with respect to previous methods:

- A series of numerical checks are made in order to be sure that other perturbing potentials, like a third body, are not present.

- Periodic variations are also taken into account during the procedure and not only secular terms.

- Realistic mean errors are computed for all fitted parameters allowing us the comparison with the theory.

- The apsidal motion period is not the only parameter to be determined, but also the eccentricity and the anomalistic period are computed.

- The whole analysis can be done completely automatized with small computers.

The equations applied are those of Kopal (1978), the equation of centre (Brouwer and Clemence, 1961) and the apparent shift of the luminosity minimum according to Martynov (1973). The computational procedure is given by the following steps:

a) Input data: observed times of minima, inclination of the orbit from the light curve and preliminary values for the eccentricity and the sidereal period.

b) Careful assignation of weight to the observed points.

c) Selection of the zero epoch time of minimum not arbitrarily, but through the weighted mean of the observed points.

d) Independent linear least-squares fitting of primary and secondary minima.

e) Initial determination of the sidereal period using Taylor expansion of the apsidal motion equation.

f) Fourier analysis of the O-C curves for primary and secondary minima with unequally spaced data to obtain a preliminary value of the apsidal period.

g) If available data are not enough or badly distributed, $T_2 - T_1$ differences are formed and the equation given by Sterne (1940) numerically applied to determine individual values of the periastron position for each epoch. Thence the apsidal period is obtained by linear least-squares fitting.

h) A differential corrector method is used to take into account minor effects considering all the observed points in the time domain instead of interpolated differences. The partial derivatives needed for the optimization are computed numerically.

i) Periodic variations in the periastron position are calculated according to the equations given by Kopal (1959).

j) The eccentricity is not a free parameter during the differential corrections and it is adjusted independently to minimize the sum of the square residuals.

As a result of the whole procedure we have five fitted parameters: eccentricity, apsidal motion period, zero epoch position of the periastron, anomalistic period and zero epoch time of minimum together with their mean errors. A computer program has been written that accomplishes all the steps in an interactive mode, in basic language, for a Hewlett-Packard mini-computer HP9845A.

With respect to the problems concerning the theoretical models, we obviously have to consider the variation of k_2 with the evolutionary state of the stars. An estimation of the age can be made, nevertheless, following the method outlined by Clausen et al (1976) as well as the chemical composition. Moreover, the membership of several candidates into open clusters can be used as a further check.

The use of different computational methods for building the theoretical models was another serious problem some years ago, but nowadays there exists some kind of convergence of all approaches to the same structures, especially for Main Sequence stars, Nevertheless, the application of different opacities like those by Cox, based on a hydrogenic model, or Carson, considering Thomas-Fermi approximation, still seems to have important influence on the computed results. It has been even proposed that these two approaches should bracket the "true" opacities according to observational conclusions from the study of apsidal motions. In fact, the internal density distribution will be much more sensitive to the variation of the opacity than the surface parameters, but also we have to keep in mind that independent checks are being made through the study of pulsational instabilities in intrinsic variables. Moreover, we have to consider that the chemical composition, especially the metal content, may introduce an effect of non-homogeneity into the comparison set of binaries.

REFERENCES

Andersen, J., Clausen, J. V. and Nordström, B.: 1979, IAU
    Symp. No. 88, *Close Binary Stars*, Toronto, Canada.
Brouwer, D. and Clemence, G. M.: 1961, *Celestial Mechanics*,
    Academic Press, New York.
Clausen, J. V., Gyldenkerne, K. and Grønbech, B.: 1976, Astron.
    Astrophys., 46, p.205.
Giménez, A. and Delgado, A. J.: 1980, IAU Comm. 27, IBVS, in
    press.
Kopal, Z.: 1959, *Close Binary Systems*, Chapman & Hall, London.
Kopal, Z.: 1978, *Dynamics of Close Binary Systems*, D. Reidel
    Publ. Co., Dordrecht, Holland.
Martynov, D. Ya.: 1973, in *Eclipsing Variable Stars*, ed. V. P.
    Tsesevich, IPST Astrophysical Library, Jerusalem.
Monet, D. G.: 1980, Astrophys. J., 237, p.513.
Popper, D. M.: 1980, Ann. Rev. Astron. Astrophys., preprint.
Sterne, T. E.: 1940, Proc. Nat. Acad. Sci., 26, p.36.

TEMPERATURE DETERMINATION OF EXCITING STARS IN HIGHLY EXCITED PLANETARY NEBULAE AND SYMBIOTIC STARS

T. Iijima

Asiago Astrophysical Observatory,
University of Padova, Italy.

ABSTRACT

An analytical method is proposed for the temperature determination of exciting stars in planetary nebulae and symbiotic stars by using relative intensities of $H\beta$, HeI $\lambda 4471$ and HeII $\lambda 4686$ emission lines. This method is applicable to the exciting stars of nebulae optically thick to the hydrogen ionizing radiation. The temperatures derived with the present method are in general higher than those obtained with the improved Zanstra's method. Close agreement is found with temperatures derived with a method based on the analysis of the relative dimensions of [NeV] $\lambda 3426$, HeII $\lambda 4686$ and $H\beta$ luminosity zones in the nebulae. Reasonable temperatures of exciting stars in symbiotic systems are also derived with this method.

1. INTRODUCTION

Temperature determination of exciting stars in planetary nebulae and symbiotic stars is a very important problem in order to understand the properties of these systems. The first results were obtained by Zanstra (1931), who derived the temperatures of central stars in planetary nebulae by comparing the intensities of hydrogen Balmer lines with the monochromatic magnitudes of the central stars at the same frequencies of respective Balmer lines. This method was extended by Harman and Seaton (1966) by using HeII $\lambda 4686$ or HeI $\lambda 4471$ emission lines. On the other hand, temperature determination methods based on intensities of nebular emission lines alone were proposed by Stoy (1933) and Ambartsumyan (1932). Stoy derived the temperature from the total intensity of forbidden lines relative to Hβ. Some improvements on this method have been made by Kaler (1967a, 1978). Ambartsumyan obtained the temperature from the intensity ratio of recombination lines of hydrogen and singly ionized helium. This method uses the asssumption that the radiation emitted by a central star lying in the frequency interval from ν_0 to $4\nu_0$ is completely absorbed by hydrogen and that the radiation beyond $4\nu_0$ is completely absorbed by singly ionized helium; ν_0 and $4\nu_0$ are Lyman limits of hydrogen and singly ionized helium, respectively. If this assumption is satisfied, the temperature of the star is given as a function of the ratio of the emitted radiation flux lying in each frequency interval, which can be deduced from the intensity ratio of the recombination lines.

Ambartsumyan's method usually gives an upper limit of the temperature, because, in general the helium secondary ionizing radiation is completely absorbed in nebulae, but the same is not necessarily true for the hydrogen ionizing radiation, and moreover the contribution of neutral helium in the absorption of stellar radiation is not taken into account. This paper will present modifications to Ambartsumyan's original method by considering the contribution of neutral helium. The optical depth of nebulae to the hydrogen ionizing radiation is qualitatively estimated with the intensity of [OI] λ6300 emission lines. An approximate analytical formula for the temperature determination is presented.

2. ANALYTICAL APPROACH

The absorption of ionizing radiation by neutral and singly ionized helium in planetary nebulae was studied by Hummer and Seaton (1964) for many cases. In this paper, however, we make the following simple assumptions:

i) Radiation emitted by an exciting star lying in the frequency intervals from ν_0 to ν_1, from ν_1 to $4\nu_0$ and from $4\nu_0$ to ∞ are completely absorbed by H, HeI and HeII respectively, where $\nu_1 = 1.809\nu_0$ is the ionization edge of neutral helium and, as mentioned in the previous section, ν_0 and $4\nu_0$ are the limits of Lyman continuum of hydrogen and singly ionized helium.

ii) In neutral helium, the triplet states occur three times more frequently than the singlet states in the recombinations.

iii) The radiation flux emitted by an exciting star can be represented by one black-body temperature.

In connection with assumption iii), it is suggested that the black-body approximation best represents the real energy distribution of high temperature stars (Kaler, 1967a; Pilyugin et al, 1978).

The intensities of radiation flux absorbed by the respective atoms and ions correlate with the intensities of their recombination lines with the following formulae (Osterbrock, 1974).

$$\frac{L\nu_f}{\int_{\nu_0}^{\nu_1} \frac{L\nu(T^*)}{h\nu} d\nu} = h\nu_{H\beta} \frac{\alpha_{H\beta}(H) \quad \pi F\nu_f}{\alpha_B(H) \quad \pi F_{H\beta}} \quad (1)$$

$$\frac{L\nu_f}{\int_{\nu_1}^{4\nu_0} \frac{L\nu(T^*)}{h\nu} d\nu} = \frac{3}{4} h\nu_{4471} \frac{\alpha_{4471}(He\ I) \quad \pi F\nu_f}{\alpha_B^3(He\ I) \quad \pi F_{4471}} \quad (2)$$

$$\frac{L\nu_f}{\int_{4\nu_0}^{\infty} \frac{L\nu(T^*)}{h\nu} d\nu} = h\nu_{4686} \frac{\alpha_{4686}(He\ II) \quad \pi F\nu_f}{\alpha_B(He\ II) \quad \pi F_{4686}} \quad (3)$$

where α_B is the total recombination coefficient for the atoms and the ion, α_λ in the effective recombination coefficient for the line with wavelength $\lambda \text{Å}$, and α_B^3 is the total recombination coefficient for the triplet states of neutral helium. $L\nu$ is the luminosity of the exciting star on frequency ν per unit frequency interval, and $L\nu_f$ is the luminosity of the star on a particular frequency ν_f. F_λ is the observed intensity of an emission line at wavelength $\lambda \text{Å}$ corresponding to frequency ν, and $F\nu_f$ is the brightness of the star on a particular frequency ν_f. T^* is the temperature of the exciting star. For the purpose of deriving the temperatures of exciting stars from intensities of nebular emission lines alone, it is necessary to obtain a formula which does not include $L\nu_f$ and $F\nu_f$. From equations (1), (2) and (3) we have

$$\frac{\int_{4\nu_0}^{\infty} \frac{L\nu(T*)}{h\nu} d\nu}{\int_{\nu_0}^{4\nu_0} \frac{L\nu(T*)}{h\nu} d\nu} = \frac{\int_{\nu_0}^{\nu_1} \frac{L\nu(T*)}{h\nu} d\nu + \int_{\nu_1}^{4\nu_0} \frac{L\nu(T*)}{h\nu} d\nu}{}$$

$$= \frac{\dfrac{\alpha_B(\text{He II}) F_{4686}}{h\nu_{4686} \alpha_{4686}(\text{He II})}}{\dfrac{\alpha_B(\text{H}) F_{H\beta}}{h\nu_{H\beta} \alpha_{H\beta}(\text{H})} + \dfrac{4}{3} \dfrac{\alpha_B^3(\text{He I}) F_{4471}}{h\nu_{4471} \alpha_{4471}(\text{He I})}} \quad . \quad (4)$$

If $L\nu$ is replaced with the black-body function we have

$$\frac{\int_{4\nu_0}^{\infty} \frac{L\nu(T*)}{h\nu} d\nu}{\int_{\nu_0}^{4\nu_0} \frac{L\nu(T*)}{h\nu} d\nu} = \frac{\int_{4x_0}^{\infty} \frac{x^2}{e^x - 1} dx}{\int_{x_0}^{4x_0} \frac{x^2}{e^x - 1} dx} \quad , \quad (5)$$

where
$$\begin{aligned}
x &= h\nu/kT* \\
x_0 &: 15.782 \times 10^4/T* \\
h &: \text{Planck constant} \\
k &: \text{Boltzmann constant} \\
T* &: \text{temperature of exciting star.}
\end{aligned}$$

Numerical values of α_B and α_λ are given by Seaton (1960) and Brocklehurst (1971), respectively; both are summarized by Osterbrock (1974). Using the numerical values, we have

$$\frac{\int_{4x_0}^{\infty} \frac{x^2}{e^x - 1} dx}{\int_{x_0}^{4x_0} \frac{x^2}{e^x - 1} dx} = \frac{2.22 \, F_{4686}}{4.16 \, F_{H\beta} + 9.94 \, F_{4471}} \cdot \quad (6)$$

Electron temperatures of 10^4K for Hβ emission region and of 1.5×10^4K for HeI $\lambda 4771$ and HeII $\lambda 4686$ emission regions are assumed, and electron number density of 10^4cm^{-3} is assumed for all regions. Numerical values of the integrations in the left-hand side of equation (6) are given by Gurzadyan (1970).

Temperatures of exciting stars in nebulae with HeII emission lines should be higher than 60000 K (Kaler, 1976a). In that temperature range, it is possible to make a quadratic approximation for the left-hand side of equation (6); namely,

$$\frac{\int_{4x_0}^{\infty} \frac{x^2}{e^x - 1} dx}{\int_{x_0}^{4x_0} \frac{x^2}{e^x - 1} dx} = 0.00266 \, t^2 - 0.0273 \, t + 0.0719. \quad (7)$$

$$t = T* \times 10^{-4}$$

Finally, from equations (6) and (7), we have a simple formula for the temperature determination of exciting stars, of the form

$$T* \times 10^{-4} = 19.38 \, \sqrt{K} + 5.13 , \quad (8)$$

$$K = \frac{2.22 \, F_{4686}}{4.16 \, F_{H\beta} + 9.94 \, F_{4471}} \cdot \quad (9)$$

The errors due to the quadratic approximation are less than ±700K in the temperature interval from 7.0×10^4 to 18×10^4K, and less than ±3000K from 6.0×10^4 to 20×10^4K. It is necessary to use the original formula, equation (6), to estimate the temperatures of exciting stars in low excited nebulae which have K in equation

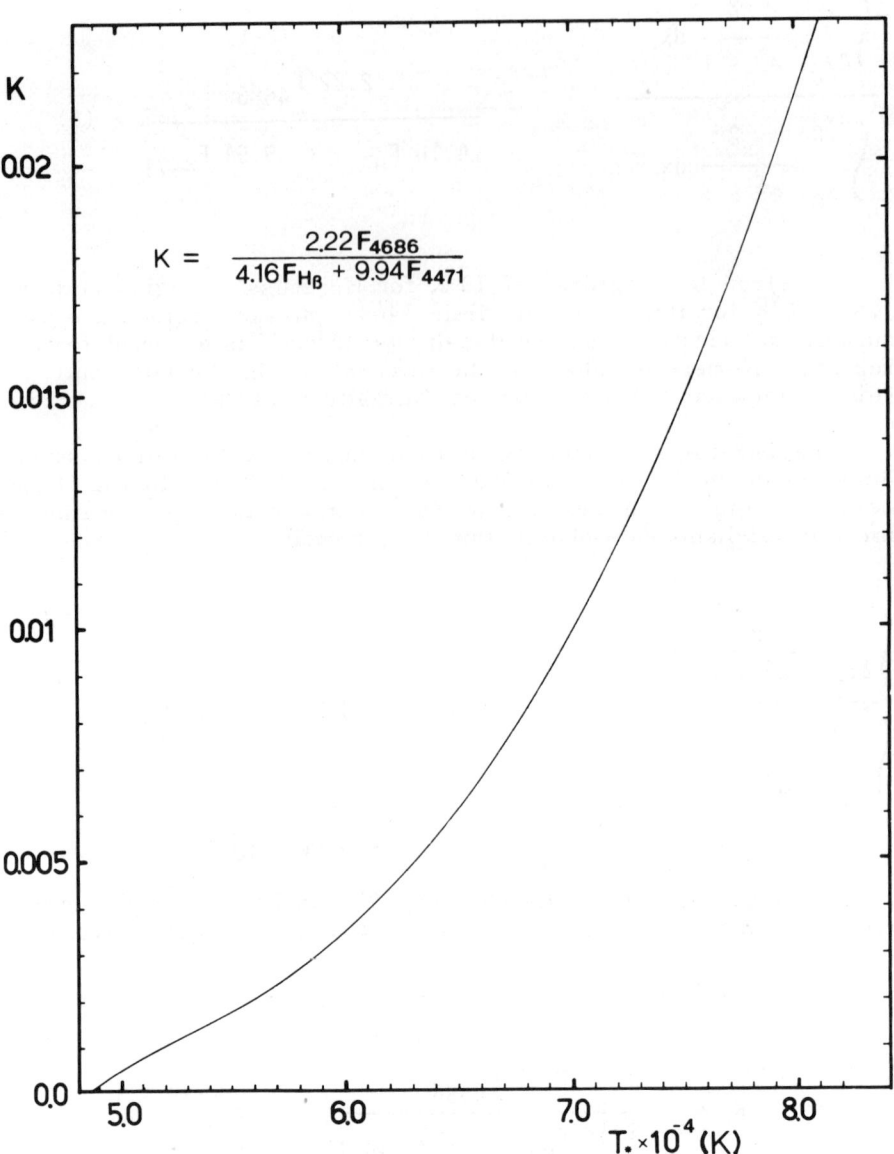

Figure 1 Relation between the intensities of the hydrogen and helium emission lines and temperature of exciting star.

(9) smaller than 0.01. This corresponds to the temperature of the exciting stars lower than $7.0 \times 10^4 K$. The original form of equation (6) is shown in Figure 1 in the temperature interval from 5.0×10^4 to $8.0 \times 10^4 K$.

In very highly excited nebulae, where the intensity of HeI $\lambda 4771$ is negligible, this method comes to the same as Ambartsumyan's method.

3. TEMPERATURE OF EXCITING STARS

A. Planetary Nebulae

The temperatures of central stars in the planetary nebulae are listed in Table 1. The names of the nebulae in the first column follow the abbreviation of the catalogue of Perek and Kohoutek (1967). The nebular numbers given in the same catalogue are shown in the second column. The temperatures derived with the method presented in this paper are shown in the third column and denoted by *He*. Two kinds of data of the relative intensities of Hβ, HeI $\lambda 4471$ and HeII $\lambda 4686$ are used to derive these temperatures. One is the mean values of the relative intensities given in Kaler's catalogue (1976b). The second is the values measured recently by Barker (1978). In both cases, the intensities corrected for interstellar extinction are used. The temperatures derived with Barker's data are denoted by a dagger. An asterisk indicates a new value. The errors of the temperatures due to the observational errors on the line intensities are an order of $\pm 1000K$ in both cases. For comparison, the temperatures derived from other methods are also summarized in Table 1; *H-S* indicates the temperatures obtained by Harman and Seaton (1966) with an improved Zanstra's method; *Gur* those of Gurzadyan (1970) derived from the intensity ratio [OIII]/[OII]; *K-T* those derived with the same modified method by Köppen and Tarafdar (1978); *Kal* the temperatures obtained by Kaler (1976a, 1978) with improved Stoy's methods; *Khr* those obtained by Khromov (1967) through the analysis of the relative dimensions of [NeV] $\lambda 3426$, HeII $\lambda 4686$ and Hβ luminosity zones in the nebulae, and finally those denoted by *UV* are derived by Pottasch et al (1978) with ultraviolet photometry made by ANS ($\lambda\lambda 1500 \sim 3300$ Å, $\Delta\lambda \sim 150$ Å). The last column of Table 1 gives the intensity of [OI] $\lambda 6300$ emission line relative to Hβ, assuming a value of 100 for Hβ. In the last column " - " means that the nebula has no [OI] $\lambda 6300$ line and "?" that the nebula was not observed at $\lambda 6300$ Å. In Table 1, low accuracy values are denoted by a colon, while a double colon indicates only an order of magnitude.

Figures 2 - 7 represent the correlation between the temperatures obtained with the present and the other methods. In the figures, nebulae with [OI] $\lambda 6300$ line are indicated by a solid circle,

Table I Temperatures of central stars in planetary nebulae derived from various methods.

NGC et al.	P.K.	$T* \times 10^{-4}(K)$							[O I]
		He	H-S	Gur	K-T	Kal	Khr	UV	
650-1	130-10°1	15.6	18.2	9.5					-
1501	144+ 6°1	17.5	7.2						-
1535	206-40°1	10.9	7.4	5.2	>10.0			4.0	34.28:
2022	196-10°1	21.1:	9.1	5.9					-
2371-2	189+19°1	18.6	10.0	4.8				>8.6	-
2392	197+17°1	14.7†	6.8	5.0			16.0	3.2	2:
2440	234+ 2°1	17.0		10.8			15.0		24.91
3242	261+32°1	11.1†	9.3	5.3				5.0	1:
3587	148+57°1	11.1	10.5	6.6				11.8	11.52:
3918	294+ 4°1	14.7*							10.21:
6058	64+48°1	17.4	7.2						?
6210	43+37°1	7.0†	5.0	4.6		5.8		4.8	2.45
6302	349+ 1°1	15.6*							14.35:
6309	9+14°1	16.4	9.6						-
6439	11+ 5°1	11.8	7.2						-
6445	8+ 3°1	14.9	18.2	>12.0	4.5				17.17
6537	10+ 0°1	18.2*							5.30
6543	96+29°1	7.1	6.6	4.6	5.0			3.8	0.28
6565	3- 4°5	10.0*							51.85
6567	11- 0°2	6.9†				5.6			3.13
6572	34+11°1	5.0:	6.2	4.6	4.9	6.0			10::
6644	8- 7°2	8.7†*							9.00
6720	63+13°1	11.6†		6.8					5::
6741	33- 2°1	14.3		6.3:	4.0				28.03
6751	29- 5°1	11.7	7.6	5.5					-
6772	33- 6°1	13.5	11.2	9.6					?
6778	34- 6°1	10.4	8.5	6.0					11.82
6781	41- 2°1	11.7	9.1	7.4					?
6790	37- 6°1	7.7		7.8:	4.8	6.6			14.64
6803	46- 4°1	7.8		4.8	5.3	6.2			19.07
6804	45- 4°1	17.9	7.2						-
6818	25-17°1	15.3	19.5	8.4			14.0		19.50
6826	83+12°1	8.3	6.9	4.6	8.0	4.5		3.4	0.1::
6833	82+11°1	6.9†		9.0:		4.9			3.80
6853	60- 3°1	13.8	13.2	7.7					-
6879	57- 8°1	12.2*							33.69
6881	74+ 2°1	16.6	6.9						-
6884	82+ 7°1	11.2		>11.0	5.4				14.76
6886	60- 7°2	14.1		6.3	4.0		16.0		13.13
6894	69- 2°1	11.5	9.8	4.0					-

Table I (continued)

NGC et al.	P.K.	He	H-S	Gur	K-T	Kal	Khr	UV	O I
				$T_* \times 10^{-4}$ (K)					
6905	61- 9°1	17.8	10.2	11.4					-
7008	93+ 5°2	17.4	9.8	10.3					-
7009	37-34°1	10.3	8.1	5.7	9.5				1.69
7026	89+ 0°1	11.0	9.8	5.8					25.12
7027	84- 3°1	14.9†		11.9:	4.8				12.14
7139	104+ 7°1	11.3	9.8	8.2					?
7354	107+ 2°1	16.4	10.2						?
7662	106-17°1	14.6	10.0	5.6	>10.0		13.0	4.8	1.92
IC 351	159-15°1	12.9†	9.1	7.1					1.25
IC 418	215-24°1	5.1†	4.3	3.8	3.3	3.4		2.95	3.37
IC 1747	130+ 1°1	11.7*							21.62
IC 2003	161-14°1	15.0†		6.9:					3.58
IC 2165	221-12°1	14.7		7.8:			14.0		4.51
IC 3568	123+34°1	6.4†	3.2	6.0		6.1		4.7	1::
IC 4406	319+15°1	8.1				7.3:			?
IC 4846	27- 9°1	5.8		4.8		6.1			-
IC 4997	58-10°1	6.1		4.2:		4.9			5::
IC 5117	89- 5°1	9.9		7.0	6.4	7.7			14.44:
IC 5217	100- 5°1	8.8†	7.4	6.1	5.4	6.8			6.77:
BD +30	64+ 5°1	10.5	4.6		2.4	3.0		3.1	2.60
CN 3-1	38+12°1	6.7†				2.5:			2.20
Hb 5	359- 0°1	17.5*							28.86
Hb 12	111- 2°1	5.4†				3.8:			6.95
Hu 1-1	119- 6°1	10.1*							85.79
Hu 1-2	86- 8°1	18.2*							16.84:
Hu 2-1	51+ 9°1	6.2†				3.6			2.21
J 320	190-17°1	6.6†		6.7		5.7			1.50
J 900	194+ 2°1	14.4*							27.84
M 1-1	130-11°1	19.4†*							1.91
M 1-2	133- 8°1	8.1†*							7.40
M 1-54	16- 4°1	11.3*							34.13
M 1-61	19- 5°1	12.1*							29.74
M 1-74	52- 4°1	5.8†*							6.80
M 2-50	97- 2°1	8.9†*							5.20
M 3-27	43+11°1	6.8†*							14.3
M 3-35	71- 2°1	<10.2†							4.64
Me 1-1	52- 2°2	8.2				6.5			25.20
Sn 1	13+32°1	7.2†*							4.77
VY 1-1	118- 8°1	9.9*							44.36:
VY 1-2	53+24°1	11.5†*							3.33

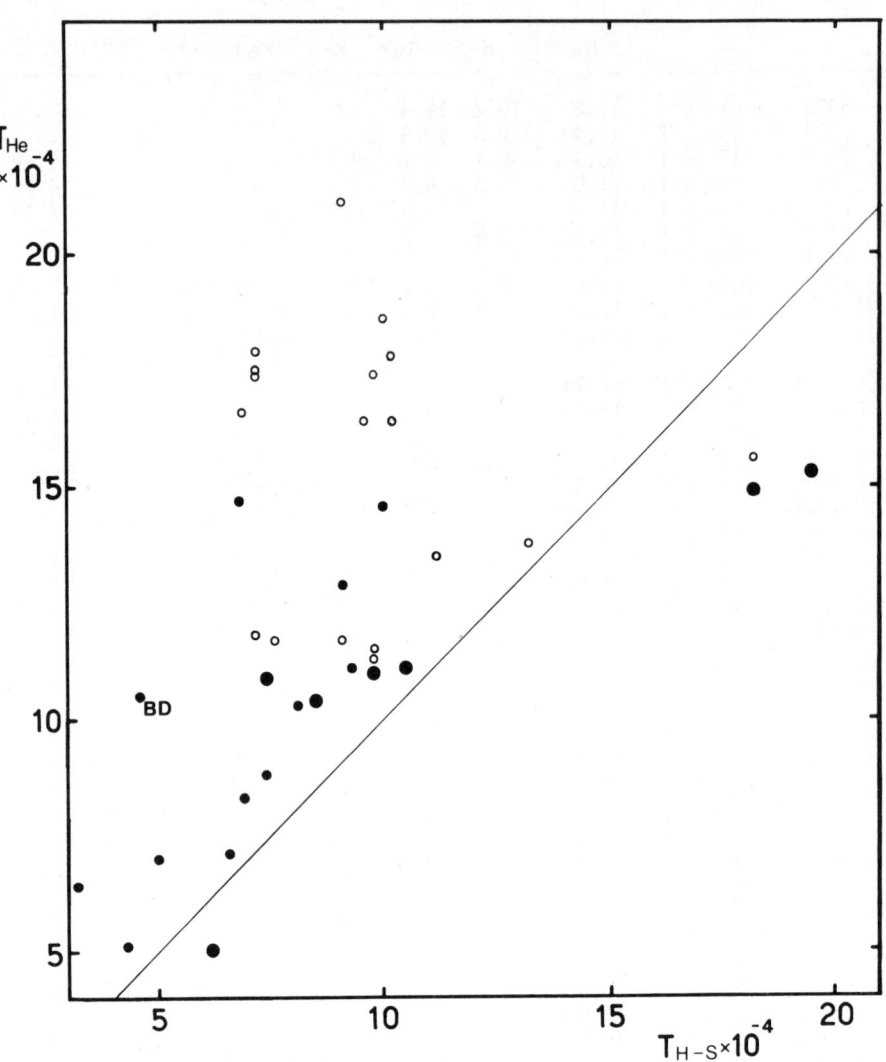

Figure 2 Correlation between T_{He} (Iijima) and T_{H-S} (Harman and Seaton, 1966). Solid circles indicate nebulae with [OI] $\lambda 6300$ line, while nebulae with strong [OI] lines are indicated with a large solid circle. Various other notations are explained in the text.

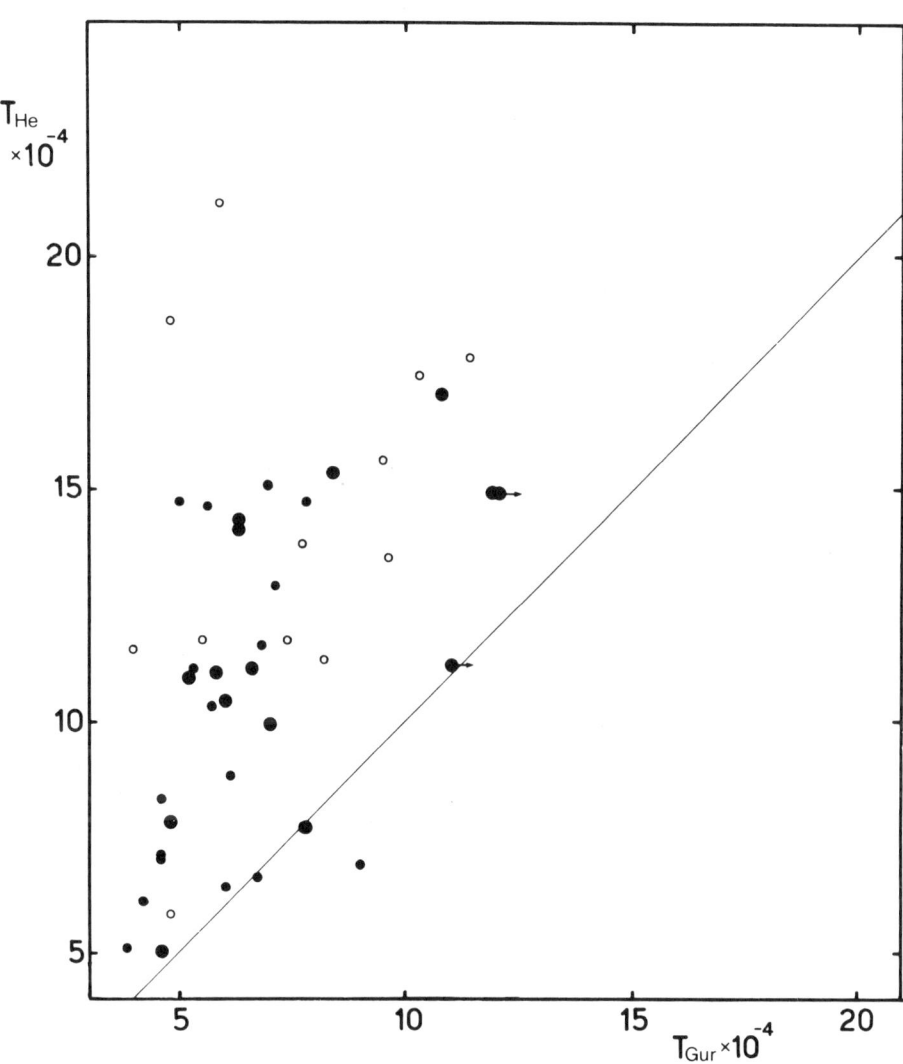

Figure 3 Correlation between T_{He} and T_{Gur} (Gurzadyan, 1970). Symbols are the same as in Figure 2.

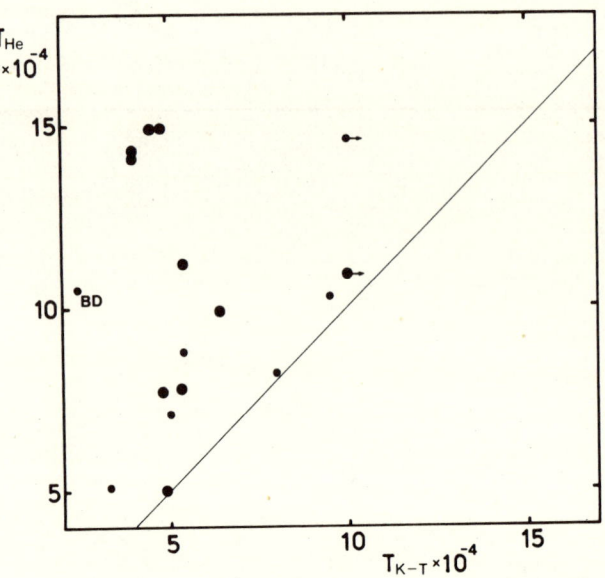

Figure 4 Correlation between T_{He} and T_{K-T} (Köppen and Tarafdar, 1978). Symbols are the same as in Figure 2.

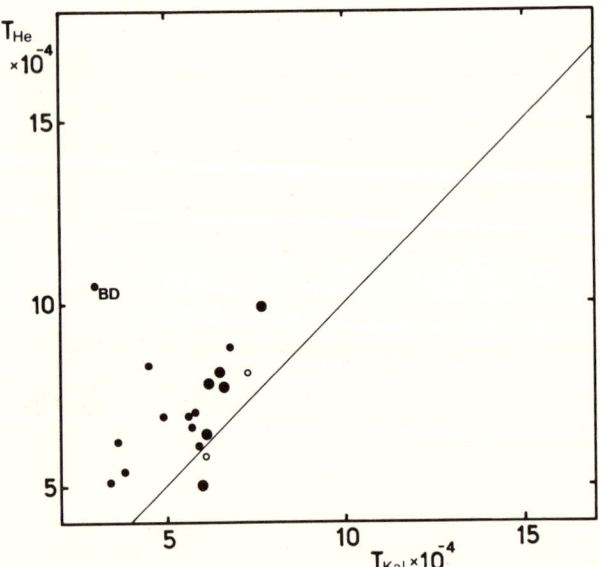

Figure 5 Correlation between T_{He} and T_{Kal} (Kaler, 1976a, 1978). Symbols are the same as in Figure 2.

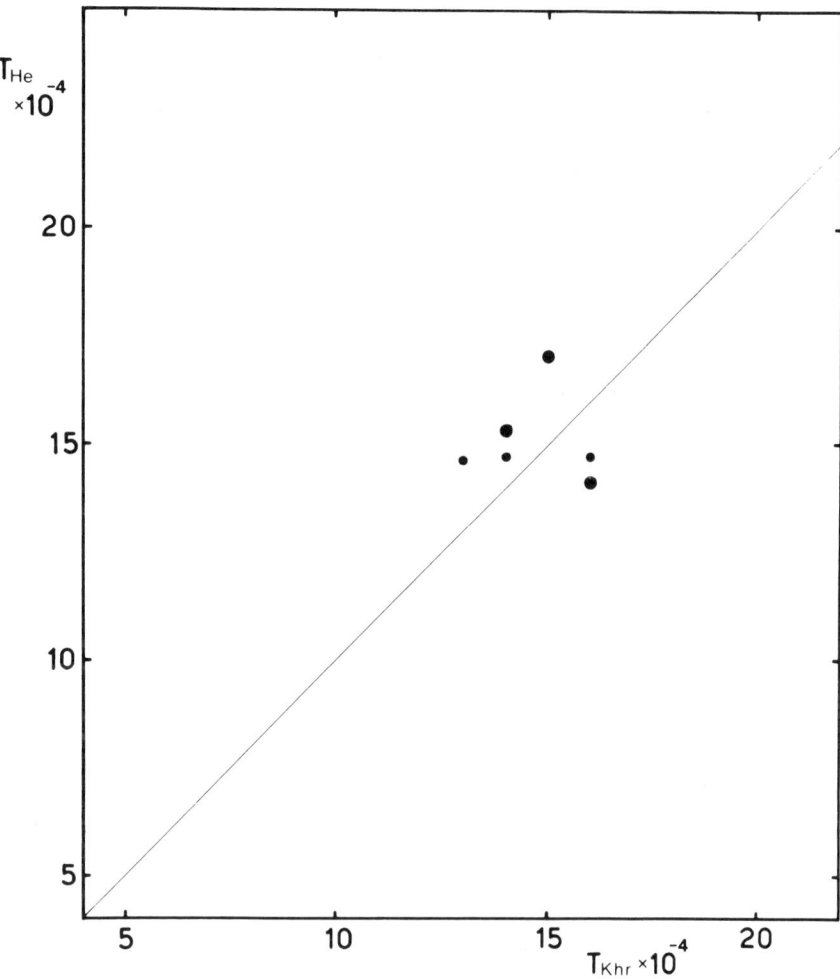

Figure 6 Correlation between T_{He} and T_{Khr} (Khromov, 1967). Symbols are the same as in Figure 2.

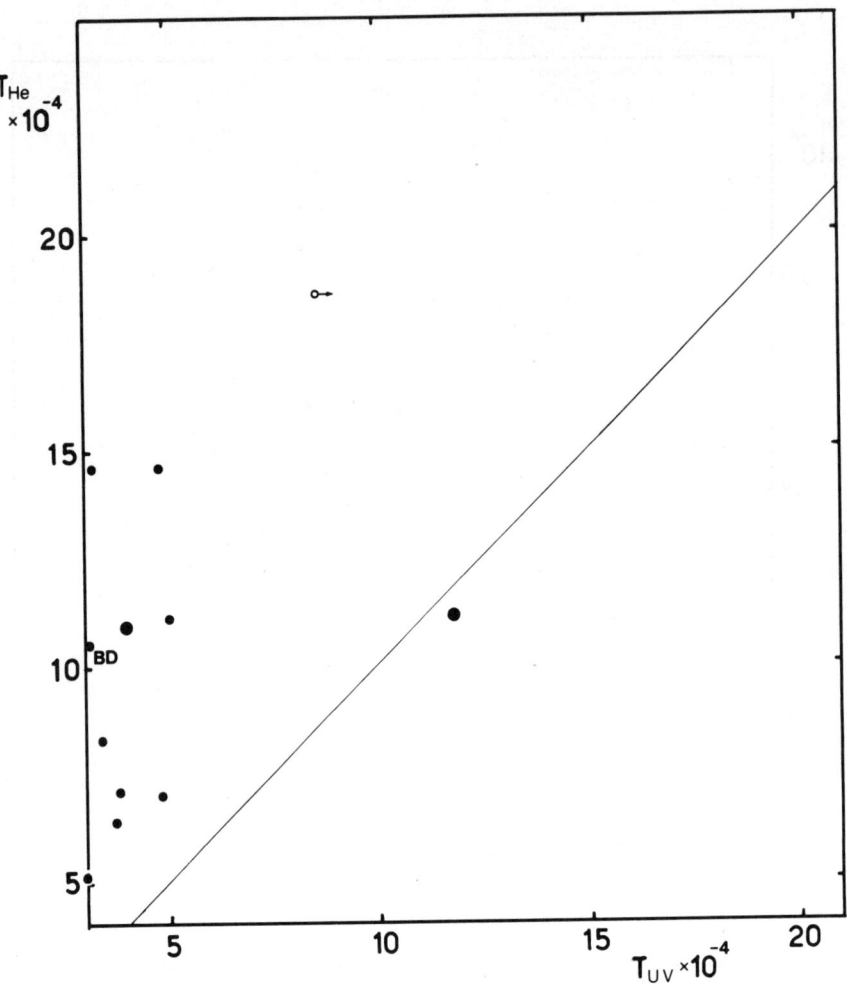

Figure 7 Correlation between T_{He} and T_{UV} (Pottasch et al, 1978). Symbols are the same as in Figure 2.

in particular those with [OI] λ6300 intensity 10% higher than Hβ are indicated by a large solid circle. In the latter case, the optical depth of the nebulae to the hydrogen ionizing radiation is expected to be much larger than unity (Khromov, 1967). The emission line on λ6300 Å is usually blended with [SIII] λ6310. This blending, however, may not affect the qualitative estimation, because, in general the intensity of [SIII] is lower than that of [OI]. Open circles indicate two kinds of nebulae denoted by " - " or " ? " in the last column of Table 1. BD +30°3639 is indicated by " BD ". Lower limits of the temperatures are indicated by an arrow. The solid line displays the equal temperature.

As shown in Figure 2, difference from the temperatures derived with the improved Zanstra's method (Harman and Seaton, 1966) may be tolerable, at least on the nebulae with [OI] λ6300 line, even though the present method gives rather high temperatures. Agreements with the temperatures derived from the intensity ratio [OIII]/[OII] are rather poor in some nebulae, as seen in Figures 3 and 4. Agreement with the temperatures derived from the intensities of forbidden lines is fairly good, except for BD +30°3639, as seen in Figure 5. Two causes may explain the disagreement on BD +30°3639. One is the existence of HeII λ4686 emission line in the spectrum of the central star itself, and the second is the high electron density in the nebula. This problem remains open here. Also a good agreement is obtained with the temperatures derived by Khromov (1967), as seen in Figure 6. To the contrary the temperatues derived by the ultraviolet observations seriously disagree, except in only one object, NGC 3587. The method based on the ultraviolet observations seems to give too low temperatures in some nebulae. For example, it may be difficult to explain the existence of [FeVII] and [NeV] emission lines in NGC 2392 and NGC 7662 (Kaler, 1976b) with the temperatures of the central stars given by the ultraviolet observations. It might be necessary to consider the possibility of binary systems for the central stars. Further efforts will be necessary to give a solution of this problem.

B. Symbiotic Stars

Exciting stars in symbiotic stars are considered to have similar properties of central stars in planetary nebulae. Some characteristics of the systems, however, prevent us applying the method proposed for planetary nebulae to symbiotic stars. For example, Zanstra's method may not be applicable because symbiotic stars are considered to be binary systems (Boyarchuk, 1975). Therefore, it is difficult to measure the brightness of the exciting stars alone. Also, the methods based on intensities of forbidden lines may not give a good approximation of the real temperature. Because nebular components of symbiotic stars have 2-3 orders of magnitude higher electron densities than those of typical planetary nebulae (Boyarchuk, 1975, Iijima, 1980a), the intensities of forbidden lines may depend not only on the temperatures of exciting stars, but also on the electron density. An applicable method to symbiotic stars was proposed by Boyarchuk (1975). His method is based on the fitting of a model calculation to Balmer discontinuity.

Table 2 lists the intensities of HeI $\lambda 4471$, HeII $\lambda 4686$ and [OI] $\lambda 6300$ emission lines of some symbiotic stars, relative to Hβ, assuming a value of 100 for Hβ. The temperatures of the exciting stars in the systems derived with the present method, denoted by He, are shown in the second column of Table 3. Those derived by Boyarchuk (1975), denoted by Boy, are shown for comparison in the third column.

The comparison of these temperatures may not be always significant, because the temperatures of exciting stars in symbiotic stars change with the activity of the systems (Boyarchuk, 1975). As seen in Table 3, however, the present method seems to give reasonable temperatures for the exciting stars in symbiotic systems. The high temperatures derived for V1016 Cyg and V1329 Cyg are consistent with the high intensities of [FeVII] and some other highly ionized emission lines (Ciatti et al, 1978; Tamura, 1977; Iijima, 1980b) in the systems. At the same time, high temperatures for V1329 Cyg explain why, when the exciting star is eclipsed, the minima on the light curve are so deep (Grygar et al, 1979; Iijima et al, 1981).

Table II Relative intensities of HeI $\lambda 4471$, HeII $\lambda 4686$ and [OI] $\lambda 6300$. (Hβ = 100).

Name	Date	HeI	HeII	[OI]	Ref.
BF Cyg	Sept 17, 1979	6.2	5.7	-	6
CI Cyg	May-Dec, 1979	10.4	36.3	-	1
AX Per	Oct 24, 1979	17.0	40.5	-	6
V1016 Cyg	May-Sept, 1977	10.7	70.8	64.6*	3
V1329 Cyg(A)	1974 - 1976	4.7	70.0	?	4
V1329 Cyg(B)	Sept 17, 1979	10.1	79.4	8.0*	2
HM Sge	Nov-Dec, 1976	3.7	1.0	4.7	5

* blended with fairly strong [SIII] $\lambda 6310$
V1329 Cyg = HBV 475.

References:

1. Iijima (1980a)
2. Iijima (1980b)
3. Ciatti et al (1978)
4. Tamura (1977)
5. Ciatti et al (1977)
6. Iijima (unpublished).

Table III Temperatures of exciting stars in symbiotic systems.

Name	$T^* \times 10^{-4}$ (K)	
	He	Boy
BF Cyg	8.3	7.9
CI Cyg	12.8	12.9
AX Per	12.7	14.1
V1016 Cyg	15.8	10.9
V1329 Cyg (A)	16.4	
V1329 Cyg (B)	16.5	
HM Sge	6.3	

4. CONCLUDING REMARKS

Although some problems remain, as mentioned in Section 3-A, we can conclude that the method proposed in this paper gives reasonable temperatures of exciting stars in optically thick planetary nebulae and symbiotic stars.

Much work has been done on symbiotic stars. Unfortunately, however, most of it was restricted to qualitative analyses, because it was difficult to estimate the temperature of exciting stars in the systems. The present method of temperature determination may give a possibility of passing to quantitative analyses. The method could be applicable also to novae in the nebular stage and to related objects.

ACKNOWLEDGMENTS

I thank the Director of Asiago Astrophysical Observatory, Professor L. Rosino, for kind hospitality, careful reading of the manuscript and encouragement. I am indebted to Dr. R. Canterna for the English version.

This work is supported by a grant from the Italian Government.

REFERENCES

Ambartsumyan, V. A.: 1932, Circ. Glav. Astron. Obs., 4, p.8.
Barker, T.: 1978, Astrophys. J., 219, p.914.
Boyarchuk, A. A.: 1975, IAU Symp. No. 67, p.377.
Brocklehurst, M.: 1971, Mon. Not. Roy. Astron. Soc., 153, p.471.
Ciatti, F., Mammano, A. and Vittone, A.: 1977, Astron. Astrophys., 61, p.459.

Ciatti, F., Mammano, A. and Vittone, A.: 1978, Astron. Astrophys., 68, p.251.
Grygar, J., Hříč, L., Chochol, D. and Mammano, A.: 1979, Bull. Astron Inst. Czechoslovakia, 30, p.308.
Gurzadyan, G. A.: 1970, *Planetary Nebulae*, D. Reidel Publ. Co., Dordrecht, Holland.
Harman, R. J. and Seaton, M. J.: 1966, Mon. Not. Roy. Astron. Soc., 132, p.15.
Hummer, D. G. and Seaton, M. J.: 1964, Mon. Not. Roy. Astron. Soc., 127, p.217.
Iijima, T.: 1980a, Astron. Astrophys., in press.
Iijima, T.: 1980b, in preparation.
Iijima, T., Mammano, A. and Margoni, R.: 1981, Astrophys. Space Sci., 75, p. 237.
Kaler, J. B.: 1976a, Astrophys. J., 210, p.843.
Kaler, J. B.: 1976b, Astrophys. J. Suppl., 31, p.517.
Kaler, J. B.: 1978, Astrophys. J., 220, 887.
Khromov, G. S.: 1967, Mon. Not. Roy. Astron. Soc., 137, p.181.
Köppen, J. and Tarafdar, S. P.: 1978, Astron. Astrophys., 69, p.363.
Osterbrock, D. E.: 1974, *Astrophysics of Gaseous Nebulae*, W. H. Freeman, San Francisco.
Perek, L. and Kohoutek, L.: 1967, *Catalogue of Galactic Planetary Nebulae*, Academia Publishing House of the Czech. Acad. Sci., Prague.
Pilyugin, L. S., Sakhibullin, N. A. and Khromov, G. S.: 1978, Astrophysics, 14, p.377.
Pottasch, S. R., Wesselius, P. R., Wu, C.-C., Fieten, H. and van Duinen, R. J.: 1978, Astron. Astrophys., 62, p.95.
Seaton, M. J.: 1960, Mon. Not. Roy. Astron. Soc., 120, p.326.
Stoy, R. H.: 1933, Mon. Not. Roy. Astron. Soc., 93, p.588.
Tamura, S.: 1977, Astrophys. Lett., 19, p.57.
Zanstra, H.: 1931, Publ. Dominion Astrophys. Obs., 4, p.209.

GAS STREAMS IN CLOSE BINARY SYSTEMS

Zdeněk Kopal

Department of Astronomy, University of Manchester, England

"Scenarios non fingo", Isaac Newton
(in *Philosophiae Naturalis Principia Mathematica*)

A number of papers presented in the past few days at our Conference have dealt - directly or indirectly - with spectroscopic phenomena observed in close binary systems which cannot be ascribed to any one of the two constituting components; and whose origin must, therefore, be sought outside the simple gravitational dipole represented by these stars. The aim of this concluding communication presented at our meetings should by no means be construed as an attempt to solve the problems still outstanding in this particular field of double-star astronomy, but rather to place these problems, and attempts at their solution, in proper perspective.

The existence of lines (both absorption and emission) in the spectra of close binary systems, whose Doppler shifts do not follow the motion of either component, has been known to us since the last century - witness the stationary lines of the so-called B5 spectrum of β Lyrae (cf. Bĕlopolsky, 1893, 1897); and - not to be outdone - the second known proto-type of close eclipsing binaries, Algol (β Persei) disclosed in its spectrum evidence of certain stationary lines of metals (mainly Fe and Mg), not sharing in the Doppler shifts of the photospheric lines of the two components; these "extra" lines (as she called them) were identified by Miss Barney (1923) in the spectra of Algol obtained by Schlesinger at Allegheny Observatory between 1907-1912. In 1930, again Carpenter found what he called an "anomalous effect" (Carpenter, 1930) in the spectra of U Cephei - an eclipsing system characterized by a (virtually) circular orbit, whose radial-velocity curve exhibited a conspicuous asymmetry (simulating the effects of a spurious eccentricity e close to 0.47) - a phenomenon to which astronomers at that time adopted an ostrich attitude (preferrring in private to explain away this effect by errors of measurements), until such a posture was made untenable by the pioneer work of Otto Struve and his school between 1940-1950.

Struve (1944) not only confirmed Carpenter's results for U Cephei, but detected at least another pair - namely RZ Scuti (cf. Neubauer and Struve, 1945), where the same effect is even more conspicuous.

As is well known, Struve and his followers (for a summary of their views cf., e.g., Struve, 1950) sought the origin of anomalous Doppler shift in *gas streams* within the respective systems; and their views - graphically represented by the "elephant trunk" diagrams so familiar in subsequent astronomical literature of lighter vintage - almost became a trademark for this line of thought. Your present speaker yields to no one in his appreciation of what Otto Struve did at the telescope in the 1940's to enrich our subject with new spectroscopic evidence. However, when it came to an interpretation of this evidence, Struve exhibited also an imagination worthy of a modern poet; and the daring reconnaissance spirit of the cavalry man he was in the days of his youth.

Was this exercise (in which Struve found many a dedicated follower) fertile? While the last word on this subject has not yet been spoken, I should become unfaithful to my self-imposed function of "devil's advocate" at this conference if I did not share with you also certain misgivings which should - in fact, must - be kept in mind as we plod along the tortuous path of progress; and which anyone can ignore only at his own risk.

First - to commence with the fundamentals - under which conditions can the Doppler shifts observed in the spectra of the components of close binary systems be regarded as true indicators of the radial velocity of their absolute space motions? The answer is in the affirmative *if - and only if - the "centres of light" of the respective components continue to project themselves constantly on their projected centres of mass.* If the stars are spherical, and the distribution of surface brightness over their apparent discs is characterized by radial symmetry, the observed Doppler shifts are indeed directly indicative of the radial velocity of the centre of mass of the respective star, due to its Keplerian motion.

For the components of close binary systems *distorted* by axial rotation and mutual tidal action this can, however, no longer remain strictly true; for their distribution of surface brightness (influenced also by gravity-darkening) will not only cease to be radially-symmetrical, but the shape of the isophotes over their apparent discs will vary with the phase. The same will, moreover, be caused by the effects of mutual irradiation ("reflection effect") even if the components could be regarded as spheres. All these effects are bound to make the *observed* radial velocities of the components of close binary systems *differ* from the Keplerian motion of their mass centres; and terms arising from their axial rotation will superpose upon the purely orbital velocity to yield a resultant which - if analyzed in the classical manner without regard to these complications - would furnish elements of spectroscopic orbits which are systematically in error.

A study of the effects, on observed radial velocities, arising from axial rotation of distorted stars was inaugurated by Sterne (1941) who as the first pointed out that effects produced by second-harmonic tidal distortion are tantamount to those produced by orbital eccentricity if the apsidal line of the respective orbit were parallel with the line of sight. Sterne's work was subsequently extended - by a more general method - by Kopal (1945) to include the effects caused by third- and fourth-harmonic partial tides. Moreover, the effects of reflection upon radial-velocity changes of spectroscopic binaries, foreseen already by Eddington (1926), were first quantitatively investigated also by Kopal (1943) to the order of approximation to which their components can be regarded as spheres. This work disclosed the extent to which the apparent centre of the irradiated star is bound to be shifted in the direction of the illuminating source - an effect diminishing the amplitudes of orbital velocity curves - as well as the degree to which the ascending branch of such curves becomes steeper than the descending one - an effect producing spurious eccentricity, reinforcing that due to limb-darkening (and lessening that due to gravity-darkening) of tidally-distorted stars.

All these effects are bound to be operative in any spectroscopic binary - regardless of whether it happens to be also an eclipsing variable. Quantitatively, however, they were found to be too *small* to give rise to an asymmetry of its radial-velocity curve which could be interpreted as a spurious eccentricity $e > 0.1$ (cf. Kopal, 1943, 1945, 1980a) - far too small to account for the anomalous spectroscopic phenomena observed in U Cephei or RZ Scuti. This remains so - mind you! - only if the distorted stars are regarded as *equilibrium* configurations, of shape appropriate for the prevalent field of force. Should, however, the components not be in equilibrium - for instance, if they exhibit *free oscillations* (and why shouldn't they?) - this would give rise to a new situation which has not yet been investigated.

The effects of distortion - whether equilibrium, or oscillatory - causing periodic displacements between the centres of mass and light on the apparent discs of the components of close binary systems - are, however, not the only reasons why their spectroscopically observed radial velocities may differ from those due to Keplerian motions of the two stars. As is well known, the observed radial velocities are deduced from the Doppler shifts of spectral lines formed in the atmospheres of the stars in question; and their use as indicators of the orbital motion of the star as a whole entails a tacit assumption that *the layers in which the spectral lines are formed are at rest with respect to the star's centre of mass*; and that this must be so is far from evident.

Regardless of what may be true if the stars are single, can this be the case in binary systems in which their components *irradiate* each other from close proximity? As is well known, such an irradiation produces a heating effect which is bound to give rise to *thermal convection* in the atmosphere - steady-state motion of gas if axial rotation of the irradiated star is synchronized with orbital revolution of the illuminating source, but non-steady motion if this source rises and sets over the illuminated star on account of asynchronism between rotation and revolution. In this connection it is well to recall that the principal components of both U Cephei and RZ Scuti - exhibiting strongly skew-symmetric velocity curves in spite of their circular orbits - rotate much *faster* than they revolve (cf. Struve, 1949) and, therefore, experience conspicuous day-and-night variations over their surfaces.

Can this be the cause (sole, or contributory) of spectroscopic anomalies shown by these two eclipsing systems? In spite of some recent investigations by Kirbiyik and Smith (1976) or Kopal (1980b), the answer is still far from certain; but the possibility exists that this may be the case. On the other hand, Struve and others sought the origin of the anomalous Doppler shifts in the absorbing effects of *gas streams* circulating between the two stars of a binary couple; and their views have been widely discussed in the past. It is not my aim to reproduce this discussion in this place; but rather to point out what remains yet to be done before such possibilities can be placed on a more reliable footing.

1. DYNAMICS OF GAS STREAMS IN CLOSE BINARY SYSTEMS.

Before, however, we approach the core of these problems, let us first consider the *dynamics* of the gas streams allegedly responsible for spectroscopic anomalies detected in close binary systems by past investigators, and partly referred to already in the preceding part of this paper. The observational significance of such anomalies is incontestable - as well as the fact that they are being observed preferentially in systems possessing contact components, and exhibiting other anomalies (irregular period changes, etc.) arguably associated with mass loss of mass escape from evolved components which, in the course of a post-Main Sequence expansion have attained their Roche limits.

The reason originally put forward in favour of such a view (cf. Kopal, 1955; Crawford, 1955) has been reduced gravity prevalent over the Roche limiting surfaces. As the star expands, the value of the total potential prevalent over each surface diminishes; and so does the gravitational acceleration. But while the surface potential should remain constant all over the star at any stage of its expansion, its gradient (i.e., the gravitational acceleration) does not. When the star has ultimately attained its Roche limit, the surface potential attains a minimum value it can possess for any closed configuration; while the gravitational acceleration, varying over the surface, and diminishing in the direction of its mate, actually vanishes at the inner Lagrangian point L_1. This fact by itself should not, to be sure, cause any matter to escape from the vicinity of such a point; for the Roche limit and any point on it (including L_1) represents a strictly *static* property of the model. However, a smallness of the gravitational acceleration in the neighbourhood of L_1 (zero at L_1) makes it certainly *easier* for any small perturbation to remove a mass element from there than from any other part of the star's surface.

This circumstance prompted in the past many investigators (Kopal, 1956, 1957, 1959; Gould, 1957, 1959; Plavec and Kříž, 1965; and others) to consider the hypothetical outflow of mass from a contact configuration as a problem of particle mechanics within the framework of the restricted problem of three bodies. In order to obtain any definite results it is, however, necessary first to specify the *boundary conditions* of the problem - i.e.,

(1) to localize the region on the star's surface from which mass is being lost;

(2) to determine the velocity with which the atoms cross the effective boundary surface; and

(3) to estimate the density (or flux) of the moving material.

As is well known, the restricted problem of three bodies (of which the two components play the role of the finite masses; and the escaping particle, that of a mass point) is one of sixth order. Consequently, six boundary conditions - i.e., the position and velocity components of the escaping particle - must be specified to render any solution determinate; and of these at best only three (i.e., the radial- velocity component, and the position of the region of escape in the plane of the celestial sphere if the system happens to be an eclipsing variable) can be indicated by the observations; the remaining ones can only be guessed at.

In order to do so, all investigators in the past considered any ejection to take place from the Lagrangian point L_1 (whose position is uniquely specified by the mass-ratio of the system) in the plane of the orbit; with an arbitrary velocity and in an arbitrary direction. The subsequent history of the particle so ejected must then be followed by numerical integration of the respective equations of motion; and the possible outcome may be (a) escape of the particle from the system; (b) its re-capture (after one or more orbits) by the body which emitted it; (c) a capture by its mate; or (d) its retention in circum-stellar space in a (simply or multiply) periodic orbit - direct or retrograde.

Most investigators in the past confined their efforts to establish initial conditions leading to the case (c) - i.e., to a mass transfer from one component to another. In particular, following an earlier suggestion by Kopal (1956), Piotrowski (1964) and Kruszewski (1963, 1964) investigated the circumstances of ejection caused by non-synchronism between axial rotation and orbital revolution - in which case a particle of finite momentum may be ejected from L_1 in the direction of a tangent to the respective branch of the Roche limit (depending on whether the rotation is faster or slower than the revolution).

In retrospect, however, this mechanical approach proved to be (at best) a detour, rather than progress, towards the solution of the underlying physical problem. The reason is the fact that the use of the equations of the restricted problem of three bodies could be physically justified *only if the mean free path of the particles ejected by the expanding star were long in comparison with the dimensions of the flow* (which are of the same order of magnitude as those of the system as a whole); so that mutual *collisions* of moving particules can be disregarded. The mean free path is, in turn, related (by the kinetic theory of gases) with the *density* of particles in the space between the stars; and this density will determine whether or not the respective assembly of particles could produce any observable effects.

In more specific terms, if N stands for the number of particles per unit volume (say, cm³) and σ, for the radius of their effective cross-section, then (cf., e.g., Jeans, 1940; p.135) the Maxwellian mean free path ℓ in neutral gas is given by the formula

$$\ell = \frac{1}{\sqrt{2}(\pi\sigma^2)N} = \frac{0.225}{N\sigma^2} \text{ cm} \quad . \tag{1.1}$$

The most common gas anywhere in the Universe (including the binary systems) is, of course, hydrogen; and for hydrodynamical purposes we shall hereafter regard the material between the components of close binary systems to consist of hydrogen alone. The effective cross-section of neutral hydrogen in ground state can, moreover, be identified with πa_o^2, where

$$a_o = \frac{h^2}{4\pi^2 m e^2} = 0.53 \times 10^{-8} \text{cm} \quad ,$$

where

$$h = 6.625 \times 10^{-27} \text{ erg sec },$$

$$m_H = 1.672 \times 10^{-24} \text{ g},$$

$$e = 4.803 \times 10^{-10} \text{ e.s.u.},$$

denote the Planck constant, mass of the hydrogen atom, and the charge of its electron. Inserting these values in the above Eq. (1.1) we find that

$$\ell = \frac{2.23 \times 10^{15}}{N_H} \text{ cm} \simeq 2 \times 10^{10}/N_H \text{ km}, \tag{1.2}$$

where N_H denotes the number of hydrogen atoms per ccm.

On the other hand, the optical depth τ of such a medium along a path of length $\tilde{r}$ will be (very approximately) given by the equation

$$\tau = K(\nu) N \tilde{r}, \tag{1.3}$$

where the coefficient $K(\nu)$ of bound-free absorption near the hydrogen ionization limit is known to be given by

$$K(\nu) = 7.91 \times 10^{-18} g_1(0) \text{ ccm}^2, \tag{1.4}$$

where the "Gaunt factor" $g_1(0)$ can be effectively set equal to 0.80. Accordingly, near the Lyman limit (ionization potential $\chi = 13.53$ eV $= 2.168 \times 10^{-11}$ erg; $\nu_c = \chi/h = 3.29 \times 10^{15} \text{sec}^{-1}$; $\lambda = c/\nu_c = 912$ Å) we should expect that, in our case,

$$K(\nu_c) = 6.33 \times 10^{-18} \text{ cm}^2 \tag{1.5}$$

for hydrogen; and $K \approx 10^{-17}$ cm² for heavier elements if present in our medium in appreciable amounts.

Moreover, if this medium is to exert noticeable effect on the light passing through it, it is necessary that its optical depth should be a quantity of zero order; and if so, a comparison of Eqs. (1.3) and (1.4) discloses that, in such a case, the length $\tilde{r}$ of the optical path should be given by

$$\tilde{r} = \frac{1.58 \times 10^{17}}{N_H} \text{ cm}; \tag{1.6}$$

i.e., about 100 times longer than the mean-free path ℓ given by Eq. (1.2). This disparity becomes greater by a further three orders of magnitude when ionization of hydrogen is taken into account (cf. next section). As, moreover, the average velocity $\tilde{v}$ of hydrogen gas (at temperature T of 10 000 deg.) is equal to

$$\tilde{v} = \left(\frac{2kT}{m_H}\right)^{\frac{1}{2}} \cong 13 \text{ km/sec} , \tag{1.7}$$

where $k = 1.380 \times 10^{-16}$ erg/deg stands for the Boltzmann constants, it follows that the mean free path ℓ given by Eq. (1.1) is, on the average, traversed in a time $t \sim 1.7 \times 10^9/N_H$ sec.

For any reasonable estimate of N_H, this value of t is by so many orders of magnitude shorter than the orbital period usually encountered in close binaries ($\sim 10^6$ seconds), that the absurdity of treating dynamical phenomena in such a medium in terms of particle mechanics - in which collisions are ignored and mean free path considered infinite - is glaringly evident. If so, it follows that the gas envelopes capable of impressing observable features in the composite spectra of such systems are no mere "exospheres" whose particles can move in ballistic trajectories, but constitute genuine "extended atmospheres". Therefore, it is *hydrodynamics*, rather than particle mechanisc, to which we must appeal in our efforts to place a study of the gas motions inside close binary systems on a sound physical basis.

The first investigator who realized this was Prendergast (1960), followed by Biermann (1971), Prendergast and Taam (1974), Sorensen et al (1975), Lubow and Shu (1975), and others. If the progress of this effort was slow, and invidiual contributions to it far between, this is due to intrinsic difficulties of the problem; and formidable difficulties are yet to be overcome before meaningful comparisons between theory and observations can be made.

Let us mention in what follows at least some of them - if alone only as a goal which we must eventually aim to accomplish. First, spectroscopic observations indicate that (at least the neutral) gas streams in close binary systems move with velocities of the order of 100 km/sec; while the velocity of sound in (hydrogen) gas at a temperature of 10 000 degrees (a typical ionization temperature indicated by the spectra) should be close to 12 km/sec. If so, however, it follows that motions of gas in between the stars and surrounding the system should be *hypersonic* and correspond to Mach numbers of the order of 10 or more. This fact, in turn, *rules out* any possibility of *linearization* of the equations of hydrodynamic motion; for in doing so we would rule out possible occurrence of *shock waves* whose effects may play an important role in any comparison between theory and observations.

Secondly, inasmuch as an overwhelming fraction of the circumstellar gas in close binary systems with at least one component of early or intermediate spectral type must be *ionized*, its viscosity μ is bound to be very high (cf. Chapman, 1954; Oster, 1957); and (in view of its low density ρ, *the kinematic viscosity* μ/ρ *may be enormous*. Therefore, it would be plainly unrealistic to treat the motions as inviscid; the terms factored by the kinematic viscosity may, in fact, be the dominant ones in the respective equations of motion.

Third, the Reynolds numbers of the respective motions are so large that such motions are probably *turbulent* as well, and characterized by high turbulent viscosity into the bargain. Therefore, fluid motions in gaseous envelopes of close binary systems may represent the case of a *hypersonic flow in viscous turbulent media* - about the worst accumulation of attributes one can ascribe to any flow - to be treated in three-dimensional space, with the time constituting the fourth independent variable (dispensable only if the flow were steady). It is only the solutions of partial differential equations (subject to appropriate boundary conditions) governing such flows which can really tell us something definite about the proportion of matter ejected by the components of close binary systems that can be captured by their mates, remain in circulation within the system, or be lost to the system altogether.

It should, furthermore, be stressed that the *forces causing ejection are internal to each star*; and before the gas will disengage itself from the gravitational field of the parent star, its companion merely stands by as a largely passive onlooker. If the velocity of ejection is of the order of 10^3 km/sec (as it is in the Wolf-Rayet stars or Novae), there is no doubt that matter so ejected will escape from the system altogether; and the same should be true of any particles carried away by "stellar winds" (cf. Siscoe and Heinemann, 1974; Baranov et al, 1976; Andriesse, 1979, 1980; and others). The role of the latter in subgiant components of semi-detached eclipsing systems, possessing external convective zones of considerable depth, should be particularly emphasized. *Hydrodynamical motions which could transfer mass from one star to another should, in general, be characterized by Mach numbers less than one*; at least that seems indicated by the outcome of extensive numerical integrations of individual particle trajectories (whose periodic orbits can represent, to be sure, only limiting cases of steady-state hydrodynamic flow, obtaining when the flow density is allowed to approach zero).

In the face of such a situation - in which emphatic nonlinearities of the problem should make us wary of simplifications which could cripple its context (remember how many consequences of linearized or inviscid hydrodynamics have in the past earned for themselves the epithets of "paradoxa"!) - it is apparent that the students of the dynamics of close binary systems (and their computing machines) are unlikely to run out of work in the foreseeable future. Also, in retrospect, it is easy to see how uncertain have been the foundations of most part of the work of the last 10-15 years on the evolution of close binary systems based on the familiar hypothesis of "mass-exchange" between the components. Even the terminology of that case was inexact; for what most of these investigators had in mind was a transfer of mass from one component to another; not a mutual exchange between them.

But what can we say about the possibility of transfer itself, or its efficiency? This question has been discussed in recent years by Plavec (1968), Kopal (1971), Paczynski (1971) and, most recently, on pp. 413-430 of my book on *Dynamics of Close Binary Systems* (1978), to which there is still nothing essential to add - except for the reiteration that the answer to the question raised above continues to be uncertain: it is concealed in the systems of equations which have so far defied attempts at their satisfactory solution; and in view of a large number of the "degrees of freedom" implied in their boundary conditions it is hopeless to guess at the outcome (the probability of any guess being right is very small indeed). It represents one aspect of the double-star problem where patience is needed - and hard work - rather than jumping to conclusions which cannot be substantiated by more than personal intuition.

2. PHYSICS OF FREE GAS IN CLOSE BINARY SYSTEMS.

Besides, all investigators of hypothetical gas streams in close binary systems avoided so far to give closer consideration to the physical state of the constituent gas, which may also profoundly influence the dynamics of its motion: namely, its *degree of ionization*.

It goes without saying that rarefied gas in close proximity of the components of a binary cannot remain in a completely neutral stare, but a certain part of it must undergo ionization. The extent of this ionization can, in turn, be ascertained by processes worked out in the theory of gaseous nebulae, initiated by Strömgren (1939), and described since in a number of sources (cf. Kaplan and Pikelner, 1970; Osterbrock, 1974; Spitzer, 1978; or Dyson and Williams, 1980) to which the reader may be referred for fuller details. In what follows we propose to apply the principles of this theory to free gas in close binary systems.

In order to so so, consider a field of hydrogen gas surrounding a star of radius R_*, effective temperature T_*, and luminosity $L\nu d\nu$ in the frequency-interval $d\nu$. Let, moreover, N_p denote the number of protons per unit volume; and N_e, the number of electrons; while N_1 stands for the number of neutral atoms (in ground state) in the assembly. The total number N_H of hydrogen atoms - neutral as well as ionized - per unit volume is then equal to the sum

$$N_H = N_p + N_1 . \qquad (2.1)$$

Let, moreover, x denote the ratio of the protons (or electrons) in the assembly to the total number N_H of hydrogen atoms in the same volume - i.e.,

$$x = \frac{N_p}{N_H} = \frac{N_e}{N_H} \quad . \tag{2.2}$$

The ratio $0 < x < 1$ signifies the degree of ionization of the respective medium; $x = 0$ corresponding to a completely neutral gas, and $x = 1$ to complete ionization.

In order to determine this quantity, let us - following Strömgren (1939) - invoke the use of Saha's equation

$$\frac{N_p N_e}{N_1} = 2 \left(\frac{2\pi m_e k}{h^2} \right)^{3/2} T_*^{3/2} e^{-\chi/kT} \times$$

$$\times \left(\frac{T}{T_*} \right)^{\frac{1}{2}} W e^{-\tau} \quad , \tag{2.3}$$

where T denotes the kinetic temperature of our hydrogen gas

$$W = \frac{\pi R_*^2}{4\pi r^2} = \frac{1}{4} \left(\frac{R_*}{r} \right)^2 \tag{2.4}$$

represents the "dilution factor" at a distance r from the star's centre; while the optical depth τ is governed by the equation

$$\frac{d\tau}{dr} = K(\nu) N_1 \, , \quad \tau(R_*) = 0 \, . \tag{2.5}$$

In these equations the numerical values of Planck's constant h, Boltzmann's constant k, and hydrogen ionization potential χ have already been quoted in the preceding section; and the mass of the electron $m_e = 9.108 \times 10^{-28}$ g; while the coefficient $K(\nu)$ on the r.h.s. of Eq. (2.5) for the optical depth τ, as given by Eqs. (1.4) or (1.5) refers to bound-free (photoelectric) absorption beyond the Lyman limit; above this limit the gas will be optically thin - as bound-bound as well as free-free absorption ("inverse bremsstrahlung") should be negligible at gas densities which we have in mind. However, this need not necessarily be the case for opacity arising from the scattering of light on free electrons (Thomson scattering), for which

$$\frac{d\tau}{dr} = \frac{8\pi}{3} \left(\frac{e^2}{m_e c^2} \right)^2 N_e = 0.665 \times 10^{-24} N_e \text{ cm}^{-1}, \tag{2.6}$$

where $c = 2.998 \times 10^{10}$ cm/sec stands for the velocity of light.

For partially-ionized hydrogen, a combination of Eqs. (2.1) and (2.2) discloses that

$$N_p = N_e = x N_H \qquad (2.7)$$

while

$$N_1 = (1 - x)N_H ; \qquad (2.8)$$

and in terms of these notations Equations (2.3) - (2.5) can be rewritten as

$$\frac{x^2}{1-x} = \frac{1.207 \times 10^{15}}{N_H(cm^3)} \left(\frac{T_*}{deg}\right)^{3/2} 10^{-\frac{68400 deg}{T}} \times$$

$$\times \left(\frac{T}{T_*}\right)^{\frac{1}{2}} \left(\frac{R_*}{r}\right)^2 e^{-\tau} \qquad (2.9)$$

and

$$\frac{d\tau}{dr} = 6.33 \times 10^{-18}(1-x)N_H \ cm^{-1}, \qquad (2.10)$$

subject to the initial condition

$$\tau(R_*) = 0 . \qquad (2.11)$$

For any given eclipsing system, the observed properties of the exciting component should permit us to estimate the values of R_* and T_* on the right-hand side of Equation (2.9); and in its hydrogen envelope the temperature T (depending on the balance between absorption and emission) should be close to 10 000 deg. If, moreover, we adopt a specific form for hydrogen density distribution $N_H(r)$ - constant or otherwise - Equations (2.9) and (2.10) subject to the initial condition (2.11) can be solved numerically for $x(r)$.

In order to do so, we find it convenient to normalize the independent variable by

$$r = R_* y ; \qquad (2.12)$$

if so, Equations (2.9) and (2.10) can be rewritten as

$$\frac{x^2}{1-x} = \frac{C}{y^2} e^{-\tau(y)}, \qquad (2.13)$$

$$\frac{d\tau}{dy} = D(1-x), \quad \tau(1) = 0, \qquad (2.14)$$

where

$$C = \frac{1.207 \times 10^{15}}{N_H} \left(\frac{T_*}{\deg}\right)^{3/2} 10^{-\frac{68400}{T}} \left(\frac{T}{T_*}\right)^{\frac{1}{2}} \qquad (2.15)$$

and

$$D = 6.33 \times 10^{-18} N_H \qquad (2.16)$$

are nondimensional constants (D represents the radius of the exciting star expressed in terms of the mean free path of the Lyman limit photons in neutral hydrogen).

By differentiating (2.13) and inserting from (2.14) for τ we find that

$$\frac{dx}{dy} = -\frac{x(1-x)}{2-x}\left\{\frac{2}{y} + D(1-x)\right\} . \qquad (2.17)$$

On the surface of the exciting star (when $y = 1$), $\tau(1) = 0$; and the corresponding value of $x(1) \equiv x_1$ follows, from (2.13) as the solution of

$$\frac{x_1^2}{1-x_1} = C , \qquad (2.18)$$

yielding

$$x_1 = \frac{1}{2}\{\sqrt{C^2 + 4C} - C\} = 1 - \frac{1}{C} + \frac{2}{C^2} - \frac{10}{C^3} + \ldots ; (2.19)$$

the expansion on the r.h.s. of (2.19) being convergent for $C > 4$.

On the other hand, at a sufficiently large distance from the exciting star (i.e., when the term $2/y$ on the r.h.s. of Eq. (2.17) can be neglected), the variables in (2.17) can be separated and the equation itself integrated to furnish an asymptotic expression of the form

$$y - y_0 = D\left\{2 \log \frac{1-x}{x} - \frac{1}{1-x}\right\} , \qquad (2.20)$$

valid when the integration constant $y_0 \gg 1$.

The problem defined by Equations (2.13) and (2.14) is not new; and has, in fact, been much studied by investigators of gaseous nebulae or physics of diffuse matter in interstellar space (for a summary of its present state cf. again Osterbrock, 1974; Spitzer, 1978; or Dyson and Williams, 1980). Since the pioneer work of Strömgren in 1939, it transpired that, up to a certain distance from hot stars, hydrogen gas that may surround them will be almost completely ionized (the "HII regions") while beyond this limit it will remain neutral ("HI regions"); a transition between HI and HII regions being quite abrupt (for bound-free absorption, its thickness is of the order of the mean free path of the photons of frequencies close to the Lyman limit).

On the other hand, the distance from the exciting source at which the transition from HII to HI state sets in depends primarily on the quality of the radiation emitted by the star, and the density of the gas through which this radiation is bound to propagate. A quantitative indicator of this distance will be the jump in the value of the ratio (2.2), obtained from a solution of Equations (2.13) - (2.14) regarded as a simultaneous system. Extensive integrations of this kind have been undertaken in the past by the students of gaseous nebulae; but their results have but little bearing on the problems confronting investigators of close binary systems.

For virtually all quantitative applications of Strömgren's theory available in the literature pertain to the cases in which the exciting stars are much hotter - and the gas surrounding them is much more rarefied - than the situations commonly encountered in close binaries. In diffuse or planetary nebulae, the spectra of the exciting stars are mostly of the O-type (of surface temperature ranging from 30 000° to 100 000°); and the density of nebular gas corresponds very largely to $N_H \sim 10^2 - 10^4$ per cm³. In close binary systems, on the other hand, the components of O-type spectra are rare (most O-type stars are spectroscopic binaries; but very few of them are near enough to us in space); the spectra of most are, in fact, later than B0 (i.e., $T_* \ll 30\,000°$). On the other hand, for any gas to be observable in such systems it must be much denser (i.e., $N_H \gg 10^4$) to produce noticeable absorption effects.

As a combined effect of both these circumstances, the extent of the HII regions in most known binaries should be by several orders of magnitude smaller than those encountered in gaseous nebulae or interstellar space; and comparable, in fact, with the size of the binary orbits (i.e., $10^{10} - 10^{11}$ cm). Under such conditions, their "Strömgren spheres" (or loops) cannot of course be resolved optically, or produce a visible nebula; but may give rise to spectroscopic phenomena which can be indirectly observed; therefore, their expected properties deserve close attention.

Extensive numerical integrations of Equations (2.13) - (2.14) for a wide range of the temperatures T_* of the exciting components encountered in close binary systems, and of gas densities around them, have recently been carried out at the University of Manchester by Salah A. A. Mahdi and Chriska H. Barzinji, as a part of their M.Sc. course under my supervision. The detailed results of their work should be published in the near future; but at least the approximate limits of the ionization zones, which the degree x of ionization drops abruptly (for not too high values of N_H) from nearly 1 to almost 0 can also be inferred from the following more elementary considerations.

Let L_ν denote the luminosity of the exciting star in a frequency ν. The number of light quanta emitted by the star at this frequency will obviously be equal to the ratio $L_\nu/h\nu$; and the total number of quanta sent out at frequencies higher than (say) ν_0 will be given by the integral

$$\int_{\nu_0}^{\infty} \frac{L_\nu}{h\nu} d\nu = 4\pi R_*^2 \int_{\nu_0}^{\infty} \frac{\pi F_\nu}{h\nu} d\nu \equiv 4\pi R_*^2 N_L , \qquad (2.21)$$

where R_* continues to stand for the star's radius,

$$L_\nu = 4\pi R_*^2 (\pi F_\nu) , \qquad (2.22)$$

and πF_ν represents the monochromatic flux of radiation emerging from unit area of stellar surface; while N_L is then the number of quanta of frequencies $\nu > \nu_0$.

If we identify now the lower limit ν_0 of integration in (2.21) with the frequency of radiation at the Lyman limit (i.e., $\nu_0 = 3.29 \times 10^{15}$ sec^{-1}, corresponding to the wavelength $\lambda = 912$ Å), all photons for which $\nu > \nu_0$ are capable of bringing about hydrogen ionization. Let us assume, in what follows, that they will do so inside a sphere of radius s_0, from which no ionizing photons escape (the medium for $r < s_0$ being regarded as *optically thick*; and the number of ionizations equal to that of recombinations).

Let, moreover, the number of recombinations in the same sphere be equal to

$$\frac{4}{3} \pi s_0^3 N_p N_e \beta , \qquad (2.23)$$

where

$$\beta = 2.45 \times 10^{-10} T^{-3/4} \text{ cm}^3 \text{sec}^{-1}$$
$$= 2.45 \times 10^{-13} \text{ for } T = 10^4 \text{ deg} \qquad (2.24)$$

represents the hydrogen recombination coefficient per unit volume and time. On the other hand, by Eqs. (2.7) and (2.8), $N_p N_e = x^2 N_H^2 \sim N_H^2$ if x is very close to 1. Therefore, by equating the number of ionizations with those of recombinations, we obtain an equation of the form

$$4\pi R_*^2 N_L = \frac{4}{3} \pi s_0^3 N_H^2 \beta , \qquad (2.25)$$

which can be solved for s_o to yield

$$s_o = \left(\frac{3R_*^2 N_L}{N_H^2 \beta}\right)^{1/3} . \qquad (2.26)$$

The number N_L of all ionizing quanta emitted by the stellar surface per unit area and time should, strictly speaking, be evaluated from the model-atmosphere of the respective star. If, however, we are prepared to accept for the monochromatic flud F_ν the Planckian approximation.

$$F_\nu = \frac{2h\nu^3}{C^2} \frac{1}{e^{h\nu/kT}-1} \quad \text{erg cm}^{-2} , \qquad (2.27)$$

where C, h and k continue to stand for the velocity of light, Planck and Boltzmann constants, respectively, it follows from (2.21) and (2.27) that

$$N_L = \frac{2\pi}{C^2} \int_{\nu_o}^{\infty} \frac{\nu^2 d\nu}{e^{h\nu/kT}-1} = \frac{2\pi\nu_o^3}{C^2} \int_1^{\infty} \frac{\eta^2 d\eta}{e^{\alpha\eta}-1} , \qquad (2.28)$$

where we have abbreviated

$$\nu = \nu_o \eta \quad \text{and} \quad \alpha = h\nu_o/kT . \qquad (2.29)$$

In most cases under consideration, $\alpha \gg 1$; and if so, the foregoing Equation (2.28) can be approximated by

$$N_L \simeq \frac{2\pi\nu_o^3}{C^2} \int_1^{\infty} e^{-\alpha\eta} \eta^2 \, d\eta = \frac{2\pi\nu_o^3}{C^2} \frac{e^{-\alpha}}{\alpha} \{1 + \frac{2}{\alpha} + \frac{2}{\alpha^2}\} . \qquad (2.30)$$

If we insert from Equations (2.24) and (2.30) in (2.26), we find that Eq. (2.26) for the radius s_o of a sphere in which hydrogen is completely ionized (while outside it remains completely neutral) can be rewritten, more concisely, as

$$s_o = K(T_*)(R_*/\odot)^{2/3} N_H^{-2/3} \quad \text{parsecs} , \qquad (2.31)$$

where the radius R_* is expressed in solar units; N_H stands for the number of hydrogen atoms per ccm; and $K(T_*)$ is a constant characteristic for each temperature T_* of the exciting star, **as** given in the following tabulation:

T_*	$\log_{10} K(T_*)$
6000°	-2.352
10000	-0.755
15000	+0.072
20000	+0.500
30000	+0.955

Should we wish to express the radii s_o in solar units ($\odot = 6.96 \times 10^{10}$ cm) rather than parsecs (the conversion factor between the two being 4.434×10^7), the above values of log K should be augmented by the factor +7.647.

The radius s_o separates (by a very sharp boundary) two regions of the same chemical composition, but of profoundly different physical characteristics. Thus the coefficient μ of viscosity - which is very low for neutral hydrogen (cf., e.g., Oster, 1957) - will for hydrogen plasma (inside $r < s_o$) be given (cf. Chapman, 1954) by

$$\mu = \frac{5\sqrt{m_H}(kT)^{5/2}}{4\sqrt{\pi}\, e^4 A_2(\xi)} \quad \frac{g}{cm\ sec} , \quad (2.32)$$

where e, k and m_H possess the same meaning as before, and

$$A_2(\xi) = \log(1 + \xi^2) - \frac{\xi^2}{1 + \xi^2} \quad (2.33)$$

with

$$\xi = (4kT/e^2) N_H^{-1/3} . \quad (2.34)$$

For $T = 10^4$ deg, Equation (2.32) yields

$$\mu = \frac{3.83 \times 10^{-5}}{\log(1 + \xi^2) - \frac{\xi^2}{1 + \xi^2}} \quad \frac{g}{cm\ sec} , \quad (2.35)$$

where

$$\xi = 2.39 \times 10^7 \, N_H^{-1/3} \quad (2.36)$$

will generally be a large number. From this it transpires that the viscosity of hydrogen plasma inside a sphere of radius s_o will be by several orders of magnitude greater than in the neutral medium surrounding it; and in spite of its low density may constitute a pretty "sticky stuff" - virtually certain to render any flow within the HII region highly turbulent.

Similarly, the mean free path ℓ of the charged particles in hydrogen plasma of temperature T will be of the order of magnitude of

$$\ell \simeq \frac{kT}{N_H e^2 \log A_2(\xi)} , \qquad (2.37)$$

which for $T = 10^4$ deg will reduce sensibly to

$$\ell \simeq 3.5 \times 10^{12}/N_H \text{ cm} \qquad (2.38)$$

by some three orders of magnitude smaller than that in neutral hydrogen of the same density (see Eq.(2.2)) - underlining the necessity to treat the motions in such a plasma as hydrodynamical flow.

3. APPLICATION TO U CEPHEI

On the background of physical preliminaries outlined in the last two sections, let us return to consider in their light the photometric as well as spectroscopic anomalies encountered in the eclipsing system of U Cephei. In doing so, it will be far from our aim to present any definitive or unified picture of this intriguing binary - this task must be left to more courageous individuals - but rather to point out the implications of the physical theory summarized in Sections 1 and 2 for the interpretation of phenomena exhibited by this (and other) eclipsing systems; and to establish the nature of the constraints which these theories are bound to impose on our interpretations.

In order to do so, let us depart from an estimate of the electron density in U Cephei deduced by Batten (1974) from different photometric anomalies of this system. In order to account for them, Batten (op.cit., p.241) assumed that the electron density in the exosphere surrounding the system corresponded to $N_e \sim 10^{12}$ to 10^{13}cm^{-3}; and if this exosphere is to be electrically neutral, it should follow that $N_p = N_e$.

The first question to which we shall seek the answer concerns the extent of the HII zone in U Cephei, in the light of the elementary theory outlined in Sec. 2. The primary component of this system, of spectral type B7 - B8, should be characterized by an effective temperature $T_* \sim 12000$ to 13000 degrees; and its absolute radius $R_* = 2.9 \odot = 2.0 \times 10^{11}$ cm. The secondary component of spectral class G8 is too cool to ionize hydrogen to any significant extent; any ions existing in the system must, therefore, be due to the primary (B7) component alone; and the HII region produced by its light should be symmetrical around that star.

If so, an appeal to Equation (2.31) discloses that the limit s_o

of the HII region should be given by

$$\log s_o = -0.03 + \frac{2}{3} \times 0.46 - \frac{2}{3}(12.5) + 7.65 = -0.41 \; ; (3.1)$$

corresponding to $s_o = 0.39 \odot$ from the star's surface, or $2.9 + 0.4 = 3.3 \odot$ from the star's centre. In other words, in this particular case the HII region should be closed around the B7 star, not too far above its visible photosphere; and the relative orbit of the secondary G8 component of radius A $14.7 \odot$ should be well outside its reach.

Is there any way in which this conclusion could be observationally verified? It goes without saying that, at ordinary optical frequencies (in fact, for $\lambda > 912$ Å) the HII region should be completely transparent; but for $\lambda < 912$ Å it should be optically thick. In other words, while the fractional photospheric radius r_1 of the B7 star is well known to be equal to

$$r_1 = \frac{2.9}{14.7} = 0.20 \qquad (3.2)$$

from an analysis of the light curve of U Cephei observed at ordinary optical frequencies, that of its HII region should be

$$r_{HII} = \frac{2.9 + 0.4}{14.7} = 0.23 \; ; \qquad (3.3)$$

i.e., significantly larger; and it is this latter value which should be obtained from an analysis of the light curve observed at $\lambda < 912$ Å.

Unfortunately, an observational verification of this difference is still outside our reach. It is true that several telescopes orbiting outside the atmosphere permit now photometric observations of eclipsing variables to be made in deep ultraviolet not accessible from the ground; but neither the International Ultraviolet Explorer, nor any other spacecraft in existence is equipped yet with light detectors sensitive to radiation beyond the Lyman limit. As long as this is so, the predictions which can be made now are not susceptible of observational verification.

Another consideration which may, however, be raised at this time concerns the distribution of circumstellar neutral hydrogen in U Cephei and similar systems. Peculiar phenomena for which neutral hydrogen alone must be held responsible were mentioned already in the first part of my address; and the primary among them is the asymmetry of the radial-velocity curves in systems describing circular orbits. Such curves are based on measurements of the Doppler shifts of the absorption lines of neutral atoms (predominantly again hydrogen); and the measured effects are no doubt due to the blending of photospheric lines with those of as yet undisclosed origin. The system of U Cephei provides again the principal example of such a situation; and a few words may be added concerning its implications.

As is well known, the principal source of information on absolute space motion of the components of U Cephei are the Doppler shifts of the Balmer lines of its hydrogen spectrum; and any blends which may cause their asymmetry require the presence of neutral hydrogen in the system. If one could disentangle these blends, one should be able to ascertain (after application of the appropriate Boltzmann factors) from their equivalent widths the number N_1 of neutral hydrogen lines along the line of sight; and from them, the degree of ionization x of the entire assembly.

Equation (3.1) used earlier in this section is based on the assumption that $x(r)$ represents a step-function with a discontinuity at $r = s_o$; x being equal to 1 for $r > s_o$, and 0 in the opposite case. In reality, $0 < x(r) < 1$ must remain continuous for $R_* < r < \infty$; and its actual value must be in harmony with the temperature T_* of the exciting star and the total density N_H of hydrogen through the medium of Saha's equation (2.13). As far as I am aware, no estimates of N_1 have so far been made from the intensity of measured spectral lines; and, for this reason, any direct appeal to Eq. (2.13) is still impossible. However, if and when this can be done, a further important constraint will be obtained which should make it possible to construct physically-consistent pictures of U Cephei and other eclipsing systems which should be more reliable than those which we have at our disposal today.

A last word for consideration of future investigators who may approach the problems exemplified in U Cephei - and that is to keep in mind a comparison with its sister-eclipsing system of U Sagittae. Geometrically - as well as by the spectra and other absolute properties (masses, dimensions) - both these systems are very much alike. Their primary minima are due to total eclipses of their early-type stars; and both components describe orbits, with similar periods, which (by the relative location of their secondary minima) are indistinguishable from circles.

Yet, in almost every other respect, U Cep and U Sge are as unlike as they can be. In the case of U Sge, the orbital period remains sensibly constant; while the period of U Cep exhibits at all times noticeable oscillations which defy any predictions. The light- as well as radial velocity-curves of U Sge are symmetrical within the limits of observational errors; while for U Cep they are both conspicuously asymmetric; and (at least for the light changes) their asymmetry varies probably with the time. Moreover, while the primary (B8) component of U Sge rotates with an angular velocity consistent with the existence of synchronism between rotation and revolution (cf. Joy, 1936); the primary component of U Cep is a very fast rotator (Struve, 1949). However, the development of a suggestion that this latter phenomenon may also be the cause of all other differences must again be left to a more courageous individual.

4. CONCLUDING REMARKS

The foregoing discussion has, moreover, by no means exhausted all problems facing us when we set out to interpret non-orbital spectroscopic phenomena, observed in close binary systems, on a physical basis. In considering these, one cannot fail again to see the gap by which the observers outdistanced the theoreticians aiming to explain the observed facts. Not that there is any shortage of anomalous phenomena to explain; nor a lack of courage with which such problems have been (and are being) attacked. Indeed, everyone seems to believe these days in the "gas streams", "rings", etc. of unspecified kind to do our bidding: the observers, because they consider them a necessary consequence of secular expansion of the star as a whole; the theoreticians, because they regard them as observed facts.

And there is indubitably something to the views of each group that deserves further consideration; the immediate task is mainly not to mix up the priorities. It is indeed very probable that stars do expand in their post-Main Sequence stage; and it is well-nigh certain that, in doing so, they divest themselves of a large fraction of their initial mass; but this happens regardless of whether the star is single or possesses a stellar companion.

But any more definite identification of escaping mass with spectroscopic anomalies is, however, still premature; and any dogmatic (or traditional) insistence may be counter-productive and inhibit further research by creating a false impression that the essential requisite clues are already in our hands. Such impressions are, we repeat, distinctly premature; and may remain so for a long time to come.

The present speaker enumerated recently several *caveats* arising in this connection (cf. p.472 of Kopal, 1978), which he should like to address once more to all workers - especially the young ones - labouring in this part of our vineyard.

1. In constructing different models in terms of which to interpret observed facts, can be proposed explanation claim uniqueness, or is it only one of many that can be invoked for the purpose? In constructing any new hypotheses of your own, always keep in mind the wisdom of "Occam's razor" not to compound hypothetical features or mechanisms of a scenario beyond necessity ("en-

tia non sunt multiplicanda praeter necessitatem"). An explanation of n observed phenomena in terms of m special assumptions represents surely nothing else but an outright guess without any real information contents if m = n; and the probability that "there is something to it" increases only as m $\ll$ n.

2. Beware of a temptation to conjure up arbitrary "scenarios" without sufficient physical reasons behind them. Conjectures are certainly free to wander far and wide over their spacious domain; but they are very unlikely to bring home conviction - let alone truth. In problems characterized by very many degrees of freedom (largely under discussion in this chapter) it is virtually hopeless to try to arrive at their solution by any kind of a guess based on mere intuition, and unsupported by adequate physical theory. It is also true that the minds of only too many theoreticians today seem to work well only in one or two "normal" modes; and are capable of a considerable amount of self-deception in their efforts to convince themselves (and others) that observations of unexpected phenomena can be made to fit into one or the other of their preconceived catagories.

3. Never forget that the only proper framework for rational interpretation of phenomena exhibited by close binary systems is *hydrodynamics* (or hydromagnetics) of radiating media, and *not* particle mechanics. If the motions of atoms or ions exhibited in the course of such events could be legitimately described by collisionless particle trajectories, the density of gas consisting of such particles would be too low by many orders of magnitude for any kind of observational detection.

4. Remember that the concept of Roche limit as defined conventionally is of significance only for *equilibrium* configurations, and becomes physically irrelevant in the presence of *motion* (or, again, of a strong field of radiation). Pay respect to the Roche limit whenever it is deserved, but do not abuse its concept by allowing it to degenerate into "folklore" in the literature on close binaries, as would happen if we were to stretch its validity to cases where it no longer applies - such as those involving gas motions, where it is the Jacobi integral which should take its place (or the Bernoulli integral in the case of fluid flow).

5. Keep in mind that a mass loss (or transfer) in close binary systems is related with any period changes by equations which are of *vector* (not scalar) form. In other words, a knowledge of the amount of mass lost (or transferred) does not, by itself, specify uniquely as yet any period change. In order to do so, we need to know, not only the magnitude of the mass involved in the process, but also the velocity and direction of escape, as well as the region of the star's surface from which the escape may take place.

6. Before you go too far conjuring up the effects which "shells", "rings", "mass-exchange", etc. may produce in the observations, ask yourself the following questions: how (and by which process) could such formations have come into being? And if they did, what is their lifetime? What maintains them, and what can prevent their dissolution? Is the contemplated formation necessary, or only optional, to produce the observed effects; or is its use invoked only because one cannot prove it to be wrong? Remember that the lack of proof of non-existence should never be confused with the demonstration that it exists! To do otherwise in binary-star astronomy could easily earn our subject the epithet of "new astrology" in the minds of our more critically-minded professional confreres. And always keep in the back of your mind the obvious thought that while n ($<\infty$) observed phenomena which may be in agreement with a given hypothesis do not yet establish its validity (they make it only more plausible), a single fact which is definitely at variance with such a hypothesis is sufficient to disprove it!

And thus, after these words of caution at the end of our school, please consider yourselves qualified to contribute through your future work your share to the common pool of human blunders; for this is how science really advances in the long run; and it happens very rarely that the first conjecture which comes to one's mind happens to hit the nail on the head! Usually it is only after gradual elimination of all mistakes of fact and judgment that we may - sometimes ourselves, but more often our descendants - arrive eventually at the genuine solution of our problems. And to all who - undaunted by such prospects, and contemptuous of jumping to conclusions on the basis of insufficient evidence - I now wish to bid "God-speed" in further pursuit of our common goals along this exciting (albeit sometimes thorny) path.

REFERENCES

Andriesse, C. D.: 1979, Astrophys. Space Sci., 61, p.205.
Andriesse, C. D.: 1980, Astrophys. Space Sci., 67, p.461; 72, p.167.
Baranov, V. B., Krasnobaev, K. V. and Ruderman, M. S.: 1976, Astrophys. Space Sci., 41, pp. 481, 491.
Barney, I.: 1923, Astron. J., 35. p.95.
Batten, A. H.: 1974, Publ. Dominion Astr. Obs., 14, p.191.
Bĕlopolsky, A. A.: 1893, Mem. Soc. Spettr. Ital., 22, p.101.
Bĕlopolsky, A. A.: 1897, Mem. Soc. Spettr. Ital., 26, p.135.
Biermann, P.: 1971, Astron. Astrophys., 10, p.205.
Carpenter, F. M.: 1930, Astrophys. J., 72, p.205.
Chapman, S.: 1954, Astrophys. J., 120, p.151.
Crawford, J. A.: 1955, Astrophys. J., 121, p.71.
Dyson, J. E. and Williams, D. A.: 1980, *Physics of the Interstellar medium*, Univ. of Manchester Press.

Eddington, A. S.: 1926, Mon. Not. Roy. Astr. Soc., 86, p.320.
Gould, N. L.: 1957, Publ. Astr. Soc. Pacific, 69, p.541.
Gould, N. L.: 1959, Astron. J., 64, p.136.
Jeans, J. H.: 1940, *Kinetic Theory of Gases*, Cambr. Univ. Press.
Joy, A. H.: 1936, Astrophys. J., 71, p.336.
Kaplan, S. A. and Pikelner, S. B.: 1970, *The Interstellar Medium*, Harvard Univ. Press.
Kirbiyik, H. and Smith, R. C.: 1976, Mon. Not. Roy. Astr. Soc., 176, p.103.
Kopal, Z.: 1943, Proc. Amer. Phil. Soc., 86, p.351.
Kopal, Z.: 1945, Proc. Amer. Phil. Soc., 89, p.517.
Kopal, Z.: 1955, Mem. Soc. Roy. Sci. Liege, (4)15, pp.684-686; Annales d'Astrophys., 18, p.379.
Kopal, Z.: 1956, Annales d'Astrophys., 19, p.298.
Kopal, Z.: 1957, in *Non-Stable Stars* (I.A.U. Symp. No. 3; ed. G. H. Herbig); pp. 123ff.
Kopal, Z.: 1959, *Close Binary Systems*, John Wiley and Chapman Hall.
Kopal, Z.: 1971, Publ. Astr. Soc. Pacific, 83, p.521.
Kopal, Z.: 1978, *Dynamics of Close Binary Systems*, D. Reidel Publ. Co., Dordrecht and Boston.
Kopal, Z.: 1980a, Astrophys. Space Sci., 70. p.329.
Kopal, Z.: 1980b, Astrophys. Space Sci., 71, p.65.
Kruszewski, A.: 1963, Acta Astron., 13, p.106.
Kruszewski, A.: 1964, Acta Astron., 14, pp. 231, 241.
Lubow, S. H. and Shu, F. H.: 1975, Astrophys. J., 198. p.383.
Neubauer, F. J. and Struve, O.: 1945, Astrophys. J., 101, p.240.
Oster, L.: 1957, Zeit. f. Astrophys., 42, p.228.
Osterbrock, D.E.: 1974, *Astrophysics of Gaseous Nebulae*, Freeman and Co., San Francisco.
Paczyński, B.: 1971, in *Ann. Reviews Astron. Astrophys.*, 9, pp. 183ff.
Piotrowski, S. L.: 1964, Acta Astron., 14, p.251.
Plavec, M.: 1968, in *Advances Astron. Astrophys.* (ed. Z. Kopal), 6, pp.202ff, Academic Press, New York.
Plavec, M. and Kříž, S.: 1965, Bull. Astr. Inst. Czechoslovakia, 16, p.297.
Pomraning, G. C.: 1973, *Radiation Hydrodynamics*, Pergamon Press.
Prendergast, K. H.: 1960, Astrophys. J., 132, p.162.
Prendergast, K. H. and Taam, R. E.: 1974, Astrophys. J., 189, p.125.
Siscoe, G. L. and Heinemann, M. A.: 1974, Astrophys. Space Sci., 31, p.363.
Sörensen, S. A., Matsuda, T. and Sakurai, T.: 1975, Astrophys. Space Sci., 33, p.465.
Spitzer, L. Jr.: 1978, *Physical Processes in the Interstellar Medium*, John Wiley, New York.
Sterne, T. E.: 1941, Proc. U. S. Nat. Acad. Sci., 27, p.168.
Strömgren, B.: 1939, Astrophys. J., 89, p.526.
Struve, O.: 1944, Astrophys. J., 99, p.222.
Struve, O.: 1949, Mon. Not. Roy. Astr. Soc., 109, p.487.
Struve, O.: 1950, *Stellar Evolution*, Princeton Univ. Press.

NAME INDEX

Abhyankar, K. D., 387
Abrami, A., 347, 359
Abramovich, G. N., 489, 508
Africano, J. L., 259, 283, 284, 450
Ahnert, P., 426, 436
Al-Naimiy, H. M. K., 156, 158, 160, 212, 214, 407, 412, 451
Alkan, H., 196, 197
Allen, C. W., 239, 250, 377, 411, 412, 415, 418
Allen, R. G., 284
Ambartsumyan, V. A., 517, 518, 533
Andersen, J., 151, 513, 516
Anderson, C., 430, 431, 432
Anderson, J. P., 253, 283, 311, 361, 362, 373, 374, 377, 379, 387
Andriesse, C. D., 543, 557
Aristotle, of Stageira, 5
Arnold, C. N., 363, 385, 387
Ashbrook, J., 417, 425, 426
Aspin, C., 509
Ausekar, B. D., 387
Awadalla, N. S., 316, 329, 330, 341, 343, 359
Ayres, T. R., 299, 303

Bachmann, P., 508
Baize, P., 256, 283
Bakoš, G. A., 347, 360, 441, 442, 445, 470
Baldwin, B. W., 469, 471
Baldwin, M. E., 417, 428, 429, 430, 431, 432
Baliunas, S. C., 302, 303
Balonek, T. J., 303
Banachiewicz, Th., 29, 30, 58
Banos, C., 442, 445, 448, 450
Baranov, V. B., 543, 557
Barker, T., 523, 533
Barney, I., 535, 557
Bartholdi, P., 283
Barzinji, C. H., 548
Batchelor, G., 489, 508
Batten, A. H., 200, 256, 283, 465, 466, 470, 489, 503, 508, 552, 557
Baumann, M., 437
Beivers, W. I., 284
Bělopolsky, A. A., 535, 557
Benoit, C., 58
Biermann, P., 467, 468, 470, 488, 508, 542, 557
Binnendijk, L., 114, 115, 120, 200, 201, 214, 215, 354, 355, 359, 417, 425, 426, 427, 429, 441, 445
Bistline, W., 508
Blanco, C., 290, 291, 292, 295, 300, 303

Bode, H., 427
Bond, H. E., 406, 409, 412
Bookmyer, B. B., 200, 201, 215, 427
Bopp, B. V., 295, 300, 303
Böhme, D., 284
Bolton, C. T., 504, 508
Born, E., 284
Born, M., 264, 283
Bortle, J., 431, 432
Boyarchuk, A. A., 531, 532, 533,
Boyer, S., 304
Braune, W., 417, 426, 427, 428, 429
Brocklehurst, M., 520, 533
Broglia, P., 200, 215
Brown, J. C., 504, 505, 508, 509
Brouwer, D., 514, 516
Bryson, A. R., 20, 58
Bucy, R. S., 20, 58
Budding, E., 196, 197, 207, 215, 313, 316, 329, 330, 341, 343, 359, 405, 406, 407, 410, 412, 473, 503, 508
Burger, M., 463, 464
Burke, E. W., 360

Caracatsanis, V. A., 196, 197
Carbon, D. F., 320, 324, 330, 498, 508
Carder, R. W., 284
Carpenter, F. M., 535, 536, 557
Carr, R., 200, 215, 345, 346, 360
Castle, K. G., 360
Castor, J. I., 461, 464
Catalano, S., 285, 286, 287, 290, 303, 335, 360
Ceraski, W., 465, 470
Cerruti-Sola, M., 290
Cester, B., 315, 330, 347, 359
Chang, Y. C., 441, 445
Chambliss, C. R., 200, 341, 343, 360
Chapman, S., 542, 551, 557
Charles, P., 304
Chisari, D., 285, 301, 303
Chochol, D., 534
Churms, J., 283
Ciatti, F., 532, 533, 534
Clausen, J. V., 160, 515, 516
Clemence, G. M., 514, 516
Cobb, C. C., 283
Code, A. D., 468
Cohen, J., 450
Cohen, M. L., 303
Collins, G. W., 7, 16
Conconi, P., 200, 215

NAME INDEX

Contopoulos, G., 442, 445, 448, 450
Cook, S., 430
Couteau, P., 283
Cowley, A., 259, 283
Cowley, P., 257, 283
Coyne, G. V., 469, 470
Cragg, T., 430, 431
Crawford, J. A., 538, 557
Crawford, R. C., 466, 467, 468, 470, 471
Cristescu, C., 442, 444, 445

Dante, Alighieri, 14
Davis, J., 284
Davis, W. D., 112, 113, 120
Dean, C. A., 315, 320, 330, 332, 335, 360
Dearborn, D. S., 412
De Biase, G. A., 226, 246, 249, 250
Debrunner, H., 284
De Campli, W. M., 302, 303
Deeming, J. T., 284
Delgado, A. J., 513, 516
De Loore, C., 463, 464
Demircan, O., 54, 125, 128, 129, 130, 131, 132, 134, 135, 137, 138,
    139, 140, 141, 142, 143, 148, 151, 152, 153, 154, 158,
    160, 164, 173, 174, 176, 180, 194, 197, 207, 208, 215,
    413, 432, 433, 434, 435
Deutsch, R., 58
De Veght, C., 264, 269, 283
Devinney, E. J., 111, 117, 120, 352, 360
Dierks, H., 271, 283
Diethelm, R., 433, 434, 435, 436, 439
Domke, K., 417, 425
Dörr, F., 425, 426
D'Orsi, A., 359
Doughty, N. A., 451
Dowers, R., 450
Dueball, J., 417, 426, 427
Dugan, R. S., 466, 471
Duinen, R. J. van, 534
Dumitrescu, A., 314, 315, 316, 320, 330, 332, 335, 360
Dunham, D. W., 283
Dworak, T. Z., 378, 387, 451
Dyson, J. E., 544, 547, 557

Eaton, J. A., 300, 301, 303
Ebbinghausen, E. G., 20, 36, 58
Edalati, M. T., 196, 197
Eddington, A. S., 253, 537, 558
Efremov, Yu. N., 417
Eggen, O. J., 199, 200, 215, 303, 314, 330

Eggleton, P. P., 288, 304, 328, 330, 378, 387
Eitter, J. J., 284
Emslie, A. G., 508
Erdélyi, A., 167, 180
Ertan, A. Y., 390, 403
Etzel, P., 11, 111, 112, 114, 117, 118, 120
Evans, D. S., 258, 262, 269, 271, 272, 283, 284, 300, 303
Evren, S., 390, 403

Fedel, B., 330
Fedorovich, V. P., 417
Fekel, F. C., 283, 284
Feldmann, P. A., 297, 303, 304
Ferland, G. J., 284
Fernandes, M., 427
Ferraz-Mello, S., 300, 304
Fieten, H., 534
Finsen, W. S., 255, 283
Fix, J. D., 469, 471
Flannery, B. P., 474, 488, 495, 508
Flin, P., 417, 432, 433
Florja, N., 425
Forbes, J. E., 335, 360
Forsythe, G. E., 479, 480, 508
Fracastoro, M. G., 285, 303
Frasinski, L., 432, 433
Frieboes-Condé, H., 341, 360, 467, 471
Friedmann, C., 300, 303
Frisina, A., 303
Frolov, M. S., 417
Fujinami, S., 271, 283
Furlani, S., 246, 251

Gadomski, J., 332, 360
Gaposchkin, S., 378, 460
Garcia-Pelayo, J. M., 155
Garrison, L. M., 468, 471
Gauss, C. F., 8
Gehlich, U. K., 264, 269, 283
Germann, R., 432, 433, 434, 435, 436, 437, 438, 439
Geyer, E. H., 346, 360
Gibson, D. M., 297, 304
Gimenez, A., 27, 155, 511, 513, 516
Gimicin, G., 330
Gingerich, O., 320, 324, 330, 498, 508
Goldstein, H., 497, 508
Goodman, I. R., 17
Gordon, K. C., 314, 330, 332, 335, 337, 360
Goudis, C., 442, 445, 448, 450
Gould, N. L., 539, 558

Greenspan, H. P., 475, 508
Gregory, P. C., 303
Grønbech, B., 516
Grygar, J., 532, 534
Güdür, N., 413, 432, 434, 435
Guinan, E., 503, 508
Gurtler, J., 301, 303
Gurzadyan, G. A., 521, 523, 527, 534
Guthnick, P., 441, 445
Gyldenkerne, K., 516

Hack, M., 453, 455, 456, 460, 463, 464
Hall, D. S., 287, 289, 300, 301, 304, 314, 329, 330, 331, 332, 354, 360, 362, 378, 387, 389, 390, 392, 397, 398, 402, 403, 466, 467, 468, 470, 471
Hampton, W., 431
Hanbury Brown, R., 255, 283
Hardie, R. H., 413, 417, 465, 471
Harland, D. M., 355, 360
Harman, R. J., 517, 526, 531, 534
Haslag, K. P., 397, 403
Hazard, C., 267, 268, 274, 283
Hazlehurst, J., 200, 215
Heinemann, M. E., 543, 558
Heintz, W. D., 254, 256, 257, 283
Henon, M., 496, 508
Herbig, G. H., 379, 387
Herczeg, T., 341, 360, 417, 441, 445, 467, 471
Herr, R. B., 257, 283
Hicks, P. D., 304
Hilditch, R., 340, 341, 344, 355, 356, 360, 412
Hiltner, W. A., 341, 360
Hipparchos, 5
Hjellming, R. M., 297, 304
Ho, Y. C., 20
Hoerner, S. von, 257, 283, 284
Honeycutt, R. K., 508
Hopp, U., 442, 444, 445
Horne, K., 448
Houck, T. E., 456, 464
Howell, A., 429
Hrič, L., 354
Huffer, C. M., 7, 16, 468
Hülscher, J., 417
Hummer, D. G., 518, 534
Hunger, K., 271, 283
Hutchings, J. B., 464
Huth, H., 345, 359, 360

Ibanoglu, C., 389, 395, 403, 431, 432

Iben, I., 378, 380, 381, 382, 383, 384, 385, 387
Iijima, T., 517, 526, 532, 534
Irwin, J. B., 15, 16
Ischi, E., 232, 240, 251

Jahn, A., 425
Jakarte, S. M., 401, 403
Jameson, R. F., 456, 464
Jaschek, C., 283
Jaschek, M., 283
Jassur, D. M. Z., 313, 329, 330, 344, 350, 360
Jeans, J. H., 540, 558
Jenkins, G., 58
Joseph, P. D., 20, 58
Joy, A. H., 555, 558
Joyce, R. R., 284
Jurkevich, I., 7, 16, 17, 20, 58, 155

Kaitchuk, R. H., 504, 508
Kaler, J. B., 518, 519, 521, 523, 528, 531, 534
Kalish, M. S., 451
Kaplan, S. A., 540, 558
Katz, J. I., 496, 508
Kemp, J. C., 505, 506, 509
Kholopov, P. M., 417
Khromov, G. S., 523, 529, 531, 534
Kietel, M., 445
Kilkenny, D., 412
Kirbiyik, H., 538, 558
Kissling, M., 437
Kizilirmak, A., 417, 432, 433, 434
Klimek, Z., 417, 433
Knipe, G. F. G., 200, 215
Koch, R. H., 378, 387, 463, 464
Kohoutek, L., 523, 534
Kondo, Y., 200, 215, 455, 464, 465, 468, 471
Koniges, A., 329, 330
Kopal, Z., 1, 7, 8, 10, 15, 16, 17, 18, 20, 24, 25, 37, 41, 42, 43,
48, 58, 111, 112, 116, 119, 120, 125, 126, 128, 131, 132,
133, 134, 135, 138, 139, 141, 142, 150, 151, 153, 154, 155,
156, 158, 159, 160, 164, 173, 177, 180, 181, 182, 183, 190,
191, 192, 194, 196, 199, 202, 203, 206, 207, 208, 209, 211,
214, 215, 283, 284, 285, 316, 325, 330, 335, 360, 371, 387,
405, 406, 411, 412, 415, 416, 418, 476, 508, 512, 514, 515,
516, 535, 537, 538, 539, 540, 544, 555, 558
Köppen, J., 523, 528, 534
Krasnobaev, K. V., 557
Kratochvíl, P., 209, 211, 215, 378, 387
Krishnan, T., 267, 283
Kristenson, H., 335, 360

Kříž, S., 539, 558
Krobusek, B. A., 417, 434, 435, 436, 437
Kron, G. E., 58, 314, 330, 337, 360
Krüger, P., 427
Kruszewski, A., 540, 558
Krzanik, L., 432
Kuiper, G. P., 415, 417
Kukarkin, B. V., 414, 417
Kukarkina, N. P., 417
Kurpinska, M., 451
Kurochkin, N. E., 417
Kurutac, M., 191, 197, 389, 397, 403, 431

Lacona, G., 285, 301, 303
Lambert, D. L., 466, 471
Lamers, H. J. G. L. M., 461, 464
Lanczos, C., 131, 154, 175, 180
Landau, L. D., 489, 508
Lang, K. R., 284
Lanzano, P., 131, 154
Larsson-Leander, G., 287, 304
Lause, F., 425
Lehmann, E., 58
Lehmann, P. B., 417, 426
Lempfuhl, F., 427
Letkowski, A., 432
Lightman, A. P., 493, 508
Liller, M. H., 448, 450
Liller, W., 412
Lin, D. N. C., 474, 495, 508
Linnell, A. P., 7, 16
Linsky, T. L., 297, 299, 303, 304
Livaniou, H., 182, 193, 197, 203, 214, 215
Livio, M., 412
Locher, K., 429, 432, 433, 434, 435, 436, 437, 438, 439
Longmore, A. J., 456, 464
Longo, G., 253
Lorenzi, F., 346, 360
Loring, D., 428
Lowden, W., 430
Lubow, S. H., 473, 474, 507, 508, 542, 558
Lucy, L. B., 200, 214, 215, 354, 360, 415, 417, 451
Lynden Bell, D., 493, 494, 508

Madirossian, F., 330
Magnus, W., 180
Mahdi, S. A. A., 544
Mallama, A. D., 417, 434, 437
Mammano, A., 533, 534
Mancuso, S., 313, 316, 323, 325, 330, 332, 360

Mannery, E. J., 412
Marcozzi, S., 359
Margon, B., 448, 450
Margoni, R., 534
Marilli, E., 290, 303
Markworth, N. L., 466, 471
Marngus, N., 503, 508
Martynov, D. Ya., 115, 120, 514, 516
Marx, H., 427, 428
Massuch, J., 427
Matsuda, T., 558
McCall, D., 479, 483, 509
McCants, M. M., 274, 283
McCluskey, G. E., 455, 464, 471
McCook, G., 508
McGraw, J. T., 283
McLean, I. S., 355, 503, 508
McMahan, P. A., 253, 283
Meaburn, J., 442, 445, 448, 450
Medvedeva, G. I., 417
Merrill, J. E., 6, 16
Messetti, M., 330
Michlovic, J. E., 360
Milano, L., 14, 313, 330, 331, 360
Milone, E. F., 350, 360, 395, 403
Mochnacki, S. W., 450, 451
Moffet, T. J., 283
Monet, D. G., 511, 516
Monske, R., 428, 429, 430, 432
Montanari, G., 5
Montemayor, T., 284
Montle, R. E., 387
Morgan, C. A., 284, 288
Morgan, J. G., 304, 328, 329, 330, 378, 387
Morgan, J. G., 304, 328, 329, 330, 378, 387
Morton, G. A., 323, 251
Moss, D. L., 200, 215
Mufson, S. L., 508
Muller, J., 425, 436

Naftilan, S. A., 469, 471
Najim, N., 406, 407, 412
Nariai, K., 487, 502, 509
Nather, R. E., 258, 259, 274, 283, 284
Nelson, B., 120, 121
Neubauer, F. J., 536, 558
Newton, Isaac, 535
Niarchos, P. G., 196, 197, 199, 203, 205, 215, 441
Nordström, B., 160, 516
Northcott, R. J., 347, 360

NAME INDEX

Oberhettinger, F., 37, 41, 42, 58, 180
Oke, J. B., 450
O'Keefe, J. A., 253, 283
Oliver, J. P., 341, 343, 345, 346, 360
Olson, E. C., 466, 467, 469, 470, 471, 509
Öpik, E. J., 254, 283
Orlovius, V., 427
Oster, L., 542, 551, 558
Osterbrock, D. E., 519, 520, 534, 544, 547, 558
Otrebski, A., 15, 16
Owen, F. N., 297, 304

Packett, W., 489, 509
Paczynski, B., 412, 451, 474, 488, 509, 544, 558
Parthasarathy, M., 469, 471
Pasinetti, L., 460, 464
Paterno, L., 220, 250, 251
Payne-Gaposchkin, C. H., 314, 324, 330, 332, 337, 360
Penford, J. E., 412
Perek, L., 523, 534
Perova, N. B., 417
Peter, H., 427, 428, 429, 432, 433, 434, 435, 436, 437, 438, 439
Petterson, J. A., 496, 509
Petty, A. F., 17, 20, 58, 155
Pickering, E. C., 5
Piirola, V., 469, 471
Pikelner, S. B., 544, 558
Pilyugin, L. S., 519, 534
Pinto, P. A., 417
Piotrowski, S. L., 7, 16, 24, 25, 58, 502, 509, 540, 558
Pitz, E., 245, 251
Plackett, R., 59
Plavec, M., 209, 211, 215, 378, 387, 463, 464, 469, 471, 539, 544, 558
Pohl, E., 413, 417, 427, 431, 432, 434, 435, 437
Polidan, R. S., 464, 469
Pomraning, G. C., 558
Popper, D. M., 112, 114, 117, 118, 119, 120, 288, 289, 303, 304, 311, 343, 346, 360, 361, 362, 373, 374, 377, 378, 379, 387, 466, 513, 516
Poss, H., 284
Pottash, S. R., 523, 530, 534
Powell, M. J. D., 129, 154
Prager, R., 441, 445
Prendergast, K. H., 474, 495, 503, 509, 542, 558
Pringle, J. E., 474, 493, 494, 495, 508, 509
Procházka, F., 441, 445
Proctor, D. D., 7, 16
Pskovsky, Yu. P., 417
Pucillo, M., 250
Purgathofer, A., 200, 215, 441, 445

Quester, W., 417, 427

Radick, B. B., 284
Rahimi-Ardabili, J., 410, 412
Rakos, K. D., 284
Rao, C. R., 59
Rao, P. V., 305, 361, 387
Rayner, P. T., 251
Reboul, H., 266, 283
Rees, M., 477, 493, 509
Rhombs, C. G., 469, 471
Ridgway, S. T., 284
Rigutti, M., 253
Robb, R. M., 360
Roberts, W. J., 496, 509
Robinson, L. J., 417, 426, 428, 429, 430
Rodono, M., 285, 287, 289, 300, 301, 303, 304, 335, 360
Roemer, E., 5
Romeo, G., 304
Rosino, L., 533
Rovithis-Livaniou, H., 181, 182, 197
Royer, A., 438
Rucinski, S. M., 200, 215, 509
Ruderman, M. S., 557
Rudolph, R., 417, 425, 426
Rudy, R. J., 505, 506, 509
Russell, H. N., 1, 6, 7, 16
Russo, G., 330, 360

Sadik, A. R., 315, 318, 329, 330, 345, 346, 351, 360
Sahade, J., 503, 509
Sakhibullin, N. A., 534
Sakurai, T., 558
Saltzman, J., 412
Sandmann, W., 284
Santin, P., 246, 251
Sareyan, J. P., 232, 240, 251
Sarma, M. B. K., 305, 341, 343, 360, 361, 387
Scaltriti, F., 303, 346, 360
Scheffe, M., 59
Scheuer, P. G., 268, 283
Schiaparelli, G. V., 8
Schiffer, F. H., 304
Schlachcic, K., 432
Schlesinger, F., 535
Schmidt, H., 444, 445
Schrick, K. W., 444, 445
Schuerman, D. W., 500, 501, 509
Seaquist, E. R., 303
Seaton, M. J., 517, 518, 520, 523, 526, 531, 534

NAME INDEX

Sedmak, G., 217, 222, 226, 227, 241, 245, 246, 250, 251
Seidl, M., 429
Semenink, I., 509
Sengouca, H., 432
Shapley, H., 6, 16
Shapley, M. B., 335, 360
Shaviv, G., 412
Shore, S. N., 329, 330, 397
Shu, F. H., 473, 474, 507, 508, 542
Simmons, J. F. L., 507, 509
Simmons, K., 431
Simon, T., 297, 303
Siscoe, G. L., 543, 558
Skillman, D. R., 417
Smith, B. W., 284
Smith, R. C., 538, 558
Smith, R. R., 479, 483, 509
Söderhjelm, S., 7, 16, 489, 509
Solazzo, C., 330, 360
Sörensen, S. A., 474, 495, 509, 542, 558
Soska, A., 432
Spitzer, L., 544, 547, 558
Stencel, R. E., 471
Stepanian, I. A., 447, 449, 450
Ştephan, N., 417, 437, 438
Šternberk, B., 314, 330
Sterne, T. E., 514, 516, 537, 558
Stoy, R. H., 517, 534
Strazzulla, G., 303, 304
Strohmeier, W., 310, 361, 362, 364, 379, 387
Strömgren, B., 544, 545, 547, 558
Struve, O., 287, 465, 503, 535, 536, 538, 555, 558
Stuhlinger, T. W., 387
Surendranath, R., 364, 387
Swandron, D., 360
Szkody, P., 450

Taam, R. E., 495, 503, 509, 542, 558
Talcott, J. C., 295, 303
Tamura, S., 532, 534
Tarafdar, S. P., 523, 528, 534
Taylor, A. R., 303, 397
Taylor, J. H., 258, 283, 403
Tomkin, J., 466, 471
Torres, C. A. O., 300, 304
Tremko, J., 470
Tsesevich, V. P., 18, 59, 164, 180, 193, 197, 208, 215, 387
Tsouroplis, A. G., 196, 197
Tulloch, M. K., 464
Tümer, O., 401, 403
Tunca, Z., 403

Ulrich, R. K., 288, 289, 304, 378, 387

Vanbeveren, D., 500, 509
Vittone, A., 313, 533, 534
Vogt, S. S., 283, 300, 302, 304

Walter, F., 297, 299, 304
Walter, K., 466, 468, 471, 502, 509
Wasow, W. R., 479, 480, 508
Watson, R. D., 251
Watts, D., 58
Wellmann, P., 371
Wells, D. C., 284
Wenzel, W. W., 447, 450
Wesselius, P. R., 534
Whelan, J. A. J., 200, 215
White, N. M., 259, 260, 261, 274, 283
Whitford, A. E., 253, 284
Wild, P. A. T., 258, 283
Williamon, R. M., 201, 215, 315, 330, 335, 360
Williams, D. A., 429, 544, 547, 557
Williams, J. D., 253, 284
Willman, W. W., 58
Wilson, R. E., 111, 117, 120, 352, 360, 455, 464
Witzigman, S., 445
Woerden, H. van, 337, 340, 341, 360, 364
Wolf, E., 364, 283
Wood, D. B., 11, 111, 117, 119, 120, 160, 335, 377
Wood, F. B., 360, 503, 509
Woodward, E. J., 451
Worley, R., 256
Wu, C. C., 534
Wuthrich, T., 437
Wyse, A. B., 15, 16, 18, 59

Young, A. T., 232, 251, 274, 284, 329, 330

Zafiropoulos, F., 161
Zanstra, H., 517, 534
Zeilik, M., 304
Zissel, R. E., 360
Zubrod, D. J., 504, 508

INDEX OF INDIVIDUAL ECLIPSING VARIABLES REFERRED TO IN THE TEXT

RT And: 313, 314, 315, 316, 317, 318, 320, 323, 324, 329, 330, 332, 333, 334, 335, 336, 337, 338, 339, 340, 357, 358, 359
AB And: 200
V 535 Ara: 200
OO Aql: 200, 413, 414, 415, 416
AC Boo: 200
TY Boo: 200
VW Boo: 200
XY Boo: 200
SS Cam: 385
SV Cam: 340, 341, 342, 343, 345, 354, 355, 356, 358, 359
WY Cnc: 305, 316, 329, 334, 342, 345, 354, 358, 359
RS CVn: 385, 389, 390, 391
RX Cas: 463
SX Cas: 463, 469
CW Cas: 200
RR Cen: 200
U Cep: 465, 466, 467, 469, 500, 504, 535, 537, 538, 553, 554, 555
VW Cep: 200, 441, 442, 443, 444
EK Cep: 21, 22, 23, 34, 39, 40, 45
RZ Com: 200
UX Com: 385
CC Com: 200
U CrB: 466
RT CrB: 390, 392
W Cru: 463, 469
CG Cyg: 329, 334, 344, 347, 348, 349, 350, 358, 359
DK Cyg: 200
V 367 Cyg: 463, 469
V 478 Cyg: 117, 118
V 1073 Cyg: 200
WW Dra: 385
BS Dra: 117
RZ Eri: 305
UX Eri: 200
YY Eri: 200
Z Her: 385, 389, 393, 394
AK Her: 200
RT Lac: 389, 395, 396
SW Lac: 200
AR Lac: 385, 389, 397, 398, 400
EM Lac: 200
UV Leo: 330
UZ Leo: 201
AM Leo: 200

β Lyr: 454, 458, 462, 465, 469
V 502 Oph: 201
V 566 Oph: 201
V 839 Oph: 201
ER Ori: 201
U Peg: 201
LX Per: 385, 389, 398, 399, 401
β Per: 504, 507, 535
AE Pho: 201
SZ Psc: 385, 389, 401, 402
UV Psc: 305, 310, 329, 334, 345, 351, 354, 358, 359
TX Pyx: 305, 311
TY Pyx: 361, 362, 363, 364, 374, 377, 378, 379, 380, 381, 386
U Sge: 500, 506, 507, 553, 554
UU Sge: 405, 406, 409, 412
ν Sgr: 460, 462
RZ Sct: 536, 537, 538
W Ser: 463, 469
RW Tau: 504
RZ Tau: 201
W UMa: 201
RW UMa: 385
XY UMa: 334, 346, 351, 354, 358, 359
AW UMa;   451
AH Vir: 201, 441, 442, 443, 444
ER Vul: 334, 347, 352, 358, 359

RAYMOND H. FOGLER LIBRARY
DATE DUE

BOOKS ARE SUBJECT TO
RECALL AFTER TWO WEEKS

DEC 14 1982

MAY 1 8 1994